Handbook of Research on Cultural Tourism and Sustainability

Claudia Ribeiro de Almeida
University of the Algarve, Portugal

Joao Carlos Martins
University of the Algarve, Portugal

Alexandra R. Gonçalves
University of the Algarve, Portugal

Silvia Quinteiro
University of the Algarve, Portugal

Maria Laura Gasparini
University of Bologna, Italy

IGI Global
PUBLISHER of TIMELY KNOWLEDGE

A volume in the Advances in Hospitality, Tourism, and the Services Industry (AHTSI) Book Series

Published in the United States of America by
IGI Global
Business Science Reference (an imprint of IGI Global)
701 E. Chocolate Avenue
Hershey PA, USA 17033
Tel: 717-533-8845
Fax: 717-533-8661
E-mail: cust@igi-global.com
Web site: http://www.igi-global.com

Library of Congress Cataloging-in-Publication Data

Names: Ribeiro de Almeida, Claudia, 1971- editor.
Title: Handbook of research on cultural tourism and sustainability / Claudia Ribeiro de Almeida, Joao Carlos Martins, Alexandra Rodrigues Goncalves, Silvia Quinteiro, Maria Laura Gasparini, Editors.
Description: Hershey, PA : Business Science Reference, [2022] | Includes bibliographical references and index. | Summary: "This publication gives a different perspective of cultural tourism with some updated theoretical frameworks of the topics in analysis, complemented with case studies, best practices and some data related to international and differentiated tourism contexts (urban, natural, rural, historical centers)"-- Provided by publisher.
Identifiers: LCCN 2021047639 (print) | LCCN 2021047640 (ebook) | ISBN 9781799892175 (hardcover) | ISBN 9781799892182 (ebook)
Subjects: LCSH: Heritage tourism--Research. | Sustainable tourism--Research.
Classification: LCC G156.5.H47 H33 2022 (print) | LCC G156.5.H47 (ebook) | DDC 338.4/791--dc23/eng/20211108
LC record available at https://lccn.loc.gov/2021047639
LC ebook record available at https://lccn.loc.gov/2021047640

This book is published in the IGI Global book series Advances in Hospitality, Tourism, and the Services Industry (AHTSI) (ISSN: 2475-6547; eISSN: 2475-6555)

British Cataloguing in Publication Data
A Cataloguing in Publication record for this book is available from the British Library.

Advances in Hospitality, Tourism, and the Services Industry (AHTSI) Book Series

Maximiliano Korstanje
University of Palermo, Argentina

ISSN:2475-6547
EISSN:2475-6555

Mission

Globally, the hospitality, travel, tourism, and services industries generate a significant percentage of revenue and represent a large portion of the business world. Even in tough economic times, these industries thrive as individuals continue to spend on leisure and recreation activities as well as services.

The Advances in Hospitality, Tourism, and the Services Industry (AHTSI) book series offers diverse publications relating to the management, promotion, and profitability of the leisure, recreation, and services industries. Highlighting current research pertaining to various topics within the realm of hospitality, travel, tourism, and services management, the titles found within the AHTSI book series are pertinent to the research and professional needs of managers, business practitioners, researchers, and upper-level students studying in the field.

Coverage

- Tourism and the Environment
- Service Management
- Service Training
- Service Design
- Sustainable Tourism
- Destination Marketing and Management
- Food and Beverage Management
- Cruise Marketing and Sales
- Travel Agency Management
- Casino Management

IGI Global is currently accepting manuscripts for publication within this series. To submit a proposal for a volume in this series, please contact our Acquisition Editors at Acquisitions@igi-global.com or visit: http://www.igi-global.com/publish/.

The Advances in Hospitality, Tourism, and the Services Industry (AHTSI) Book Series (ISSN 2475-6547) is published by IGI Global, 701 E. Chocolate Avenue, Hershey, PA 17033-1240, USA, www.igi-global.com. This series is composed of titles available for purchase individually; each title is edited to be contextually exclusive from any other title within the series. For pricing and ordering information please visit http://www.igi-global.com/book-series/advances-hospitality-tourism-services-industry/121014. Postmaster: Send all address changes to above address.

Titles in this Series

For a list of additional titles in this series, please visit: www.igi-global.com/book-series

Gastronomy, Hospitality, and the Future of the Restaurant Industry Post-COVID-19 Perspectives
Ana Pinto Borges (ISAG, Portugal & Research Center in Business Sciences and Tourism (CICET-FCVC), Portugal) António Lopes de Almeida (ISAG - European Business School and Center in Business Sciences and Tourism (CICET - FCVC), Portugal) Elvira Pacheco Vieira (ISAG - European Business School and Center in Business Sciences and Tourism (CICET - FCVC), Portugal) Rui Rosa Dias (ISAG - European Business School and Center in Business Sciences and Tourism (CICET - FCVC), Portugal) and Paula Rodrigues (Lusíada University. COMEGI and CICET- FCVC, Porugal)
Business Science Reference • © 2022 • 315pp • H/C (ISBN: 9781799891482) • US $215.00

Optimizing Digital Solutions for Hyper-Personalization in Tourism and Hospitality
Nuno Gustavo (Estoril Higher Institute for Tourism and Hotel Studies, Portugal) João Pronto (Estoril Higher Institute for Tourism and Hotel Studies, Portugal) Luísa Carvalho (Polytechnic Institute of Setúbal, Portugal & and CEFAGE, University of Évora, Portugal) and Miguel Belo (Lisbon University, Portugal & Portuguese National Funding Agency for Science, Research, and Technology, Portugal)
Business Science Reference • © 2022 • 379pp • H/C (ISBN: 9781799883067) • US $195.00

Challenges and Opportunities for Transportation Services in the Post-COVID-19 Era
Giuseppe Catenazzo (ICN Business School, France)
Business Science Reference • © 2022 • 320pp • H/C (ISBN: 9781799888406) • US $225.00

Handbook of Research on Digital Communications, Internet of Things, and the Future of Cultural Tourism
Lídia Oliveira (University of Aveiro, Portugal)
Business Science Reference • © 2022 • 587pp • H/C (ISBN: 9781799885283) • US $325.00

New Governance and Management in Touristic Destinations
Marco Valeri (Niccolò Cusano University, Italy)
Business Science Reference • © 2022 • 357pp • H/C (ISBN: 9781668438893) • US $225.00

Global Perspectives on Literary Tourism and Film-Induced Tourism
Rita Baleiro (University of the Algarve, Portugal & CiTUR, Portugal) and Rosária Pereira (University of the Algarve, Portugal & CinTurs, Portugal)
Business Science Reference • © 2022 • 371pp • H/C (ISBN: 9781799882626) • US $215.00

IGI Global
PUBLISHER of TIMELY KNOWLEDGE

701 East Chocolate Avenue, Hershey, PA 17033, USA
Tel: 717-533-8845 x100 • Fax: 717-533-8661
E-Mail: cust@igi-global.com • www.igi-global.com

Editorial Advisory Board

List of Contributors

Table of Contents

Detailed Table of Contents

other researchers in the field of sustainable development studies of cultural tourism using an interpretive approach and a systematic meta-combination method. Then, based on the hidden content analysis strategy, the extracted codes are identified in the form of categories. Finally, the authors present a conceptual model. In this research, valid scientific databases were reviewed, and selected articles were identified using CASP criteria. Based on this, 104 primary codes in the form of 29 categories of primary and secondary categories are identified, and the four central codes are components, effective factors, consequences, and challenges of sustainable development of cultural tourism. Also, a conceptual model of sustainable development of cultural tourism and tourist destinations in need was presented. Finally, the validity of the research results with Kappa Cohen index in SPSS software was confirmed based on the opinions of experts outside the authors of the research with a size of 0.712.

The chapter describes the theoretical and methodological approach followed to develop the MED sustainable path and cultural route (MED S&C Path) model as a practical tool to monitor sustainable tourism development along cultural routes. In the framework of BEST MED (Beyond European Sustainable Tourism MED Path) project, the model aims at improving the management of cultural routes by assessing their sustainability level using a set of criteria and indicators. The implementation process of the model proposes a multi-stakeholder approach which contributes to higher levels of cooperation at local, regional, and transnational level, improving the governance of cultural routes. The MED S&C Path model responds to the concrete need of increasing the tools to measure cultural routes' social, economic, and environmental impact, in order to raise awareness among policy makers about their strategic role in sustainable local development, especially in rural and less-known areas.

This chapter highlights the research-action practices developed under the Phase 4 of the Interreg BEST MED Project (Beyond European Sustainable Tourism Med Path, Grant Number: 5987), "Testing." In this project, stage was determinant to prototype and test the best practices towards sustainability, developed by the project's partners, detailing its achievements towards Mediterranean tourism sustainability, resulting from the ground-level operationalization of a theoretical and methodological framework. Here stakeholder participation is determinant to achieve breakthrough advances on more enduring and resilient touristic territories, addressing seasonality and lack of cooperation among tourist actors. This data gathering was developed among public officials, private operators, academics, third sector associations, and citizens in eight Mediterranean areas. Despite the common sustainable and cultural path model, different outputs were found from the participation and auscultation practices.

sustainable cultural heritage tourism management in Turkey. The study aims to be inspirational for both theoreticians and practitioners to create brighter cooperative projects.

Foreword

It gives me great pleasure to write the Foreword for this dynamic new publication on cultural tourism and sustainability. From my first encounter with Professors Cláudia Ribeiro de Almeida, Sílvia Quinteiro, Alexandra Rodrigues Gonçalves and colleagues at a symposium in Faro in 2019 on literary tourism and cultural routes, hosted by the University of the Algarve, I was struck by the group's intellectual curiosity and energy devoted to the field of cultural heritage and linked tourism initiatives. Their work is always exciting to read and worth the academic effort.

Of course, cultural heritage and tourism runs into a dead end without involving itself in concerns about sustainability, authenticity and community involvement, among other important issues. This is why this current book *Handbook of Research on Cultural Tourism and Sustainability* is useful as it aims to tackle such issues theoretically, as well as illustratively, by drawing on various case studies. Furthermore, it is not one-sided when it comes to the spread of contributors – stakeholders, consumers and theoreticians in cultural heritage are all called upon to add their voices to the volume.

The range of geographical locations from which these voices arise is also impressive and speaks of an effort on the part of the editors to spread the intellectual net as widely as possible. Though the northern hemisphere is predominant both in terms of authors' academic and official affiliations, the range is intercontinental: contributions are sourced from Colombia, Norway, Italy, Spain, Greece, Albania, Turkey, Hong Kong and, of course, Portugal. Coming from the southern hemisphere, I would be interested to see such debates brought to bear even more widely but that is work for another book. Suffice it to say, this is an important publication for both academics and stakeholders when looking at the pressing issue, particularly, of sustainability within the contemporary world in which we live.

The book is well organised leading the reader into 3 sections: the first begins with concepts and methodologies which inform and remind us of the best current practice in social innovation in tourism, together with the complexities and concerns that accompany any cultural heritage efforts. Many of the 8 chapters in this section stress the importance of citizen participation and co-creation of knowledge if sustainability long-term is to be assured. This is a complicated matter when realising the non-homogeneous nature of many communities and the rural-urban divide which some chapters rightly emphasise. Section 2 has stakeholders' analysis as focus and includes fascinating references to specific areas such as the Algarve and the local tourism council, the terraced fields of La Gomera in the Canary Islands, and 'orange tourism' as potential stimulus for tourism in a Colombian state. Section 3 continues the analysis of cultural tourism using case studies from an academic point of view – the potential for literary tourism in Albania, tourism 'pilgrimage' walks, cultural tourism efforts in the Cyclades, favela tours and how these impact on communities and tourists alike, Macao's festivals and pressures plus others. There is a wide range of subject matter and approach here; enough for a broad range of interests.

The contemporary focus of the book is also pleasing: the debates referred to are current and applied to recent research efforts as in the MED sustainable path and cultural route model; certainly, the adverse impact of the recent and ongoing covid pandemic on tourism is referred to specifically in one chapter, whilst current climate issues and changing farming practices form the subtext of another.

Again, and again, the emphasis falls on the importance of "local people", the communities. I would add, from my experience in running a research project on literary tourism and trails in South Africa, that another vital ingredient when it comes to sustainability of cultural heritage efforts is the involvement of local authorities and tourism councils. That conversation is imperative and it has its echoes in these papers as well. Even the best devised routes and sites need ongoing support from those council networks and municipalities within which they fall. Good relationships between stakeholders, local communities and even the academics mean there can be a vision for the future. A good cultural heritage 'product' – be it a literary trail, a gastronomic tour, a musical festival, a slum tour – must harness the efforts of local people, in tandem with those that develop these preservers of intangible heritage and indigenous knowledge, together with the authorities, ideally. Then long-term sustainability is possible.

This volume is a step along the path to thinking through these issues and promoting best practice among those already engaged in the various fields that fall under the umbrella of cultural heritage and tourism. For that reason, it is a worthy academic effort and deserves our attention.

Lindy Stiebel
University of KwaZulu-Natal, South Africa

Preface

The United Nations World Tourism Organization (2020), considering the present uncertain tourism context, states that Cultural Tourism can be a determinant for economic recovery in Tourism, through a double strategy. In the present, to search for more and better data, develop a strong connection with tech and digital economy, ensuring sustainable and resilient tourism and its workforces, implementing participative governance models and practices to attract new audiences, particularly in the national market. To the future, a shift towards more qualitative standards of tourism study, a diversification of tourism products, a boost of community's participation in Tourism, customization of tourism offers and the improvement of cultural tourism entrepreneurship and innovation.

Sustainable Development represents a critical answer to an economic and social model, a critical reply to a maximized and extractives' vision of natural and human resources, combined with the improvements in the productive techniques. But in fact, the expected global improvement did not occur, and even some societies have lowered their quality of life, undermining the importance of social dimensions, namely the cultural aspects. (Swanson & DeVereaux, 2017; Soini, & Dessein, 2016; Hawkes, 2001). In a global context of increasing access to diverse cultures, new questions arise as authenticity, creativity, distinctiveness, and innovation, highlighting the constituents of identity and sense of belonging. We assist to an increasing interest in cultural aspects by tourists, being culture at the centre of all changes, particularly on territories specialized in leisure appropriations (Farrell, B & Twining-Ward, 2004; UNESCO, 2001).

In this sense, is determinant to ensure three main Sustainable Development dimensions: Firstly, Socio-Cultural dimensions, acknowledging the importance of Cultural Tourism, Heritage and Local products in visiting communities, strengthening their sense of identity and belonging. In this process, is essential to promote arts and cultural experimentation, a stronger and more intimate relation between tourists, visitors, consumers and the local population, its culture, and social practices; Secondly, to understand management practices, with the dynamization of the local ecosystem of stakeholders, towards resilient participative governance of touristic territories; Thirdly, to understand long term economic results, particularly towards tourism territories, avoiding massification and other unauthentic leisure experiences developed in the past.

The successful recovery of Tourism as an industry worldwide, as an economic, socio-cultural and environmental driver of sustainable local communities, but also as a social phenomenon of individual evasion, artistic and cultural amusement, at the local, national, and global levels is strictly dependent on the efficient development of new models of leisure. It is fundamental that all tourism actors, namely tourism operators, local and national governments, local associations and residents, generate robust trust among leisure consumers (Fang et al. 2021). In addition, human-induced climate change has emerged as another challenge for the mid-term and long-term future of leisure industry at the local, national, and global levels (Prideaux, Thompson & Pabel,2020).

With all of these problematic in mind, the editors developed *The Handbook of Research on Cultural Tourism and Sustainability.* Organized in three sections, the Handbook presents 18 chapters incorporating case studies from different geographical regions and realities, thus helping to understand cultural tourism and sustainability issues.

Section 1 focuses on concepts and methodologies for cultural tourism and sustainability presented in eight chapters.

Chapter 1 (Social Innovation and Cultural Tourism: A Review of Co-creation Strategies for Sustainable Tourism) intends to explain and understand how the enhancement of culture for the development of tourism is approached through processes of Social Innovation, as a new form of social action that is capable of transforming the frameworks of action and proposing new formulas for the development of cultural tourist destinations.

Chapter 2 (Cultural Routes as a Tool for Sustainable Tourism and Socio-Economic Development) discusses the importance of cultural routes as a way to connect various attractions and tourist destinations within a country or across many countries, bringing to light the cultural legacy of the area, creating opportunities for the social and economic development.

Chapter 3 (Dialogue of Knowledge for the Consolidation of Territories Through Cultural Tourism) reflects on how the identification of ancestral and local agents' wisdom can promote processes of social appropriation of knowledge, by creating spaces for discussion between academia and local actors, recognizing the socio-cultural diversity of local communities.

Chapter 4 (Innovation for Sustainability in World Heritage Destinations: Opportunities and Challenges of the Idealistic Paradigm) presents the narratives from two cultural World Heritage destinations in Norway to illustrate potentials and challenges of using the idealistic paradigm in experience-based transformations.

Chapter 5 (Conceptual Model of Sustainable Development of Cultural Tourism With a Hybrid Approach) discusses some of the main findings of the studies conducted by other researchers in the field of sustainable development studies of cultural tourism and also presents a conceptual model of sustainable development of cultural tourism and tourist destinations.

Chapter 6 (Monitoring Sustainability Along Cultural Routes: The MED Sustainable Path and Cultural Route Model) describes the theoretical and methodological approach followed to develop the MED Sustainable Path and Cultural Route (MED S&C Path) model as a practical tool to monitor sustainable tourism development along cultural routes.

Chapter 7 (A New Framework for Tourism Sustainability and Its Prototyping in Pilot Areas: Insights From BEST MED Testing Phase) highlights the research-action practices developed under the Phase 4 (testing) of the Interreg BEST MED Project, determinant for the prototype and test of the best practices towards sustainability in the Mediterranean area.

Chapter 8 (Mapping the Cultural Tourism Experience: Role of Senses, Emotions, and Memories in Cultural and Heritage Experience and Gastronomic Experience) focuses on the role of senses, emotions, and memories model (SEMs) to mapping a cultural tourism experience.

Section 2 present four chapters that introduce the stakeholders' analysis in different geographic contexts.

Chapter 9 (Moving Towards Sustainable Cultural Tourism Experiences: A Participatory Journey – The stakeholder Perspective) presents a discussion of the stakeholders' perception of the development of sustainable cultural tourism experiences in a council located in the Algarve, Portugal.

Chapter 10 (Actor Networks and Development of Cultural Tourist Destinations) addresses the relational structure of the networks of actors of three cultural tourist destinations using Social Network Analysis (SNA).

Chapter 11 (The Tourism Value of Terraced Fields Landscapes: The "Bancales" of the Island of La Gomera, Spain) discusses the strategy to develop tourism on the agricultural terraces in the north of the island of La Gomera, part of the territorial identity of the island

Chapter 12 (Answers to the Crisis in the Tourism Sector in a COVID Environment: The Orange Economy Cluster initiative in the State of Boyacá, Colombia) presents the Orange Cluster Initiative for the State of Boyacá, Colombia, focusing on the tourism sector's capability to promote sustainable tourism based on local resources

Section 3 presents six case studies from different countries.

Chapter 13 (Developing Literary Tourism to Support Local Culture and Tourism: The Case of Shkodra) discusses the potential of Shkodra (Albania) for the development of literary tourism and the multiple benefits coming from the literature as part of the cultural heritage and tourist offer in this city.

Chapter 14 (Promoting the Development of Local Communities Through "Touripilgrimage": The Portuguese Inner Way of Santiago de Compostela) discusses the importance of "touripilgrimage" in the Portuguese Way of the Interior of Santiago de Compostela and how the different experiences along the route can help to promote the development of the local communities.

Chapter 15 (Local Perspectives on Cultural Tourism and Cultural Sustainability: The Case of the Cyclades, Greece) explores cultural tourism perceptions, practices, concerns and prospects, among local residents and business representatives in the Cyclades and discusses their implications for local cultural sustainability.

Chapter 16 (Favela Tour Experience: The Impacts in The Host Communities) aims to verify the tangible and intangible impacts of the slum tours in the host communities of Rocinha and Santa Marta (Rio de Janeiro, Brazil).

Chapter 17 (Challenges in Safeguarding Traditional Festivals Facing Tourism Pressure: A Case Study of Macao SAR) discusses the challenges in safeguarding two of Macao's and China's nationally recognized intangible cultural heritage festivals which are facing tourism pressures, the Macao Feast of the Drunken Dragon and the Lion Dance Gala and The Beliefs and Customs of Na Tcha.

Chapter 18 (Recognition of Tangible and Intangible Cultural Heritage for the Sustainability of Tourism: Turkey [Türkiye] Case) describes how sustainability for cultural heritage tourism management can be ensured through recognition of the tangible and intangible cultural heritage, with a focus on the Turkey case.

The book editors believe that readers will be made aware of the importance of cultural tourism, not only through the different conceptual and methodological chapters but also with the different case studies presented, comprising both urban and rural destinations. The knowledge and diversity of examples in analysis will help the creation of new perspectives about the importance of cultural tourism, as well as the need for involvement and coordination of different stakeholders and local communities in order to create sustainable and innovative cultural tourism offers.

Cláudia Ribeiro de Almeida
University of the Algarve, Portugal

João Carlos Martins
University of the Algarve, Portugal

Alexandra R. Gonçalves
University of the Algarve, Portugal

Sílvia Quinteiro
University of the Algarve, Portugal

Maria Laura Gasparini
University of Bologna, Italy

Acknowledgment

Editing a new book is always a great challenge, not only for the editors, but for the ones that work with us.

Saying this, the editors would like to acknowledge the help of all the people involved in this handbook and, more specifically, to the authors and reviewers that took part in the review process.

Without their support, this book would not have become a reality.

To all the authors, our sincere gratitude to their contributions, time and expertise.

To all the reviewers, our appreciation to their valuable contributions that allowed the improvement of quality, coherence and content presentation of the chapters.

To Professor Lindy Stiebel, for her kind words in the foreword.

And to all the others that directly or indirectly helped us during the project, with some recommendations and technical work in order to bring to light this new publication.

To all… THANKS!

Cláudia Ribeiro de Almeida
University of the Algarve, Portugal

João Carlos Martins
University of the Algarve, Portugal

Alexandra R. Gonçalves
University of the Algarve, Portugal

Sílvia Quinteiro
University of the Algarve, Portugal

Maria Laura Gasparini
University of Bologna, Italy

Section 1

Concepts and Methodologies

Chapter 1
Social Innovation and Cultural Tourism:
A Review of Co-Creation Strategies for Sustainable Tourism

Maria D. Pitarch-Garrido
University of Valencia, Spain

ABSTRACT

The main objective of this chapter is to explain and understand how the enhancement of culture for the development of tourism is approached through processes of social innovation as a new form of social action that is capable of transforming the frameworks of action and proposing new formulas for the development of cultural tourist destinations. Initiatives classified as social innovation are characterised by mobilising local and external resources to change the logic of collective action and provide new responses to common problems. It is based on citizen participation, shared leadership, co-creation of knowledge, initiatives, and policies. Cultural heritage is understood in this way as linked to the society that generated it, maintains it, and gives it its own identity. The tourist using this heritage must favour its conservation and sustainability. This chapter aims to review the literature concerning social innovation in tourism and good practices in order to generate a theoretical basis so that the social innovation processes already validated can be replicated in other countries.

INTRODUCTION

Culture and heritage are part of the essence of every human group. However, neither one nor the other has always been equally valued by society, which leads to problems for their conservation, for their transmission to future generations and for their enhancement through activities such as tourism. In more recent history, the value of culture and heritage has almost always been granted, or at least ratified, by governmental bodies at different levels. The consolidation of a certain approach that we can call "top-down" makes the identification of the population with its history, its culture, and therefore its heritage,

DOI: 10.4018/978-1-7998-9217-5.ch001

less efficient. Local governance processes have not always considered culture as one of its objectives, so it has not always been correctly used for the development of local societies.

The author understands social innovation from a territorial orientation, i.e., as new social, organisational or institutional arrangements or configurations established to modify social relations, to transform a framework of action or to propose new cultural orientations in a territory (Klein et al., 2009; Lévesque, Fontan & Klein, 2014; Klein & Pitarch, 2020). The territorial development process is stronger if it is the product of an innovative process that involves the local population and all internal and external stakeholders with an interest in the territory. "Bottom-up" development, based on participation and dialogue between citizens and between institutions, the enhancement of endogenous and exogenous resources, as well as the consolidation of shared leadership and the co-construction of knowledge that allows for seeing emerging development options, generate a more solid social capital, positive synergies between different sectors and sustainable and integrated development.

The concept of social innovation is closely related to notions such as solution and transformation, which ultimately means policy and practice with the capacity to influence action for a new, fairer, sustainable and ultimately free and responsible society (Pitarch-Garrido et al., 2019). In terms of tourism, it is especially important to guarantee an attractive, authentic and respectful experience for tourists, which is increasingly demanded and valued.

Today, the notion of cultural heritage refers to all assets, whether man-made or natural, tangible or intangible, without limit of time or place, collected or inherited by previous generations and preserved for transmission to future generations. It is noteworthy that it emphasises the value given by the subject within a community (which links it directly to the aspects of social innovation mentioned above), as an integral part of a social group that appropriates the elements that form part of its culture and history, but from the present moment on. This is a new categorisation of heritage as memory and the importance of cultural diversity as an unequivocal notion associated with it (Alba-Pagán & Pitarch-Garrido, 2018).

Social innovation, and the consequent co-creation process, makes the heritage and culture of each territory available for the creation of value and facilitates new business opportunities and novel concepts, especially in tourism. Some of its results have great indirect potential. An innovation process from the grassroots, from society, can offer new ideas, which are especially useful for the creation of future tourism products and services linked to heritage and local resources.

For the tourism industry, tools and content co-created through so-called social innovation processes, with a bottom-up perspective, are fundamental in adding value to the world's tourism destinations. This is important because the current crisis caused by the COVID-19 pandemic has irreparably affected the tourism sector, making it less competitive, globally and locally.

The tourists of the future will be more informed and specialised (Kuzior, et al., 2021; Wachyuni & Kusumaningrum, 2020). They will be looking for unique experiences, or at least experiences that cannot be easily replicated elsewhere. Cultural tourism is unique to each territory or destination and is capable of attracting a more informed, interested and spending tourist. In areas with competitiveness problems, and with an interesting heritage, tourism is the engine of development. However, there is also the danger of overstepping the limits and breaking the difficult balance between profit and saturation, which is why control, sustainability and local citizen initiative guarantee an adequate enhancement of heritage for visitors. Sustainability is, then, a key concept (Kuzior, et al., 2021; Ribes & Baidal, 2018).

The complexity of the implementation of social innovation initiatives linked to cultural tourism means that they are neither very abundant nor very well known. However, it can be affirmed that their success is enormous, overcoming any reluctance they may generate, especially among political decision-makers.

The main benefits or advantages of their implementation are linked to aspects that are difficult to quantify, which also favours their invisibility, such as the linking of the population and visitors with the heritage, that is, with the territory and its culture, the indirect generation of businesses in the creative economy (artists, craftsmen...) whose products or services tourists can purchase, the strengthening of social relations and solidarity networks between residents and tourists, which are particularly evident in periods of crisis but which remain strong for a long time, the conservation (respect) of the cultural heritage, etc.

There are examples in the world of social innovation initiatives that have led to the enhancement of local culture for tourism, which is of great interest because of its enormous value, its uniqueness, and because it can be an opportunity to overcome, through social innovation, a situation of social and economic crisis.

BACKGROUND: SOCIAL INNOVATION AND TOURISM

The concept of social innovation is one of the concepts that has generated the most literature in recent years, particularly in sociology and political science. In order to define a process or initiative as innovative from a social point of view, it must meet a series of characteristics that present certain discrepancies among academics. Without entering into controversy, social innovation is defined here as new social, organizational or institutional arrangements or configurations established to modify social relations, to transform a framework of action or to propose new cultural orientations in a territory (Klein et al., 2009; Lévesque, Fontan & Klein, 2014; Klein & Pitarch, 2020, Mosedale & Voll, 2017).

Social innovation, like any other kind of innovation, sometimes does not go beyond being a one-off experience that does not end up being consolidated. The characteristics of the environment are what determine society's capacity to generate and accept an innovation. Studies on this subject have shown that the existence of an open innovation ecosystem in which the participation of all social actors in the territory is promoted, favours the success of a product or service. As well, its acceptance by the customer or user is higher (Polese, et al., 2018).

The evolution of global capitalism since the 1980s has forged conditions that favour the social and territorial concentration of both wealth and poverty, which also leads to growing differences at both international and inter-regional levels. The latest international financial crisis, the pandemic crisis of 2020 and the current environmental crisis have made it clear that the dominant local or regional development models do not always contribute to reducing social differences or to achieving sustainable development. The current form of political, economic and social organisation generates resistance and alternative forms of action, which serve as a starting point for change, particularly in territories where they become innovations that demonstrate a different way of tackling social problems and finding solutions to them (Fontan et al., 2014). In short, social actors react when policies that do not contribute to solving their problems become evident. It is then that society claims what we could call "the right to the territory," giving rise to changes in the support structures for local development and the emergence of new forms of sharing responsibilities that involve both social actors and public entities. In this confrontation, new forms of governance that transform the relationship between the political decision-maker and civil society are experimented with (Moulaert et al., 2007). These experiences replace the priority traditionally given to economic growth and launch initiatives and projects that follow guidelines that are more respectful of the environment and quality of life.

This phenomenon is not new. Since the 1980s, and even before, the mobilisation of social actors has contributed to the establishment, in many places in Europe and America, of new forms of governance in the framework of public management that favour the integration of civil society in decision-making (Stöhr & Taylor, 1981). In all these cases, the territory, and its heritage, has become an identity reference that favours collaboration between actors belonging to different social groups. In some cases, local actors explore counter-hegemonic and innovative paths (Klein & Pitarch-Garrido, 2020).

In this context, we assume that social innovation implies social transformation. Thus, we understand social innovation in tourism as a complex process that incorporates new ideas (products, services or models) capable of responding to social needs and aspirations in a more appropriate way, while at the same time creating social relations or forms of collaboration between people that can be characterised as novel in order to achieve sustainable and respectful tourism development. Social innovation in tourism is capable of transforming the frameworks for action and proposing new formulas for tourist development that are not limited by the dominant institutional and economic situation. This context includes cultural heritage as one of the key elements that generate identity, value, personal and social ties, and that is at the basis of a territory's tourism development.

CULTURAL TOURISM: A FIELD OF DEVELOPMENT OF SOCIAL INNOVATION INITIATIVES

Social Innovation, Culture and Sustainability in Tourism

The main challenges of cultural heritage today are: to disseminate, communicate, promote and transmit cultural values. This leads to a desirable identification between society and the heritage that is part of its cultural legacy, and for which it is also responsible, as it is responsible for the actions of guardianship, conservation and dissemination (Alba-Pagán & Pitarch-Garrido, 2018). This transfer of knowledge must achieve different levels of communication with the aim of covering the widest possible social register from an inclusive and non-elitist perspective.

In recent years, UNESCO has stipulated that, in planned actions, heritage stakeholders should be aware that, on certain occasions, the level of participation in decision-making and the capacity of local actors to make real contributions are often limited. For this reason, it is necessary to involve the weakest actors (children, women, marginalized groups, youth, etc.), seeking to broaden the social spectrum of participation beyond the necessary collaboration that often takes place with cultural associations or groups interested in history, art or culture in general (Labadi, 2013). Thus, an effective and innovative participatory approach provides reciprocal benefits between cultural assets and society. One of the most effective ways of achieving this is through tourism.

These new approaches, in short, respond to the need to adapt to current conceptual changes, even with regard to seeing and understanding the world. These processes are framed within the issues related to social innovation, and understood, as noted above, as a process of creation, implementation and dissemination of new social and cultural practices in very different areas of society. It is therefore a matter of finding new ways of satisfying new social needs, in which the importance of the public sphere has grown exponentially, or of producing the behavioural changes needed to solve society's major challenges, empowering citizens and generating new social relations and new models of collaboration.

In short, it is a question of finding ways of responding to the great challenges of tourism: safety, authenticity, experience, sustainability, involvement of local society, etc. In this, the development of the culture of innovation finds its raison d'être in values such as creativity, risk-taking and the training of people in new knowledge and skills. Tourism, in this context, necessarily involves a change in social relations and a strengthening of citizen empowerment, both on the part of the host society and of visitors.

The existence of heritage in a territory makes a place unique. It implies the presence of a possible differentiating argument with respect to other cities or regions with other characteristics. The idea of betting on the activation of cultural heritage is one of the most important arguments to diversify and differentiate a territory from the point of view of the economy (particularly tourism). According to UNESCO, there are currently a total of 1,121 sites (869 cultural, 213 natural and 39 mixed) in 167 countries that have been declared World Heritage Sites. This recognition is more than just a "label," as the 1972 Convention concerning the Protection of the World Cultural and Natural Heritage points out; it is an effective tool for conserving sites and cultural forms of outstanding value to humanity, as well as protecting them from problems associated with climate change, uncontrolled urbanization, mass tourism and natural disasters, among others.

Throughout history, territories have undergone different learning processes that respond to a vision and strategy shared by local actors. This vision is the product of the social and economic structures that have taken shape over the centuries, all of which make up the culture of that territory. Sharing it with the rest of the world makes it valuable; at the same time, tourism contributes to making local culture visible. In so doing, it makes it part of everyone's heritage.

In some parts of the world, the lack of social appreciation of heritage has led to the loss and disappearance of representative elements of it. Tangible and intangible heritage bear witness to people's lives, to technology and to social relations throughout history.

From the point of view of culture, it is possible to apply a social innovation approach to sustainable development and the achievement of socially intelligent territories. This approach, together with the co-creation of a long-term strategy, including the collaboration of all social agents, is the basis of the new orientation that the European Union wants to promote at local and regional levels (European Commission, 2018). It is necessary to set in motion a process of teaching and co-learning to achieve an informed civil society capable of making decisions about its own future, based on scientific evidence that overcomes the barriers not only due to a lack of information, but also to preconceived ideas or ideas based on ideologies that do not correspond to reality.

Sustainable tourism is, according to the WTO (2001), tourism that takes full account of current and future economic, social and environmental impacts to meet the needs of visitors, of industry, of the environment and of host communities. The very definition of sustainable tourism includes the participation of social actors, as well as strong leadership by the public administration to ensure not only consensus, but also constant monitoring of actions aligned with the achievement of sustainable development objectives. Furthermore, the perspective of sustainable development in tourism has been especially important to fight against the problems generated by the negative impacts of this activity and to ensure the long-term viability of tourism activity. Sustainability in tourism is closely related to concepts such as resource conservation, empowerment of the host society, respect for the environment, participation, etc. (Edgell, 2019; Liu, 2003; Coccossis, 2008; Sharareh & Badaruddin, 2013; WTO, 2001; WTO, 2002; Mosedale & Voll, 2017). All of these, in particular the issues related to participation, are similar to, if not confused with, the objectives that define social innovation in tourism. Sustainability and social innovation are two concepts that interact and reinforce one another. The sustainable development of tourism is nowadays a priority, both for the WTO and for regional and local governments. In the search for sustainable tourism strategies, the management of cultural heritage and local resources has become a central concern (Coccossis, 2008).

As mentioned in the introduction, the existence of a favourable environment for social innovation will also be positive for the development of actions that guide tourism to achieve sustainable and integrated development. It should be noted that various authors speak of an innovative ecosystem to refer to technological innovations; however, this term has also been used to refer to spaces of interrelation between social actors, whose cooperation generates a suitable environment for achieving far-reaching social transformations. In tourism, Polese et al. (2018) have called these environments smart tourism ecosystems. These are socio-economic environments capable of generating social transformation, i.e., transforming innovation in tourism (often technical) into social innovation. This requires time and social maturity, which implies the existence of a solid relational social capital, characterized by strong bonds of trust and a well-established democratic and participatory culture. In those territories with an important cultural heritage, where tourism is a defining activity, this innovative ecosystem can help this tourism become the engine of sustainable development.

In this sense, it is essential to initiate a process of social innovation that focuses on coordinating different types of agents present in the territory in order to achieve joint learning, both horizontally, i.e., with respect to the rest of the local partners, and vertically, with partners from other territories that collaborate to create synergies and common paths. The university has a liaison function and also a function of knowledge transmission and transfer of science to society, while at the same time participating in the co-creation process, like the rest of the stakeholders (the quadruple helix). In this way, opportunities are generated to carry out processes of revitalisation of local heritage resources, as well as culture, environment and relational social capital networks. This revitalisation aims to establish a medium and long-term tourism development process to generate sustainable economic activity and social welfare, based on heritage and culture.

Social innovation in tourism is closely related to community-based tourism. It arises as a response to mass tourism, based on initiatives adapted to the potential of the territories and contributing to the development of their capacities. Community-based tourism, combining nature tourism and cultural tourism, is often a real and easy-to-implement alternative for local communities, usually in rural environments, to carry out their own development in contexts of crisis regarding traditional activities (Alkier, Milojica & Roblek, 2017; Arboleda Jaramillo et al., 2020). Tourism initiatives, which include an express will to achieve positive changes in society, are especially abundant in rural territories, where there seem to be fewer obstacles to a process of participation and co-creation by social agents with an active role in tourism. Therefore, these types of territories are especially likely to contribute to the development and consolidation of social innovation initiatives, even without intending to, spontaneously, as a consubstantial part of the development of community-based tourism.

Some cases of successful social innovation initiatives, not all of them in rural areas, are analysed below.

Some Keys to Understanding Social Innovation in Cultural Tourism Based on Case Studies

There are different case studies on initiatives, called social innovation in tourism, in different parts of the world, especially in developing countries (Alkier, Milojica & Roblek, 2017, Aksoy et al., 2019, Elias & Barbero, 2020, Polo de Lobatón, Expósito & Lobatón Polo, 2020, Gómez Carreto, et al., 2017, Ortega-

Hoyos & Marín-Verhelst, 2019, Quandt, et al., 2016, Díaz & Castro, 2017, Arboleda Jaramillo, et al., 2020, Martini, Buffa & Notaro, 2017). There are still no studies of tourism ecosystems of social innovation that help to create a theory in this regard. However, case studies are helping to better understand the impact of social innovation in the sustainable development of tourism destinations. More analyses of the reality of tourism territories are needed to help detect more and better good practices and, above all, to help define the key aspects that contribute to consolidating innovative ecosystems with social impact in favour of sustainable tourism development.

As noted in previous sections, the so-called quadruple helix (some already speak of the quintuple helix), which includes academia, public administration, companies and individuals (civil society) (Castillo-Vergara, 2020), is the basis for the generation and transfer of innovation. The synergies between actors from these social spheres consolidate and reproduce innovative processes with repercussions in all spheres. Moreover, each of them, individually, can develop initiatives that end up having a positive social impact. However, in this study, the objective has been the detection of social innovation initiatives developed by private entities, i.e., by companies of any type, particularly social enterprises or initiatives of social economy (Alkier, Milojica & Roblek, 2017). The justification for such a decision is practical: in tourism, it is private initiative that mainly generates the product, particularly in countries where the state has a minor, purely regulatory, role. In any case, the role that the administration has cannot be ignored. Local administration and the research centres, in particular the universities, have a role of development catalyst, to propitiate the necessary conditions that strengthen an innovative territorial system in which tourism can become an activity that generates innovative solutions to social problems. Entrepreneurship is basic to producing empowerment and sustainability in the tourist destination. Social innovation can be found in the philosophy of the company and also in the action or tourism product developed (Díaz & Castro, 2017). The role of the rest of the propellers is also fundamental, particularly in a second phase: beyond the immediate environment, they can contribute to disseminating the tourism model generated, becoming the predominant one. In the case of social innovation in cultural tourism, there is still no evidence of this second phase, which can become particularly relevant especially in cultural tourism, where the tourist experience can be linked to awareness-raising policies on issues related to heritage conservation, aspects of promoting inclusive tourism (awareness of social and functional diversity), etc.

Methodology

The aim of this section is to learn about social innovation initiatives in cultural tourism and how they can contribute to the sustainable development of the tourist destination. Qualitative methodology in social sciences, and in particular in tourism studies, is appropriate for the purpose of this research, as it is possible to collect information beyond the merely quantitative (Riley & Love, 2000; Gouthro, 2010; Hillman & Radel, 2018). The methodology followed here to gain in-depth knowledge of the possibilities of social innovation in cultural tourism was based on an online survey of a series of different types of initiatives, in order to gather information that is not publicly available on the internet. Nowadays, any type of tourism offer uses the internet and social networks for its commercialisation. All socially innovative tourism initiatives have a website and accounts on the most common social networks. The questionnaire was carried out in English and Spanish in order to reach the maximum possible number of companies of this type. The list of innovative initiatives was compiled using the information provided on the website of the Social Entrepreneurship Competition in Tourism, an event organized every year that brings together numerous entrepreneurs from around the world in order to reward, make visible and help grow companies in the tourism sector that generate a positive social impact. The lists of winners and finalists of the different editions were reviewed. In addition, a detailed internet search was carried out in order to include other companies that consider themselves to be social innovation initiatives. Once the list was compiled, it was sifted through to leave only those that have a cultural component. The result does not exhaustively reflect all initiatives of this type; however, they are a good example worldwide of socially innovative experiences in tourism. We have tried to make it a diverse sample from the point of view of the territory (different countries) and the tourism product offered. Not all of them responded to the survey. Of those that did not, the information was obtained through their websites.

The surveys are anonymous and, in their analysis, priority was given to the elements they have in common and that correspond to what we have called social innovation (table 1). The questionnaire was kept short, with open-ended questions, and was launched via Google-forms, sent directly to the email addresses of the websites of each of the initiatives detected. A total of 55 questionnaires were sent out, 20% of which were answered.

The questions in the questionnaire were as follows:

- Name of the initiative or company
- Type of company
- Number of employees (or participants in the initiative, including managers)
- Do you think that the company you represent can be considered a social innovation initiative?
- What is the main innovation that your company brings to cultural tourism? (Please explain)
- What is the main problem your company is currently facing? (Please explain)

Finally, they were also asked if they would like their company's or initiative's website to appear and if they would like to make any additional comments.

Results

Most of the initiatives that responded to the questionnaire indicated that their main problems were related to the limited possibilities of growth and expansion, especially due to the Covid-19 pandemic and to the reduced demand that still exists for this type of responsible tourism offer linked to a social objective. The latter, they point out, may be related to problems of effective marketing aimed at attracting a type of traveller that is still a minority.

The initiatives are small; all of them have fewer than 10 workers and half of them have fewer than 5.

The offer they make for the cultural tourist is always in small groups, normally aimed at people of mature age, and the product offered is of high quality and very personalised.

The features of the trip include the possibility of interacting with local communities, guided visits by experts, the possibility of learning and understanding the local culture from the inside. In addition, all initiatives make clear to tourists the positive impact their trip will have on the destination they are visiting. In general, tourism is considered as a way to make specific social groups visible, such as ethnic communities, people with disabilities, LGTBI people, etc. It is, therefore, what we could call conscious tourism. In fact, many of these initiatives try to avoid the word tourist by using the word traveller instead, which is already a clear stance for a very specific type of tourism. The customer consciously chooses this type of trip, even more so when the price of most of this offer is medium-to-high.

The initiatives analysed can be divided into two types: companies that deal directly with tourists and platforms to facilitate the placing of companies, accommodation, etc. on the tourism market. Although the former are more numerous in the study carried out here, the platforms have the advantage of being able to bring together under one roof a large number of entrepreneurs, all of whom share the same business philosophy.

Finally, it should be noted that companies and platforms incorporate terms such as ecotourism, slow tourism, sustainable tourism, social innovation, authenticity, learning, community and inclusion. Some include innovative forms in the product they offer, such as apps, storytelling and the possibility of booking a practically personalised product via the internet. However, others do not innovate in the product, but rather in the impact of the trip. Both elements, technological innovation and social impact, combine to offer the customer a unique, highly attractive product that raises awareness.

At least half of the initiatives analysed indicate that they have counted on the participation of different social agents in the design and promotion of their product. None of them have been part of a complete co-creation process, with the quadruple helix, but they have been part of a partial co-creation process with some of the social agents, especially with civil society and the local administration.

It is clear that the initiatives are socially innovative, and that is how they define themselves, but the client, the tourist or traveller is also a social innovator in the sense that they choose a type of experience aware of the impact it will have on the environment and the local and global society. The fight against climate change and the path towards the SDGs are among the objectives that some of these initiatives demonstrate.

CURRENT SITUATION AND CONSEQUENCES

Tourism initiatives that can be considered innovative from a social point of view involve clear improvements in the relationship between tourism and the host society. The design of the tourist product is carried out with the participation, effective or not, of the social group that will be favoured by the tourist development. Co-creation is an important part of a remarkable number of this type of initiatives. Therefore, from the supply side, social innovation in cultural tourism contributes to a more sustainable and integrated development.

On the demand side, multiple studies confirm that the tourists of the future will be more informed and specialized. They will be looking for unique experiences, or at least experiences that cannot be easily replicated elsewhere. Heritage is capable of attracting a more informed, more interested and more spending tourist. The originality and authenticity of social innovation initiatives meets the expectations of this profile of tourist or traveller, so demand is likely to be increasingly high.

The complexity of the implementation of social innovation initiatives linked to heritage means that they are neither very abundant nor very well known. However, it can be affirmed that their success is enormous, overcoming any reticence they might generate. The main benefits or advantages of their implementation are linked to aspects that are difficult to quantify, which also favours their invisibility, such as the linking of the population with their heritage, that is, with their territory and their culture, the indirect generation of businesses in the creative economy (artists, artisans...), not always well recognised by certain social groups, the strengthening of social relations and solidarity networks, which are particularly evident in periods of crisis, but which remain strong for a long time, the conservation (respect) of cultural heritage, etc.

Table 1. 20 examples of social innovative initiatives in Tourism based on culture and heritage.

Name	Country	Summary
Khusvegi English and Cultural Camp	Mongolia	It's community-led and designed to focus on slow travel. Khusvegi is also a women of color-led initiative that aims to create equity and inclusion.
Triple F Photo Tours	Canada, Uganda	Triple F Photo Tours is a female-led tourism company providing competitively priced, high quality, "non-touristy" trips with a philanthropic edge: participants learn photo skills while supporting or conducting workshops for local young women who are training to become photojournalists, all led by an experienced Asian-Ugandan-Canadian founder with a desire to improve the lives of others.
Black Tourism Talent Directory	United States	The Directory is designed to create broader diversity, equity and inclusion in the tourism industry by connecting tourism players directly with qualified black tourism talents.
GetAboutAble	Australia	This social enterprise focuses on improving travel and leisure options for people with disabilities.
Khwela Womxn	South Africa	Khwela Womxn is an organisation which uplifts young womxn in the tourism industry by providing various tools and resources for personal growth.
Local Purse	Sweden	Local Purse is a web-based platform that supports cultural artisans and local guides using live video shopping experiences to virtually host travellers around the world.
Get up and go Colombia	Colombia	Get Up And Go Colombia is an NGO, which promotes tourism in regions affected by the armed conflict through activities and projects with bilingual university students, young victims of the armed conflict, young people belonging to the LGBTI community and former combatants of armed groups, building peace, inclusion and diversity through tourism and languages.
Ecopreneur Evolution	Costa Rica	Ecopreneur Evolution creates opportunities for future entrepreneurs by providing the know-how and resources for them to solve today's most pressing social and environmental challenges.
Rai Kuen Rang's Farmstay	Thailand	Their mission is to enhance and preserve community culture through community-based tourism by using slow tourism as a tool.
Socialbnb	Germany	Socialbnb is an online platform that connects travellers with social and ecological organisations, which list their unused accommodations to gain financial independence from donations.
Bonding through Caring	Mexico	The project supports the local population of ten nearby villages to develop community ecotourism projects and also promote sustainable development.
Espresso yourself	El Salvador	The project aims to make sustainable coffee cultivation a playful experience for different target groups.
Impact Travel Collective	Brazil	The collective combines three interconnected platforms that redefine the way people give and take tours.
Know Your Tourist	Rwanda	Know Your Tourist is an education and market intelligence platform that strives to attract more tourists from their own country and Africa.
Locals From Zero	Slovenia	Locals From Zero is an online booking platform through which travellers can book authentic and yet undiscovered experiences.
Memopin	Armenia	Memopin helps tourism players deliver online experiences to travellers by leveraging a map and storytelling tools in one application to help people discover new places in the most personal way.
Mucha Miga	Spain	Mucha Miga aims to be a platform that connects rural destinations in Spain and potential tourists, taking into account the differentiated needs of each user.
Community Homestay Network	Nepal	This social enterprise works with grassroots communities to develop attractive tourism experiences through a value chain approach.
WildyNess	Tunisia	WildyNess offers a complete digital solution via an online platform and mobile application that connects travellers with hidden tourist attractions of Tunisia.
YESD Social Enterprise	Viet Nam	YESD assists homestay owners in managing their own business, trains local guides, provides free English classes and organises community development projects.

Source: (Own elaboration from https://socialtourismcompetition.com/)

The importance of social entrepreneurship in cultural tourism lies in the potential to create social value while generating economic benefit, in the implementation of social innovation activities that enhance the active participation of society and aim at sustainable development. The tourism system is, in itself, complex, as it involves multiple stakeholders, and dynamic, i.e. changing, interactions.

There are examples, as highlighted in this chapter, in the world of social innovation initiatives that have led to the enhancement of local culture. Innovative thinking in tourism goes beyond technology. All these initiatives can be considered as an opportunity to overcome a situation of economic, environmental and social crisis, and therefore deserve the recognition and institutional support of governments and international entities that advocate a more responsible and sustainable tourism, which contributes to generating well-being and combating the consequences of climate change, among others.

FUTURE RESEARCH DIRECTIONS

Social innovation in cultural tourism is still a very small phenomenon and very little known by academia. Practices and examples of social innovation are scarce and not always successful in the long term, which makes it difficult to study. It is necessary to continue analysing this alternative tourism offer, especially in order to define more adequately what we mean by social innovation in tourism and, on the other hand, to highlight the lack of data and statistics on this phenomenon, which seems to be growing and which is undoubtedly increasingly in demand.

A better understanding of the diversity of social innovation initiatives in cultural tourism is needed. Some authors (Quandt, et al., 2016) have already pointed out that social innovation in tourism territories does not necessarily respond to theoretical models that arise from social innovation in other types of territories. Tourism needs deeper analyses because of its enormous diversity and because social innovation in tourism, especially in cultural tourism, needs different and specific ways to measure its impact. The definition of social innovation as an innovative solution to human needs (Mulgan, 2006) implies, in the case of tourism, including key aspects closely linked to this type of activity, for example, the environment. Furthermore, it is necessary to count on the active participation of segments of society that are not linked to the tourist territory, except at the time of travel, i.e., tourists. Bearing in mind that the impacts of tourism are global, even if the specific activity is developed at a local level, the scope of the benefits of developing socially innovative tourism initiatives goes beyond the closest territorial environment, pointing towards good sustainable tourism practices for the entire planet.

To carry on with the previous idea, the proposal of methodologies for measuring the social impacts of tourism is a line of research that has not yet been developed. It is essential to agree on how the social innovation of tourism influences the reduction of poverty and inequality. On the other hand, there is a lack of research on the forms of governance of the territories in which socially innovative cultural tourism initiatives have been developed. In short, rigorous studies are needed on innovative ecosystems in tourism territories, on the networks and social capital that form the basis of social innovation (Elias & Barbero, 2020). Knowledge about how these innovative ecosystems or territories function, in which cultural tourism is the basis, clearly contributes to one of the most interesting characteristics of social innovation: replicability. In other words, it contributes to mutual learning and, with it, to achieving a change in the tourism model towards more sustainable forms of tourism that are more respectful of society and the planet.

CONCLUSION

The work presented here is exploratory, as there are few studies on this topic due to the limited data available, which explains why most of the available analyses are case studies. The relationship between social innovation and cultural tourism is even less well known. A review of the content of some initiatives of this type based on the information available on the internet allows us to obtain information on their objectives and the social value they promote. This information helps to qualify some of the cultural tourism companies as social innovation initiatives. This is not an exhaustive analysis but rather a sample that helps to understand how these initiatives work and what impact they have. Even so, the scarce information collected advises caution and further research on this topic, which is still new and has little to offer.

However, despite said caution, some interesting conclusions can be drawn. Notably, the process of tourism development in a destination should be based on collaboration, participation and co-creation of knowledge and policy. This depends on the ability of citizens, whether they are regular residents or not, to imagine their own future and take the initiative to transform the current scenario into the desired utopia. Given that not everyone can dream of the future in the same way, it is necessary to converge on common points and to cooperate, that is, to promote the participation of all the social agents involved in order to reach agreements that facilitate collective strategic decision-making and that, in turn, generate the quality knowledge necessary to continue evolving and taking advantage of the positive synergies from which all the stakeholders benefit.

Smart innovative tourism ecosystems are a novel way of understanding tourism territories, bringing together the definition of social innovation ecosystem (Klein, et al., 2009; Levesque, Fontan & Klein, 2014) and that of smart territory. The United Nations defines smart territories as innovative territories that use information and communication technologies and other means to improve the quality of life, efficiency and competitiveness of urban operations and services, ensuring that they meet the needs of present and future generations, as well as respecting cultural aspects. Intelligence, understood as technology development, is not at odds with social intelligence. The intelligence (knowledge) and innovative capacity of a territory is found in the people who inhabit it and, therefore, in the culture. It is this type of intelligence that social innovation aims to revitalise. In many cases, these are areas in crisis, physically and socially distant from the most important urban centres, where the most competitive activity is carried out, including tourism. The recovery of social intelligence is one of the challenges facing these territories, which can be called peripheral. Social innovation initiatives in regions or localities in central countries with a long tradition of tourism, such as European countries, are more invisible due to the large existing tourism offer. However, there is an increase in demand linked to the quality of the product offered and its originality. The process of co-creation definitely contributes to this success, although these are initiatives that remain small-scale and deeply rooted in the territory and its cultural resources.

Social innovation, and the consequent co-creation process, uses heritage and culture (in a broad sense) for value creation, and this facilitates new tourism business opportunities adapted to the challenges faced by this sector of activity. Some of its results have a great indirect potential for regional development, supporting new creative approaches that could become new innovative services, leading the change of traditional business models based on heritage or local culture.

A process of innovation from the grassroots, from society, can offer new ideas, especially useful for the creation of future products for the development of cultural and sustainable tourism. For this reason, institutional support and the creation of incentives for companies and territories that are committed to socially integrated and territorially sustainable tourism are particularly important.

This research received no specific grant from any funding agency in the public, commercial, or not-for-profit sectors.

REFERENCES

Aksoy, L., Alkire, L., Choi, S., Kim, P. B., & Zhang, L. (2019). Social innovation in service: A conceptual framework and research agenda. *Journal of Service Management*, *30*(3), 429–448. doi:10.1108/JOSM-11-2018-0376

Alba-Pagán, E., & Pitarch-Garrido, M. D. (2018). Las propuestas culturales socio-participativas en la ciudad de Valencia. Innovación social y cultura en los barrios: la emergencia necesaria. Actas del Congreso Internacional de Ciencia Regional.

Alkier, R., Milojica, V., & Roblek, V. (2017). Challenges of the social innovation in tourism. *Tourism in South East Europe.*, *4*, 1–13. doi:10.20867/tosee.04.24

Arboleda Jaramillo, C. A., Arias Arciniegas, C. M., Pérez Sánchez, E. O., & Correa Janne, P. (2020). Innovación social como estrategia para fortalecer el turismo rural comunitario en Colombia. *Revista Venezolana de Gerencia*, *25*(89), 92–104.

Castillo-Vergara, M. (2020). La teoría de las N-hélices en los tiempos de hoy. *Journal of Technology Management & Innovation*, *15*(3), 3–5. doi:10.4067/S0718-27242020000300003

Coccossis, H. (2008). Cultural heritage, local resources and sustainable tourism. *International Journal of Services Technology and Management*, *10*(1), 8–14. doi:10.1504/IJSTM.2008.020340

Díaz Zarco, V., & Castro Morales, F. (2017). El turismo emocional creativo: La ciudad como taller inclusivo. *International Journal of Scientific Management and Tourism*, *3*(1), 357–371.

Edgell, D. L. (2019). *Managing Sustainable Tourism: A Legacy for the Future*. Routledge. doi:10.4324/9780429318122

Elias, S., & Barbero, A. C. (2021). Social innovation in a tourist coastal city: A case study in Argentina. *Social Enterprise Journal*, *17*(1), 44–62. doi:10.1108/SEJ-02-2020-0011

European Commission. (2018). *A New European Agenda for Culture*. https://eur-lex.europa.eu/legal-content/EN/TXT/PDF/?uri=CELEX:52018DC0267&from=EN

Fontan, J. M., Klein, J.-L., Caillouette, J., Doyon, M., Lévesque, B., Tremblay, D.-G., Tremblay, P.-A., & Trudelle, C. (2014). Vers de nouveaux modèles d'action en développement territorial: L'expérimentation à l'échelle locale de la transition vers le «buen vivir». *Economie et Solidarites*, *44*(1-2), 84–102.

Gómez-Carreto, T., Zarazúa Escobar, J. A., Guillen Cuevas, L. A. & Castellanos Albores, A. C. (2018). Innovación social, turismo rural y empresas sociales. Evidencias desde el Sur-Sureste de México. *El periplo sustentable, 34*, 44-81.

Gouthro, M. B. (2010). Qualitative method research and the 'tourism experience': a methodological perspective applied in a heritage setting. In *Tourist Experience* (pp. 221–234). Routledge.

Hillman, W., & Radel, K. (2018). *Qualitative Methods in Tourism Research*. Channel View Publications.

Klein, J.-L., Fontan, J. M., Harrisson, D. & Lévesque, B. (2009). *L'innovation sociale au Québec: un système d'innovation fondé sur la concertation*. Cahier du CRISES, Collection Études théoriques, n. ET0907.

Klein, J.-L., & Pitarch-Garrido, M. D. (2020). La innovación social. In *Manual de Economía Social*. Tirant lo Blanch.

Kuzior, A., Lyulyov, O., Pimonenko, T., Kwilinski, A., & Krawczyk, D. (2021). Post-Industrial Tourism as a Driver of Sustainable Development. *Sustainability*, *13*(15), 8145.

Labadi, S. (2013). *UNESCO, Cultural Heritage, and Outstanding Universal Value: Value-based Analyses of the World Heritage and Intangible Cultural Heritage Conventions*. Rowman & Littlefield.

Lévesque, B., Fontan, J. M. & Klein, J.-L. (2014). *L'innovation sociale: les marches d'une construction théorique et pratique*. Presses de l'Université du Québec.

Liu, Z. (2003). Sustainable tourism development: A critique. *Journal of Sustainable Tourism*, *11*(6), 459–475. doi:10.1080/09669580308667216

Martini, U., Buffa, F., & Notaro, S. (2017). Community Participation, Natural Resource Management and the Creation of Innovative Tourism Products: Evidence from Italian Networks of Reserves in the Alps. *Sustainability*, *9*(12), 2314. doi:10.3390u9122314

Mosedale, J., & Voll, F. (2017). Social Innovations in Tourism: Social Practices Contributing to Social Development. In P. J. Sheldom & R. Danielle (Eds.), *Social Entrepreneurship and Tourism* (pp. 101–115). Springer. doi:10.1007/978-3-319-46518-0_6

Moulaert, F., Martinelli, F., González, S., & Swyngedouw, E. (2007). Social Innovation and Governance in European Cities Urban Development Between Path Dependency and Radical Innovation. *European Urban and Regional Studies*, *14*(3), 195–209. doi:10.1177/0969776407077737

Mulgan, G. (2006). The Process of Social Innovation. *Innovations: Technology, Governance, Globalization*, *1*(2), 145–162. doi:10.1162/itgg.2006.1.2.145

Ortega Hoyos, A. J., & Marin Verhelst, K. (2019). La innovación social como herramienta para la transformación social de comunidades rurales. *Revista Virtual Universidad Católica del Norte*, (57), 87–99. doi:10.35575/rvucn.n57a7

Pitarch-Garrido, M.-D., Sales-Ten, A., & Salom Carrasco, J. (2018). Políticas de nueva generación para la innovación social: el caso de la ciudad de Valencia. *XVII Congreso Internacional de Investigadores en economía Social: La Economía Social: transformaciones recientes, tendencias y retos de futuro*. http://ciriec.es/eventos/xvii-congreso-internacional-de-investigadores-en-economia-social-y-cooperativa/comunicaciones

Polese, F., Botti, A., Grimaldi, M., Monda, A., & Vesci, M. (2018). Social Innovation in Smart Tourism Ecosystems: How Technology and Institutions Shape Sustainable Value Co-Creation. *Sustainability*, *10*(1), 140. doi:10.3390u10010140

Polo de Lobatón, G., Expósito, M. Y., & Lobatón Polo, D. (2020). Gerencia social: Estrategia de innovación social con enfoque economía solidaria, para el turismo comunitario en las poblaciones palafíticas de la Ciénaga Grande de Santa Marta. *Revista Loginn: Investigación Científica y Tecnológica, 4*(1), 51–65. doi:10.23850/25907441.3137

Quandt, C., Ferraresi, A., Kudlawicz, C., Martins, J., & Machado, A. (2017). Social innovation practices in the regional tourism industry: Case study of a cooperative in Brazil. *Social Enterprise Journal, 13*(1), 78–94. doi:10.1108/SEJ-12-2015-0038

Ribes, J. F. P., & Baidal, J. I. (2018). Smart sustainability: a new perspective in the sustainable tourism debate. *Investigaciones Regionales-Journal of Regional Research*, (42), 151-170.

Riley, R. W., & Love, L. L. (2000). The State of Qualitative Tourism Research. *Annals of Tourism Research, 27*(1), 164–187.

Sharareh, A. D., & Badaruddin, M. (2013). Local Perception of Tourism Development: A Conceptual Framework for the Sustainable Cultural Tourism. *J. MGMT & Sustainability, 31*, 31–40.

Stöhr, W., & Taylor, F. (Eds.). (1981). *Development from above or below? The Dialectics of Regional Planning in Developing Countries*. Wiley & Sons Ltd.

Wachyuni, S. S., & Kusumaningrum, D. A. (2020). The Effect of COVID-19 Pandemic: How are the Future Tourist Behavior? *Journal of Education. Society and Behavioural Science, 33*(4), 67–76.

WTO. (2001). *The concept of sustainable tourism*. http://www.worldtourism.org/sustainable/concepts.htm

WTO. (2002). *World tourism stalls in 2001*. https://www.world-tourism.org/

ADDITIONAL READING

Ciro, A., Toska, M., & Nientied, P. (2019). Social innovation and sustainable economic development: Participatory tourism destination management. In *The Role of Public Sector in Local Economic and Territorial Development* (pp. 173–192). Springer. doi:10.1007/978-3-319-93575-1_10

Garau, C. (2017). Emerging Technologies and Cultural Tourism: Opportunities for a Cultural Urban Tourism Research Agenda. In Tourism in the City (pp. 67-80). Springer.

Hjalager, A. M. (2009). Cultural Tourism Innovation Systems–The Roskilde Festival. *Scandinavian Journal of Hospitality and Tourism, 9*(2-3), 266–287. doi:10.1080/15022250903034406

Klein, J.-L., & Harrisson, D. (2007). *L'innovation sociale. Émergence et effets sur la transformation des sociétés*. Presses de l'Université du Québec.

MacCallum, D. (Ed.). (2009). *Social Innovation and Territorial Development*. Ashgate Publishing, Ltd.

Martínez-Pérez, Á., Elche, D., García-Villaverde, P. M., & Parra-Requena, G. (2019). Cultural Tourism Clusters: Social Capital, Relations with Institutions, and Radical Innovation. *Journal of Travel Research, 58*(5), 793–807. doi:10.1177/0047287518778147

Peeters, L. W., & Ateljevic, I. (2017). Women Empowerment Entrepreneurship Nexus in Tourism: Processes of Social Innovation. In Tourism and Entrepreneurship (pp. 94-109). Routledge.

Sørensen, F. (2007). The Geographies of Social Networks and Innovation in Tourism. *Tourism Geographies*, *9*(1), 22–48. doi:10.1080/14616680601092857

KEY TERMS AND DEFINITIONS

Co-Creation: Process by which agreements or consensus are reached to make decisions that have an impact on society as a whole or on a specific group.

Cultural Heritage: This term refers to all assets, whether man-made or natural, tangible or intangible, without limit of time or place, collected or inherited by previous generations and preserved for transmission to future generations.

Innovative Tourism Ecosystem: Social, economic, and environmental context, generally linked to a territory, that contributes positively to generating and consolidating innovative tourism initiatives from which, in turn, they benefit positively. This ecosystem includes a solid relational social capital, a highly qualified population, a mature democratic and participatory culture, and adequate regulation (legislation) to promote sustainable tourism and participatory governance.

Social Innovation: New social, organizational, or institutional arrangements or configurations established to modify social relations, to transform a framework of action or to propose new cultural orientations in a territory.

Sustainable Tourism: Tourism that takes full account of current and future economic, social and environmental impacts to meet the needs of visitors, industry, the environment and host communities.

Tourism Stakeholders: Social actors who have an interest in tourism activities or in the impacts generated by tourism.

Community-based tourism: The set of tourism activities generated and managed by the local community, of endogenous origin, and whose benefits accrue to the territory in which it is generated. It is always linked to local tourism resources, both natural and cultural.

Chapter 2
Cultural Routes as a Tool for Sustainable Tourism and Socio-Economic Development

Ruhet Genç
Turkish-German University, Turkey

ABSTRACT

Connecting various attractions and tourist destinations within a country or across many countries, cultural routes rely on a region's cultural legacy to give learning opportunities and tourist delight. These heritage-based routes that exist in all parts of the world, on various scales, features, and purposes, are considered an important tool for the social and economic development of tourism destinations via cultural tourism and feature in tourism marketing that offers various interests such as gastronomy, arts, linguistics, architecture, spirituality, religion, and sports. This chapter examines the contribution of cultural routes to sustainable tourism and socio-economic development due to the significance they have. The methodology will be literature review and critical evaluation of the findings.

INTRODUCTION

Cultural route is a recently developed concept for tourism development. Connecting various attractions and tourist destinations within a country or across many countries, it offers the discovery of history and culture associated with the region. It is an innovative tool for revitalizing, promoting and preserving cultural heritage (McCormack, 2017). As complex and multidimensional as it is, what a cultural route represents is broader than physical pathways. It means a network of places united by a common theme, seen in various forms with the characteristics of each place (Berti, 2015). According to the Council of Europe (CoE), a cultural route refers to a cultural project that intends to develop and promote an itinerary or a network based on a historic route, a cultural concept, figure, or phenomenon which involves transnational importance to demonstrate common European values (2010).

Cultural routes may appeal to various interests such as gastronomy, arts, linguistics, architecture, spirituality, religion, and sports. (Puczkó & Rátz, 2000, 2007). Thus, they are attractive travel destinations

DOI: 10.4018/978-1-7998-9217-5.ch002

with creative activities that serve multiple purposes. They promote sustainable tourism and simultaneously improve cultural and socio-economic development.

This chapter will examine the contribution of cultural routes to sustainable tourism and socio-economic development due to the significance they have. The methodology will be literature review and critical evaluation of the findings. First, characteristics and types of cultural routes will be presented based on their extent, purpose, duration, and creation, namely whether they are naturally evolved or purposively designed for tourism attraction. Second, the context and some examples of cultural routes will be reviewed. Then, the ways and objectives of creating them will be studied along with the significant aspects of their planning, development, and management in the tourism industry. Finally, the potentially beneficial aspects, the structure of activity they bring into the tourism industry, and the associated socio-economic outcomes will be analyzed.

BACKGROUND

Characteristics and Types of Cultural Routes

As Timothy & Boyd (2014) classified, cultural routes may be assessed according to their formation in two broad perspectives: organically evolved cultural routes and purposive cultural routes. The former is developed in ancient days and used by native people for various needs whereas the latter is, built purposefully in modern times for recreational and touristic use and has significant developmental value. Both are cultural routes by their human-made nature and human activities.

Organic Cultural Routes

Today following the earliest footprints of a true historic trail or a modern or an intentional path, which are often founded on ancient indigenous hunting or migratory routes (Goodey, 1975; MacLeod, 2004), organic cultural routes are cultural in nature. Some may be based on linear-shaped natural resources like rivers, escarpments, or coasts (Schutt, 1997; Timothy, 2009).

Some ancient hunting, commerce, and migratory pathways have developed into today's roads and motorways, some still exist in rural regions and serve footpaths for recreational activities (Timothy & Boyd, 2014). For instance, as accounted by Murray and Graham (1997) Europe's Camino de Santiago pilgrimage road Roman Catholic pilgrims created over centuries has changed geographically, temporally, and purposefully from its original function to a more generic tourism role. As the support infrastructure is built along the road such as marked paths, bridges, hostels, inns, and the new entry points to the road are included to provide accessibility, the function of the route has evolved to have additional uses with increased mobility and more diverse users. Eventually, with the introduction of additional facilities, such as accommodation, food, and tourist information that are made available to trail users, the sites along the route have become more significant to visitors than just reaching the destination which was the original function of the road. Thus, the route has become a path for tourism services in a more extended scope, with its multi-directional use, numerous means of mobility such as walking, horseback riding, bicycles, and motor vehicles, and product diversification with market growth efforts (Timothy & Boyd, 2014).

Organic cultural routes come in a variety of forms that are frequently used and marketed as tourist trails, as presented in Table 1. International organizations such as the World Tourism Organization or

Table 1. Varieties of organic cultural routes

Trade routes	• The Silk Road • The Frankincense Route
Explorer, settler, and migration routes	• The Mormon Trail • The Chilkoot Trail • Historic Route 66 • The Slave Route
Religious routes or pilgrimage trails	• The Mormon Trail • The Way of St James (Camino de Santiago)
Intentionally-built linear resources	• Historic railways • Canals and towpaths • Political borders

Source: (Adapted from Timothy & Boyd, 2014)

UNESCO, focusing on a number of traditional and famous trade routes, attempt to build long-distance cultural routes in line with the economic activities that formerly characterized them. For example, the Silk Road that has numerous interconnected branches and sub-routes (Misra, 2011; Tang, 1991) has been an enduring project of The World Tourism Organization (UNWTO) (1996) and UNESCO (Shackley, 2003) in terms of making it a cultural tourism product. Many governmental bodies manage and preserve these trails as in the case of the Chilkoot Trail which is in the control of Parks Canada and the US National Parks Service (Jackson et al., 2003; Masson, 2001) or the Historic Route 66 attracting an increasing number of cultural heritage tourists which influence the Route 66 Corridor Preservation Program enacted by US Congress (DuVal, 2001; Wallis, 1992). The Slave Route that begins in West Africa's villages, slave forts, and monuments (Teye & Timothy, 2004; Timothy & Teye, 2004; Yankholmes & Boakye, 2010), and offers many itineraries across countries, locations, and organizations was created by UNESCO in 1994 to honor and reveal the tragic events of the trans-Atlantic slave trade and thus, to fight racism.

Purposive Cultural Routes

Referring to trails that have been created by design for modern touristic and recreational use, purposive cultural routes are established based on a predetermined geographic region that comprised nodes of similar thematic appeal. Their thematic content that connects locations, businesses, and communities with similar histories and goods that appeal to similar demands, rather than the historical affiliation with a specific route, is what defines these designed cultural routes. It is stated by Logan (2002) that it is essential to view the places along the path holistically as a whole thematic compilation rather than separately. For instance, the theme of 'Highway 89 was designed as a scenic mountain route in northern California to promote the surrounding villages altogether for collective benefits, although the related rural communities did not possess the financial means to promote themselves independently (Smith et al., 1986).

Purposive routes that exist in various areas of the world last for decades in some cases. In addition, new routes will very certainly be included in the tourism sector. However, many trails disappear after some time as the associated region or community loses the motivation to work together for common benefits. It is common for failed routes that certain tourism enterprises or locals neglect their responsibilities in promoting the trail's aims and purposes, which in turn would cause the created network to fail

Table 2. Varieties of purposive cultural routes

Short trails at historic sites and gardens	• Paths at the archaeological site of Tulum, Mexico
Maritime routes	• Route of the Bounty, Tahiti • The South Pacific Whaling Route • The Maine Maritime Heritage Trail, US
Urban heritage trails:	• The ancient walls of Girona, Spain • Rothenburg ob der Tauber, Germany • Freedom Trail, US • Little India Heritage Trail, Fort Canning Trail, Chinatown Trail, Bukit Timah Trail, Peranakan Trail, Kampong Glam Trail and the Civic District Trail, Singapore
Literary, film and music trails	• The Catherine Cookson Trail, UK • The Cromwell Trail, New Zealand • The Robert Burns Trail, Scotland • Fairy Tale Route, Germany
Industrial trails	• The Route of the Clockmakers, France • The Pyrenean Iron Route, Spain • The Iron Road, EU • The Otago Goldfields Heritage Trail, New Zealand • The Colliery Route, Canada
Agriculture trails: • Food and wine routes • Beer and whiskey trails • Food trails	• The Niagara Wine Route • The Bairrada Wine Route, Portugal • The Ale Trail, Canada • The Bacon & Ale Trail, Canada • Lorraine's Beer Route, France • The Brewery Trail, Germany • Malt Whiskey Trail, Scotland • Ruta de Tequila, Mexico • The Charlevoix Flavour Trail, Canada • Bellarine Taste Trail and the Taste of the Tropics Trail, Australia
Religious trails (inorganic)	• St Patrick's Trail, Ireland • the Jesus Trail, Israel • the Gospel Trail, Israel

Source: (Adapted from Timothy & Boyd, 2014)

(Scott et al., 2008). Even so, they remain a valuable tourism resource by facilitating access to important tourist attractions.

Similar to organically evolved cultural routes, purposive routes come in various types (Timothy & Boyd, 2014), most prominent of which are presented in Table 2, along with their most well-known examples.

Besides being classified according to their creation, the cultural routes can also be classified in several other categories as shown in Table 3. Basic distinguishing factors related to the cultural route are as follows: the route's structure and physical substratum, historical data about its use, any physical structures associated with its material purpose and functionality, communication elements, and the existence of cultural manifestations of shared origin along the route (ICOMOS, 2008)

Purposes of Cultural Routes

With constantly growing popularity as significant tourism resources, cultural routes deliver public enjoyment, preservation of heritage, and image enhancement of region. They have a wide range of touristic, social, economic, and even political functions.

Table 3. Types of cultural routes according to several other indicators

territorial extent	local, national, regional, continental, intercontinental
cultural scope	within a specific cultural region or across different regions that continue to form mutual cultural values
purpose	social, economic, political, cultural, mixed
time span	obsolete, contemporary
structural configuration	linear, circular, cross-shaped, radial, network.
natural environment	land, water, air, mixed

Source: (Based on ICOMOS charter on cultural routes, 2008)

Cultural routes principally serve to preserve historic values and to protect cultural resources including migration routes, exploration trails, railways, and logging pathways (Hayes & MacLeod, 2007; Ramsay & Truscott, 2003; Snowball & Courtney, 2010; Timothy, 2014). By delivering details about and memorializing events, places, and peoples from the past, they improve the territorial image and social, psychological, and political empowerment of locals (Timothy, 2007), and, in turn, advance efforts to protect it (Timothy & Boyd, 2003). Furthermore, they help to build a sense of place by inviting outsiders to explore local cultures (Hayes & MacLeod, 2007).

Another advantage of cultural routes is to create a positive image for the place, which will bring social, psychological, and political benefits to local communities (Timothy, 2007). By supporting a cultural identity for the locals and a sense of place for their communities, cultural routes promote the social development of regions (Markwell et al., 2004). Moreover, they are important means for economic development (Garcés Feliú, 2011) through promoting the growth of heritage and cultural tourism (Hitrec, 1996; Moulin & Boniface, 2001). The economic prospects of cultural routes in terms of bringing revenue to the area, bringing new job opportunities for residents, and generating additional tax earnings is a substantial motivation for their establishment. Both the people work mainly in the management of the cultural routes and the organizations that provide retail activities or other related services benefit from the economic opportunities offered during this process. The economic stimulation is prevalent particularly around purposive cultural trails where individual tourism-related enterprises are set up such as wineries (Timothy & Boyd, 2014).

Additionally, cultural trails, similarly to other cultural heritage assets, are occasionally cultivated to serve as political tools or other means of exerting influence and persuasion (Timothy & Boyd, 2003, 2006). Because the people or institutions that construct a particular route simultaneously define the features of the future events it intends to memorialize, cultural routes become political accordingly by their sheer nature of maintaining and shedding light on parts of human history. Travelers are directed along a predetermined trail that portrays a specific narrative or image of a location, which mostly gives priority to the dominant culture or powerful people when designing the content (Shaw and MacLeod, 2000). For instance, the Sun at-sen Heritage Trail in Hong Kong aims to honor the renunciation of colonial control in the former British colony and by deliberately avoiding most references to the colonial era and by favoring landmarks in Hong Kong that were pivotal to Sun Yat-revolution, which in turn it manifests a strong political message (Teather & Chow, 2003).

There are many cultural routes focusing on the peoples, places, and events that have been known through unofficial sources of history, with the aims of portraying a more comprehensive truth about history and culture. As shown in the works of Shaw and MacLeod (2000) Clerkenwell Historic Trail placed in London highlight the marginal places and immigrant peoples that are largely ignored or unknown in official history of the city. Similarly, in the southeastern US, to counteract the asymmetry in the ways of presenting the history with inclusive views of slavery, the misery of African Americans in the southern US, and the important role African Americans' have played in the formation of the American nation, various black history, Civil War, and civil rights routes have been created (Ebony, 1990; Geiger & Werner, 2009; Kammen, 1993; Mahoney, 1999; Sevigny, 1992). Also, as seen in the case of the European Cultural Routes Programme with its intention of promoting unified Europe, pan-European interaction, and the "European ideal" (Moulin & Boniface, 2001) and strengthening the collective European identity through a network of cultural routes (Hitrec, 1996; Moulin & Boniface, 2001), these trails have become a political means to form a sense of regional, national, or supranational identity.

FINDINGS AND DISCUSSION

Cultural Routes as a Cultural Tourism Products

World Tourism Organization (UNWTO, 2017) defines cultural tourism as excursions with the primary or secondary motivation of learning, discovering, experiencing locations and events of cultural and historical significance that have become part of a community's cultural legacy. Contrary to mainstream mass tourism which mostly does not appreciate the unique ambiance of travel destinations and does not support the sustainability of the local economy, cultural tourism has as its core element the idea of a particular sense in the visited places known for their authentic nature, history, architectural attractions, and cultural elements.

Over the recent years an abundance of studies, conferences have been delivered about cultural tourism on account of elaborating the features, effects, and implications of this type of tourism (Timothy & Boyd, 2003; Smith, 2003; Pedersen 2002; McKercher & du Cros, 2001; Shackley, 1998; Richards, 1996; Ashworth, 1995; Swarbrooke, 1994; ICOMOS&WTO, 1993, Cohen, 1988). Upon the improvement of relevant literature and the realization of prospects cultural tourism may provide, the stakeholders in the tourism sector devoted much effort in making various touristic locations favorable for potential tourists who are in quest of interesting and memorable experiences around original environments. Accordingly, investments in resources as cultural routes have become essential as they serve as innovative and sustainable cultural tourism products and simultaneously contribute to the socio-economic development of tourism destinations. Examples of newly developed cultural routes and corridors on a national or international scale have appeared frequently during this course. For instance, the Punta Gorda Trail was found In Belize for preserving traditional Garifuna and Maya cultures, including architecture, agriculture, cuisine, and language (Boucher, 1990). Also, to boost the demand for cultural tourism in Serbia and the Balkans, a cluster of cultural routes was established in 2013 including The Valley of the Dragons, The Nikola Tesla and Stanojević Ways, and The Mining Route (Graf, 2013). Another example is Romania where two significant plans have been implemented to build new cultural routes namely, the Transhumance Route project and the Route of Fortified Churches (Oprică, 2015). The Council of Europe, numerous other national and international organizations have promoted these enterprises which increased the tourist flow in the area who are in search of discovering authenticity of before-mentioned countries.

Cultural Routes of the Council of Europe

The Cultural Routes Programme was initiated by the Council of Europe in 1987 with the aims of preserving the common European culture and heritage and promoting the knowledge of Europe's history. Since the Programme have focused on management of the cultural heritage sustainably and responsibly, it has brought many other benefits for communities inhabited around the routes such as increased social welfare and economic development. The main concept is to develop networks connecting cultural places that incorporate a regions' whole diversity of creative assets (Richards, Russo & Grossman 2008). Consequently, the development of cultural corridors paved the way for long-term international cultural heritage promotion programs between European neighbors. The program also launched some major projects to conserve natural and cultural heritage resources. Moreover, the programme correlates directly to current European cultural tourist development patterns. It demonstrates that creation of special interest tourism sites may render people engaged in the preservation of cultural heritage, natural resources, and environmental management. Recently established cultural routes exhibited great potential for SME development, clustering, networking between different cultures.

The Committee of Ministers of the Council of Europe establishes the rules for issuing the Cultural Route of the Council of Europe certification and the applicant countries should meet some basic conditions. Furthermore, the European Cultural Routes Institute offers guidance and professional support for cultural route development and provides training opportunities for institutions work in cultural route management.

The Council of Europe has established 45 cultural routes so far, with different themes that present European memory, history, and heritage, which are demonstrated in Table 4 and there are 16 new applicant countries. Today, these cultural routes host hundreds of cultural events, educational tours, and projects.

Sustainable Development of Cultural Routes

There are a multitude of methods on the creation, operation, and management of cultural routes around the world as it incorporates numerous participants actively such as NGOs, public associations and universities, public and private investors, tour operators, and the local community itself (Aspridis et al., 2018). The process of designing and developing the cultural route necessitates cooperative work by local authorities and entrepreneurs, tour companies and operators and residents (MacLeod 2016).

Research indicates that cultural routes and the tourism industry cooperate in different ways with the aim of promoting cultural tourism (Hospers, 2002; Meyer, 2004; Briedenhann & Wickens, 2004; Jansen-Verbeke, Priestley & Russo, 2008; Khovanova-Rubicondo, 2012; Graf, 2013). Activities that aim to advance the prominence of cultural routes are put into practice such as organizing contemporary festivals and events in related places, using interactive guidelines and advertisements in the media, and marketing through multiple actors in the tourism sector. Branding of the cultural route is considered essential to realize its full potential in terms of attracting new visitors (Vana & Malaescu 2016). The prominent use of brochures, maps, and audio recordings are important for assisting visitors in navigating the destinations, as they serve as sources of information, interpretation, and marketing (Markwell et al., 2004; O'Brien & Curtis, 1979).

Last but not least, since there are many stakeholders in the fields of culture and tourism, the management of cultural routes is especially significant. Creation, promotion, and operation processes of cultural routes can lead to the development of new innovative products. Additionally, for cultural routes

Table 4. Cultural routes of Council of Europe

Cultural Route	Certification Date	Countries With Network Members	Theme
Santiago de Compostela Pilgrim Routes	1987	Belgium, France, Italy, Lithuania, Netherlands, Poland, Portugal, Spain	Pilgrimage
The Hansa	1991	Germany, Latvia, Netherlands, Poland, Russian Federation, Sweden, United Kingdom	Trade & Historic
Viking Route	1993	Denmark, Estonia, Finland, France, Iceland, Irelan, Latvia, Netherlands, Norway, Poland, Spain, Sweden, United Kingdom	Historic
Via Francigena	1994	France, Italy; Switzerland, United Kingdom	Pilgrimage
Routes of El legado andalusí	1997	Egypt, Italy, Jordan, Lebanon, Morocco, Portugal, Spain, Tunisia	Artistic
Phoenicians' Route	2003	Albania, Belgium, Croatia, Cyprus, France, Greece, Italy, Lebanon, Malta, Spain, Tunisia	Historic & Navigation
Iron Route in the Pyrenees	2003	Andorra, France, Spain	Industrial
European Mozart Ways	2004	Austria, Belgium, Czech Republic, Germany, Greece, Switzerland, Ukraine, United Kingdom, The United States of America	Artistic
European Route of Jewish Heritage	2004	Austria, Azerbaijan, Belgium, Bosnia and Herzegovina, Croatia, Czech Republic, France, Georgia, Germany, Italy, Lithuania, Poland, Portugal, Republic of Belarus, Romania, Serbia, Spain, Turkey, United Kingdom	Historic & Pilgrimage
Saint Martin of Tours Route	2005	Belgium, Croatia, France, Germany, Hungary, Italy, Luxembourg, Netherlands, Slovak Republic, Slovenia, Austria	Pilgrimage
Cluniac Sites in Europe	2005	France, Germany, Italy, Poland, Spain, Switzerland, United Kingdom	Pilgrimage
Routes of the Olive Tree	2005	Albania, Algeria, Croatia, France, Greece, Italy, Lebanon, Montenegro, Morocco, Portugal, Republic of North Macedonia, Slovenia, Spain, Tunisia, Turkey	Landscape & Trade
Via Regia	2005	France, Germany, Poland, Republic of Belarus, Ukraine	Historic
The Romanesque Routes of European Heritage	2007	Austria, France, Germany, Italy, Portugal, Romania, Serbia, Slovak Republic, Spain	Architecture
Iter Vitis Route	2009	Armenia, Azerbaijan, Croatia, Cyprus, France, Georgia, Greece, Israel, Italy, Luxembourg, Montenegro, Portugal, Republic of Moldova, Republic of North Macedonia, Romania, Russian Federation, Slovenia, Spain, Tunisia, United Kingdom	Landscape
European Route of Cistercian Abbeys	2010	Belgium, Czech Republic, Denmark, France, Germany, Italy, Netherlands, Poland, Portugal, Spain, Sweden, Switzerland, United Kingdom	Architecture
European Cemeteries Route	2010	Austria, Bosnia and Herzegovina, Croatia, Denmark, Estonia, France, Germany, Greece, Ireland, Italy, Netherlands, Norway, Poland, Portugal, Romania, Serbia, Slovenia, Spain, Sweden, United Kingdom	Historic
Prehistoric Rock Art Trails	2010	Azerbaijan, Finland, France, Georgia, Italy, Norway, Portugal, Spain	Artistic
European Route of Historic Thermal Towns	2010	Austria, Azerbaijan, Belgium, Croatia, Czech Republic, Estonia, France, Georgia, Germany, Greece, Hungary, Italy, Luxembourg, Poland, Portugal, Spain, Turkey, United Kingdom	Historic & Landscape
Route of Saint Olav Ways	2010	Denmark, Finland, Norway, Sweden	Pilgrimage
European Route of Ceramics	2012	Austria, Azerbaijan, France, Germany, Italy, Netherlands, Poland, Spain, Turkey	Artistic
European Route of Megalithic Culture	2013	Denmark, France, Germany, Italy, Netherlands, Spain, Sweden, United Kingdom	Historic
Huguenot and Waldensian Trail	2013	France, Germany, Italy, Switzerland	Historic & Religious
Architecture of Totalitarian Regimes of the 20th Century in Europe's Urban Memory	2014	Albania, Bulgaria, Croatia, Italy, Romania	Architecture
Réseau Art Nouveau Network	2014	Austria, Belgium, Cuba, France, Germany, Hungary, Italy, Latvia, Norway, Portugal, Romania, Serbia, Slovenia, Spain, Switzerland	Artistic
Via Habsburg	2014	Austria, France, Germany, Switzerland	Historic
Roman Emperors and Danube Wine Route	2015	Bulgaria, Croatia, Hungary, Romania, Serbia	Historic & Landscape
European Routes of Emperor Charles V	2015	Algeria, Belgium, France, Germany, Italy, Morocco, Netherlands, Panama, Portugal, Spain, Tunisia	Historic
Destination Napoleon	2015	Belgium, Croatia, Czech Republic, France, Germany, Greece, Italy, Poland, Portugal, Spain	Historic
In the Footsteps of Robert Louis Stevenson	2015	France, Germany, United Kingdom	Landscape & Artistic
Fortified Towns of the Grande Region	2016	France, Germany, Luxembourg	Historic & Architecture
Impressionisms Routes	2018	Belgium, Croatia, Denmark, Finland, France, Germany, Greece, Hungary, Lithuania, Netherlands, Poland, Russian Federation, Slovenia, Spain	Artistic
Via Charlemagne	2018	Belgium, France, Germany, Italy, Luxembourg, Spain, Switzerland	Historic & Religious

continued on following page

Table 4. Continued

Cultural Route	Certification Date	Countries With Network Members	Theme
European Route of Industrial Heritage	2019	Austria, Belgium, Bulgaria, Croatia, Czech Republic, Denmark, Finland, France, Germany, Greece, Hungary, Ireland, Italy, Latvia, Liechtenstein, Luxembourg, Netherlands, Norway, Poland, Portugal, Serbia, Spain, Sweden, Switzerland, Turkey, Ukraine, United Kingdom	Industrial
Iron Curtain Trail	2019	Austria, Belgium, Croatia, Czech Republic, Germany, Greece, Hungary, Lithuania, Serbia, Slovak Republic, Turkey	Historic
Le Corbusier Destinations: Architectural Promenades	2019	Argentina, Belgium, France, Germany, Japan, Switzerland	Architecture
Liberation Route Europe	2019	Belgium, Czech Republic, France, Germany, Luxembourg, Netherlands, United Kingdom, United States of America	Historic
Routes of Reformation	2019	Austria, Czech Republic, Germany, Hungary, Italy, Poland, Slovenia, Switzerland	Artistic & Pilgrimage
European Route of Historic Gardens	2020	Georgia, Germany, Hungary, Greece, Italy, Poland, Portugal, Spain	Historic & Landscape
Via Romea Germanica	2020	Austria, Germany, Italy	Pilgrimage & Landscape
Aeneas Route	2021	Albania, Greece, Italy, Tunisia, Turkey	Archaeology & Artistic
Alvar Aalto Route	2021	Denmark, Estonia, Finland, France, Germany	Architecture
Cyril and Methodius Route	2021	Bulgaria, Czech Republic, Greece, Hungary, Slovak Republic, Slovenia	Religious
European Route d'Artagnan	2021	Belgium, France, Netherlands, Portugal, Spain	Historic &Artistic
Iron Age Danube Route	2021	Austria, Croatia, Hungary, Slovenia	Archaeology & Landscape

Source: (Council of Europe, 2021)

to become a successful tool for sustainable development by enriching the tourist offer, cooperation of different institutions, associations, and organizations on multiple levels is of utmost importance (Terzić et al., 2014a, 2014b).

Impacts of Cultural Routes on Socio-Economic Development

Cultural routes as a tourism product affect the lives of communities in the destination (Campolo et al., 2016). One of the most significant impacts is that they benefit the residents economically by creating a positive image for the area which cannot become apparent as an important tourism destination prior to the creation of cultural route (Stoddart and Rogerson 2009). Additionally, they create new job opportunities and improve the tourism infrastructure including hotels and roads (Káposzta et al. 2016). For example, cultural routes in Alberta, Canada, bring income and create jobs for locals besides allowing the visitors to interact with Aboriginal Canadian culture (McCormack, 2017). Moreover, cultural routes can eliminate seasonality by lessening visitor concentration in certain areas (Butler & Ivanovic, 2016).

In terms of urban recreation and spatial design, cultural trails are critical since most of them also entail driving and public transportation. They have play an important role for transportation and traffic flow, infrastructure investment, economic growth, and conservation (Javaheri, 2011; Orbaşli, 2000; Wang & Hji-Avgoustis, 2011), not to mention their capacity to assist in place-making exercises (Markwell et al., 2004).

According to Berki & Csapó (2008), main conveniences in developing a cultural route are:

- From the supply side, costs are minimal so that they require small investments,
- Diversity can captivate new groups of tourists to the given attraction,
- Economic effects can motivate the enterprises among the local community,

- Investments in infrastructure, buildings and tourism-related services can be utilized by the residents,
- Overall increase in tourism flow affects positively the living standards of locals.

CONCLUSION

Cultural routes organized around a specific historical, artistic, or social theme have become essential tourist destinations. Despite the fact that their core idea is not mainly aimed toward tourist usage, they do constitute a tourism product in addition to the cultural dimension. Both organic and purposive cultural routes can increase a destination's competitiveness and attractiveness by promoting local traditional products, natural and cultural heritage, food, and drink resources. This increases regional visitation and encourages investment in cultural and creative tourism which will contribute to the social, psychological, political, and economic empowerment of local communities, which is vital for sustainable development.

REFERENCES

Ashworth, G. J. (1995). Heritage, Tourism and Europe: a European Future for a European Past. In D. T. Herbert (Ed.), *Heritage, Tourism and Society* (pp. 68–83). Mansell Publishing Ltd.

Aspridis, G., Rossidis, I., Katsimardos, P., & Bouas, K. (2016). Modulation of conditions and infrastructure for the integration of organizational changes in tourist sector. In *Tourism, culture and heritage in a smart economy*. Springer.

Berki, M., & Csapó, J. (2008). The Geographical Basis for the Development of Thematic Routes. In Progress in Geography in the European Capital of Culture 2010. University of Geography, University of Pécs.

Berti, E. (2015). Defining the Cultural Routes of the Council of Europe. *Cultural Routes Management: from theory to practice; Step-by-step guide to the Council of Europe Cultural Routes.*

Boucher, B. J. (1990). Punta Gorda to create ecotourism trail. *The Belize Review*, *2*(9), 12–13.

Briedenhann, J., & Wickens, E. (2004). Tourism routes as a tool for the economic development of rural areas - vibrant hope or impossible dream? *Tourism Management*, *25*(1), 71–79. doi:10.1016/S0261-5177(03)00063-3

Butler, G., & Ivanovic, M. (2016). Cultural heritage tourism development in post-apartheid South Africa: Critical issues and challenges. *Cultural Tourism in Southern Africa*, 58-75.

Campolo, D., Bombino, G., & Meduri, T. (2016). Cultural landscape and cultural routes: Infra- structure role and indigenous knowledge for sustainable development of inland areas. *Procedia: Social and Behavioral Sciences*, *223*, 576–582. doi:10.1016/j.sbspro.2016.05.350

Cohen, E. (2004). *Contemporary Tourism – Diversity and Change*. Elsevier.

Council of Europe. (2010). *Resolution CM/Res (2010)53. Establishing an Enlarged Partial Agreement on Cultural Routes (Adopted by the Committee of Ministers at its 1101st meeting of the Ministers' Deputies).* https://search.coe.int/cm/Pages/result_details.aspx?ObjectID=09000016805cdb50

Council of Europe. (2021). *Cultural Routes of the Council of Europe Programme Activity Report.* https://rm.coe.int/cultural-routes-of-the-council-of-europe-programme-activity-report-202/1680a245ec

DuVal, L. (2001). Historic Route 66 is still kicking in Albuquerque. *Arizona Republic, 17*(June), T3.

Ebony. (1990). Georgia's Black History Trail. *Ebony, 45*(4), 167-172.

Garcés Feliú, E. (2011). Cultural tourism in fragile institutional contexts: Project for cul-tural routes in Tierra del Fuego. *Journal of Tourism Research, 3*(2), 35–43.

Geiger, B. F., & Werner, K. A.National Voting Rights Walk and Selma Antebellum Historic District. (2009). A guided walking trail to explore the Martin Luther King Jr. National Voting Rights Walk and Selma Antebellum Historical District. *International Journal of Heritage Studies, 15*(5), 467–476. doi:10.1080/13527250903072807

Goodey, B. (1975). Urban trails: Origins and opportunities. *The Planner, 61*, 29–30.

Graf, M. (2013). Industrial Cultural Routes. Journal "Energy, Economy, Ecology", 15(2), 328-335.

Hayes, D., & MacLeod, N. (2007). Packaging places: Designing heritage trails using an experience economy perspective to maximize visitor engagement. *Journal of Vacation Marketing, 13*(1), 45–58. doi:10.1177/1356766706071205

Hitrec, T. (1996). Kulturni itinereri: Važan segment europske turistčke ponude. *Turizam, 44*(3), 47–60.

Hospers, G. J. (2002). Industrial heritage tourism and regional restructuring in the European Union. *European Planning Studies, 10*(3), 397–404. doi:10.1080/09654310220121112

ICOMOS. (2008). *Charter On Cultural Routes.* https://www.icomos.org/images/DOCUMENTS/Charters/culturalroutes_e.pdf

ICOMOS & WTO. (1993). *Tourism at World Heritage Cultural Sites: The Site Manager's Handbook* (2nd ed.). World Tourism Organization.

Jackson, S. A., Haider, W., & Elliot, T. (2003). Resolving inter-group conflict in winter recreation: Chilkoot Trail National Historic Site, British Columbia. *Journal for Nature Conservation, 11*, 317–323.

Jansen-Verbeke, M., Priestley, G. K., & Russo, A. P. (2008). *Cultural resources for tourism: patterns, processes, and policies.* Nova Science Publishers.

Javaheri, H. (2011). Optimal tourism route in historical urban areas: A case study of Tehran's down town, Iran. *South Asian Journal of Tourism and Heritage, 4*(1), 34–43.

Kammen, C. (1993). Gateway to Freedom: A black history trail. *Heritage, 9*(4), 16–21.

Káposzta, J., Nagy, A., & Nagy, H. (2016). The impact of tourism development policy on the Hungarian regions. *Региональная икономика,* (1), 10-17.

Khovanova-Rubicondo, K. (2012). Impact of European Cultural Routes on SMEs' innovation and competitiveness – Study. Provisional Edition, Council of Europe.

Logan, W. S. (2002). Vietnam's Highway No.1: Corridor of power and patrimony. *The Historic Environment, 16*(2), 23–26.

MacLeod, N. (2016). Self-guided trails – a route to more responsible tourism? *Tourism Recreation Research, 41*(2), 134–144.

MacLeod, N. E. (2004). Seeing place: A critique of the heritage trail as a visual interpretation of place. *LSA Publication, 84*, 61–78.

Mahoney, C. (1999). Heritage tourism: Strategies for implementation and the model of The Civil War Discovery Trail. *History News, 54*(1), 21–23.

Markwell, K., Stevenson, D., & Rowe, D. (2004). Footsteps and memories: Interpreting an Australian urban landscape through thematic walking tours. *International Journal of Heritage Studies, 10*(5), 457–473.

Masson, G. (2001, May). *The Chilkoot Trail National Historic Site: A Canadian cultural route management approach* [Paper presentation]. ICOMOS Making Tracks Conference, Alice Springs, Australia.

McCormack, P. (2017). Walking the land: Aboriginal trails, cultural landscapes, and archaeological studies for impact assessment. *Archaeologies*.

McKercher, B., & du Cros, H. (2002). *Cultural Tourism: The Partnership between Tourism and Cultural Heritage Management*. Haworth Hospitality Press.

Meyer, D. (2004). *Tourism routes and gateways: Examples and a selection of key issues for development of tourism routes and gateways and their potential for pro-poor tourism*. ODI.

Misra, M. (2011). Shadow of the Silk Road. *The International Journal of Environmental Studies, 68*(3), 400–404.

Moulin, C., & Boniface, P. (2001). Routeing heritage for tourism: Making heritage and cultural tourism networks for socio-economic development. *International Journal of Heritage Studies, 7*(3), 237–248.

Murray, M., & Graham, B. (1997). Exploring the dialectics of route-based tourism: The Camino de Santiago. *Tourism Management, 18*, 513–524.

O'Brien, R. & Curtis, C. (1979). The Donkin Heritage Trail. *Planning and Building Developments, 40*, 13–26.

Oprică, M. A. (2015). European Cultural Routes in Romania: Opportunities for Developing Sustainable Tourism and Architectural Heritage. *Acta Technica Napocensis: Civil Engineering & Architecture, 58*(4), 119–127.

Orbaşli, A. (2000). *Tourists in Historic Towns: Urban Conservation and Heritage Management*. E & FN Spon.

Pedersen, A. (2002). Managing Tourism at World Heritage Sites: A Practical Manual for World Heritage Site Managers. In World Heritage Manual No. 1. Paris: UNESCO World Heritage Centre.

Puczkó, L., & Rátz, T. (2007). Trailing Goethe, Humbert and Ulysses Tourism: Cultural Routes in Tourism. In G. Richards (Ed.), *Cultural Tourism: Global and Local Perspectives*. Haworth Press.

Ramsay, J., & Truscott, M. C. (2003). Tracking through Australian forests. *The Historic Environment, 16*(2), 32–38.

Richards, G. (Ed.). (1996). *Cultural Tourism in Europe*. CABI Publishing.

Richards, G., Russo, A. P., & Grossman, M. (2008). *Cultural Corridors in South East Europe: Refinement of concept and development of pilot projects*. Council of Europe.

Schutt, A. M. (1997). *The Bruce Trail, Ontario Users and their Economic Impact*. Trent University.

Scott, N., Baggio, R., & Cooper, C. (2008). *Network Analysis and Tourism: From Theory to Practice*. Channel View Publications.

Sevigny, R. (1992). Florida's black heritage trail. *Historic Preservation Forum, 6*(4), 6-9.

Shackley, M. (Ed.). (1998). *Visitor Management – Case studies from World Heritage Sites*. Butterworth Heinemann.

Shackley, M. (2003). The Frankincense Route: A proposed cultural itinerary for the Middle East. *The Historic Environment, 16*(2), 12–17.

Shaw, S., & MacLeod, N. (2000). Creativity and conflict: Cultural tourism in London's city fringe. *Tourism, Culture & Communication, 2*(3), 165–175.

Smith, M. K. (2003). *Issues in Cultural Tourism Studies*. Routledge.

Smith, V. L., Hetherington, A. Q., & Brumbaugh, M. D. D. (1986). California's Highway 89: A regional tourism model. *Annals of Tourism Research, 13*, 415–433.

Snowball, J. D., & Courtney, S. (2010). Cultural heritage routes in South Africa: Effective tools for heritage conservation and local economic development? *Development Southern Africa, 27*(4), 563–576.

Stoddart, H., & Rogerson, C. (2009). Tourism routes, local economic promotion and propoor development: The case of the crocodile ramble. *Africa Insight, 38*(4), 10–26.

Swarbrooke, J. (1994). The future of the past: heritage tourism in the 21st century. In A. V. Seaton (Ed.), *Tourism, the State of the Art* (pp. 222–229). John Wiley & Sons.

Tang, M. D. (1991). Evaluation of the tourist resources of the northern Silk Road, Xinjiang. *Chinese Journal of Arid Land Research, 4*(4), 329–337.

Teather, E. K., & Chow, C. S. (2003). Identity and place: The testament of designated heritage in Hong Kong. *International Journal of Heritage Studies, 9*(2), 93–115.

Terzić, A., Bjeljac, Ž., & Jovanović, R. (2014b). Protection, revitalization and use of cultural heritage through cultural route formation system: Evaluation of the Roman Emperor's Path master plan. *Kultura, 143*, 319–335.

Terzić, A., Bjeljac, Ž., Jovičić, A., & Penjišević, I. (2014a). Cultural Route and Ecomuseum Concepts as a Synergy of Nature, Heritage and Community Oriented Sustainable Development. *European Journal of Sustainable Development*, *3*(2), 1–16.

Teye, V. B., & Timothy, D. J. (2004). The varied colors of slave heritage in West Africa: White American stakeholders. *Space and Culture*, *7*(2), 145–155.

Timothy, D., & Boyd, S. W. (2014). *Tourism and trails: Cultural, ecological and management issues*. Channel View Publications.

Timothy, D. J. (2007). Empowerment and stakeholder participation in tourism destination communities. In A. Church & T. Coles (Eds.), *Tourism, Power and Space* (pp. 199–216). Routledge.

Timothy, D. J. (2009). River-based tourism in the USA: Tourism and recreation on the Colorado and Mississippi Rivers. In B. Prideaux & M. Cooper (Eds.), *River Tourism* (pp. 41–54). CABI.

Timothy, D. J. (2014). Contemporary cultural heritage and tourism: Development issues and emerging trends. *Public Archaeology*, *13*(1-3), 1–12.

Timothy, D. J., & Boyd, S. W. (2003). *Heritage Tourism*. Prentice Hall.

Timothy, D. J., & Boyd, S. W. (2006). Heritage tourism in the 21st century: Valued tradi- tions and new perspectives. *Journal of Heritage Tourism*, *1*(1), 1–17.

Timothy, D. J., & Teye, V. B. (2004). American children of the African diaspora: Journeys to the motherland. In T. Coles & D. J. Timothy (Eds.), *Tourism, Diasporas and Space* (pp. 111–123). Routledge.

UNESCO. (2012). *The Slave Route*. https://en.unesco.org/themes/fostering-rights-inclusion/slave-route

Vana, M. V., & Malaescu, S. (2016). Cultural thematic travel itineraries: Mediators of success. *Procedure Economics and Finance*, *39*, 642–652.

Wallis, M. (1992). *Route 66: The Mother Road*. St. Martin's Press.

Wang, S., & Hji-Avgoustis, S. (2011). Evaluating costs and benefits of a tourism project: A case study of the Indianapolis Cultural Trail. *Tourism Today (Nicosia)*, *11*, 172–181.

World Tourism Organization. (2017), *General Assembly - Twenty-second session*. UNWTO.

Yankholmes, A. K. B., & Boakye, K. A. (2010). 'Awusai Atso': Community attachment to and use of transatlantic slave trade resources in Danish-Osu, Ghana. *Journal of Heritage Tourism*, *5*(1), 49–67.

ADDITIONAL READING

Olsen, D. H., & Toronto, A. (Eds.). (2018). *Religious Pilgrimage Routes and Trails: Sustainable Development and Management*. CABI. doi:10.1079/9781786390271.0000

Palmi, P., Esposito, M., & Prete, M. I. (2021). Change in Perspectives in Cultural Tourism: A Sustainable Managerial Model for Cultural Thematic Routes Creating Territorial Value. In P. Demartini, L. Marchegiani, M. Marchiori, & G. Schiuma (Eds.), *Cultural Initiatives for Sustainable Development: Management, Participation and Entrepreneurship in the Cultural and Creative Sector*. Springer. doi:10.1007/978-3-030-65687-4_10

Chapter 3
Dialogue of Knowledge for the Consolidation of Territories Through Cultural Tourism

Milena Santiago Escobar
https://orcid.org/0000-0001-8774-435X
Centro Regional de Gestión para la Productividad y la Innovación de Boyacá, Colombia

Diana María Dueñas Quintero
https://orcid.org/0000-0002-1004-1290
Centro Regional de Gestión para la Productividad y la Innovación de Boyacá, Colombia

ABSTRACT

The dialogue of knowledge through cultural tourism in search of the consolidation of territories can help in the identification of the sociocultural diversity of local communities. The research focuses on identifying and interpreting the social, cultural, and tourist practices present in the territories in a way that facilitates the understanding of the individual and collective dynamics for the creation of strategies that establish guidelines of belonging in the different institutional interventions and provide spaces for the dialogue through strategies of social appropriation of knowledge. The content analysis reveals that there are some aspects of the initiatives studied that may be successful, but without the full participation of all the actors involved, discrepancies may occur in the results obtained, because they tend from assumptions of homogeneity of the populations, even overlooking the differences that the context provides in terms of the development of economic, social, and tourist activities.

INTRODUCTION

The dialogue of knowledge through the social appropriation of knowledge in local communities allows a social, cultural, and environmental integration to be established through the recognition of the cultural dynamics established by the critical experiences and the contributions of the communities. Indeed, the objective of this research is to integrate the context of the territories and the realities of the local com-

DOI: 10.4018/978-1-7998-9217-5.ch003

munities, starting from intercultural spaces for training, education and knowledge transfer adapted to the environment in which peasant activities take place so that there is a construction collective that confronts the lines of thought with the local, regional, or universal reality with the interaction of knowledge.

The research focuses on the identification of ancestral knowledge, peasant senses and cultural heritage, in order to recognize alternatives that manage to promote processes of social appropriation of knowledge, through the dialogue of knowledge that allows working on the relationship between cultural diversity and territory, considering spaces for meeting and discussion between the academy and regional and local agents, so that a methodological proposal can be built that questions the concepts and practices of development to transform tourism processes in the territories, so that they can rediscover knowledge that allow influencing the ways in which the current situation of the communities under study is understood, understood and dealt with.

The processes of support and intervention of the communities on the part of public or private institutions focus on the fulfilment of national, regional, or local development plans with a low participation of the community. These processes of collective construction, based solely on official politics, have caused a disproof between official representations and the needs and / or expectations of the cultural agents present in the territory. This situation means that the participation of these agents is dispersed and that their symbolic repertoires and their cultural frameworks remain outside the scope of public plans and policies. The value given to heritage, however fragile they may seem, and to the agents of the production of meaning in the territories, will allow greater relevance in community interventions and in the scope of development plans and public policies. Therefore, there is evidence of the need to identify alternatives that manage to promote processes of social appropriation of pertinent and inclusive knowledge insofar as they can recognize agents, patrimonies and territories from institutional plans and programs to impact in the medium and long term in the communities and in cultural dynamics.

Finally, the appropriation of ancestral and peasant knowledge and a knowledge that has been given a new meaning by the recognition and potential of cultural, economic, and symbolic resources, will allow rural communities of Alto Ricaurte to empower and legitimize themselves as communities to know. From the recognition of some narratives associated with representations and imaginaries of the countryside and the peasant, the appropriation of the natural and cultural landscape of rural communities can be claimed as territories of high culture density.

The main objective of the research is to strengthen the processes of social appropriation of knowledge through the recognition of cultural diversity and dialogue of knowledge in the Province of Alto Ricaurte in the Department of Boyacá - Colombia. To meet this objective, three specific objectives are proposed: first, to interpret sociocultural and tourism practices that facilitate the understanding of individual and collective dynamics in the territories; second, to establish guidelines of relevance in the institutional interventions of the territories; and third, to promote spaces for the dialogue of knowledge through strategies of social appropriation of knowledge in interventions present in the territories.

The article begins with a review of the existing literature on dialogue of knowledge and the social appropriation of knowledge in the context of cultural tourism through sociocultural practices and ancestral knowledge of the communities; subsequently, a description of the methodology used throughout the investigation is made and the different sources used and the results found through them are presented; then an approach is made to the guidelines for future research that can be carried out as a continuation of this research; and ends with the conclusions of the investigation.

BACKGROUND

Since the beginning of humanity, the human being creates and recreates culture, symbols, tools and discourses, generating concepts with which to define the world, live in it and transform it (Gómez et al., 2015), for this reason, in the different communities there are and have been knowledge that become practices developed through their stories, ways of being, doing, thinking and living in their territories. Knowledge, as it is conceived today, is a progressive and gradual process developed by man to understand his world and realize himself as an individual and a species (Ramírez, 2011).

Knowledge can mean an activity or state of the subject or individual, as well as an active process by which a person comes to grasp a reality (Villoro, 1982) but at the same time, it can also designate the result of that activity; the state of the subject that also subsists in him as a disposition determined by that reality. Villoro (1982) ensures that the knowledge process cannot be reduced to a theoretical activity, detached from practice: practice is a condition of knowledge, which is always directed by concrete interests, and in this sense, it occurs in real individuals who seek to satisfy their practical needs to guide your actions in the world to ensure their success.

In this context, the dialogue of knowledge can be understood as a process of recovery, deconstruction, resignification, and recreation of knowledge in which the interaction characterized by dialogue, recontextualizes, facilitates reflexivity and the configuration of meanings in processes, actions, knowledge, histories, and territorialities (Cano, 2018) of the people and the communities to which they belong. In this sense, territories are those spaces where all the relationships established in a community are perceived to have materialized, and in it political, social, economic and cultural actions will be displaced under conditions of appropriation (Llanos-Hernández, 2010) and they find the relations of power that are organized at a given time present.

There is a social construction of the territory when the voice of the inhabitants is incorporated to understand the network of meanings and the concrete interactions with which they build and rebuild a society and a territory, in a continuous process of construction of a shared knowledge that includes references territorial, ways of identifying the territory, appropriating it, making it a place or many places, that is, loading it with symbolic codes (Barrionuevo, 2012). Territories as tourist destinations are the scene where cultural diversity becomes visible (Santiago-Jiménez & Morfin-Herrera, 2011), being a source of encounters that celebrate diversity, while enriching knowledge.

At present, the processes of support and intervention towards the communities from different public or private institutions focus on the fulfillment of national, regional, or local development plans with a low participation of local communities. These processes of collective construction based solely on official policies have caused a break in the balance between the official representations and the needs and / or expectations of the cultural and tourist agents present in the territories. It should also be mentioned that, because of the desire to comply with development plans, these intervention processes are based on the assumptions of homogeneity of populations, territories, and heritage. So, the mistake of overlooking these differences within their context in terms of the development of economic, social, and cultural activities distances this type of initiatives from a real relevance by not considering relativity and cultural diversity.

Consequently, this situation causes the participation of the agents to be dispersed, generating that their symbolic repertoires and cultural frameworks remain outside the scope of public plans and policies. Therefore, the value given to culture and heritage, however fragile they may seem and to the agents of the production of meaning in the territories, will allow generating greater relevance in the interventions developed in the communities and in the scope of the development plans and public policies. Therefore, the need to identify alternatives that manage to promote processes of social appropriation of pertinent and inclusive knowledge becomes evident insofar as agents, patrimonies and territories can be recognized from institutional plans and programs that can impact in the medium and long term. term in the communities and in the cultural and tourist dynamics.

The identification of ancestral and peasant knowledge and cultural heritage, in order to recognize alternatives that manage to promote processes of social appropriation of knowledge, through the dialogue of knowledge that allows working on the relationship between cultural diversity and territory, considering spaces of meeting and discussion between the academy and the different regional and local agents, through a "social learning process", which is achieved when individual and experiential knowledge is codified and socialization is achieved in a community, developing capacities and abilities in people that allow them to respond successfully to permanent changes in their environment. So, learning is the fundamental process that leads from knowledge to innovation and social change. The two complementary processes of appropriation of knowledge lead to a dynamic conception of the relationships that exist between knowledge, the subject who knows and the environment on which the subject acts based on that knowledge (Chaparro, n.d.).

According to Santiago-Jiménez (2011, 2016), the dialogue of knowledge allows communities to recognize their identity by fostering an internal local dialogue to define: what to rescue, what to modify, what to reinvent or what to co-create, in order to define which parts of their culture will be shared with the rest of society (Santiago-Jiménez & Morfin-Herrera, 2011). In this sense, the dialogue of knowledge provides open spaces to interpret the territory, allowing socialization, learning and the defense of personal and community identities. Although, these dialogues form a basis for the formulation of strategies with sustainable ends and with an inclusive vision where the communities face the future from their own conceptions of development and in this way contribute to a consolidation of territorial identity, becoming a new shared vision for the future (Bermúdez et al., 2005).

Cultural diversity provides tourism with the fundamentals of the territory (causal and intentional). Which form numerous products resulting from everything that society is like identity, images, expectations, or denials. As well as its practices, processes, capacities, knowledge, ideologies, and taboos. It is important to mention that not all the cultural environment is involved in the occurrence of the tourist phenomenon, but only certain constructions (Navarro, 2010). In the field of tourism, cultural diversity constitutes the portion of the intangible reality built by man collectively and over time (Navarro, 2010). In this sense, it is important to promote processes of social appropriation of knowledge to consolidate local territories that enable tourism actors to recognize socio-cultural aspects that mark the identity of a territory and allow innovation and social change in them.

The influence of modernity in the development of the territories in this case of tourism is established by the forms of interaction between various actors in specific spaces that respond to the needs and challenges of specific communities. As a form of organization, knowledge is part of these processes involving interaction, relationship, mutual transformation, co-dependency, and co-evolution; that is to say, it has a direct link in the cultural, social, political, environmental, and economic changes in the development of communities.

These changes lead to confronting the scientific, productive, organizational, or other paradigms. Knowledge and consciousness as a transversal axis for Boisier (2001) are due to the need to involve improvement processes in companies, governments, development promotion agencies, communities as well as any individual. Just as the conception of scientific knowledge is established, the definition itself of social, traditional, or popular knowledge is also complex according to the experience of research oriented to participatory research and rural praxis carried out by Fals (1981) and for which Habbermas in the sociology of knowledge refers to the importance of the origin of ideas and the interests that govern knowledge.

The validity of this approach lies in the importance of social knowledge with respect to scientific knowledge, due to the establishment of intercultural relationships and the existence of "technique, technology and tradition", therefore, techno-productive knowledge in the social context, it is a space of human endeavor that can be analyzed without the need to contrast it or confuse it with science (Díaz et al., 2004).

Commonly, the motivations for changing social knowledge bring with them fields or forms of meaning. The importance of social knowledge is established in maintaining and strengthening the capacity of the actors to use it according to their expectations, motives, purposes, social practices constituted by groups that carry out actions oriented to determined ends and using specific means and that are part of a dynamic system with the following elements: a) A set of agents with capacities and with common purposes that interact and have transactions with the environment, immersed in a culture and a social system. b) A medium where agents interact and have transactions with objects. c) A set of objects that are part of the environment and d) actions that involve intentions, purposes, ends, projects, tasks, representations, beliefs, values, norms, rules, value judgments and emotions (Guber, 2004; Olivé, 2009).

In the Language of Links, Denisse Najmanovich (1995) explains that this subjectivity allows the transformation of knowledge and practices, creates new areas, weaves connections, and establishes an ethical-political rethinking of the knowledge that dialogic creation and expansion implies. in the production, transmission and validation of knowledge determined by a framework of social and cultural relations suitable for its use.

This context implies societies in dialogue, permanent communication and interaction between community actors; is its correlation with the environment involves the recognition of the place that local or autochthonous knowledge occupies in knowledge societies whose development models considerably value the forms of codification of knowledge due to cultural and linguistic diversity, highlighting the problem of access to knowledge because it is inseparable from the conditions in which they occur (Bindé, 2005; Saso & Oliver, 2003).

Interculturality as part of the dialogue of knowledge is characterized by the effort to communicate and interrelate between individuals, groups, and culturally different knowledge, which encourage cooperative actions to learn, work and act; identifies common issues as well as developing attitudes of responsibility and solidarity. It implies the establishment of dialogues, interaction, and critical communication between the actors from their own realities, recognizing the way of articulating popular knowledge with the scientific knowledge of researchers (Abatedaga et al., 2015; Levalle, 2013; Najmanovich, 1995; Walsh, 2005).

In tourism, the dialogue of knowledge is commonly carried out through tourist animation as a means of interaction between communities and tourists, where the identity, values, norms, and behaviors of the territory where the tourist activity takes place are shared. It is also considered as a mechanism for the defense of material and immaterial resources mediated by dialogical encounters between communities and tourists with new forms of exchange that invite us to reflect on social practices, their imaginaries and the ways in which territories are appropriated (Cammarata, 2006; Santiago-Jiménez & Morfin-Herrera, 2011).

The interactions between the actors to determine the value offer in tourism from the dialogue of knowledge, implies establishing the scope of generating knowledge in tourism, its meaning and interest that is commonly based on commodification rather than on how to take advantage of the generation and co -construction of knowledge to modify the realities of tourist communities seen in the ways of life mainly due to social, economic and environmental conditions (Tobar et al., 2015).

In tourism, the way in which knowledge is built depends on the researcher's trajectory and the dynamics of reality based on their ideologies, practices, and traditional knowledge. This conjugation of elements allows us to reflect on how to verify and verify the statements made in the dialogue of knowledge. As mentioned by Tobar et. al (2015) the diversity of theories arranged in this economic activity refers to areas of knowledge such as sociology, anthropology, management, economics, philosophy, among others, understanding that the framework of the relationships that occur around tourist services is complex.

Commonly, the complexity in tourism could be explained from social practices understood as the actions carried out by individuals in community in the territories, which start from inherited local imaginary, and which also co-evolve according to their context and local culture, trying to maintain their identity, what Cammarata (2006) refers to as a network of effective relationships, which must be constantly woven in order to guarantee its existence, its continuity. Interest is placed on individuals, tourists and the social conditions that enable this condition, a framework in which they make sense (Cammarata, 2006).

It is understood the scope of these elements in the dialogue of knowledge in tourism and for this research dimensions of analysis are established on the way in which communities interact in the creation, recognition, transfer, and co-creation of knowledge as principles to improve their living conditions through tourism as an economic alternative. Taking advantage of the constructs developed for this activity, tourism production establishes the articulation between systems of actors, images, and spaces with high cultural significance that for the dialogue of knowledge not only translates into commercial processes but also into the possibility of taking advantage of tangible and intangible elements. arranged in the territories that value their resources and identify new alternatives to strengthen the ways of life for the benefit of the community with respect to their social demands that can be resolved collectively.

Finally, regarding tourist entertainment, Santiago-Jiménez & Morfin-Herrera (2011) suggests considering it as a pedagogical strategy that can be useful in the dialogue of knowledge because it contains resilient elements if there are some of these conditions:

1. A community willing to accept the support that external actors offer it.
2. A group of conscious "experts" who will use Tourist Animation as a tool to share and acquire knowledge for fair or solidarity tourism, as a comprehensive strategy for the reconstruction and preservation of social and ecological diversity.
3. A community or a group of inhabitants willing to strengthen and / or rescue their uses and customs, which define their cultural identity.
4. Active participation of local actors in the design of these tourism strategies, built through dialogues between experts and the community, through an exchange.

In Figure 1 you can see reflected the conceptual framework in which it is detailed how, starting from the societies present in the territories, the local communities can be identified, who, being possessors of traditional knowledge, perceptions and diverse sociocultural practices through the dialogue of knowledge, its needs, perceptions and experiences can be identified, which will allow the development of a social appropriation of knowledge that allows the growth of communities to strengthen social and tourist integration.

Figure 1. Conceptual framework
Source: Own processing

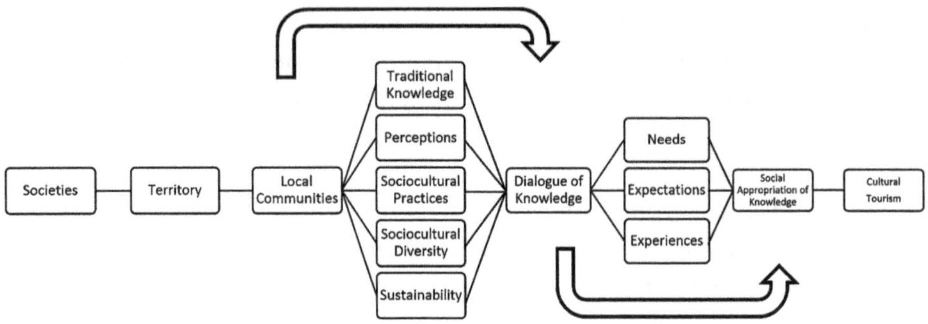

STUDY AREA

The scope of work is in the Province of Alto Ricaurte is in the Department of Boyacá in central Colombia, in the region recognized as the Cundiboyacense highlands as it can be seen in Figure 2. The province is divided into seven municipalities: Villa de Leyva, Gachantivá, Ráquira, Sáchica, Santa Sofía, Sutamarchán and Tinjacá; municipalities characterized by their cultural and natural wealth, as well as their geological and paleontological, archaeological, architectural, and civil, urban, furniture, historical and intangible documentary heritage typical of the Colombian colonial era.

The province is characterized by having beautiful landscapes and cultural and ecological attractions that suggest tourism as a real alternative for development from an economic, social and cultural perspective (Vélez, 2007). The economic activities of this province are focused on intensive agricultural production, mining, tourism and the transformation of some agricultural products such as milk and handicrafts (Casas et al., 2017), each of these activities is characterized by its ancestral knowledge, its traditions and its unique socio-cultural practices in its territories.

RESEARCH METHODOLOGY AND DESIGN

The research uses a qualitative descriptive methodology in which it seeks to strengthen the processes of social appropriation of knowledge through the recognition of cultural diversity and the dialogue of knowledge in the territories of the Province of Alto Ricaurte, through the identification of alternatives that allow Recognize agents, patrimonies and territories through the valuation of the sociocultural elements that offer the possibility of helping in the identification of guidelines that allow achieving institutional interventions with greater relevance in the territories.

The method used is ethnography, since it is necessary to coexist with the human group under study to achieve a better understanding of their reality (Campos, 2020), and at the same time to know first-hand their behaviours, how they arise, how they develop, what is their dynamics, how they change and how they are modified (Ortiz, 2015), which is characterized by the realization of field work in which the interpretation of the meanings, the analysis of the social structure must play a very important role. and of the roles of the population studied (Álvarez, 2008).

Figure 2. Generalities of the study area
Source: Own processing, map elaborated with ArcGIS 10.5

GENERALITIES		
	BOYACÁ	
	Extension	23.189 km^2
	Department Capital	Tunja
	Provinces	13
	ALTO RICAURTE	
	Extension	746,66 km^2
	Municipalities	7

The development of the ethnographic method was developed in several stages, taking into account the research question and the objectives in addition to the steps proposed by Ortiz (2015): first, the analysis and selection of the territories was carried out; second, identification of the tourism projects that were being developed at the time in the Alto Ricaurte Province, in order to establish the participation of the different agents in the support and intervention processes developed in their territories. Third, selection of techniques taking into consideration that the interpretation of sociocultural and tourist practices is sought and the relevance of institutional interventions, which is why participant observation, in-depth interviews, focus groups and the field diary were selected.

Once the initiatives under study had been identified, workshops were held to identify actors and their role in the territories, to gain access to the territories. Of all the actors involved in the initiative, 16 individuals were selected among actors, community, and representatives of the instructions, who agreed to participate in the research. The information collected has been analyzed through a system by categories that has been generated following the proposals of Romero (2005), Fernández (2006), Hernández (2014) and Rivas (2017). Nine categories were established: sociocultural practices, territories, heritage, community empowerment, social appropriation of knowledge, characteristics of the actors, perceptions, institutions, and problems.

An exhaustive review of the plans, programs and projects that were being developed in Alto Ricaurte was carried out to identify those related to tourist activity. A single, descriptive, and exploratory case study was selected from this review (Stake, 2010). The selected study unit presents unique characteristics that allowed the identification and analysis of sociocultural practices and the relevance of said intervention with the territory. The use of the ethnographic method was chosen because it is the discipline that studies and describes the ways of life of certain social groups (Maldonado, 2018), and will allow a description of the social and cultural events that take place in life. communities studied, with essential consideration of the social structures and the behavior of the subjects as members of the group, as well as their interpretations and meanings of the culture to which they belong.

The selected sample consisted of 16 informants, among partners, territorial entities, and enterprises of the selected initiatives. Among the criteria developed for the selection of said sample, it was sought that they were participating actors and that they belonged to different municipalities of the province and dedicated to various activities related to tourism. To obtain the data, semi-structured interviews were applied. The aspects addressed during the interviews were these: i) the characteristics of the actors and their relationship with the selected initiative; ii) identification of sociocultural practices; iii) territory and social construction; iv) heritage, symbolic repertoires, and ancestral knowledge; v) individual and community empowerment; vi) perceptions of culture, environment, and own initiative; vii) public institutions in each the territory; and viii) perceived social and economic problems. The data were recorded in a field journal (Espinosa & Ríos, 2017; Martínez, 2007; Universidad De Costa Rica, 1987; Valverde, 1993) and were later systematized.

DIALOGUE OF KNOWLEDGE FOR THE CONSOLIDATION OF LOCAL TERRITORIES

The territory is a terrestrial extension that includes a relationship of power or possession by an individual or a social group, which contains limits of sovereignty, property, appropriation, discipline, surveillance and jurisdiction, and transmits the idea of enclosure (Rodríguez, 2010). On the other hand, the concept implies ambivalence; firstly, it appears as the space of the Nation-States (national territory) and secondly, from a culturalist dimension of the territory (Gouëset, 1999). Territoriality is not only a matter of appropriation of a space by a state or by any power group, but also of relevance to a territory through a process of identification and representation, either collective or individual, which is often unknown. classical political or administrative borders (Rodríguez, 2010). All the relationships established in a community are perceived as materialized, and in it political, social, economic, and cultural actions will be displaced under conditions of appropriation (Llanos-Hernández, 2010). There is a social construction of the territory when the voice of the inhabitants are incorporated to understand the network of meanings and the concrete interactions with which they build and rebuild a society and a territory, in a continuous process of construction of a shared knowledge that includes references territorial, ways of identifying the territory, appropriating it, making it a place or many places, that is, loading it with symbolic codes (Barrionuevo, 2012).

In the territories are social and economic actors, who have important characteristics and influences on the development of their communities. By actors, reference is made to individuals, groups, or institutions whose system of action coincides with the limits of local society, that is, it is about those actors whose scope of action coincides with the local scene. In each territory, the participation of its actors in the construction of the development of their territories is sought through the incidence of individuals and social groups in the different stages in which matters of public interest are resolved, that is, in consultation, discussions, proposing proposals, and all kinds of activities that involve State institutions and other agents (Quispe et al., 2018). The foregoing is based on the empowerment of the actors, which tries to turn people and communities into protagonists of their own change, giving them control and dominion over their reality, making them responsible for the actions in which they are involved and therefore those that are affected, to achieve a positive transformation of the same, improving their quality of life.

Empowerment is related to community participation, collective decision-making, good governance, full citizenship, and democracy (Soler et al., 2014) and is related to a process of growth, strengthening, empowerment and development of trust of individuals and communities to promote positive changes in the context, gain power, authority, decision-making capacity and change both individually and collectively (Titterton & Smart, 2008). All this implies the self-realization and emancipation of individuals and communities, the recognition of groups / communities and social transformation.

The experiences of social groups are related to the different forms of appropriation and use of geographic space, highlighting the material and symbolic dimensions from which the links are established between the cultural actions developed throughout the trajectory of a community and the territories that materialize them (Carvalho & Moquete, 2011). And the knowledge of the communities has a series of characteristics that make it a common good, a heritage, which is usually defined as an inherited good, which can be subjected to a range of uses and which can be subjected to a variety of uses and that it must be delivered, duly conserved, inventoried, enriched and disseminated to future generations who will be in charge of applying "their" way of conceptualizing this inheritance and proper management to the people who will come and so on (Galí et al., 2000) . In other words, it is the cultural heritage of the past of a community with which it lives today and that it transmits to present and future generations.

However, heritage is not limited only to monuments and collections of objects, but also includes traditions or living expressions inherited from ancestors and transmitted to their descendants, such as oral traditions, performing arts, social uses, rituals, festive events., knowledge and practices related to nature and the universe, and knowledge and techniques related to traditional crafts (UNESCO, n.d.). In other words, life, identities, and collective memory are represented in the cultural heritage of a community.

As part of the heritage, there is what has been called traditional knowledge to traditional and popular knowledge and practices in very different ways, in which different knowledge enters and is transformed, generating others. Traditional knowledge can be understood as that knowledge that has been generated, preserved, applied, and used by traditional communities and peoples (Valladares & Olivé, 2015). This knowledge is used and transmitted, sometimes orally, others through practices, and others through learning sessions, and they result in knowledge that has been fixed in some way (Olivé et al., 2018), and are the result of the ability to connect knowledge and knowledge generated in social practices and representations that help to understand the diversity of life, learning from their own interaction with the environment.

Within each territory, problems can also be found that refer to a condition that affects a significantly considerable number of people, in a way that is considered inconvenient and which, according to it is believed, should be corrected through collective social action (Suárez, n.d.). Social problems are modified over time, either because the standards of desirability are more demanding, because of a growth in collective consciousness, or vice versa because a policy, deliberate or not, reduces aspirations and expectations. Traditional knowledge constitutes an important part of peoples and has enormous potential for understanding and solving different social problems (Valladares & Olivé, 2015), as well as modifying behavior patterns derived from tradition, reciprocity, complementarity. and relationship between actors, who contribute, safeguard, promote, confront, and negotiate their particular knowledge, until it is regularized and transformed into concrete processes in addition to being expressed in a dialogue and consensual way (Acevedo et al., 2005; Argueta, 2012; Díaz et al., 2004; Ingold et al., 2001; Olivé, 2009, 2010).

The importance of the traditional social knowledge of the communities lies in their dynamic expressions of their perception and understanding of the world, since they have made and continue to make a valuable contribution to science and technology, being necessary to preserve, protect, research and promote their heritage and its knowledge (Argueta, 2012), as well as its great potential for economic and social development, but also in many other ways (Olivé et al., 2018), which can even be incorporated into innovations and contribute to social development in different ways (Splinder, 2013). Regarding tourism, this traditional knowledge would lead to the appreciation of local knowledge by the tourist and not to look at the otherness established in the destinations as the exotic, but by having a close experience, the perception of the other would cease to be in the plane of indifference to move to the level of the ethics of otherness, unleashing solidarity and fair exchange (Santiago-Jiménez & Morfin-Herrera, 2011).

A large amount of traditional knowledge, as explained by Olivé et al. (2018) has to do with the sustainable use of biodiversity, the use and preservation of forests and jungles, with medicine and health, with the improvement of species for food and with the manufacture of objects of interest. cultural and useful in everyday life. In other words, traditional knowledge has great potential, which is why it is important to develop social mechanisms for its preservation, promotion, and proper exploitation for the benefit of those who have developed and preserved it, but also the rest of society.

The motivations that have generated change in social knowledge bring with them various fields of significance: common sense as the field of specialized knowledge, linked to solving problems due to its interest in the particular and concrete, its practical and immediate nature, and its way. to proceed related to the experience; theory, characterized by its interest in knowledge centered on its productive activities; and interiority, for relating the subject to himself (Díaz et al., 2004). Perceptions are found along the same lines, which refer to the mental image that is formed in the mind of an individual with the help of experiences and needs, it is the result of a process of selection, interpretation, and correction of sensations. Individual perceptions are subjective, selective, and temporary. Subjective since the reactions to the same stimulus vary from one individual to another. Selective, since perception is a consequence of the subjective nature of the person who cannot perceive everything at the same time and selects her perceptual field based on what he wants to perceive. And temporary, since it is a short-term phenomenon, since it evolves as experiences are enriched, or their needs or motivations vary.

The concept of the knowledge society should be understood in a broad sense, where the generation of wealth and well-being is not conceived only in terms of scientific and technological knowledge but must also include the use of traditional knowledge (Olivé et al., 2018). The knowledge society is considered as a new type of social, amplifying and innovative organization, in which each actor can give rise to new knowledge (Marín, 2012); However, this can only be achieved to the extent that knowledge is renewed and disseminated, in a way that contributes to the adaptation of individuals to the changing and unstable environment, and allows them to have a vision of the future; since the speed of the transformations of society makes it impossible for the mediating discourse used by one generation to continue to be effective for the next (Pabón, 2018).

In this sense, the search for the knowledge society has given rise to what has been called social appropriation of knowledge, where scientific communities project strategies aimed at socializing the knowledge obtained from their research so that the different actors in society achieve understand and assimilate them. In this topic, a series of factors, areas, professions, and disciplines converge, not only those that serve for the creation of knowledge, but also those that in one way or another contribute to the better use and benefit of the community (Marín, 2012), since the world of knowledge cannot only be in the heads of the scientific community but must be a common heritage of the people (Ahumada & Miranda, 2003).

It is, then, about promoting problem-solving networks, in which various agents can participate to constitute social networks of innovation that require attention from public policies to work simultaneously on several fronts. Part of this interaction is the transfer of knowledge. Various authors as Mora (2012) and Perkins & Salomon (1992) define it as the use of knowledge in a situation different from that which has been learned, that is, it refers to the process of recovery and use of knowledge in moments after it was learned and improves its related performance in another context. It is necessary to emphasize that the subjects that participate have diverse knowledge, ways of learning, motivations, and contexts where their different activities take place.

In this sense, the social appropriation of knowledge is understood as a process that implies the disposition of scientific and technological knowledge in a common setting and language for society; as well as those individuals make such knowledge their own as useful and necessary elements for their benefit and benefit. Social appropriation of knowledge, from the perspective of the knowledge society, means, then, the democratization of access and use of scientific and technological knowledge, as a strategy for its adequate transmission and use among the different social actors, which will result in the improvement of the quality of life of the communities and their members (Marín, 2012).

Dialogue of Knowledge and Cultural Tourism

Societies are characterized by being immersed in interculturality, understood as a deep reflection on the opportunities offered by cultural diversity as an element of fruitful exchange of values and attitudes, breaking prejudices and initiating spaces for interrelation and exchange, thus creating ties and interpretations between diverse cultural groups that coexist in the same reality. Therefore, the appropriation of knowledge requires that societies interrelate and enhance in an integrated way development, education and culture (Ahumada & Miranda, 2003), focused in such a way that it allows to examine and contribute to the solution of the problems that they concern the population, who through their active participation in the processes, must reach a balanced understanding of what knowledge, science and technology mean and can contribute to their well-being.

Interculturality is essential, since it recognizes the differences in origin, practices, customs, knowledge and knowledge, in this sense the dialogue of knowledge protects, revitalizes and promotes ancestral knowledge and traditional knowledge; As well as the possibility of establishing a new management of the creative potential of cultural diversity requires the configuration of innovative institutional and social platforms that facilitate the diffusion of power to the base of society, particularly to local communities (Valladares & Olivé, 2015).

Interculturality is part of social interaction and is a relevant aspect in the transfer of knowledge, characterized by the effort to communicate and interrelate between individuals, groups and culturally different knowledge, represented by their cultural practices, generators of identity, if they produce concrete subjects, and in that sense, there is no social action without representation, and in the same way, there is no representation that is not put into practice. In other words, the social world is built from what has already been built in the past, its social forms are reproduced, appropriated, displaced and transformed in the practices and interactions of the daily life of the actors, affirming that the interdependence between action and representation, and cultural practice and identity (Rizo, 2004).

Interculturality gives way to social and cultural practices, which are the different activities that are carried out on a daily, constant, and repeated basis within a given community, which are considered as the fruit of tradition and the passage of historical time, since in the same community these practices vary as their notions of morality, society, identity, among others, change. Social practices can be understood as the set of actions produced by specific social agents in significant situations for them (Jociles, 2018), they are a recurring way of carrying out a certain activity, shared by all the members of a community

(Pérez & Gardey, 2013). Within them you can find jobs, exercises, customs, practices, or procedures that serve as an element of interconnection between communities. Sociocultural practices are a form of link between the individual and the community and differ from one culture to another. Likewise, cultural practices involve tradition, celebrations, folklore, national stories, and group identity. Their practices that reinforce the idea of belonging to a cultural group and tradition, that is, to a particular history. Through culture, people and groups define, identify, and adjust to the shared values of society, that is, these practices are generators of identity (Rizo, 2004) of a town or community, they can change to give themselves a new meaning in its relationship with time and space (Itchart & Donati, 2014).

According to Boullón (2006), the tourist practice is projected in natural or urban spaces through planning and territorial ordering, seeking to meet the needs of the tourist demand in terms of mobility and accommodation, as well as the assessment and adaptation of areas for recreation for visitors. In this way, Carvalho & Moquete (2011) refers to the concept of tourist as the space arises, which can be understood as the juxtaposition of material elements and symbolic relationships that are projected in the geographical space and condition the flow of visitors. In other words, tourist territories result from tourist practice, since by commercializing cultural attractions, tourism gives rise to a dialectical process of creation and recreation of places.

Social and cultural practice is part of the historical process of each of the communities, as Cammarata (2006) affirms, that thanks to its dialectical nature and its active relationship with the world, places and landscapes of the territory, and with other individuals, transforms nature and he transforms himself, building a qualitatively new world of meanings, values and human works, and realizing himself as a human being who appropriates and consolidates himself in his territory, involving economic, social and way of life territorial relations.

Territorial Dynamics and the Dialogue of Knowledge

Since 1960, the struggles of the peasant-indigenous movement were the determining factor in society in the vindication of their cultural differences and their right to development. In the Latin American context, some intellectuals began to criticize the system, stating the need to establish a dialogue with popular knowledge to establish the bases of a social science at the service of the people. Freire (2005) and Fals (2009) with the participatory action research proposal expressed the need to build science -and politics- on a subject-subject relationship, in addition to the constitution of a point of departure for scientific work (Levalle, 2013) in various educational, social, administrative and cultural aspects (Ortega-Ibarra et al., 2019).

The economic, political, and social development is framed in the trends of globalization due to changes in the use of resources, there is priority in the use and access of knowledge as a competitive strategy with respect to the use of capital, labor, and land. That is, the dynamics are sustained in knowledge societies by the involvement of the actors that are part of a community, locality or territory in which capacities are built and links are generated for the acquisition and development of knowledge. The impact of this type of society lies in understanding its problems, taking advantage of other types of knowledge (traditional, local, or scientific - technological) for its solution, as well as different ways of thinking, problematizing, and exploring a territory in permanent transformation. It depends on the interaction between agents and / or actors to link encounters, narratives with structured forms and languages (Colciencias, 2005; Najmanovich, 1995; Olivé, 2010). In this sense, an empathetic attitude towards others about their life experiences and an authentic participation must be adopted for the development of a responsible, democratic, and participatory science (Fals, 2009).

Within this framework of interaction between agents, there are university - company - State - society relations determined by the achievement of project results in alliance with research groups, companies, or productive sectors. However, the relevance and impacts of this relationship are measured in the majority only in the short term by ambiguous results and operational indicators without considering those related to factors of change and improvement of the actors in the transfer of knowledge, making evident the lack of strategies that guarantee teaching - learning processes among the carriers of knowledge and recipients of it.

In the Latin American case and especially Colombia, to reinforce this interaction model, initiatives, programs, norms, and policies have been developed that mainly promote science, technology, and innovation in the territories. In said intentions, the universities have been fundamental actors due to their institutional capacities that organize processes and determine lines of work. In this sense, the coordinated research through centers and groups of research and technological development can be highlighted, and its economic environment in the development of consultancies, consultancies, internships, and assistance programs.

In the framework of the appropriation of knowledge, it is highlighted that in the documents of policy and frameworks the importance of the articulation of the actors in the National System of Science, Technology and Innovation - SNCTI is mentioned, referring to the development of activities with actors social and non-profit and for-profit organizations for the transfer and appropriation of knowledge, not to mention directly the exchange of knowledge where research also strengthens their work; Higher Education Institutions - IES are generally selective in the discrimination of topics and results; and the emphasis on integration is directed primarily at the international academic community. There is clarity in the respect and recognition of traditional knowledge, but not in the possibility of the existence of a real dialogue of knowledge.

This is based on the way these processes are measured and monitored. Currently, the validation of scientific knowledge in research is related to the impact of measurement indicators and the application of models or techniques for transfer without considering whether the way it is transferred is relevant to communities, companies, unions, to mention a few. In the framework of the National Strategy for the Social Appropriation of Knowledge for Colombia, it is established that appropriation is found between the production, distribution, and consumption of knowledge, which promotes innovation as a social process through the exchange of knowledge among multiple actors. Additionally, it is suggested to carry out programs and projects supported by lines of action that seek "effective dialogue and knowledge exchange, in which a complex vision of science and technology is evidenced, techno-scientific problems are presented in their dimensions. political and social, and realize the role that different interest groups have in structuring events" (Colciencias, 2018; Lozano & Maldonado, 2010).

Due to the dynamics and reflections promoted in recent years, the need has been seen to involve and empower different social groups within the framework of university social responsibility, trying to link the results of the research with the extension. Generally, the integration and transfer mechanisms are established by the dynamics of the communities, productive sectors and companies and then try to codify and standardize, in the case of the university, within the framework of the accreditation processes, as well as recognition and measurement of research centers and groups, which on many occasions ends up being the object of the transfer regardless of whether the communities actually generated learning processes in the medium and short term.

The selected initiative is developed in the municipalities of Sáchica, Gachantivá, Ráquira, Villa de Leyva, Sutamarchán, Tinjacá and Santa Sofía in the Province of Alto Ricaurte, located in the Department of Boyacá - Colombia. Those territories are characterized by having a great diversity of climates and terrain, in addition to having a great natural and cultural wealth. The initiative seeks the positioning and promotion of tourism, handicraft and agroindustry products and services of the local communities, highlighting their culture and tradition.

The meetings with local agents made it possible to show that the dialogue of knowledge is a scenario that encourages reflection, discussion and the exchange of experiences and knowledge about communication practices and processes for social change in communities, as well as being a commitment policy that favours the collective construction of the valuation of the sociocultural elements that can be considered as elements of approach in the operationalization of public policies.

The sociocultural practices identified are pluri-directional, conjunctive, supportive and conducive to changes in emotion, understanding, expression and action. These individual and collective movements start from something, they have an origin, a starting point, which in the case of practices is the context, experiences, knowledge, and their life stories. Knowledge is conceived as platforms for ideas, experiences, beliefs, images, reflections, opinions that are built throughout life, both individually and collectively. Knowledge accounts for the ways in which people participate and act in their social groups, communities, and institutions. Throughout life, according to the context and environment, the actors weave diverse knowledge around conflicts and the way of dealing with them leads them to act and to understand in a singular way what happens with them.

The underlying issue has to do with the complex relationship between sociocultural factors and tourism, the internal fractures between the actors present in the territories and who seek to become relevant participants in processes that transcend time and allow the improvement of living conditions. of the communities in the territories from the transfer and appropriation, through the strengthening of capacities and productive practices, taking advantage of traditional knowledge as a basis for cooperation and solidarity.

These initiatives are relevant if one considers the need to understand the behavior of the territories and the role of the agents that intervene in it, so that these initiatives do not end up being ephemeral actions in time due to the lack of adaptation and appropriation of the communities of the Provinces of Alto Ricaurte, but promote and guarantee the right of the communities to preserve their identity, flourish, and decide on their symbolic and material heritage, and participate in decision-making (Olivé, 2010).

The understanding the complexity of the interaction in the transfer and appropriation of knowledge, it is understood that there must also be the interest of the interested actors in promoting institutional, legislative, and structural transformations that achieve the reduction of gaps and visible barriers in the territories. Due to the dynamics of rural communities, dialogue that favors cultural, social, and environmental integration through the recognition of cultural dynamics established by experiences, criticisms and contributions is necessary. Indeed, the contribution of this research is to integrate the context of the territories and the realities of the communities, starting from intercultural spaces for training, education and knowledge transfer between the academy and regional and local agents, adapted to the environment in which Peasant activities are developed so that there is a collective construction that confronts the lines of thought with the local and regional reality with the interaction of knowledge.

Tourism acts as an incentive to recognize socio-cultural aspects, in the endogenous search for local images themselves: search for originalities, roots, as well as accelerated reconstruction of the symbolic past that creates new metaphors. The past is created from the present with a non-ideological utilitarian sense, according to the primary interests of location and potential for visitors, to consolidate the territory of life, enhancement and development based on an inclusion strategy of certain spaces, different in the consideration of the tourist and operators of the tourist activity.

FUTURE RESEARCH DIRECTIONS

This study highlights the need to promote shared management in the approach of programs, plans and projects not only from official institutions, but also the community and the actors present in the territories must be taken into account, so that they can be Recognize agents, patrimonies and territories, highlighting their differences, as well as their local values, so that development projects can be carried out that increase the quality of life for all the actors involved in this activity.

The particularities of the territories can add value to the process of creating tourism products, promoting greater articulation between local agents in the management of destinations, as well as greater interaction between the global and the local to achieve the sustainability of tourism projects.

It is important to highlight that the dialogue of knowledge in the territories, in this project, seeks a holistic sense and has as main objectives:

1. offer a vision of socio-cultural and tourist practices, a way of life or represent a particular style of the province under study.
2. create spaces for exchange of knowledge and organized dialogue.
3. integrate individuals and promote their social, cultural, economic, and personal adaptation.
4. enable greater knowledge and knowledge.
5. plan processes for creating strategies that allow the involvement of different actors through workshops that integrate diversity.

The following are areas in which this research could be followed up:

1. Recognition of knowledge, heritage, and territorial dynamics in different regions of the country, to identify if the conclusions identified here are repeated in other regions or are isolated facts.
2. Appropriation of socio-cultural practices seen from the point of view of local communities

CONCLUSION

This research suggests the valuation of the sociocultural and tourist elements that can be considered as elements of approach in the operationalization of public policies, the identification of guidelines that allow achieving institutional interventions of greater relevance to the territories, as well as the strengthening of the dialogue of knowledge through practices social, cultural and tourist according to the conditions of the territories.

This need lies in the criticism that arises from the various models of the transfer of knowledge and appropriation between institutions, non-profit organizations, and regional and local actors in which the strategic role of economic positioning is overvalued, ignoring the characteristics of the territories. such as the knowledge, interests and motivations of the actors immersed in the productive sectors and in the sociocultural dynamics, with benefits oriented to the institutions and measured from indicators, without any assessment of the impact of the appropriation in the communities involved.

Under this position, the discussion of the dialogue and dialectical transfer of knowledge / consciousness are established, questioning the scope of the transfer and appropriation of knowledge to turn it into ideals according to foreign logics (Mora, 2012). With the above, it is not being ignored that this knowledge has no value; rather, the need to understand traditional knowledge, its relevance and logic of why it is still valid in the communities is highlighted (Mora, 2012).

Due to the importance of achieving processes that transcend over time and allow the improvement of the living conditions of the communities in the territories from the transfer and appropriation, they relate the search for answers to social problems, and their relationship between science, technology and society, mentioning that the scope is to achieve the strengthening of productive capacities and practices, taking advantage of traditional knowledge (Olivé, 2009, 2010), which helps to better understand problems and can be used to propose and carry forward alternative solutions: "Go from paradigms to figures of thought, problematize and explore a fluid territory in permanent transformation" (Najmanovich, 1995) in the language of links (Najmanovich, 1995; Olivé, 2009, 2010).

These initiatives are relevant if one considers the need to understand the behavior of the territories and the role of the agents that intervene in it, so that these initiatives do not end up being ephemeral actions in time due to the lack of adaptation and appropriation of the communities of the Provinces of Alto Ricaurte, but promote and guarantee the right of the communities to preserve their identity, flourish, and decide on their symbolic and material heritage, and participate in decision-making (Olivé, 2010).

The understanding the complexity of the interaction in the transfer and appropriation, it is understood that there must also be the interest of the interested actors in promoting institutional, legislative, and structural transformations that achieve the reduction of gaps and visible barriers in the territories. Due to the dynamics of rural communities, dialogue that favors cultural, social, and environmental integration through the recognition of cultural dynamics established by experiences, criticisms and contributions is necessary.

Indeed, the contribution of this research is to integrate the context of the territories and the realities of the communities, starting from intercultural spaces for training, education and knowledge transfer between the academy and regional and local agents, adapted to the environment in which Peasant activities are developed so that there is a collective construction that confronts the lines of thought with the local, regional, or universal reality with the interaction of knowledge.

It can be concluded that the factors that influence the transformation of the territory and the appropriation of tourist activities are i) social organization, ii) mechanisms of inclusion and exclusion, iii) relations with the external environment and iv) learning collective (González et al., 2019). In the dialogue of knowledge for the consolidation of territories, tourism implies the framework of a set of actors and institutions (Cammarata, 2006) that forms an organizational base with historical depth and cultural roots, at the same time that it represents a limiting factor for the implementation of the economic logic of tourism in rural communities (González et al., 2019).

In rural communities, the dialogue of knowledge has provided a great opportunity to consolidate spaces for discussion that have been evidenced by the absences registered from the imposition of plans or programs far from the reality of their territories, since it has become evident that from the institutional framework point of view, the homogeneity of the populations, territories and heritage has been assumed for the development of economic, social and tourist activities, distancing this type of initiatives from a real relevance by not taking into account relativity and cultural diversity. The dialogue of knowledge is fundamental both for the collective construction of joint visions, and for the exchange and adaptation of socio-cultural and tourist practices.

The dialogue of knowledge for the consolidation of territories is something that goes beyond an exchange of opinions and beliefs, it is a staging of meanings of life, cultural identities, territories, and heritage that has its expression in daily life and in recognition of knowledge that the communities have in an integral way in the common areas, since they are part of life and are existential knowledge that has been exposed to the knowledge of the entire community.

The work carried out with the communities through the dialogue of knowledge has served to reflect on issues not only related to the initiatives studied, but they have also been spaces that have made it possible to address urgent and necessary stories, in relation to memories and territories, especially with those issues that have been presented in some communities on the lack of collective construction in the processes raised from the local and regional governments. A commitment is proposed that favors the collective construction of strategies, consciousness, and knowledge that intends to strengthen, both the communicative mediation processes that advocate for socio-cultural and tourist diversity.

When it comes to tourist activity, Carvalho & Moquete (2011) conclude that it is possible for it to take over the territories, reinterpreting them and providing them with a new dynamic and rationality regarding market expectations. Thus, the experience of the actors and local community in relation to the tourist space is opposed to the place of social coexistence and reciprocity. The development of the tourist activity causes changes in the dynamics of the host populations, and the presence of tourists interferes in the network of relationships of social actors.

The proposed rationale sought to show the dynamics of territorial appropriation from the identification and valuation of the sociocultural aspects of the communities through cultural tourism, reflecting on the impacts of the projects or initiatives in the lives of the actors involved in them, determining that in the territory tourism produces new sociocultural configurations that must be considered in the planning of the territories.

The topic addressed in this research has served to help broaden the references on the relationship between cultural diversity, tourism, and territory as a joint part of learning to identify local community organizations as regulatory entities of new tourism dynamics, but at the same time they arise as new scenarios of some local groups over others. It is observed that tourism is presented as an activity with access barriers built based on the possession of different forms of capital (economic, physical, cultural, and social) by local actors.

The reflections presented here were generated through dialogues with strategic actors of the Province of Alto Ricaurte, in Boyacá - Colombia, together with the members of the research project "Dialogue of knowledge through the relationship between diversity and tourism for the consolidation of the territory". It should be noted that the research is still under development, it is not a completed project. At this moment, it is in the process of collecting information in other provinces of the Department of Boyacá belonging to other initiatives, and in the design of strategies that manage to promote processes of social appropriation of pertinent and inclusive knowledge.

The work has the limitation of being based on the perspective of local actors linked to a single initiative regarding tourism, so future research should be approached from the point of view of other social, individual, and institutional actors, who are directly associated with the territory.

ACKNOWLEDGMENT

This research was supported by MinCiencias and CREPIB under the framework of the Call for Postdoctoral Researchers Fellow Position 848-2019.

REFERENCES

Abatedaga, N., Haiquel, M., & González, V. (2015). ¿Cómo permear la extensión universitaria desde la práctica social? *Actas Del XVII Congreso de La Red de Carreras de Comunicación Social y Periodismo de Argentina.*

Acevedo, M., González, O., Zamudio, L., Abello, R., Camacho, J., Gutiérrez, M., Barreto, E., Ochoa, J., Torres, G., Quintero, M., & Baeza, Y. (2005). Un análisis de la transferencia y apropiación del conocimiento en la investigación de universidades colombianas. *Investigación & Desarrollo, 13*(1), 128–157.

Ahumada, J., & Miranda, F. (2003). Ciencia, tecnología y Sociedad: Algunas Reflexiones. In Documento preparado para la Organización de los Estados Américanos. doi:10.2307/j.ctvn5tw9m.5

Álvarez, C. (2008). La etnografía como modelo de investigación en educación. *Gazeta de Antropología, 24*(1). http://hdl.handle.net/10481/6998

Argueta, A. (2012). El diálogo de saberes, una utopía realista. *Integra Educativa, V*(3), 15–29.

Barrionuevo, C. A. (2012). *El territorio como construcción social : una pregunta que importa. El caso de Rincón de Las Perlas*. Universidad Nacional de la Plata.

Bermúdez, O. M., Mayorga, M. L., Jacanamijoy, B., Seygundiba, A., & Fajardo, T. (2005). *El diálogo de saberes y la educacíon ambiental*. Academic Press.

Bindé, J. (2005). Hacia las sociedades del conocimiento. In *Informe Mundial*. UNESCO.

Boisier, S. (2001). *El lenguaje emergente en el desarrollo territorial*. CATS.

Boullón, R. C. (2006). *Planificacion del espácio turístico*. http://prepacihuatlan.sems.udg.mx/sites/default/files/planificaciondelespacioturisticorobertoc.boullon.pdf

Cammarata, E. B. (2006). El turismo como práctica social y su papel en la apropiación y consolidación del territorio. *América Latina: Cidade, Campo e Turismo*, 351–366.

Campos, E. (2020). Etnografía. In D. Páramo Morales, S. J. Campo Sierra, & L. M. Maestre Matos (Eds.), *Métodos de investigación cualitativa. Fundamentos y aplicaciones* (pp. 163–203). Editorial UniMagdalena.

Cano, A. M. (Ed.). (2018). Diálogo de saberes, memorias y territorios (1a ed.). Sello Editorial Universidad de Medellín.

Carvalho, K. D., & Moquete, S. (2011). El turismo en la dinámica territorial ¿Lógica global, desarrollo local? *Estudios y Perspectivas en Turismo, 20*, 441–461.

Casas, L. F., Pineda, M., Andrade, M., & Pulgarin, Y. (2017). *El Alto Ricaurte e Iguaque territorio de patrimonio natural y cultural, acciones para un manejo sostenible.* La Imprenta Ltda. https://www.natura.org.co/publicaciones/alto-ricaurte-e-iguaque-territorio-patrimonio-natural-cultural-acciones-manejo-sostenible/

Chaparro, F. (n.d.). *Apropiación Social del Conocimiento.* Aprendizaje y Capital Social.

Colciencias. (2005). Encuentro Internacional de Investigadores "Colombia en la Sociedad del Conocimiento." In *Memorias del Encuentro de Investigadores Colombianos Residentes en el País y en el Exterior.* Panamericana formas e impresos S.A.

Colciencias. (2018). Política de Ética de la Investigación, Bioética e Integridad Científica. In *Documento de Política Nacional de Ciencia, Tecnología e Innovación.* Colciencias.

Díaz, M. G., Ortíz, P., & Núñez, I. (2004). *Interculturalidad, saberes campesinos y educación. Un debate con la diversidad cultural.* El Colegio de Tlaxcala, A. C.

Espinosa, R. A., & Ríos, S. (2017). El diario de campo como instrumento para lograr una práctica reflexiva. *Congreso Nacional de Investigación Educativa - COMIE,* 11.

Fals, O. (2009). La investigación Acción en Convergencias Disciplinarias. *Revista Paca, 1,* 7–21.

Fals Borda, O. (1981). Investigación participativa y praxis rural nuevos conceptos en educación y desarrollo comunal. Mosca Azul Editores.

Fernández, L. (2006). Fichas para investigadores ¿Cómo analizar datos cualitativos? *Bulletí LaRecerca, 7,* 1–13.

Freire, P. (2005). *Pedagogía del Oprimido* (2nd ed.). Siglo XXI Editores.

Galí, N., Majó, J., & Vidal, D. (2000). Patrimonio Cultural Y Turismo: Nuevos Modelos De Promoción Vía Internet. *Cuadernos de Turismo, 6*(julio-diciembre), 73–88.

Gómez, E., Vásquez, G., Betancur, V., Maartínez, D., Ocampo, M., Uribe, E., Soto, M., Ramírez, M., Jafeth, E., Karupia, W., Lectamo, J. N., Medina, B., Atehortúa, O., Valdés, D., & Arcos, A. (2015). *Diálogo de saberes e interculturalidad. Indígenas, afrocolombianos y campesinado en la ciudad de Medellín* (C. Gaviria & M. Orozco, Eds.). Pulso & Letra Editores.

González, I., Thomé, H., & Osorio, R. (2019). Apropiación turística de territorios indígenas: Cambios y continuidades en una comunidad Matlatzinca. *Turismo e Sociedade, 25*(junio), 113–120. doi:10.18601/01207555.n25.06

Gouëset, V. (1999). El territorio colombiano y sus márgenes. La difícil tarea de la construcción territorial. *Territorios, 1*(Enero), 77–94.

Guber, R. (2004). *El salvaje metropolitano. Reconstrucción del conocimiento social en el trabajo de campo.* Paidós. www.paidosargentina.com.ar

Hernández, R., Fernández, C., & Baptista, P. (2010). Metodología de la investigación (5a ed.). McGraw-Hill.

Hernández, R. M. (2014). La investigación cualitativa a través de entrevistas: Su análisis mediante la teoría fundamentada. *Cuestiones Pedagógicas*, *23*, 187–210.

Ingold, T., Hornborg, A., Pálsson, G., Descola, P., Ellen, R. F., Howell, S., Rival, L., Hviding, E., Arhem, K., Hell, B., Knight, J., Papagaroufali, E., Nothnagel, D., Richards, P., & Ruivenkamp, G. (2001). Naturaleza y Sociedad. Perspectivas antropológicas. In P. Descola & G. Pálsson (Eds.), *Naturaleza y sociedad: perspectivas antropológicas*. Siglo XXI Editores.

Itchart, L., & Donati, J. I. (2014). Practicas culturales. In Textos Iniciales (3ra ed.). Academic Press.

Jociles, M. I. (2018). La observación participante en el estudio etnográfico de las prácticas sociales. In Revista Colombiana de Antropologia (Vol. 54, pp. 121–150). Instituto Colombiano de Antropologia e Historia. doi:10.22380/2539472x.386

Levalle, S. (2013). El pensamiento latinoamericano puesto en práctica: descolonización y construcción dialógica del conocimiento en las experiencias pedagógico-investigativas del Centro Indígena de Investigaciones Interculturales de Tierradentro de Colombia. *X Jornadas de Sociología*, 26.

Llanos-Hernández, L. (2010). El concepto del territorio y la investigación en las Ciencias Sociales. *Agricultura, Sociedad y Desarrollo*, *7*(Septiembre-diciembre), 207–220.

Lozano, M., & Maldonado, O. J. (2010). *Estrategia Nacional de Apropiación Social de la Ciencia, la Tecnología y la Innovación*. Colciencias.

Maldonado, J. E. (2018). Metodología de la Investigación Social. Paradigmas: cuantitativo, sociocrítico, cualitativo, complementario. In Educación. Ediciones de la U.

Marín, S. A. (2012). Apropiación social del conocimiento: Una nueva dimensión de los archivos. *Revista Interamericana de Bibliotecología*, *35*(1), 55–62.

Martínez, L. A. (2007). La Observación y el Diario de Campo en la Definición de un Tema de Investigación. *Perfiles Libertadores*, *4*(80), 73–80.

Mora, D. (2012). Diálogo y transferencia dialéctica de saberes/conocimientos. *Integra Educativa*, *V*(2), 31–75.

Najmanovich, D. (1995). *El lenguaje de los vínculos. De la independencia absoluta a la Autonomía Relativa*. Academic Press.

Navarro, D. (2010). *Turismo: portal de la diversidad cultural. El turismo receptivo como espacio para el encuentro multicultural*. Academic Press.

Niño, V. M. (2019). Metodología de la Investigación. Diseño, ejecución e informe. In A. Gutiérrez (Ed.), *Educación* (2nd ed.). Ediciones de la U.

Olivé, L. (2009). Por una auténtica interculturalidad basada en el reconocimiento de la pluralidad epistemológica. In *Pluralismo epistemológico* (pp. 19–30). Muela del Diablo Editores.

Olivé, L. (2010). Multiculturalidad, interculturalismo y el aprovechamiento social de los conocimientos. *Recerca. Revista de Pensament i Anàlisi*, *10*, 45–66. doi:10.16309/j.cnki.issn.1007-1776.2003.03.004

Olivé, L., Argueta, A., & Puchet, M. (2018). Interdisciplina y transdisciplina frente a los conocimientos tradicionales. *Revista Iberoamericana de Ciencia. Tecnología y Sociedad, 13*(38), 135–153.

Ortega-Ibarra, I. H., Ortega-Ibarra, E., & Hernández-Jiménez, A. (2019). La investigación de Acción Participativa: ¿una alternativa de investigación o una estrategia de solución? *Universita Ciencia, 20*, 61–70.

Ortiz, A. (2015). Enfoques y métodos de investigación en las ciencias sociales y humanas. In *Educación*. Ediciones de la U.

Pabón, R. (2018). Apropiación social del conocimiento: Una aproximación teórica y perspectivas para Colombia. *Educación y Humanismo, 20*(34), 116–139. doi:10.17081/eduhum.20.34.2629

Páramo, D., Campo, S. J., & Maestre, L. M. (2020). Métodos de Investigación Cualitativa. Fundamentos y aplicaciones. In D. Páramo Morales, S. J. Campo Sierra, & L. M. Maestre Matos (Eds.), *Adminsitración y Negocios*. Editorial UniMagdalena.

Pérez, J., & Gardey, A. (2013). *Definición de Práctica Social.* https://definicion.de/practica-social/

Perkins, D. N., & Salomon, G. (1992). Transfer of learning. In International Encyclopedia of Education (2nd ed., pp. 443–467). Pergamon Press. doi:10.1037/h0071273

Quispe, G., Ayaviri, D., & Maldonado, R. (2018). Participación de los actores en el desarrollo local en entornos rurales. *Revista de Ciencias Sociales, 24*(3). https://www.redalyc.org/jatsRepo/280/28059580008/html/index.html

Ramírez, S. (2011). Nueva sociedad. *Enero-Febrero 2011*. https://nuso.org/articulo/el-giro-de-la-politica-exterior-colombiana/

Rivas, L. A. (2017). Capitulo 6. La definición de variables o categorías de análisis. In Elaboración de tesis: estructura y metodología (p. 284). Editorial Trillas. doi:10.13140/RG.2.1.3446.6644

Rizo, M. (2004). *Prácticas culturales y redefinición de las identidades de los inmigrantes en El Raval (Barcelona): aportaciones desde la comunicación.* Universitat Autònoma de Barcelona.

Rodríguez, D. (2010). Territorio y territorialidad Nueva categoría de análisis y desarrollo didáctico de la Geografía. *Unipluriversidad, 10*(3), 12.

Romero, C. (2005). La categorizacón un aspecto crucial en la investigación cualitativa. *Revista de Investigaciones Cesmag, 11*(Jun), 113–118.

Santiago-Jiménez, M. E., & Morfin-Herrera, M. del C. (2011). Diálogo de saberes a través de la actividad turística. *El Periplo Sustentable,* (21), 31–59. doi:10.21854/eps.v0i21.897

Santiago-Jiménez, M. E., & Morfin-Herrera, M. del C. (2016). *Animación Turística para el Diálogo de Saberes*. Fundación Universidad de las Américas.

Saso, E., & Oliver, E. (2003). Las comunidades de aprendizaje: Un modelo de educación dialógica en la sociedad del conocimiento. *Revista Interuniversitaria de Formación Del Profesorado, 17*(3), 91–103.

Soler, P., Planas, A., Ciraso-Calí, A., & Ribot-Horas, A. (2014). Empoderamiento en la comunidad. El diseño de un sistema abierto de indicadores a partir de procesos de Evaluación Participativa. *Pedagogía Social. Revista Interuniversitaria, 24*, 44–77. doi:10.7179/PSRI_2014.24.03

Splinder, E. (2013). The history of sustainability. The origins and efects of a popular concept. In I. Jenkins & R. Schröder (Eds.), *Sustainability in tourism: A multidisciplinary approach* (pp. 9–31). Springer Gabler.

Stake, R. E. (2010). *Qualitative Research: Studying how things work*. The Guilford Press.

Suárez, F. M. (n.d.). *Problemas Sociales y Problemas de Programas Sociales Masivos*. CEPAL.

Titterton, M., & Smart, H. (2008). Can participatory research be a route to empowerment? A case study of a disadvantaged Scottish community. *Community Development Journal: An International Forum, 43*(1), 52–64. https://www.jstor.org/stable/44258060

Tobar, L. E., Castillo, M., Mendoza, R., & Tamayo, A. L. (2015). La visión crítica en el construir del conocimiento turístico. *Journal of Tourism Analysis, 19*, 33–44.

UNESCO. (n.d.). *¿Qué es el patrimonio cultural inmaterial?* Patrimonio Cultural Inmaterial. https://ich.unesco.org/es/que-es-el-patrimonio-inmaterial-00003

Universidad De Costa Rica. (1987). *Instructivo para la elaboración del Diario de Campo*. Author.

Valladares, L., & Olivé, L. (2015). ¿Qué son los conocimientos tradicionales? Apuntes epistemológicos para la interculturalidad. *Cultura y Representaciones Sociales, 19*, 61–101.

Valverde, L. (1993). El Diario de Campo. *Revista de Trabajo Social, 12*(10), 308–319. https://www.binasss.sa.cr/revistas/ts/v18n391993/art1.pdf

Vélez, M. (2007). Papel de la población local en procesos de valoración turística. Caso: Turismo Rural en la Región del Alto Ricaurte (Boyacá): 2002-2003. *Anuario Turismo y Sociedad, 8*(Noviembre), 144–161.

Villoro, L. (1982). *Creer, Saber y Conocer*. Siglo XXI Editores.

Walsh, C. (2005). La Interculturalidad en la Educación. In O. Chávez (Ed.), Journal of Chemical Information and Modeling (Vol. 53, Issue 9). UNICEF. doi:10.1017/CBO9781107415324.004

ADDITIONAL READING

Cammarata, E. B. (2006). El turismo como práctica social y su papel en la apropiación y consolidación del territorio. *América Latina: Cidade, Campo e Turismo*, 351–366.

Espeso-Molinero, P., Carlisle, S., & Pastor-Alfonso, M. J. (2016). Knowledge dialogue through Indigenous tourism product design: A collaborative research process with the Lacandon of Chiapas, Mexico. *Journal of Sustainable Tourism, 24*(8–9), 1331–1349.

Gómez, E., Vásquez, G., Betancur, V., Maartínez, D., Ocampo, M., Uribe, E., Soto, M., Ramírez, M., Jafeth, E., Karupia, W., Lectamo, J. N., Medina, B., Atehortúa, O., Valdés, D., & Arcos, A. (2015). *Diálogo de saberes e interculturalidad. Indígenas, afrocolombianos y campesinado en la ciudad de Medellín* (C. Gaviria & M. Orozco, Eds.). Pulso & Letra Editores.

Levalle, S. (2013). El pensamiento latinoamericano puesto en práctica: descolonización y construcción dialógica del conocimiento en las experiencias pedagógico-investigativas del Centro Indígena de Investigaciones Interculturales de Tierradentro de Colombia. *X Jornadas de Sociología*, 26.

Rodríguez, D. (2010). Territorio y territorialidad Nueva categoría de análisis y desarrollo didáctico de la Geografía. *Unipluriversidad*, *10*(3), 12.

Santiago-Jiménez, M. E., & Morfin-Herrera, M. del C. (2011). Diálogo de saberes a través de la actividad turística. *El Periplo Sustentable,* (21), 31–59. doi:10.21854/eps.v0i21.897

KEY TERMS AND DEFINITIONS

Actors or Stakeholders: It is an agency, organization, group, or individual that has an interest (direct or indirect) in the planning process, project, or its programs or that affects or is positively or negatively affected by their execution and results.

Dialogue of Knowledge: Communicative process in which two different logics are brought into interaction: that of scientific knowledge and that of everyday knowledge, with a clear intention to understand each other; implies the recognition of the other as a different subject, with different knowledge and positions.

Social Appropriation of Knowledge: It is a process where civil society is empowered through knowledge. This process is the foundation of any form of innovation that involves the interaction of different social groups.

Sociocultural Practices: It is the process or phenomenon related to the social and cultural aspects of a community or society, where a participatory attitude is promoted in the members to contribute in this way to social and cultural development.

Territoriality: The territories that are created from a certain intention and, as a human construction, are constantly redefined and adapted to the needs of social groups.

Tourism Practices: They are those behaviours and activities that are carried out in visited destinations, which include economic, cultural, political, and technological aspects.

Transfer of Knowledge: Process by which research results, discoveries, scientific findings, or knowledge flow between different stakeholders.

Chapter 4
Innovation for Sustainability in World Heritage Destinations:
Opportunities and Challenges of the Idealistic Paradigm

Olga Høegh-Guldberg
Nord University, Norway

Evelina Maziliauske
Nord University, Norway

Dorthe Eide
Nord University, Norway

Anne Wally Ryan
Nord University, Norway

ABSTRACT

Experience-based sectors are increasingly used to transform rural communities into attractive places to live, work, and visit. However, such experience-based transformations (EBT) can bring sustainability challenges. The fields of sustainability and innovation have merged and developed into three paradigms to innovation for sustainability (IFS): skeptic, pragmatic, and idealistic. This chapter explores and develops the idealistic paradigm by using the ecological economy (EE) and community-based sustainability (CBS) approaches. Narratives from two cultural world heritage destinations in Norway are used to illustrate potentials and challenges of using the idealistic paradigm in EBT. The comparison of EE and CBS shows some overlaps but also differences, hence possible complementarity. Holistic sustainability and broad partnerships are central, but also ambitious, complex, and still rare, calling upon methods for involvement and conflict solving. Further research exploring these and other approaches and empirical research are needed to further concretize the idealistic paradigm.

DOI: 10.4018/978-1-7998-9217-5.ch004

INTRODUCTION

Innovation concepts and assumptions largely originate from studies of firms within manufacturing and technology sectors. However, as service sectors like tourism became the dominating employers in Western worlds, researchers have increasingly emphasized that mainstream theory has significant knowledge gaps regarding understanding and measuring service innovations (Djellal & Gallouj, 2010). Later, similar issues have been argued about experience-based innovations (Alsos et al., 2014).

Innovation means the exploration and exploitation of new or significantly changed ideas. Traditionally, innovation has been a new product, process, market, or organizational form (OECD, 2005). Innovations in experience-based sectors are, however, complex, since they are often developed incrementally over time and imply intertwinement across traditional innovation forms (Hjalager, 2010). This is due to the entirely or partly intangible nature of experience products, their user-value, and the importance of co-creation with users and other organizations (Fuglsang & Eide, 2013).

Many rural communities face grand challenges due to declining populations and reduction in traditional industries, e.g., fishing or farming (Bullvåg et al., 2020). Tourism and other experience-based sectors can be vital strategic tools for developing thriving local societies (Higgins-Desbiolles, 2020), by promoting attractive communities to live in, work in, and travel to (Jørgensen et al., 2021). However, experience-based transformations (EBT) are not always successful (Soares, et al., 2021). One reason is that the growth of tourism before Covid-19 increased pressure and tendencies of over-tourism (Epler Wood et al., 2019). Later, Covid-19 increased other socio-economic challenges (Gössling et al., 2020). Both call for more radical innovations for sustainability, such as to "rebuild better" (Higgins-Desbiolles, 2020).

The world's increasing sustainability challenges have bridged innovation and sustainability as research fields and called for cross-disciplinary approaches, novel innovation types with stronger sustainability and partnerships (Bocken et al., 2019; Schot & Geels, 2008). A number of terms are used, such as sustainable innovation (Schot & Geels, 2008), eco-innovation, and sustainability innovation (Klewitz & Hansen, 2014), sustainability-oriented innovation (Adams et al., 2015), responsible innovation (Stilgoe et al., 2013), green and transformative social innovations (Haxeltine et al., 2016), and innovation for sustainability (Leach et al., 2012). There is little consensus on whether it is "sustainability" or "innovation" that constitutes the primary purpose of development processes, what is meant by "sustainability," and who should be involved. This chapter uses "innovation for sustainability" (IFS), as an umbrella concept (Bocken, et al., 2019), with sustainability as the purpose and innovation as the means. Research on IFS has increased in the last decade (Albareda & Hajikhani, 2019) and vary according to the following factors: (1) innovation types; (2) innovativeness (e.g., radical, incremental); (3) value types (economic, environmental and/or social); and (4) analytical level (product, business model, firm, value chain, multi-sector/stakeholder, and systems) (Bocken et al., 2019). In his generic systematic literature review, Ritala (2019) summarizes IFS into three paradigms, "skeptic," "pragmatic" and "idealistic" based on different ontological assumptions about the four above-mentioned factors, as well as views on reality and actors. The concept of paradigm is commonly used when describing how scientific disciplines work based on the basic assumptions about ontology (reality and human nature), epistemology, and methodology (Kuhn, 1962). A paradigm shift means a radical change in basic assumptions and is rare, as most changes are within a worldview. More ambitious holistic and partnership approaches are needed; the notion of IFS paradigms can help explain why radical system transformations are complex.

This conceptual chapter explores the research question, "What are the core, potentials, and challenges of the idealistic paradigm to IFS in rural communities involved in experience-based transformations

(EBT)?" The idealistic paradigm shares the vision of the UN's development sustainability goals (SDGs) about holistic sustainability and broad partnerships. The chapter explores and elaborates on this paradigm by literature within the approaches of Ecological Economy (EE) and Community-Based Sustainability (CBS). In addition, two cultural World Heritage destinations in Norway are used to illustrate the potentials and challenges of using the idealistic paradigm. This chapter contributes by developing the framework of the idealistic paradigm more concretely and suggests implications for empirical research.

PARADIGMS ON INNOVATION FOR SUSTAINABILITY (IFS) WITH PRIMARY FOCUS ON THE IDEALISTIC

The *skeptic paradigm* to IFS represents the mainstream (neoclassical economics) paradigm with the focus on firms, markets, growth, and economic sustainability. Innovations usually take place within the firm, or more openly in business networks or triple helix relations (Fuglsang & Eide, 2013). The *pragmatic paradigm* combines economic value with one of two other values: environment or social. Most typically, it is the environment value, as the green shift and green economy increases in popularity by firms and politicians (Hall, 2018); for example, the idea that resource efficiency is important for both the environment and businesses (Simonis, 2011; UNEP, 2011). The main stakeholders are still mainly from industry, often across sectors, and involving technology (Engelmann & Al-Sadi, 2019); examples are green energy, such as windmills and electric transportation. The *idealistic paradigm* is ambitious since all three sustainability values are to be balanced, which is complex and rare in research and practice. Inspired by Ritala (2019), Bocken et al. (2019), and Alberada and Hajikhani (2019), the three paradigms are illustrated in Figure 1 with two main dimensions: "sustainability value focus" and "partnerships of stakeholders" in innovations, both gradual rather than strict dichotomies.

The sustainability focus varies from narrow economic value focus, medium combining two value foci (in the figure illustrated with economy and environment), to strong holistic sustainability (combining the three values). Partnerships vary from narrow focus on firms, medium—adding cross-sectoral partnerships, to strong with multi-stakeholder and multi-level partnerships, where local residents are central. Interpreting the dimensions in the experience-based contexts, we also find alternative skeptic approaches, which differ from the one focused on economy described above, such as a strong focus on environmental or cultural preservation in e.g., national parks or museums. There are also alternative pragmatic approaches, for example, when environmental or cultural preservation is extended with some focus on economy, e.g. when national parks or museums work to increase quality for visitors to earn money for preservation (Bayer et al., 2017).

The idealistic paradigm is ambitious: it aims at a "fundamental transformation towards more sustainable modes of production and consumption" (Markard et al., 2012, p. 955), including long-term system changes (Seyfang & Haxeltine, 2012). The paradigm can be used in business model innovations, for example, through localism and place development when local conditions are considered in business decisions (Dybdahl, 2019). Innovation can also mean making intentional changes to a community's philosophy and values, as well as to its products, processes, or practices to create social and environmental values in addition to economic returns (Adams et al., 2016). The success of the idealistic paradigm depends on value co-creation through multi-level and transformative innovations, which involve diverse stakeholder types and levels (Albareda & Hajikhani, 2019; Bocken et al., 2019). Top-down and bottom-up perspectives can depend on each other for success, where governance, management, and inte-

Figure 1. Three paradigms in innovation for sustainability
Source: authors

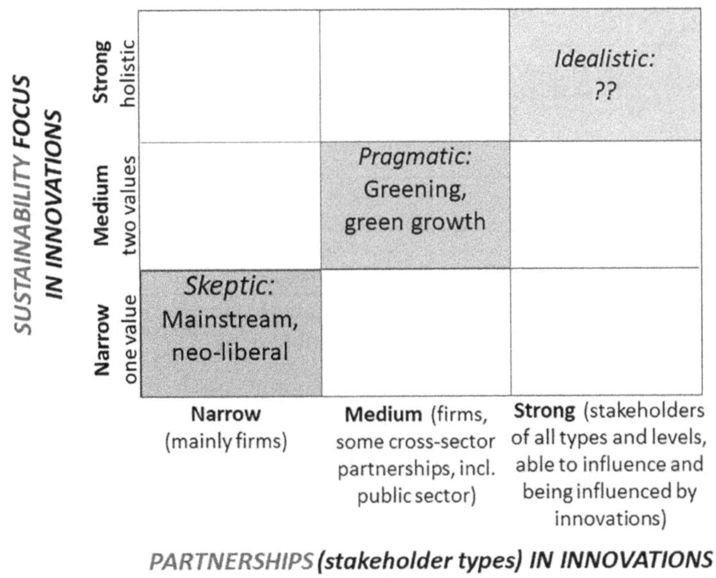

grated partnerships are central (Forrest & Wiek, 2014). However, the idealistic paradigm is challenging due to its scale (system innovations) and complexity (Smith et al., 2010), and the knowledge gaps are many due to its novelty. Local communities struggle with operationalizing and balancing sustainability dimensions and the SDGs, as there is a significant lack of research and practices balancing the three main value domains (Albareda & Hajikhani, 2019; Oftedal et al., 2021). Additionally, the complexity and operationalization of IFS depends on engaging different stakeholders (Tura et al., 2019). Empirical research and success cases are needed to show how IFS can take place over time, involve also SMBs (Tura et al., 2019), address the role of localism, local food, and development of place (Dybdahl, 2019), as well as involve action research (Bocken et al., 2019).

In the further exploring of the research question, and concretizing the idealistic framework, the theoretical approaches of Ecological Economics (EE) and Community-Based Sustainability (CBS) were chosen. These are summed up in Table 1.

Ecological Economy (EE)

The EE approach builds upon different streams of research within social, humanistic, and natural sciences, thus making it a transdisciplinary approach that crosses boundaries of two or more different disciplines. Studies related to ecological worldview date back to the 1930s. "The Conservation Ethic" by Leopold (1933, pp. 182), for example, highlights that conservation efforts had failed because humanity's relationship to land had been dominated by economic expediency: "The land-relation is still strictly economic, entailing privileges but not obligations." Later, the research on human ecology emerged e.g., Duncan (1961) suggested that it is time for human ecology to shift its interest "from social system to ecosystem." Subsequently, the ecological framework was used in other fields, including planning, where it could help understand and utilize new principles about the systems in which we try to intervene (Holling &

Table 1. Literature review of the ecological economy and community-based sustainability approaches

Approaches Within Idealistic Paradigm/Characteristics	EE	CBS
Searching for TITLE-ABS-KEY in Scopus, search words, number of hits	1) "Ecological economics": 3236 2) "Ecological economics" + "tourism": 76	1) "Community-based sustainability": 28 2) "Community-based sustainability" + "tourism": 1
Majority of/increase in publications	1996-2021	2009-2021
Disciplines	Environmental science; Economics, Econometrics and Finance; Agriculture and Biological Science; Social Science; Business, Management and Accounting	Social sciences, environmental sciences, energy, business, management, accounting, and engineering
Issues to resolve/application contexts	Community Resilience, Environmental Challenges, Social-, Ecological Well-Being, Ecosystem Services, Sustainability, De-Growth	Community gardens, solidarity purchasing groups, food cooperatives, community energy, recycling, and mobility initiatives
Development of sustainability dimensions/values	From primary focus on ecology and economy (Costanza, 1996) to holistic approach to sustainability focusing on integrated social-ecological systems (Plessis & Brandon, 2015).	Primary focus on socio-cultural sustainability (Axon, 2020), but also economic (Richins, 2009) and environmental (Hidayat & Stoecker, 2021)
Central characteristics of the approach	Radical shift from understanding the world as a machine to seeing it as an integrated organic network of all living and non-living entities	Community involvement in prioritizing, planning and implementing community developments
Geographic context	North Americas, Western Europe and increasingly China	Spread from Northern America and Europe to Japan, Singapore, and Australia

Source: authors

Goldberg, 1971). Simultaneously, "The Limits to Growth" by Meadows et al. (1972) shifted focus to various aspects of our energy and resource future.

Still, ecology and economics have been separate disciplines throughout their recent histories in the 20th century. After numerous trials and joint meetings between economists and ecologists, the International Society for Ecological Economics (ISEE) was formed in 1988 and the journal "Ecological Economics" was initiated and published its first issue in February 1989 (Costanza, 1996). As a result, the search term "Ecological Economics" currently generates over 3200 publications ranging from the 1980s until today (Scopus). It includes such first publications as "Toward an Ecological Economics" by Stanfield (1983, pp. 16), who argued that "Conventional, economic analysis cannot foot the whole bill and must therefore be supplemented with an economics of a different scope and method." In more recent studies, EE worldview is applied in fields such as business, social science and environmental studies in the North Americas, Western European countries and, increasingly, in China (Scopus). The number of EE studies has significantly increased since 1998s and thus helped formulate it as a distinct scientific transdisciplinary field, where various scholars exchange ideas, tools, and knowledge to gain deeper understanding and manage the connections between the economy, ecosystems, and the social structures that sustain human well-being (Zografos & Howarth, 2010). Currently, EE overlaps with several other approaches and has inspired concepts of economic models. For example, some circular economy streams find their roots in ecological economics (McDonough & Braungart, 2002). Another stream inspired by

Figure 2. The worldviews of ecological economics and neoclassical economics
Source: Temesgen et al., 2019, p. 45-47

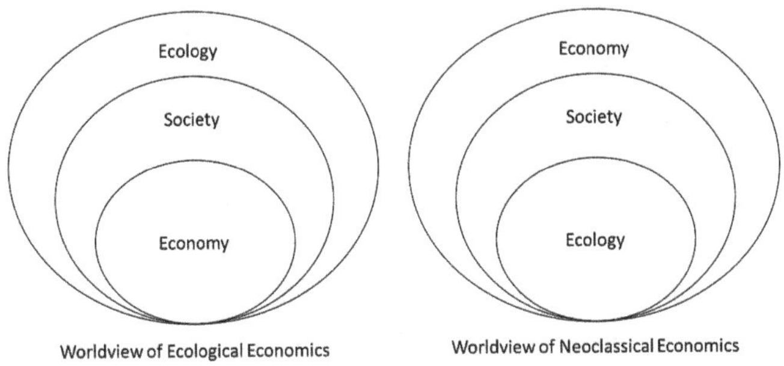

EE is the so-called doughnut economy, where the goal of economic activity is to "meet the needs of all" while respecting planetary boundaries (Raworth, 2017).

Traditionally, a key assumption of the widely used neoclassical economic view is that natural capital and created capital are perfect substitutes. Natural resources are "essentially limitless due to technical progress and infinite substitutability" (Costanza et al., 2014, pp. 58). According to the neoclassical view, nature, society, and culture do not possess an intrinsic value, they are merely instruments used to achieve maximum economic utility. It also employs a mechanistic way of thinking, where the highest possible output must be achieved with the means that are available (Liening, 2013). The neoclassical view of sustainability aims to assign natural resources and ecosystems an economic value with the expectation that market mechanisms will solve any occurring problems (Liening, 2013). It assumes that sustainability is achieved provided total capital stock is maintained.

Contrarily, the worldview of EE asserts that "wicked" problems originate in the ever-expanding size of the economy (Grierson, 2009). Nature, society, and culture each have both instrumental value and their own inherent value, which cannot be converted into economic value (Jakobsen, 2018). EE recognizes the instability, multi-dimensionality, and increasing complexity of natural and social systems, such as the environment, technology, population, and culture (Grierson, 2009). The view on sustainability implies that social, cultural, and environmental (nature) dimensions cannot be viewed as a mere stock of resources, rather they are a set of complex systems consisting of evolving elements (Grierson, 2009). A key point is that the economy is the means to achieve social, environmental, and cultural sustainability, and not vice versa (Figure 2). Moreover, this worldview resists the mechanistic way of thinking and a rejection of the assumption of human self-importance in the larger scheme of things (Grierson, 2009). Emphasis is placed on the whole, and the view is described as holistic, organic, ecological, and systemic. Reed (2007, pp. 675) suggests that this sustainability view goes beyond current notions of increased resource efficiency and reduced impact while meeting basic needs, to being based on the idea of whole or living systems' thinking, in which the "purpose of sustainability is sustaining life-enhancing conditions." Additionally, ecological worldview moves from competition to cooperation and partnerships as a necessary means for finding solutions to common sustainability challenges (Jakobsen, 2018). For example, dialogues and utopia-workshops are well-tested ways of shedding light on assumptions rarely critically examined (Jakobsen & Storsletten, 2019).

It has been argued that the trends in tourism development emphasizing scientific and technological success, consumerism, and materialism, will continue to produce undesired sustainability issues, such as local culture dilution, pollution, conflicts with locals, damage to the natural environment, and so on

(Mason, 2020). Thus, new knowledge from different spheres, such as EE, with focus on complex adaptive systems, a shift from mechanistic to systems thinking, integration of human and natural systems, and a new understanding of sustainability, are of great relevance to current tourism studies (Farrell & Twining-Ward, 2004). The tourism industry is complex with interdependent factors and activities, thus systems thinking can be an effective and powerful tool to clarify and integrate isolated problems associated with the industry, and provide a system for group learning and decision making to achieve desirable (sustainable) outcomes (Thanh & Bosch, 2010). Some argue that tourism is not merely an industry but a social force with deep transformative capacities for culture and environmental sustainability (Johannesson & Van der Duim, 2015). Empowerment and sustainable transformation through tourism occurs when tourism is implemented through a collaborative process, where residents are able to influence tourism development. Thus, co-creation and collaboration in innovation can be tools for sustainability in tourism development, when offerings are developed from inside of the community following a systematic process that inspires actual participation, capacity-building, cultural development, creativity, and innovation (Espeso-Molinero, 2018).

To sum up, the main potentials of EE lie in the holistic sustainability focus, the need for systemic changes, broad and local partnerships, application of inter-and transdisciplinary knowledge, and the importance of thriving and quality of life of humans and non-humans. When it comes to challenges, negotiating between biophysical reality and plural constructed social realities is the main question. In addition, mediating among different actors and competing regulating propositions can be difficult (Wironen & Erickson, 2020). Therefore, it requires innovative, complex, and multi-layered actions.

Community-Based Sustainability (CBS)

Community-based sustainability (CBS) is another approach relevant within the idealistic paradigm. Research on CBS is scarce, with too few empirical examples, and existing publications are rather heterogeneous in nature. Most of the publications have been produced between 2009 and 2021, with only two publications before the millennium, in 1995 and 1998. The two latter publications discuss CBS in different geographic contexts in such contrasting developments as communities taking up "the ownership and maintenance of their water sources" in the light of water supply system breakdown (Okuni & Rockhold, 1995, pp. 32), and community gardens management as opposite to being managed by local governments and how that can contribute to physical, ecological, sociocultural, and economic sustainability (Stocker & Barnett, 1998). Such contrasting publications, nevertheless, both emphasize the role of innovative technology, training, and stakeholder engagement.

In papers published after 2000, CBS has been applied in contexts such as "community gardens, solidarity purchasing groups, food cooperatives, community energy, recycling, and mobility initiatives" (Celata & Sanna, 2019, pp. 939), spread geographically from Northern America and Europe to Japan, Singapore, and Australia (Scopus). Community involvement stands central in different developments: from economic (Richins, 2009) and environmental (Hidayat & Stoecker, 2021) to a primary focus on the socio-cultural sustainability (Axon, 2020). In addition to Rydin (2007), O'Brien and Sakris (2014), and Richins (2009) are among the most cited authors published after 2000; newer publications attract scholarly attention by discussing meaningful resident engagement with CBS projects and transformational change (Axon, 2020). Rydin (2007) discusses the role of local sustainability indicators in governing sustainable development, and related to the issues of subjectivity, central-local relations, and constructing community. O'Brien and Sakris (2014) focus on sustainability in higher education and how CBS projects can be a part of deep learning. Richins (2009, pp. 785) discusses CBS in a tourism destination context with the focus on "long-range destination planning and implementation," where visitor commitment to the community's lifestyle, natural features, and contribution to the economy is important.

In the tourism context, Saarinen (2006) discusses CBS as one of the three traditions to understand sustainable development; the two others are resource- and activity-based sustainability. The discourse is rooted in the Brundtland report with ecological, social-cultural, and economic sustainability, as well as principles of futurity, equity, and holism, on one hand, and the necessity to limit mass tourism, on the other. As a result, the three traditions of Saarinen (2006) differ in their focus and relation to resources of a destination, and destinations' carrying capacities offering more time and space-specific answers to the destinations (Butler, 1999; Eckert et al., 2019).

The CBS approach is introduced as an alternative to the limitations of the resource-based tradition, "grounded in positivism and natural sciences" describing the limits of growth as objective and measurable, as well as the activity-based tradition, with its relativist ontology emphasizing the needs of the tourism industry, its changing activities, and capacities as central for limits of growth (Saarinen, 2006, pp. 1126-1128). As destination resources often can be overstepped, sustainability negotiation cannot be left to the industry alone, and there is a need of integrating tourism development with the host communities, and tangible and intangible resources in their destinations (Saarinen, 2014). According to Saarinen and Gill (2019, pp. 19), the CBS "creates a broader stakeholder context and collective action for sustainable tourism planning and development in which the limits to growth are based on negotiations and participation." Thus, CBS means a major system shift from an industry-dominated focus on growth and profits (Becker, 2016; Epler Wood et al., 2019) to a *holistic sustainability* focused on communities' life quality and well-being. Therefore, the tangible and intangible destination resources, represented by respective protection and preservation organizations, as well as other local stakeholders affected by tourism development, but often absent from discussions or marginalized (Heslinga et al., 2019; Sørensen & Balsby, 2020), should be involved in dialogues and negotiating tourism futures. Stakeholder participation needs to be enabled through community empowerment in planning and decision making; understanding contexts, shared knowledge and values; trust building; and acknowledgment of community situatedness in larger management and governance systems (Saarinen, 2019). Therefore, facilitations and the use of common meeting arenas, such as living labs, are essential in enabling sustainability in EBT (Høegh-Guldberg et al., 2022). Living labs can be defined as a "tool to utilize residents' participation and cross-stakeholder involvement" (Thees et al., 2020, pp. 1).

The CBS understanding has implications for open innovation processes, where tourism and experience sectors become a means to create attractive local communities and where visitor management is used to increase sustainability (Høegh-Guldberg et al., 2021; Plichta, 2019). The involvement of all stakeholders - affecting and being affected by the EBT, depends on broad partnership and participation (Eide & Hoarau-Heemstra, 2022), as well as co-creation and cross-sector (both public/private and across industries) multi-level innovations (Bosak, 2019). Involved residents can also co-create with visitors to find superior solutions (Soares et al., 2021).

Such open processes and holistic sustainability are, however, not without challenges. These include increasing risk of tensions and power conflicts, where certain firms, individuals, or other stakeholders with administrative, cultural, and economic power can take advantage and hinder community-based development (Cooper et al., 2009; Heslinga et al., 2019). Besides different stakeholders may have conflicting interests and values, and their engagement varies (Lindberg et al., 2019). To organize and make sure that relevant partners get involved and shared community arenas and methods function can also be resource demanding for both organizers and participants; it can be difficult to reach out to and engage the participants (Axon, 2020). In addition, stakeholders might miss essential knowledge or awareness to contribute to community discussions (Moscardo, 2008; Ratten et al., 2020). Therefore, more knowledge is needed about CBS challenges, and how CBS can be implemented in practice, as well as management and coordinated actions.

METHODS

This is a conceptual paper, and therefore literature review is the main method used. Exploration of the EE and CBS perspectives was inspired by the existing reviews of sustainability research (e.g., Albereda & Hajikhani, 2019; McCool & Bosak, 2019), emphasizing the need for stronger sustainability perspectives. Literature review of the EE and CBS publications was conducted by means of systematic literature search in the Scopus database, with the last update in November 2021. See the search words and the number of hits in Table 1. The search results were filtered based on the following criteria: limited to publications in English, article, conference paper, review, book chapter, and book types of documents. The authors then analyzed and selected the highest cited publications, as well as newer publications to learn about the evolution of EE and CBS approaches. In addition, a search to identify tourism publications was done among the EE and CBS publications (Table 1). Given only one tourism publication within CBS among the Scopus search results, the search was later broadened by using the snowballing method on Google Scholar.

Given that the research has also been inspired by the observation of conflicts in rural Norwegian communities, the authors chose to use empirical illustrations. Hence, the empirical methods are briefly introduced. Data from two cases were used to illustrate some potentials and challenges with experience-based transformations (EBT) seen from an idealistic paradigm. Both cases are Norwegian rural local communities of Røros and Vega. The main data gathering took place in June 2021–August 2021, with semi-structured interviews as the main method. It was supplemented with observations, documents, and photos, and some interviews in 2017 and the fall of 2021. Informants belong to different types of stakeholder groups within experience-based organizations, tourism, local food industry, NGO's, municipality organizations, voluntary interest groups, county governance, etc. Twenty-two and 15 interviews were conducted, respectively in Vega and Røros, during the main period.

With the aim to illustrate the idealistic framework; a narrative approach (Czarniawska, 1997; Webster & Mertova, 2007) was used when interviewing, analyzing, and writing. Narratives depend on a scene (here: contexts) and a plot (here: activities, potentials and challenges) (Connelly & Clandinin, 1990); the material was interpreted and constructed based on this. Due to the need to limit the length of the narratives, without losing the contextual situatedness, the cases were analyzed and described at three main levels: 1) broadest with the local community, 2) narrowing to the destination and role of experience scapes or events, and 3) core plot by focusing on a sub-destination (Sundsvoll) or one event (Elden-festival). There are relations and interwovenness of the levels, as illustrated by the circles in the Figure 3.

Experience scape means the physical, social, and symbolic environment where the experience and customer-interactions happen (Eide & Mossberg, 2019; O'Dell, 2005). They can be outdoor and/or in-door, involve interior, exterior, nature, and cultural landscapes, and therefore also influence sustainability.

Events are activities planned for a special purpose and usually involve many people, e.g., a meeting, party, trade show, conference, or festival. To use Leiper's (in Page & Getz, 2019) analogy of events tourism, events have become central in many destination systems where accommodation, attractions, transport, and ancillary services have been utilized or specifically developed.

Figure 3. Two different design approaches in the illustrations of EBT
Source: Authors

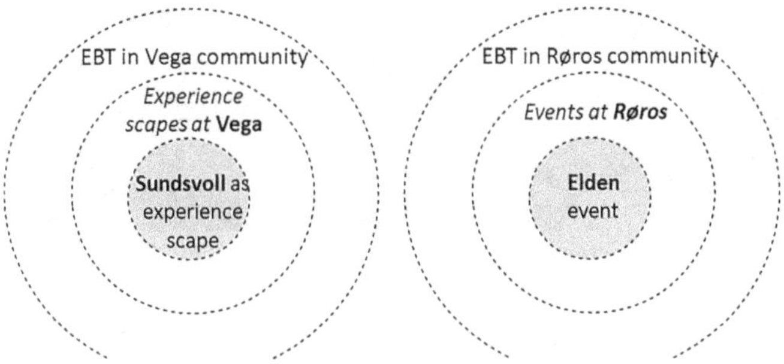

VEGA ILLUSTRATIONS

Vega as a Community and World Heritage (WH) Site

Vega municipality is situated on the coastline of Nordland County, with about 1200 residents and about 6500 islands and skerries. The first people settled there during the Stone Age about 10.000 years ago (Næss, 2013). The largest island is called Vega; most of the residents live there, and it hosts the main service infrastructure.

People who lived on many of the islands practiced fishing and had small farms to sustain their livelihoods. Traditionally, males from the age of 14 were away for weeks fishing in other regions like Lofoten. A central part of the female culture was eider duck care - building houses with nests and providing security from predators during the nesting period (Figure 4). For hundreds of years, families could generate the same income selling eider duck eggs and down feathers, as from fishing (Næss, 2013). Vega achieved UNESCO Heritage status in 2004 based on the fishing and farming culture, with a special focus on women's practices with eider ducks (Visit Vega, 2021).

Following the attainment of the World Heritage status, different innovations spurred and interacted, such as new education on the eider duck practices; the eider duck culture museum (E-house); restaurant (Gåkka) on a small island, with local food and storytelling; a "living museum" (Utværet Lånan) established at another small island where the eider duck culture is practiced. The living museum offers guided tours, produces and sells physical products (duvets, baby shoes, etc.) (see e.g., Eide & Mossberg, 2019).

However, there are also community challenges related to an increase in tourism, such as tensions between residents about modern fish-tourism, ferry queues, and pressure on infrastructure (financed based on number of citizens, but not visitors). The largest challenge for the community is the decline of primary sectors and hence a decrease in population. There is a need for new workplaces and promotion of population growth. In addition, there is a large conflict locally, due to two fish farms, one in a nature protected area and the other just outside of Sundsvoll. Some want them for economic and work reasons, others argue against them due to nature and cultural value.

Figure 4. Eider duck nesting
Source: Dorthe Eide

Destination Development at Vega

As the numbers of tourists with no family connection increased, more fishing huts opened for accommodation, and modern fish-tourism developed. A hotel was established with a restaurant famous for its local food, attracting both holiday and conference markets. Newer slow adventure tourism became popular, like kayaking and biking, including the concept of "jumping between" islands that is central in branding the region of the Helgeland coastline. Around 2009, the attention of this region increased significantly, as the TV-series "Skye blue" was released, using one of the islands in Vega municipality as a main location (Eide, 2011). Visits from Norwegians and other Nordic countries increased, causing the first indications of over-tourism in the form of pressure on ferries and narrow roads.

For a long time, July has been the busiest month, while the rest of the year has been characterized by under-tourism. Therefore, the tourism manager and others have encouraged innovations of products that could improve the spring and fall seasons. One example is local food and meals, farm cafés, and farm shops, where farmers got involved in multi-sector activity. Another example is new active adventures, such as hunting, fishing concept "hunt & release," as well as Via Ferrata and the Vega steps at Sundsvoll.

In 1995, the number of citizens on the islands was CA 1500 (SSB, 2021), while in July the number of people on the islands almost tripled (own observations). In 2019, the number of visitors was assumed to be about 50,000. In 2020, with the Covid-19 crisis and almost no international visitors, cruise and/or bus travel, domestic visitors increased significantly, making the total visitors around 65,000 and that is about the same for 2021 (Informants Vega, 2021).

While working with WH, the preservation, mitigation, and reporting increased the focus on sustainability. It was a natural next step, in 2008, to become one of the pilot projects in the Norwegian sustainable destination certification program. A core team with complementary competences has been the main stakeholderresponsible for managing sustainability at the destination level, by also involving and motivating others (Eide & Hoarau-Heemstra, 2022). One example is the invitation of all the residents in a sub-destination to participate in the living labs with dialogues about destination strategies:

We put them together and talk in meetings. What do we want? Do we want tourism in our area? Where, when why/why not? But at some areas, we must have a new meeting, because it has gone from low activity to very high in a very short time (informant Vega, 2021).

Figure 5. Photo collage of changes in Sundsvoll
Source: Dorthe Eide

Another example is a video shown at the ferry, informing about the use of mobile homes in Vega, where the sub-destination Sundsvoll has gotten the largest pressure related to mobile home tourism.

Transformations of Sub-Destination Sundsvoll

Before 2019, the community of Sundsvoll included two farms with farmland, and a few houses and second homes. Most parts of the sub-destination consisted of untouched nature, except for the brief signs of cultural landscape with trails shaped by sheep, hikers, and berry pickers. It was a raw, pristine nature, with mountains, stone beaches, and moorlands (left side of Figure 5). It also attracted a few tourists for its visual lookout to a main photo icon, i.e., the small island "Søla." Berry pickers and hikers were mostly locals and second house owners with family relations at the islands. As tourism increased, the municipality bought some of the land to make it a part of the nationally regulated outdoor area. The photo collage below shows some of the changes in the last four to five years.

Around 2016 to 2017, firms in a business network for modern fishing tourism and fishing huts explored how to develop the spring and fall seasons into more active seasons in Vega e.g., suggesting a Via Ferrata (guided climbing) at Sundsvoll (informants Vega, 2017). They discovered a need for steps to get down, which inspired another innovative idea, i.e., the Vega steps. The local sport club and the municipality became central in developing the steps, using crowdsourcing to get parts of it financed. Sculptures were set up along the way. Parking and a service building with toilets and garbage disposal areas were built. The Via Ferrata and the steps became a popular attraction and activity for locals and visitors. Visitors were both from the traditional segments of culture and slow adventure, and new segments within more active adventure tourism. It resulted in access to nature in new ways, changing some of the old experience scapes and creating new ones.

The summer of 2020, i.e., during Covid-19, the island had a record number of visitors, further increasing pressures on parking, narrow roads, toilets, and waste management at Sundsvoll in particular, causing traffic problems and much wild camping (Figure 5). Mobile homes blocked the view to the iconic Søla, with up to 80 mobile homes counted in the area (Brønnøysundsavis, 2021). Stakeholders saw an increasing need for visitor governance and regulations of wild camping and numbers of mobile homes.

It called upon innovations for sustainability, e.g., behavioral advice like the video motivating people to leave their cars on the mainland and offering bus service at the island to visitor attractions. Voluntary visitor tax was introduced in 2021 to cover service functions. Moreover, before the summer of 2021, two new firms were established in Sundsvoll. One, mobile camping site close to the Vega steps, as the free camping at Sundsvoll was closed by physical barriers. Another firm offered accommodations in small cabins formed as E-houses. Currently more innovative ideas are being considered as the high pressure on the roads and ferries remains.

The experience innovations implemented in Sundsvoll have changed the experience scapes as well as the image of Vega from primarily being a cultural and soft adventure destination to a more active adventure island. However, the unsolved situation with the planned fish farm just outside of Sundsvoll, which will change the visual image of Søla and Sundsvoll, may become the next major change in the area.

RØROS ILLUSTRATIONS

Røros as a Community and World Heritage Site

Røros, a mining town taking form in 1644, currently has around 5,600 inhabitants and covers 1,956 square kilometers of a sparsely populated mountain region with a mix of forests, mountains, agricultural lands, and small settlements (Destination Røros, n.d.). Røros has been on UNESCO's World Heritage List since 1980 due to many years of mining and agriculture. These activities have resulted in a unique land-use system and interaction between resource use and landscape change, and a multitude of well-preserved buildings of great cultural value. From 2010, the Heritage Site has been extended to include the surrounding area, the circumference (UNESCO, 1992–2010). The Røros Museum (Figure 6) is one of the key actors in Røros municipality and the management of the World Heritage site.

Before the establishment of the mining industry, the area of Røros was mainly used for reindeer herding among the southern Sámi population (indigenous people), and as mountain pasture and seasonal dwelling for the few farmers in the area. Later, mining, agriculture, and extensive use of backcountry resources, such as hunting, fishing, and forestry, intertwined and resulted in a diversified economy. As the last mine closed in 1977, tourism, manufacturing, agriculture, and public services became the main sectors of industry in Røros. The South Sámi culture is still in the area, with more than 20 Sámi families actively working with reindeer husbandry. Despite this, the Sami culture and people were not incorporated into the application process for WH status (informants Røros, 2021). Moreover, there are still a number of challenges regarding Sámi culture, such as increasing the visualization and dissemination of the culture in an authentic way to avoid exoticism and the misrepresentation of the indigenous culture.

Destination Development at Røros

Today, Røros is advertised as a place where you can get "a glimpse of the past"—a unique place with character and authenticity, "a living museum"—and this heritage is widely used for branding the city and as a source for creating events, jobs, and branding value for products (Lillevold & Haarstad, 2019). For example, it has developed a strong identity connected with local and organic food and is often referred to as "the local food capital of Norway" in the media. The town has its own brewery, dairy, and rain deer

Figure 6. The Røros Museum located in a former smelting house
Source: Anne Wally Ryan

produce, while firms have for long been active participants in a food network. Røros has been marked the "Best Cultural Municipality" of Norway for seven consecutive years (2011–2018).

Additionally, tourism has become an important industry in Røros, with approximately one million visitors every year. It was one of the pilot projects in a sustainable destination certification program and has won two awards for sustainable tourism in Norway (in 2011 and 2012). Cultural experiences are highly important for tourism development in Røros. Here they are mostly based on tangible, preserved artifacts, mining infrastructure, old wooden houses, and the Sami culture, and are well-known as a center for arts, crafts, and local food. Nature activities are also very central in destination development. During winter, the selection of activities includes dogsledding, ice fishing, reindeer experiences, snowmobile safaris, etc. While in the summer, the region offers hiking trips in the mountains and activities such as biking, kayaking, etc.

Røros is also known for being an "evental city" (Pløger, 2016) by having many different events that attract visitors, year-round income, and vitality, e.g., outdoor theatre, music festivals, and markets. Most of them have links to the history of the place and making use of the old urban structures. For example, Rørosmartnaen, Vinterfestspelet, and Femundsløpet are large and well-known events, nationally and

Figure 7. The two pictures give a glimpse from the Elden festival
Source: Elden, 2021

internationally. Such a vast selection of events contributes to the strongly balanced seasonality of the destination. There is a major event happening in Røros each month of the year except for the month of May (Informant Røros, 2021). However, not all local stakeholder groups e.g., local people, minorities, etc., feel included in the organizing of these events, spurring some communal conflicts e.g., challenges with over-tourism, the disturbance of local everyday practices, increased costs, noise, traffic, and waste pollution.

Thus, the Destination Management Organization (DMO) is engaged in collaborating efforts with local community members and local businesses in conflict management practices and mitigation efforts between different interests and viewpoints. In addition, other organizations, such as The World Heritage Center, The Elden Foundation, The Røros Museum, local church, etc., play a significant role in the involvement of various stakeholders of the community in the destination development efforts and practices.

Elden Event

One of the most popular events, and the second biggest of this kind in Norway, is Elden, a historical musical outdoor theatre set up in spectacular surroundings in the middle of the World Heritage site of Røros (Figure 7). Elden premiered in 1980, and the brutality of war, the fear of the civilian population, and love are the main themes in the play. On average, it attracts over 12,000 spectators yearly, which is double the population of Røros (Elden, 2021). Therefore, it plays a significant role in the community of Røros, including different industries and stakeholders, as well as its impacts and contributions.

For instance, in terms of the economy, Elden contributes 6.5 million Norwegian krone (approximately $747, 000 US) directly back into the community each year (informant Røros, 2021). It tries to involve different social groups of the community in various activities pre, during, and after the event. Each year the management of Elden tries to be more environmentally sustainable, promoting locality and green practices. Additionally, Elden plays a vital role in passing on the history of the place and its people. It aims at keeping the local culture alive by continuously innovating not only the management of the event, but also the content of the historic play—adapting, modernizing, and promoting cultural history for present and future generations, as well as local and international visitors.

The quote "Elden is for all" (Informant Røros, 2021), illustrates the philosophy behind the event. It is about the involvement of all—firms, residents, visitors, destination management organizations, participants, etc. By including various stakeholders in the planning and organizing of the event, managers aim to strengthen the holistic sustainability of the place, in terms of economy, environment, culture, and social wellbeing. It also addresses the continual development of Røros as a well-known brand, where collaboration is important.

By using workshops and innovation labs, that involve various stakeholder groups, the management team of Elden can gather different opinions, ideas, and suggestions. They are focusing on more holistically sustainable practices—local and sustainable food options, locally-sourced souvenirs, better systems for waste management and pollution, volunteer wellbeing, involvement of socially vulnerable groups, and promotion of the cultural heritage of the place. For instance, Elden is highly dependent on voluntary work, in which the people of Røros are highly engaged. This, in turn, creates meeting places and a strong sense of community. One informant puts it like this, *"We need the volunteers. They are gold. Without volunteers, we can shut down immediately. So, we must involve them in all the processes"* (Informant Røros, 2021). Furthermore, local businesses and organizations are invited to participate in the event. They emphasize that Elden is important for them to market and sell their products.

However, some groups in the destination community do not feel the benefits of the event and feel like they have no voice in the planning and performing stages. For instance, according to some local Sami people, the story of the play is not authentic and lacks the representation of the indigenous population. By inviting non-Sami to perform Sami songs on stage, it feels highly inauthentic and insensitive. In addition, it is often criticized for noise pollution and crowds, thus motivating the organizers to consider new and innovative solutions for such issues in the future. As this event attracts a vast number of spectators, crowd control tactics and other logistical solutions must be considered.

DISCUSSION

The chapter suggests two approaches, EE and CBS, to explore the research question: "*What are the core, potentials and challenges of the idealistic paradigm to IFS in rural communities involved in experience-based transformations (EBT)?*" To further concretize and discuss the idealistic framework, the approaches are compared and summarized through the following categories: ontology; core disciplines; sustainability values; innovation focus; partnerships; methods; organizing; potentials; and challenges (columns 2, 3 Table 2). The discussion is further developed by drawing upon the narratives from Vega and Røros (columns 4, 5 Table 2).

EE and CBS – Conceptual Comparison

1. **Ontology**, or the nature of reality: EE's ontology ranges from positivism to relativism. Ecology is based on natural reality that has intrinsic value and where the constitutive parts of nature cannot be isolated, and economy that has a given collection of social realities that emerge at a certain time (Temesgen et al., 2019). The CBS ontology, in contrast, is socially constructed where interpretations, understandings, and practices are situated in interactions of a community.
2. **Disciplines:** The complexity in social and natural systems, and sustainability challenges, require new ways to understand reality and, thus, a shift to inter- and trans-disciplinarity (Temesgen et al., 2019). EE can strengthen the ability to explain interdisciplinary methods, as it requires the knowledge of more than one system. CBS shares this assumption, as communities often need local solutions to wicked problems that are complex. However, communities are often driven by

"social or economic principles of sustainability" (Zizka et al., 2021) that can be conflicting with environmental values and can be short-term oriented.

3. **Sustainability Values:** Both EE and CBS include holistic sustainability, nested systems, and a view where economy is a means to achieve other values. According to EE, the future vision of a sustainable planet lies on a high quality of life and, hence, thriving, for all its members (humans and other species) within material and ecological constraints. CBS is also holistic, but prioritizes social sustainability of the communities with their time and space-specific challenges and community life quality, with a less pronounced link to the common good of broader geographic spaces.
4. **Innovation:** Since EE questions the hard core of mainstream economics, it requires system changes and innovations (Jakobsen, 2018), with new solutions being ecologically and economically beneficial as well as socially and culturally balanced (Pastoors et al., 2017). This will require transformative social innovations with changes in social relations involving new ways of doing, organizing, knowing, and framing (Haxeltine et al., 2016). Research illustrates a wide spectrum of forms and types of CBS innovations as contextual solutions to local or greater societal challenges. Environmental, social, and economic goals stand high on the CBS agenda, while the community-based initiatives in Europe contribute to "*strengthening social ties, the skills of their beneficiaries*" (Celata and Sanna, 2019, p. 943).
5. **Partnerships:** As EE calls for interdisciplinary and transdisciplinary knowledge, there is a need for strong broad partnerships to collaborate and co-create instead of competing. Similarly, CBS highlights the necessity of multi-stakeholder involvement, with a particularly important role for the local residents in taking on responsibility and ownership to the development processes. Both prioritize involvement, cooperation, networking, and multi-level interactions.
6. **Methods:** EE calls for collaborations and co-creation of transdisciplinary knowledge. This is facilitated by involving various stakeholders in the co-creation in so-called utopia-workshops, with scenarios enabling dialogical processes, releasing visions and energy into transformative processes, individually and collectively (Jakobsen & Storsletten, 2019). CBS also facilitates dialogues and co-creation through different types of labs, particularly, living labs, as innovation methods (Van der Walt et al., 2009; Zavratnik et al., 2019).
7. **Organizing:** The EE approach requires transformative, multi-layered, and complex actions and interactions to achieve systemic changes. To enable such processes, political, governmental, and managerial, i.e. top-down, actions are necessary (Wironen & Erickson, 2020). Given the community focus and strong role of local residents in the CBS, the bottom-up initiatives seem vital. Nevertheless, these initiatives can often only be realized when they meet respective, encouraging policies and action from the top, and someone has to be the boundary broker and facilitate meeting arenas and processes (Wenger, 2003).
8. The main **potentials** of both the EE and CBS are holistic sustainability, strong partnerships, and common good. However, negotiating between biophysical reality and plural constructed social realities is also one of the main **(9) challenges**. Difficulties with mapping the relevant stakeholders, engaging them into the dialogues, negotiating different interests and values, and continuously following up the development processes (Wironen & Erickson, 2020) is demanding and requires a change in community practices (Nicolini & Monteiro, 2017), which is a long-term endeavor.

Cross-Case Reflections on Idealistic Paradigm to EBT

The illustrative EBT cases showed different positive and negative sustainability impacts at different levels, including community, sub-destination, events, local firms, and individuals. In Sundsvoll (Vega) the new activities and experience scapes involve changes in the physical, social, and symbolic environment, involving different stakeholder types, both creating and destroying sustainability values. The intangible and sensory elements are more addressed in Elden (Røros), as the natural environment is maintained, but staged with aesthetics and symbolic elements.

Both cases show that a new use of areal can create significant changes in the symbolic meaning of landscapes and other experience scape elements (Fitchet et al., 2021; Ryan & Wollan, 2013). The perceptual symbolic, sensing, and aesthetic dimensions of landscapes are reasons to visit and have an intrinsic value of their own. At the same time, more functional and service-related aspects, such as parking places, waste disposal, roads, transportation, etc., must also be addressed. Stakeholder involvement in destination development and decision-making takes place through living lab-like arenas, some by getting locals and non-locals involved, which means that not only multi-stakeholder and multi-level partnerships get involved, but also different disciplines and combinations of practical and more research-based knowledge.

Both case narratives have core elements, potentials, and challenges revealed in the EE and CBS literature, but there is limited integration and overall system thinking. There is some facilitation done, but both cases have conflicts hampering communication and thriving of the communities. However, none seem to have the legitimacy or courage to address them constructively, pointing to the need for negotiating and legitimate orchestrators across sectors.

Development of the Idealistic Paradigm Framework and Further Research

The calls for rethinking sustainability before and during Covid-19, included in this debate, have intensified the need of paradigm nudging or shifts (Weaver, 2009). Although paradigm shifts require a radical change in thinking and practice, which are complex and require further interactive research and innovation. There is a need for a better combination of strong elements from EE and CBS in the idealistic paradigm. Namely, the categories describing the core of EE and CBS (Table 2) suggest how this can be done.

The ontologies and epistemologies of EE and CBS can be seen as complementary in the idealistic paradigm to IFS, as also illustrated by Vega and Røros ontics. In particular, the systems thinking of EE and contextualization of CBS, with holistic sustainability in focus, demonstrate the potential of combining abstract and applied knowledge. As the cases show, the transdisciplinarity of EE could lift CBS practices by involving researchers and residents with sound sustainability background and e.g. creating community-based indicators (Boyd & Charles, 2006; Schmitz et al., 2010). However, practicability of the holistic umbrella of the idealistic paradigm is challenging while leading to changes across levels and industries, certain sustainability values are contextually prioritized and hence shaping what innovation types are involved. The combination of the EE and CBS innovation methods could potentially make up for this challenge by combining broad stakeholder involvement in the earlier visionary development phases of innovation processes through utopia-workshops and stronger engagement of selected stakeholder in the concretization and implementation phases through living labs. As illustrated by the cases, common vision on the community level is also important to make sure that local developments are not purely building on economic values.

Table 2. The core, potentials, and challenges of EE and CBS, with respective illustrations from the cases

Core	EE	CBS	Vega	Røros
(1) Ontology or Ontic	Organic, collectivism, anti/ reductionism.	Social constructivism, dependent on knowledge, dialectical, negotiated, power-laden.	Rural community, local tourism destination, sub-destination Sundsvoll in particular, and experience scapes.	Rural community, local tourism destination, events, and Elden, in particular.
(2) Discipline	Inter- and transdisciplinary. Natural, social, and humanistic science combined.	Social sciences primarily, in the times of Anthropocene. Locals can bring other sciences in.	Tourism, culture, nature preservation, governance, business, subsector-related (farming, food, sport club, outdoor life, culture+), another local knowledge.	Communication, journalism, tourism, art, nature management, and history (in addition local firms, public workers, unemployed, youth).
(3) Sustainability Values	Holistic, but environment is most important, then the social. Economy as a means to achieve the common good.	Holistic through social lenses, community life quality.	Holistic with focus on social values (work, quality of life), some put culture, nature, or economy first.	Holistic with focus on social values and an inclusive policy, balanced use and conservation of local resources and maintaining quality of life for the residents.
(4) Innovation (focus)	System innovations, social innovations. Multi-level, Managerial, and governance innovations. More radical innovations.	Directed at solving local challenges (of very heterogeneous nature), troubling issues. Grassroot and multi-level.	Experience and service products and scapes; visitor governance; responsible marketing; culture; transport; beach garbage; season development. New methods for co-creation.	Methods for co-creation; Local food menu; Visualizing the local in high-quality branding products. Aesthetics of the performance and venue.
(5) Partnership	Multi-stakeholders from different disciplines, both theoretical and practical knowledge.	Multi-stakeholder, with resident participation as a prerequisite.	Multiple: experience-providers, residents, DMO, municipality, voluntary organizations, county.	Multiple: private or public experience-providers, residents, voluntary organizations, municipality, county, DMO.
(6) Methods	Dialogue cafes; Utopia-workshops (scenarios and dialogues).	Living labs and other labs/ workshops.	Destination and sub-destination workshops/ living labs, dialogue, mitigations (part of WH and certification).	Workshop involving different stakeholders to achieve an integrated and creative involvement of the locals.
(7) Organizing	Often political, governance, managerial levels. Combination of top down-bottom up approaches.	Often public bodies are connecting others and facilitating processes.	Core team in certification, mostly tourism manager (tourism), preservation manager (national parks).	Event manager, tourist manager (DMO), WH coordinator, CEO of Museum.
(8) Potentials	Diverse theoretical and practical knowledge/opinions to solve grand challenges and handle more radical innovations.	Local initiatives, tacit knowledge, specific, operationalizable, contribute to community quality of life.	Increased cooperation and value creation synergies across levels and sectors. New firms, work, pride.	Involvement of the ingenious, authentic performances and branding. Development of cooperation with local firms and use of local competence.
(9) Challenges	Complex due to many perspectives, higher risk of misunderstanding and conflicts. Too utopian?	Missing knowledge since mainly local and mainly social science. Changing existing routines. Stakeholder engagement and power, resource demanding.	Tensions and conflicts across sectors. Lack cross-industry dialogue, negotiation, and coordination. Lack proactive coordination of innovations and their consequences.	Controlling noise pollution and crowds. Solving conflicts with the Sami. Maintaining and renewing an attractive event. Keeping the volunteers' enthusiasm and engagement.

Source: Authors

The idealistic paradigm to IFS calls for a combination of the bottom-up and top-down approaches, that echoes into Giddens' (1984, p. 121) claim of the "duality of structure": *"the structural properties of social systems are both medium and outcome of the practices they recursively organize."* Thus, community benefit sharing (Heslinga et al., 2019) requires power redistribution to avoid coercive and induced resident participation, where community interests are only heard symbolically (Saarinen, 2019). Besides, EE and CBS are rather limited in discussing the involvement of non-human stakeholders, such as animals and plants. The authors suggest strong partnerships need to be more explicit in this regard, and other approaches, such as the actor-network theory (Latour, 1996) and the quintuple helix (Carayannis & Campbell, 2018) can be helpful. In addition, one can argue in line with stakeholder theory that the definition of primary and secondary stakeholders can depend on the problem at stake (Bonnafous-Boucher & Rendtorff, 2016).

Although having used empirical illustrations, there is still a need for empirical research using and further developing the idealistic paradigm, not only related to firms, but also in local communities, and by involving different analytical levels. In particular, our illustrations point to the need for studies and participative innovations exploring and coordinating how experience-based and other sector innovations are not isolated, but are influenced by, and influencing each other, throughout time. As broad partnerships are so central, one needs to study and innovate more on methods and resources (financing, time, competence) for participation, and who, how, and when to involve. Finally, as many communities face not only sustainability challenges but also value conflicts, there is a need for new knowledge on how and who can be responsible in negotiating conflicts within and across sectors in communities at different levels.

CONCLUDING REMARKS AND LIMITATIONS

The authors have discussed the idealistic paradigm as an alternative to the narrower skeptic and pragmatic sustainability paradigms confined to one or two sustainability values and limited involvement of stakeholders. This does not imply that the idealistic paradigm can dismiss the accumulated knowledge; instead, it should bridge, combine, and balance the different aspects as long as the economy remains the means of achieving common good on different levels. Exploring the EE and CBS approaches, the chapter demonstrates that both approaches harmonize with the idealistic paradigm, have similarities and differences and, most importantly, can complement one another when concretizing the framework and using the idealistic paradigm. The main limitations of the chapter are naturally related to its conceptual nature and restricting it to the EE and CBS approaches to the idealistic paradigm.

REFERENCES

Adams, R., Jeanrenaud, S., Bessant, J., Denyer, D., & Overy, P. (2016). Sustainability-oriented Innovation: A Systematic Review. *International Journal of Management Reviews, 18*(2), 180–205. doi:10.1111/ijmr.12068

Albareda, L., & Hajikhani, A. (2019). Innovation for Sustainability: Literature Review and Bibliometric Analysis. In N. Bocken, P. Ritala, L. Albareda, & R. Verburg (Eds.), *Innovation for Sustainability. Business Transformations Towards a Better World* (pp. 35–57). Palgrave Macmillan. doi:10.1007/978-3-319-97385-2_3

Alsos, G. A., Eide, D., & Madsen, E. L. (2014). Introduction: Innovation in tourism industries. In G. Alsos, D. Eide, & E. L. Madsen (Eds.), *Handbook of Research on Innovation in Tourism Industries* (pp. 1–24). Edward Elgar Publishing.

Axon, S. (2020). The socio-cultural dimensions of community-based sustainability: Implications for transformational change. *Journal of Cleaner Production, 266*, 121933. doi:10.1016/j.jclepro.2020.121933

Bayer, J., Fehringer, A., Lehar, G., Jurgeit, F., & Leitner, T. (2017). The relevance of visitor's national park affinity for effective visitor management in protected areas. In J. N. Albrecht (Ed.), *Visitor Management in Tourism Destinations* (pp. 75–90). Cabi. doi:10.1079/9781780647357.0075

Becker, E. (2016). *Overbooked: The exploding business of travel and tourism.* Simon and Schuster.

Bocken, N., Ritala, P., Albareda, L., & Verburg, R. (2019). *Innovation for Sustainability: Business Transformations Towards a Better World.* Palgrave Macmillan. doi:10.1007/978-3-319-97385-2

Bonnafous-Boucher, M., & Rendtorff, J. D. (2016). *Stakeholder theory: A model for strategic management*. Springer International Publishing. doi:10.1007/978-3-319-44356-0

Bosak, K. (2019). The tourism system. In S. F. McCool & K. Bosak (Eds.), *A Research Agenda for Sustainable Tourism* (pp. 14–20). Edward Elgar Publishing. doi:10.4337/9781788117104.00009

Boyd, H., & Charles, A. (2006). Creating community-based indicators to monitor sustainability of local fisheries. *Ocean and Coastal Management, 49*(5-6), 237–258. doi:10.1016/j.ocecoaman.2006.03.006

Brønnøysundsavis. (2021). *Ny bobilparkering på Sundsvoll* [New motorhome parking at Sundsvoll]. https://www.banett.no/nyheter/2021/06/28/Ny-bobilparkering-på-Sundsvoll-Har-hatt-besøkende-siden-vi-åpnet-24177

Bullvåg, E., Nyvold, C. E., Ovesen, S., Kvarum, M. K., Finsveen, R., & Steffensen, T. (2020). *Indeks Nordland, Nummer 17 - Årgang 2020, En rapport om utviklingen gjennom 2019 og utsiktene for 2020*. Indeks Nordland. https://indeksnordland.no/

Butler, R. W. (1999). Sustainable tourism: A state-of-the-art review. *Tourism Geographies, 1*(1), 7–25. doi:10.1080/14616689908721291

Carayannis, E. G., & Campbell, D. F. (2018). *Smart quintuple helix innovation systems: How social ecology and environmental protection are driving innovation, sustainable development and economic growth*. Springer.

Celata, F., & Sanna, V. S. (2019). A multi-dimensional assessment of the environmental and socioeconomic performance of community-based sustainability initiatives in Europe. *Regional Environmental Change, 19*(4), 939–952. doi:10.100710113-019-01493-9

Connelly, F. M., & Clandinin, D. J. (1990). Stories of experience and narrative inquiry. *Educational Researcher, 19*(5), 2–14. doi:10.3102/0013189X019005002

Cooper, C., Scott, N., & Baggio, R. (2009). Network position and perceptions of destination stakeholder importance. *Anatolia, 20*(1), 33–45. doi:10.1080/13032917.2009.10518893

Costanza, R. (1996). Ecological Economics: Reintegrating the study of humans and nature. *Nature Sciences Sociétés, 5*(2), 978–990.

Costanza, R., Cumberland, J. H., Daly, H., Goodland, R., Norgaard, R. B., Kubiszewski, I., & Franco, C. (2014). *An Introduction to Ecological Economics*. CRC Press. doi:10.1201/b17829

Czarniawska, B. (1997). *Narrating the organization. Dramas of institutional identity*. University of Chicago Press.

Destination Røros. (n.d.). *Information about Røros*. Destination Røros. https://www.roros.no/en/ information/

Djellal, F., & Gallouj, F. (2010). The innovation gap and the performance gap in the service economies: A problem for public policy. In F. Gallouj & F. Djellal (Eds.), *The Handbook of Innovation and Services* (pp. 653–675). Edward Elgar Publishing. doi:10.4337/9781849803304.00041

Duncan, O. D. (1961). From Social System to Ecosystem. *Sociological Inquiry, 31*(2), 140–149. doi:10.1111/j.1475-682X.1961.tb00518.x

Dybdahl, L. M. (2019). Business model innovation for sustainability through localism. In N. Bocken, P. Ritala, L. Albareda, & R. Verburg (Eds.), *Innovation for Sustainability. Business Transformations Towards a Better World* (pp. 193–211). Palgrave Macmillan. doi:10.1007/978-3-319-97385-2_11

Eckert, C., Zacher, D., Pechlaner, H., Namberger, P., & Schmude, J. (2019). Strategies and measures directed towards overtourism: A perspective of European DMOs. *International Journal of Tourism Cities, 5*(4), 639–655. doi:10.1108/IJTC-12-2018-0102

Eide, D. (2011). *Turisme basert på fortellinger, bøker og film: en litteratur og case studie*. UiN-rapport nr. 2-2011. University in Nordland.

Eide, D., & Hoarau-Heemstra, H. (2022). Innovation for sustainable destinations: The role of certification and partnership. In I. Booyens & O. Brouder (Eds.), Handbook of Innovation for Sustainable Tourism. Edward Elgar Publishing.

Eide, D., & Mossberg, L. (2019). Toward a framework of experience quality assessment in cultural tourism. In D. A. Jelinčić & Y. Mansfeld (Eds.), *Creating and Managing Experiences in Cultural Tourism* (pp. 101–120). World Scientific. doi:10.1142/9789813233683_0007

Elden. (2021). *About Elden*. https://elden-roros.no/utendorsteateret/historiene/

Engelmann, J., & Al-Sadi, M. (2019). Business-Driven Ecological Innovations in Green Growth Strategies. In N. Bocken, P. Ritala, L. Albareda, & R. Verburg (Eds.), *Innovation for Sustainability. Business Transformations Towards a Better World* (pp. 97–113). Palgrave Macmillan. doi:10.1007/978-3-319-97385-2_6

Epler Wood, M., Milstein, M., & Ahamed-Broadhurst, K. (2019). *Destinations at Risk: The Invisible Burden of Tourism*. The Travel Foundation. https://www.thetravelfoundation.org.uk/invisible-burden/

Espeso-Molinero, P. (2018). Collaborative capacity building as a resilience strategy for tourism development in indigenous Mexico. In J. Cheer & A. A. Lew (Eds.), *Tourism, Resilience, and Sustainability: Adapting to Social, Political and Economic Change* (pp. 184–201). Routledge.

Farrell, B. H., & Twining-Ward, L. (2004). Reconceptualizing tourism. *Annals of Tourism Research, 31*(2), 274–295. doi:10.1016/j.annals.2003.12.002

Fitchett, J., Lindberg, F., & Martin, D. M. (2021). Accumulation by symbolic dispossession: Tourism development in advanced capitalism. *Annals of Tourism Research, 86*, 103072. doi:10.1016/j.annals.2020.103072

Forrest, N., & Wiek, A. (2014). Learning from success—Toward evidence-informed sustainability transitions in communities. *Environmental Innovation and Societal Transitions, 12*, 66–88. doi:10.1016/j.eist.2014.01.003

Fuglsang, L., & Eide, D. (2013). The experience turns as 'bandwagon': Understanding network formation and innovation as practice. *European Urban and Regional Studies, 20*(4), 417–434. doi:10.1177/0969776412448090

Giddens, A. (1984). Elements of the theory of structuration. In G. M. Spiegel (Ed.), *Practicing History* (pp. 121–142). Routledge.

Gössling, S., Scott, D., & Hall, C. M. (2020). Pandemics, tourism and global change: A rapid assessment of COVID-19. *Journal of Sustainable Tourism*, *29*(1), 1–20. doi:10.1080/09669582.2020.1758708

Grierson, D. (2009). The shift from a mechanistic to an ecological paradigm. *International Journal of Environmental, Cultural, Economic and Social Sustainability*, *5*(5), 197–206. doi:10.18848/1832-2077/CGP/v05i05/54670

Hall, C. M. (2018). Resilience theory and tourism. In J. Saarinen & A. M. Gill (Eds.), *Resilient Destinations and Tourism* (pp. 34–47). Routledge. doi:10.4324/9781315162157-3

Haxeltine, A., Avelino, F., Pel, B., Dumitru, A., Kemp, R., Longhurst, N., . . . Wittmayer, J. M. (2016). A framework for transformative social innovation. *TRANSIT Working Paper, 5*, 2-1.

Heslinga, J., Groote, P., & Vanclay, F. (2019). Strengthening governance processes to improve benefit sharing from tourism in protected areas by using stakeholder analysis. *Journal of Sustainable Tourism*, *27*(6), 773–787. doi:10.1080/09669582.2017.1408635

Hidayat, D., & Stoecker, R. (2021). Collective Knowledge Mobilization Through a Community University Partnership. *Journal of Higher Education Outreach & Engagement*, *25*(2).

Higgins-Desbiolles, F. (2020). Socialising tourism for social and ecological justice after COVID-19′. *Tourism Geographies*, *22*(3), 610–623. doi:10.1080/14616688.2020.1757748

Hjalager, A. M. (2010). A review of the innovation research in tourism. *Tourism Management*, *31*(1), 1–12. doi:10.1016/j.tourman.2009.08.012

Høegh-Gulberg, O., Eide, D., & Yati, Y. (2022). Towards sustainable tourism through lab-driven innovations: a systematic literature review. In I. Booyens & O. Brouder (Eds.), Handbook of Innovation for Sustainable Tourism. Edward Elgar Publishing.

Høegh-Guldberg, O., Seeler, S., & Eide, D. (2021). Sustainable Visitor Management to Mitigate Overtourism – What, Who, and How. In A. Sharma & H. Azizul (Eds.), *Over-tourism as Destination Risk: Impacts and Solutions* (pp. 167–186). Emerald Publishing Limited. doi:10.1108/978-1-83909-706-520211012

Holling, C. S., & Goldberg, M. A. (1971). Ecology and Planning. *Journal of the American Institute of Planners*, *37*(4), 221–230. doi:10.1080/01944367108977962

Jakobsen, O., & Storsletten, V. M. L. (2019). Beyond the green shift–Ecological economics. In J. S. Methi, A. Sergeev, M. Bieńkowska, & B. Nikiforova (Eds.), *Borderology: Cross-disciplinary Insights from the Border Zone* (pp. 173–183). Springer. doi:10.1007/978-3-319-99392-8_13

Jakobsen, O. D. (2018). *Anarchism and ecological economics: A transformative approach to a sustainable future*. Routledge. doi:10.4324/9780429486876

Johannesson, G. T., Ren, C., & van der Duim, R. (2015). *Tourism encounters and controversies: Ontological politics of tourism development*. Ashgate Publishing, Ltd.

Jørgensen, M. T., Vorre, H. A., Sørensen, F., Fuglsang, L., Sundbo, J., & Jensen, J. F. (2021). Collective tourism social entrepreneurship: A means for community mobilization and social transformation. *Annals of Tourism Research, 88*, 103171. doi:10.1016/j.annals.2021.103171

Klewitz, J., & Hansen, E. G. (2014). Sustainability-oriented innovation of SMEs: A systematic review. *Journal of Cleaner Production, 65*, 57–75. doi:10.1016/j.jclepro.2013.07.017

Kuhn, T. S. (1962). *The Structure of Scientific Revolutions*. University of Chicago Press.

Latour, B. (1996). On actor-network theory: A few clarifications. *Soziale Welt, 4*, 369–381.

Leach, M., Rockström, J., Raskin, P., Scoones, I., Stirling, A., Smith, A., Thompson, J., Millstone, E., Ely, A., Arond, E., Folke, C., Olsson, P., Scoones, I., Stirling, A., Smith, A., Thompson, J., & Olsson, P. (2012). Transforming innovation for sustainability. *Ecology and Society, 17*(2), art11. doi:10.5751/ES-04933-170211

Leopold, A. (1933). The conservation ethic. *Journal of Forestry, 31*(6), 634–643.

Liening, A. (2013). The breakdown of the traditional mechanistic worldview, the development of complexity sciences and the pretence of knowledge in economics. *Modern Economy, 04*(04), 305–319. doi:10.4236/me.2013.44034

Lillevold, K., & Haarstad, H. (2019). The deep city: Cultural heritage as a resource for sustainable local transformation. *Local Environment, 24*(4), 329–341. doi:10.1080/13549839.2019.1567481

Lindberg, F., Fitchett, J., & Martin, D. (2019). Investigating sustainable tourism heterogeneity: Competing orders of worth among stakeholders of a Nordic destination. *Journal of Sustainable Tourism, 27*(8), 1277–1294. doi:10.1080/09669582.2019.1614188

Manual, O. E. C. D. (2005). *Guidelines for collecting and interpreting innovation data. OECD/Eurostat*. OECD Publishing.

Markard, J., Raven, R., & Truffer, B. (2012). Sustainability transitions: An emerging field of research and its prospects. *Research Policy, 41*(6), 955–967. doi:10.1016/j.respol.2012.02.013

Mason, P. (2020). *Tourism impacts, planning and management*. Routledge. doi:10.4324/9780429273544

McCool, S. F., & Bosak, K. (2019). *A Research Agenda for Sustainable Tourism*. Edward Elgar Publishing. doi:10.4337/9781788117104

McDonough, W., & Braungart, M. (2010). *Cradle to cradle: Remaking the way we make things*. North Point Press.

Meadows, D. H., Meadows, D. L., Randers. J., & Behrens, W. W. (1972). *The Limits to Growth*. Earthscan.

Moscardo, G. (2008). Sustainable tourism innovation: Challenging basic assumptions. *Tourism and Hospitality Research, 8*(1), 4–13. doi:10.1057/thr.2008.7

Næss, I. E. (2013). *Øylandet. Dagligliv på Helgelandskysten*. Orkana.

Nicolini, D., & Monteiro, P. (2017). The practice approach: For a praxeology of organisational and management studies. In A. Langley & H. Tsoukas (Eds.), *The Sage handbook of process organization studies* (pp. 110–126). Sage.

O'Brien, W., & Sarkis, J. (2014). The potential of community-based sustainability projects for deep learning initiatives. *Journal of Cleaner Production*, *62*, 48–61. doi:10.1016/j.jclepro.2013.07.001

O'Dell, T. (2005). Experiencescapes: Blurring borders and testing connections. In T. O'Dell & P. Billing (Eds.), *Experiencescapes -Tourism, Culture, and Economy* (pp. 11–33). Copenhagen Business School Press.

Oftedal, E. M., Bertella, G., Lanka, S., Grzegorczyk, M., & Molthan-Hill, P. (2021). Perspectives of Sustainability. *Revista de Administração Contemporânea*, *25*(3), 1–8. PMID:34334863

Okuni, P. A., & Rockhold, J. P. (1995). *An approach for community-based sustainability*. Paper presented at the 21st WEDC Conference: Sustainability of Water and Sanitation Systems, Kampala, Uganda.

Page, S. J., & Getz, D. (2019). *Event studies: Theory, research and policy for planned events*. Routledge.

Pastoors, S., Scholz, U., Becker, J. H., & Van Dun, R. (2017). *Towards Sustainable Innovation: A five step approach to sustainable change*. Tectum Wissenschaftsverlag.

Plessis, C. A., & Brandon, P. (2015). An ecological worldview as basis for a regenerative sustainability paradigm for the built environment. *Journal of Cleaner Production*, *109*, 53–61. doi:10.1016/j.jclepro.2014.09.098

Plichta, J. (2019). The co-management and stakeholder's theory as a useful approach to manage the problem of overtourism in historical cities–illustrated with an example of Krakow. *International Journal of Tourism Cities*, *5*(4), 685–699. doi:10.1108/IJTC-12-2018-0107

Pløger, J. (2016). The Evental City. *Space and Culture*, *19*(3), 260–274. doi:10.1177/1206331215595729

Ratten, V., Braga, V., Álvarez-García, J., & del Rio, M. D. L. C. (2020). Tourism innovation: The role of technology, sustainability and creativity. In V. Ratten, V. Braga, J. Álvarez-García, & M. d. l. C. del Rio (Eds.), Tourism Innovation: Technology, Sustainability and Creativity (pp. 1-13). Routledge.

Raworth, K. (2017). *Doughnut economics: Seven ways to think like a 21st-century economist*. Chelsea Green Publishing.

Reed, B. (2007). Forum: Shifting from 'sustainability' to regeneration. *Building Research and Information*, *35*(6), 674–680. doi:10.1080/09613210701475753

Richins, H. (2009). Environmental, cultural, economic and socio-community sustainability: A framework for sustainable tourism in resort destinations. *Environment, Development and Sustainability*, *11*(4), 785–800. doi:10.100710668-008-9143-6

Ritala, P. (2019). Innovation for Sustainability: Scepital, Pragmatic and Idealist Perspectives on the Role of Business as a Driver for Change. In N. Bocken, P. Ritala, L. Albareda, & R. Verburg (Eds.), *Innovation for Sustainability. Business Transformations Towards a Better World* (pp. 21–34). Palgrave Macmillan. doi:10.1007/978-3-319-97385-2_2

Røros Museum. (n.d.). *About the Røros Museum.* Rorosmuseet. https://rorosmuseet.no/en/about-the-museum

Ryan, A. W., & Wollan, G. (2013). Festivals, landscapes, and aesthetic engagement: A phenomenological approach to four Norwegian festivals. *Norsk Geografisk Tidsskrift-Norwegian Journal of Geography, 67*(2), 99–112. doi:10.1080/00291951.2013.784352

Rydin, Y. (2007). Indicators as a governmental technology? The lessons of community-based sustainability indicator projects. *Environment and Planning. D, Society & Space, 25*(4), 610–624. doi:10.1068/d72j

Saarinen, J. (2006). Traditions of sustainability in tourism studies. *Annals of Tourism Research, 33*(4), 1121–1140. doi:10.1016/j.annals.2006.06.007

Saarinen, J. (2014). Critical sustainability: Setting the limits to growth and responsibility in tourism. *Sustainability, 6*(1), 1–17. doi:10.3390u6010001

Saarinen, J. (2019). Communities and sustainable tourism development: Community impacts and local benefit creation in tourism. In S. F. McCool & K. Bosak (Eds.), *A research agenda for sustainable tourism* (pp. 206–222). Edward Elgar Publishing. doi:10.4337/9781788117104.00020

Saarinen, J., & Gill, A. M. (2019). *Resilient destinations and tourism. Governance Strategies in the Transition Towards Sustainability in Tourism.* Routledge.

Schmitz, C. L., Stinson, C. H., & James, C. D. (2010). Community and environmental sustainability: Collaboration and interdisciplinary education. *Critical Social Work, 11*(3).

Schot, J., & Geels, F. W. (2008). Strategic niche management and sustainable innovation journeys: Theory, findings, research agenda, and policy. *Technology Analysis and Strategic Management, 20*(5), 537–554. doi:10.1080/09537320802292651

Seyfang, G., & Haxeltine, A. (2012). *Growing grassroots innovations: Exploring the role of community-based initiatives in governing sustainable energy transitions.* SAGE Publications.

Simonis, U. E. (2011). Greening urban development: On climate change and climate policy. *International Journal of Social Economics, 38*(11), 919–928. doi:10.1108/03068291111171423

Smith, A., Voß, J. P., & Grin, J. (2010). Innovation studies and sustainability transitions: The allure of the multi-level perspective and its challenges. *Research Policy, 39*(4), 435–448. doi:10.1016/j.respol.2010.01.023

Soares, J. R. R., Casado-Claro, M. F., Lezcano-González, M. E., Sánchez-Fernández, M. D., Gabriel, L. P. M. C., & Abríl-Sellarés, M. (2021). The Role of the Local Host Community's Involvement in the Development of Tourism: A Case Study of the Residents' Perceptions toward Tourism on the Route of Santiago de Compostela (Spain). *Sustainability, 13*(17), 9576. doi:10.3390u13179576

Sørensen, F., & Balsby, N. (2020). Brokers and saboteurs: Actor roles in destination innovation network development. *Tourism Planning & Development, 18*(5), 547–572. doi:10.1080/21568316.2020.1830300

SSB. (2021). *Alders- og kjønnsfordeling i kommuner, fylker og hele landets befolkning (K) 1986-2021.* Statistics Norway. https://www.ssb.no/statbank/table/07459/

Stanfield, J. R. (1983). Toward an Ecological Economics. *International Journal of Social Economics*, *10*(5), 27–37. doi:10.1108/eb013942

Stilgoe, J., Owen, R., & Macnaghten, P. (2013). Developing a framework for responsible innovation. *Research Policy*, *42*(9), 1568–1580. doi:10.1016/j.respol.2013.05.008

Stocker, L., & Barnett, K. (1998). The significance and praxis of community-based sustainability projects: Community gardens in western Australia. *Local Environment*, *3*(2), 179–189. doi:10.1080/13549839808725556

Temesgen, A. K., Storsletten, V. M., & Jakobsen, O. D. (2019). Circular economy–Reducing symptoms or radical change? *Philosophy of Management*, *19*, 1–20.

Thanh, V. M., & Bosch, O. J. H. (2010). Systems Thinking Approach as a Unique Tool for Sustainable Tourism Development: A Case Study in the Cat Ba Biosphere Reserve of Vietnam. In *54th Annual Conference of the International Society for the Systems Sciences (ISSS) 2010: Governance for a Resilient Planet, 18-23 July 2010*. Waterloo: ISSS.

Thees, H., Pechlaner, H., Olbrich, N., & Schuhbert, A. (2020). The Living Lab as a tool to promote residents' participation in destination governance. *Sustainability*, *12*(3), 1120. doi:10.3390u12031120

Tura, N., Mortimer, G., & Kutvonen, A. (2019). Exploring the Pitfalls of Systemic Innovations for Sustainability. In N. Bocken, P. Ritala, L. Albareda, & R. Verburg (Eds.), *Innovation for Sustainability* (pp. 157–175). Palgrave Macmillan. doi:10.1007/978-3-319-97385-2_9

UNEP. (2011). *Towards a Green Economy: Pathways to Sustainable Development and Poverty Eradication*. UN environment programme. http://www.unep.org/greeneconomy/Portals/ 88/documents/ger/ger_final_dec 2011/Green%20EconomyReport_Final Dec_2011.pdf

UNESCO. (1992-2010). *Røros mining town and the circumference*. UNESCO World Heritage Centre. http://whc.unesco.org/en/list/55

Van Der Walt, J. S., Buitendag, A. A., Zaaiman, J. J., & Van Vuuren, J. J. (2009). Community living lab as a collaborative innovation environment. *Issues in Informing Science and Information Technology*, *6*(1), 421–436. doi:10.28945/1070

Visit Vega. (2021). *Vega World Heritage Centre*. Vega verdensarv-senter. https://www.visitvega.no/ no/vega-verdensarvsenter

Weaver, D. B. (2009). Reflections on sustainable tourism and paradigm change. In S. Gössling, C. M. Hall, & D. Weaver (Eds.), *Sustainable tourism futures* (pp. 53–60). Routledge.

Webster, L., & Mertova, P. (2007). *Using narrative inquiry as a research method: An introduction to using critical event narrative analysis in research on learning and teaching*. Routledge. doi:10.4324/9780203946268

Wenger, E. (2003). Communities of Practice and Social Learning systems. In D. Nicolini, S. Gerardi, & D. Yanow (Eds.), *Knowing in organizations: A practice-based approach* (pp. 76–99). ME Sharpe.

Wironen, M. B., & Erickson, J. D. (2020). A critically modern ecological economics for the Anthropocene. *The Anthropocene Review*, *7*(1), 62–76. doi:10.1177/2053019619884485

Zavratnik, V., Superina, A., & Stojmenova Duh, E. (2019). Living Labs for Rural Areas: Contextualization of Living Lab frameworks, concepts and practices. *Sustainability*, *11*(14), 3797. doi:10.3390u11143797

Zizka, L., McGunagle, D. M., & Clark, P. J. (2021). Sustainability in science, technology, engineering and mathematics (STEM) programs: Authentic engagement through a community-based approach. *Journal of Cleaner Production*, *279*, 123–715. doi:10.1016/j.jclepro.2020.123715

Zografos, C., & Howarth, R. B. (2010). Deliberative ecological economics for sustainable development. *Sustainability*, *2*(11), 3399–3417. doi:10.3390u2113399

Chapter 5
Conceptual Model of Sustainable Development of Cultural Tourism With a Hybrid Approach

Omid Salek Farokhi
University of Girona, Spain

Seyedeh Yasamin Hosseini
https://orcid.org/0000-0002-4186-496X
University of Girona, Spain

ABSTRACT

In this study, an attempt has been made to systematically review the findings of the studies conducted by other researchers in the field of sustainable development studies of cultural tourism using an interpretive approach and a systematic meta-combination method. Then, based on the hidden content analysis strategy, the extracted codes are identified in the form of categories. Finally, the authors present a conceptual model. In this research, valid scientific databases were reviewed, and selected articles were identified using CASP criteria. Based on this, 104 primary codes in the form of 29 categories of primary and secondary categories are identified, and the four central codes are components, effective factors, consequences, and challenges of sustainable development of cultural tourism. Also, a conceptual model of sustainable development of cultural tourism and tourist destinations in need was presented. Finally, the validity of the research results with Kappa Cohen index in SPSS software was confirmed based on the opinions of experts outside the authors of the research with a size of 0.712.

INTRODUCTION

Sustainable development is a development that meets the needs of the present generation without consuming the resources that belong to the future generations. The most common definition is given by

DOI: 10.4018/978-1-7998-9217-5.ch005

the World Commission on Environment and Development, which states that it is a development that meets current needs without reducing the ability of future generations to meet their needs (Cerin, 2006). Tourism is one of the greatest international blessings and in the meantime, cultural tourism is growing rapidly. From the 1980s onwards, cultural tourism emerged as the primary source of economic development for many destinations. According to the World Tourism Organization in 2017, one billion and 326 million international tourists have been relocated worldwide (UNWTO,2017). Travel has always been an important topic, as evidenced by the development of grand tours in the 16th century (Towner, 1995). The importance of this issue is such that UNESCO, the only UN agency with special authority in the field of culture, has declared culture to be the fundamental issue of the 21st century for the phenomenon of tourism. Therefore, this international cultural agency encourages the development of cultural tourism model in countries to protect and preserve cultural and natural heritage. Cultural and natural heritage as a source of development of countries always attract a large number of tourists. Although tourism can be an opportunity for cultural and natural heritage, it is a threat if it is poorly managed. Although tourism can be an opportunity for cultural and natural heritage, it is a threat if it is poorly managed. Although tourism can be an opportunity for cultural and natural heritage, it is a threat if it is poorly managed. For many experts, sustainable tourism and cultural tourism are synonymous. This research aims to show the necessity of using the concepts of sustainable development in cultural tourism by using a meta-combined approach, a systematic review of research literature, and a conceptual model for sustainable development of cultural tourism in order to better plan and line this issue in offering cultural heritage destinations.

According to research, three main criticisms can be made in this regard:

- **Lack of systematic research**; Most studies have not taken a systematic approach. Therefore, there is a vacuum of research that systematically reviews the past research (Ismagilova et al., 2015; Loulanskia & Loulanski, 2011; Underberg-Goode, 2014).
- **Lack of qualitative exploratory research**; Most of the past researches have only described the components, factors or consequences of the sustainable development of cultural tourism in different destinations and have used less qualitative exploratory or meta-combined approach in their studies ((Bill) Carter * & Beeton, 2004; Chaplin, 2003; FRANC & IŞTOC, 2002; Pereira Roders & Van Oers, 2011; Xu et al., 2009).
- **Lack of social and holistic vision;** According to the research that has been done so far, A holistic view of the sustainable development of cultural tourism has not yet been provided to guide the planners to manage cultural tourism destinations with a sustainable approach (du Cros, 2001; Farid, 2015; Vargas-Sánchez et al., 2011).

In view of the above, as detailed below, the present researcher seeks to provide a comprehensive and integrated approach using the hyper-combined qualitative research method, with the aim of systematically combining qualitative studies. This study is based on the main question "What is the conceptual model of sustainable development of cultural tourism?" Search in scientific databases indicates that so far, no research has been presented to identify, categorize, analyze and interpret categories and present them in the form of a conceptual model in this regard, which is the result of this type of innovation. In the end, this study presents the sustainability approach as a comprehensive and holistic approach to the development of tourism in cultural destinations.

Purpose and Research Questions

The first purpose of this research is to identify the components, effective factors, consequences, and challenges of sustainable development of cultural tourism in order to provide a conceptual framework and model. Therefore, in this study, we seek to answer the following two questions:

1. What are the main components, effective factors, consequences and challenges of sustainable development of cultural tourism?
2. What is the conceptual model of sustainable development of cultural tourism?

Research Methodology

One of the studied scientific methods for studying and combining the research done by other researchers is meta-study. The meta-study is an in-depth analysis method of research work carried out in a specific field, which can be divided into four distinct methods depending on the nature of the type of study and the nature of the type of study: Meta-analysis (quantitative analysis of the content of early studies), Fararoosh (analysis of the meta-analytical method of early studies), meta-theory (analysis of theories of early studies), and meta-composition (qualitative analysis of the content of preliminary studies) (Bench & Day, 2010). This research started as a qualitative study routine with a hybrid strategy following the question: "What is the conceptual model of sustainable development of cultural tourism?" First, the factors and components of sustainable cultural tourism should be identified. Considering that the nature of the collected data is qualitative, therefore, it was done in a meta-study with a meta-combined method. This method is based on a research question that aims to create a creative and integrated interpretation of qualitative findings. Generally, the methodological study of this research based on the Saunders and Barroso research onion model shows that this research in terms of interpretive research paradigm, with inductive approach, content analysis strategy, and qualitative tactics, is a cross-sectional study based on preliminary studies and documents in qualitative data format (Text) (Sandelowski & Barroso, 2003). Meta-composition; It is a qualitative study that examines studies and findings derived from other qualitative studies conducted by other researchers in relation to the subject and similar. The meta-composition focuses on qualitative studies that do not cover a broad theoretical basis and creates an interpretive combination of the findings rather than a comprehensive summary of the findings (Zimmer, 2006). Therefore, in this study, we have used meta-combined strategy to review and interpret past research using a systematic approach. In this study, which aims to investigate the components, effective factors, consequences and challenges of sustainable development of cultural tourism, the method of content analysis or topic has been used. In this method, theoretical saturation is the criterion of sample adequacy. The sample for meta-composition is selected from qualitative studies and based on their relationship with the research question to achieve theoretical saturation.

The meta-composition study approach has its own variations and ambiguities in terms of executive procedures, and in a way, all the authors of meta-composition texts acknowledge the complexity and challenges of this method. The first method for the systematic study of qualitative research was presented by (Noblet and Hare,1988) 5 steps. Afterwards, Peterson et al., 2001 presented a five-step process with a trans-study approach as shown in Figure 1. The critical interpretive mix of (Dixon-Woods et al. (2006) was also presented in 5 steps. (Popay et al., 2006) presented a narrative composition based on 5 steps. The schematic composition was provided by Thomas & Harden, (2008) with 4 steps and finally,

Figure 1. The process of meta-composition
Source: Sandelowski and Barroso (2007)

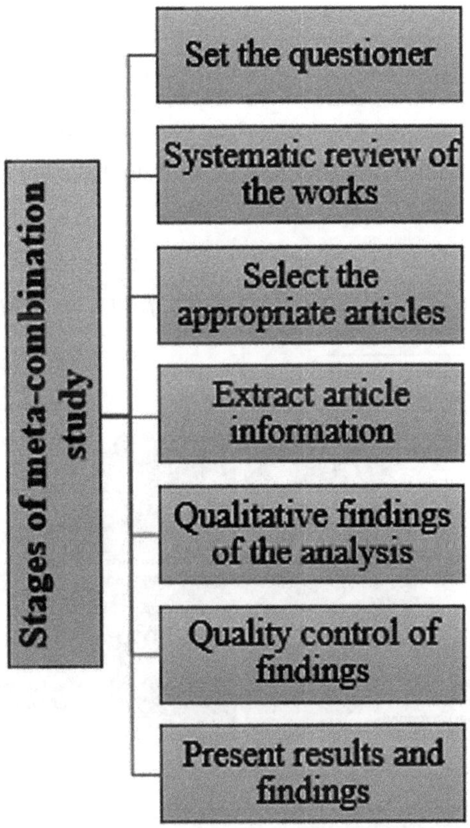

the Edwards & Kaimal quadruple combination was presented in 2016. According to many researchers, including Chenail, the method of (Sandelowski & Barroso, 2003), one of the most comprehensive methods for conducting research is meta-combination, and ultimately, it provides more comprehensive results (Chenail, 2009). The selective approach of the present study to perform meta-study with the meta-combined technique is the seven-step model of Sandelowski and Barroso.

Research Findings

In order to achieve the goals of the researcher and answer the questions, each executive steps was performed based on the framework of the seven-step model of Sandelowski and Barroso (2003) (Figure 1).

Step 1: Set up the Research Question; in this step, as the first step of studying meta-composition, the basic question or questions are set. To answer the research question in this method, parameters such as "What is the subject?", "What is the study community?", "What is the time limitation?", and finally "how?" It is suggested that for the correctness of the process of the meta-combination method and the completion of the researchit is necessary to give them an appropriate answer. In Figure 2, the research questions are identified along with the studied parameters. Based on the evaluation, articles were included in this study that addressed or referred to the components, effective factors, consequences and challenges of sustainable development of cultural tourism. Obviously, articles that address other issues related to the sustainable development of cultural tourism are set aside from this study. The main subjects of the study are scientific-research articles, books, dissertations. The final selected materials were identified and analyzed based on the CASP evaluation form using the content analysis technique (Appendix 1). The extracted materials then formed categories, concepts, and categories. The whole process of data analysis was done with MAXQDA version 2018 update software.

Figure 2. Parameters and research questions based on the first step of the meta-combination method

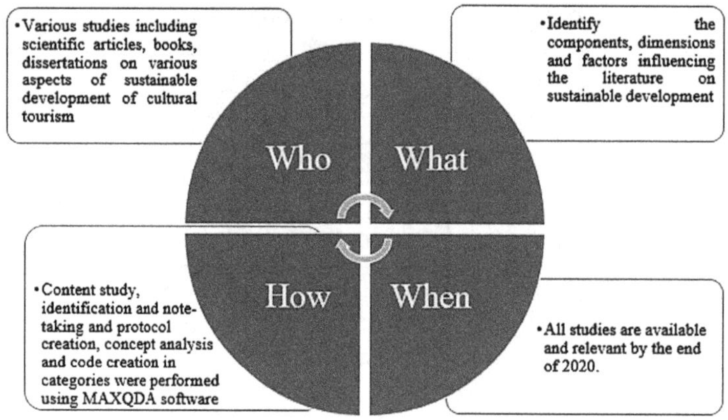

Step 2: Systematic Review of Studies; in this step, all texts and articles related to the research question were reviewed. Eligible studies were selected to enter the meta-analysis. Then, secondary data performed with a qualitative method are used in the field of research. Therefore, the statistical population of this study includes all studies conducted in this field. Systematic search for resources by keywords "Sustainable Development," "Cultural Tourism," "Sustainability of Cultural Tourism," and "Model of Sustainable Development of Cultural Tourism" were carried out in scientific databases. A variety of keyword selections were performed to further retrieve the resources. To search for resources in databases Science Direct, Springer, Emerald, ProQuest, Google Scholar and the like were mentioned. Scientific reports from the World Tourism Organization and UNESCO sites were also considered in this regard. In general, it was an attempt to select sources of scientific quality. For example, scientific-review articles and low credit conferences were excluded in this study. Based on extensive research in the mentioned scientific databases, 348 documents for meta-composition analysis of the subject; Sustainable cultural tourism resources were stored. In the qualitative research, the sample size based on the researcher's opinion continues until the theoretical saturation is reached. In this form of the study, it is normal to reach a sample size of 12 articles or less. If the obtained sample is selected and examined accurately and systematically in accordance with the principles and criteria of qualitative studies, it will certainly cover all the information that the researcher is looking for (Guba & Lincoln, 1994).

Figure 3. Search algorithm and selection of appropriate sources for the present study

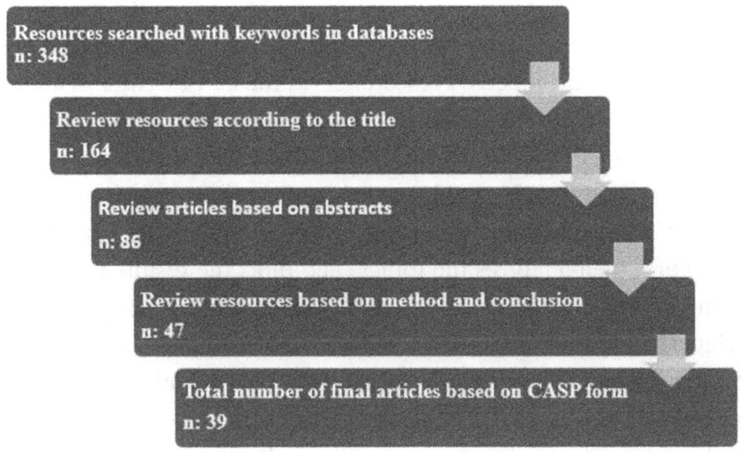

Step 3: Selecting right articles; At this stage, the researcher must evaluate the quality of stored resources based on comprehensive and keyword selection in scientific databases. The purpose of completing the step-by-step study process, which is also an important step, is to eliminate low-validity studies. A useful tool that is often used to evaluate the quality of early research studies is the Critical Assessment Skills Program. This program helps the researcher assess the accuracy, validity, and importance of qualitative research studies by asking ten questions. The questions asked in the CASP evaluation form are: 1) Research objectives, 2) Method logic, 3) Research design, 4) Sampling method, 5) Collecting data, 6) Reflectivity, 7) Ethical considerations, 8) Accuracy of data analysis, 9) Clear expression of findings, and 10) The value of research.

At this stage, the researcher gave a small score to each of the questions and based on the sum of the scores given to each document, a final review and evaluation was performed. Based on the 50-point Critical Assessment Skills Form Scale, any degree that scored below a good level (below 30) was excluded from the field of study. Therefore, in this study, only sources and documents that scored excellent and very good (with a score above 30) were selected. After identifying the appropriate sources based on the CASP evaluation form in the Sandelowski & Barroso (2003) model, the next step is to refine and load them into the MAXQDA software to extract information. Based on the extensive research in the mentioned scientific databases, 348 documents for meta-composition analysis of the subject; Sustainable development of cultural tourism was stored. By reviewing the titles of the articles in accordance with the subject, question, purpose and approach of the research, 164 retrieved articles were selected for the initial review. In the next step, 86 items were selected by examining the key parts of the keywords. In the next review, 47 items were selected based on the method and conclusions of the selected articles. Finally, after a thorough study of the content of the articles and the CASP scoring form, 39 cases were finalized and selected as the basis of this study (Appendix 1). Figure 3 shows the documentation selection algorithm and the process of refining stored resources and their final selection.

Step 4: Extract textual information; At this stage of the qualitative trans-study method (meta-combination), the researcher should choose one of the nine methods according to the nature of the subject and the collected resources. In this research, content analysis has been used as a method used to extract information from final sources. Content analysis is used as one of the methods of qualitative data analysis, especially (but not exclusively) along with the combination of quantitative data. This method was first used and developed by researchers at the EPPI Center to address questions about "what works" primarily in relation to health promotion interventions. Recently, this method has been used in combining findings from multiple qualitative studies in systematic reviews. Thomas and Harden developed a new approach called thematic composition which incorporates the principles of basic qualitative research and other methods. This method identifies the topic or repetitive topics in the parent literature, then analyzes the topics, and plots the results in a systematic review. The purpose of this method is to develop analytical themes through a descriptive synthesis and to find explanations related to a specific case. This method is somewhat similar to the two methods of data foundation theory and trans-ethnographic ethnography (Ring et al., 2011). The importance of this step is very valuable for the correct completion of the next steps of the research. Accordingly, to increase the quality of the research, a summary of the documents was prepared, and all resources were uploaded to the MAXQDA software (version 2018). Then we started to study and analyze the content of the documents on the components, effective factors, consequences and challenges of sustainable development of cultural tourism as central codes. Since the study is systematic, with the help of content analysis technique, codes were extracted from texts and in return, the source codes and the authors of the works were given (Tables 1,2,3 and 4). Then, at a higher level than the primary codes, categories (primary and secondary) having a more abstract concepts were defined and identified to provide the basis for interpreting the research results. Since content analysis is done covertly, the frequency of the extracted studies will not be the basis of performance. However, in the tree diagram drawn for each central code using MAXQDA software, the thickness of the lines indicates the frequency of the relevant code in the sources (Figure 4).

Step 5: Analysis and Composition of Qualitative Findings; In the fifth step, after extracting the information in the previous step, based on the content analysis method, the qualitative findings of the prediction studies are analyzed and interpreted in order to provide an integrated and systematic interpretation with a new approach Sandelowski and Barroso (2007). In this way, all identified codes are classified into categories with similar concepts according to their concepts, and categories are created that have a more abstract surface than the code. By doing this, an interpretive process is provided beyond any of the studies included in the meta-combination on the subject, which will include all of them, so that the effect of the each of the initial studies can be sought in this whole. Based on the analysis performed using the content analysis method on 39 cases of finalized work, 104 initial codes were identified, and the list of these initial codes extracted from the initial studies is given in Tables 1, 2, 3 and 4. Then, considering the first category of codes, categories (primary and secondary) were formed, and finally the central codes are obtained. The categories created in the form of 29 categories in the form of 4 central codes are components, effective factors, consequences, and challenges of sustainable development of cultural tourism. That is the innovation of the research done in categorizing codes for final analysis, interpretation and presenting a conceptual model.

Step 6: Quality Control of Findings; In the step of analyzing the selected resources based on the method of coding and classifying information with content analysis technique, there is always an attempt to review and re-codify the data extracted by the research team members. All these activities have

Table 1. Codes and articles on the components of sustainable development of cultural tourism

Axial Code	Category	Codes	Code Extraction Resources
Components of sustainable development of cultural tourism	**Economic**	Economic benefits of cultural tourism for the host and destination community (employment, income, etc.)	(Durovic & Lovrentjev, 2014) (UNWTO, 2017) (Padin & Svensson, 2013) (Khlaikaew, 2015) (Amir et al., 2015) (Brcar & Colaric-Jakse, 2017) (Hamimi Omar, 2013) (Wan & Li, 2013) (Nocca & Fusco Girard, 2018) (Pakdeepinit, 2017) (Ko, 2005) (Loulanskia & Loulanski, 2011) (UNWTO, 2017)
		Sustainability of tourist satisfaction	
		Cultural facilities	
		Institutional settings	
		Seasonality of tourism activity	
		Transportation related to tourism	
		Cultural routes	
	Social	Protecting the cultural identity of the local community	(Padin & Svensson, 2013) (Torggler & France, 2015) (Amir et al., 2015) (Brcar & Colaric-Jakse, 2017) (Hamimi Omar, 2013) (Wan & Li, 2013)
		Public local safety	
		Protection of cultural heritage	
		Destination social tolerance capacity	
		Social and cultural effects of tourism on the host community	
		Quality of life in general	
	Environmental	Preservation of natural ecosystems around cultural monuments	(Nocca & Fusco Girard, 2018) (Seyfi et al., 2018) (Ko, 2005) (Loulanskia & Loulanski, 2011) (Durovic & Lovrentjev, 2014) (UNWTO, 2017) (Padin & Svensson, 2013) (Khlaikaew, 2015) (Amir et al., 2015) (Brcar & Colaric-Jakse, 2017) (Hamimi Omar, 2013) (Nocca & Fusco Girard, 2018) (Pakdeepinit, 2017) (Loulanskia & Loulanski, 2011)
		Energy Management	
		Water availability and management	
		Wastewater treatment	
		Waste management	
		Air pollution	
		Visual impact management of facilities and infrastructure	
		Intensity of use	
		Environmental management	
	Protection	Legal basis for protection	(UNWTO, 2017) (Gurira & Ngulube, 2016) (Pakdeepinit, 2017) (Agyeiwaah et al., 2017)
		Determining the effects	
		Financing for protection	
		Subject Specifications	
		Location and environmental conditions	
		Threat to the integrity and jurisdiction of the property	

Source: Research Findings

been done to ensure the quality of research findings. Another key tool adopted to ensure the theoretical validity and practicality of the research is the use of expert opinion. In order to evaluate the reliability of the research, researchers have used the opinion of two experts outside of the research

Table 2. Codes and articles on the components of sustainable development of cultural tourism

Axial Code	Main Category	Subcategory	Codes	Code Extraction Resources
Factors affecting the sustainable development of cultural tourism	Economic perception		Sources	(Liu & Chen, 2014) (Ismagilova et al., 2015) (Awang, 2010) (Hunt et al., 2015) (UNWTO, 2004) (Swanson & DeVereaux, 2012)
			Fund	
			Technological innovations	
			Work force	
	The volume of tourists	compression ratio		(Camus et al., 2012) (Buckley, 2012)
		Revenue from tourism	Tax revenue from tourism	(Ko, 2005) (du Cros, 2001) (Liu & Chen, 2014) ((Bill) Carter * & Beeton, 2004)
			Average tourist expenses per person	
			Average tax rate	
		Degree of tourists' satisfaction with cultural product innovation	Rate coefficient of tourism satisfaction	(Padin & Svensson, 2013) (Vargas-hernández, 2012) (Brcar & Colaric-Jakse, 2017)
			Degree of tourist satisfaction with the destination density ratio	
			The degree of general satisfaction of tourists	
			The rate of cultural product innovations	
			Coefficient of technology	
	The extent of local community participation in the tourism process	Number of human resources in the tourism industry	Volume of non-native participation in tourism	(Pakdeepinit,2007) (Loulanskia & Loulanski, 2011) (Hamimi Omar, 2013) (Awang, 2010) (Gurira & Ngulube, 2016)
			The ratio of cultural elites in society	
		Local community income from tourism	Average wages from various tourism businesses	
		Local community participation rate		
	The value of cultural heritage tourism	The value of cultural heritage		(Ismagilova et al., 2015) (Amir et al., 2015)
		The amount of value added from the invested capital ratio	Financial investment	(Awang, 2010) (Torggler & France, 2015) (Camus et al., 2012) (UNWTO, 2017) ((Bill) Carter * & Beeton, 2004) (Bowitz & Ibenholt, 2009) (Nocca & Fusco Girard, 2018) (Carter et al., 2015)
			Tax exemption for capital invested in the tourism industry	
			Investment in the tourism industry	
			Donated capital	
			Investment capital ratio	
		Value added value of human resources coefficient		
		Value added rate of cultural innovation coefficient		
	Stakeholders involved in cultural tourism			(Khlaikaew, 2015) (Pakdeepinit,2007)
	Geographical view			(Manrique-de-Lara & Guerra-Baez, 2016)
	Sociological perspective	**Innovation in cultural products**		(Brcar & Colaric-Jakse, 2017)
	Cultural changes			(UNWTO, 2004)
	Private sector initiatives			(UNWTO, 2004)
	Local community policies			(UNWTO, 2004)
	Public sector development policies			(UNWTO, 2004) (Pakdeepinit,2007)

Source: Research Findings

Table 3. Sustainable consequences of cultural tourism development

Axial Code	Main Category	Subcategory	Codes	Code Extraction Resources
Factors affecting the sustainable development of cultural tourism	Tourists		International visitors	(Buckley, 2012) (Hamimi Omar, 2013) (Wan & Li, 2013) (Borges et al., 2011) (Buckley, 2012) (Carter et al., 2015) (Brcar & Colaric-Jakse, 2017)
			Satisfaction of visitors	
			National visitors	
			Number of Visitors	
			Familiarity of tourists with the local culture	
	Environmental Implications		Fluctuations in weather conditions	(Su & Wall, 2015) (Wan & Li, 2013) (Nocca & Fusco Girard, 2018) (Hunt et al., 2015) **Adams** (Underberg-Goode, 2014) (Dragouni, 2017) (Padin & Svensson, 2013) ((Bill) Carter * & Beeton, 2004) (Hamimi Omar, 2013) (Yun & Zhang, 2017)
			Development of festivals and local food	
			Rational use of resources	
			environmental effects	
			Tourism damage to the environment	
	Local community	Protection of cultural heritage	Holding a national tourism marketing campaign	(Hamimi Omar, 2013) (Buckley, 2012) (Underberg-Goode, 2014) (Padin & Svensson, 2013) ((Bill) Carter * & Beeton, 2004) (Seyfi et al., 2018) (Borges et al., 2011) (Pakdeepinit, 2017) (Manrique-de-Lara & Guerra-Baez, 2016) (du Cros, 2001) (Amir et al., 2015) (Brcar & Colaric-Jakse, 2017)
			Development of archeological sites	
			Development of tourist routes	
			Conservation activities	
			Maintenance and protection	
		Economic resources	Poverty alleviation	(Hamimi Omar, 2013) (FRANC & IŞTOC, 2002) (Su & Wall, 2014) (Seyfi et al., 2018) (Awang, 2010) (Yun & Zhang, 2017) (Farid, 2015) (Dragouni, 2017) (UNWTO,2017) (Amir et al., 2015) (Wan & Li, 2013) (Nocca & Fusco Girard, 2018) (Underberg-Goode, 2014) (Padin & Svensson, 2013) (Bowitz & Ibenholt, 2009) (Amir et al., 2015) (Brcar & Colaric-Jakse, 2017) ((Bill) Carter * & Beeton, 2004) (Borges et al., 2011) (Camus et al., 2012)
			Changes in oil prices	
			Economic value	
			Economic Development	
			Monetization	
			Economic prosperity and large-scale recession	
			Development of local products, arts and crafts and other attractions	
			The local community has gained new benefits, including infrastructure	
			Tourism raises the prices	
			The power to earn foreign currency	
		Socio-cultural effects	Educate local people for cultural tourism	(Hamimi Omar, 2013) (FRANC & IŞTOC, 2002) (Su & Wall, 2014) (Wan & Li, 2013) (Nocca & Fusco Girard, 2018) (Underberg-Goode, 2014) (Loulanskia & Loulanski, 2011) (Buckley, 2012) ((Bill) Carter * & Beeton, 2004) (Borges et al., 2011) (Yun & Zhang, 2017) ((Bill) Carter * & Beeton, 2004) (Carter et al., 2015) (Awang, 2010) (Underberg-Goode, 2014) (Padin & Svensson, 2013) (Dragouni, 2017) (Liu & Chen, 2014) (Vargas-hernández, 2012) (Seyfi et al., 2018)
			Pilot strikes	
			Language retrieval	
			Strengthen cultural identity	
			Host-guest community interaction	
			Changing the attitude of the host community	
			Evolution of social and cultural effects	
			Change language	
			Product change and preservation of ethnicities	
			Changing the nature of the host community	
			Changing the nature of the tourism industry	
			Socio-cultural impact	
			Tourism increases crime rates at cultural destinations.	
			Political or social unrest	
			Tourism increases population density and transportation	
			Improve recreational opportunities	
			Facilities for local participation	
			Increases the identity of the local community	
			Increase the value of heritage, brand, marketing and infrastructure	(Buckley, 2012)

Source: Research Findings

Table 4. Codes and categories of challenges for sustainable development of cultural tourism

<table>
<tr><th>Axial Code</th><th>Category</th><th>Codes</th><th>Code Extraction Resources</th></tr>
<tr><td rowspan="8">Challenges of sustainable development of cultural tourism</td><td rowspan="2">Local community</td><td>Threatened by the urbanization process</td><td rowspan="2">(Carter et al., 2015)
(Pereira Roders & Van Oers, 2011)</td></tr>
<tr><td>Indigenous community involvement with sustainability issues</td></tr>
<tr><td rowspan="6">Structural and managerial</td><td>Threat of heritage resources Development and land use</td><td rowspan="6">(Carter et al., 2015)
(Awang, 2010)</td></tr>
<tr><td>Lack of governance and clarity about access to cultural resources</td></tr>
<tr><td>Insufficient financial capacity of government and non-governmental organizations</td></tr>
<tr><td>Limited human resources</td></tr>
<tr><td>Use advice for challenges</td></tr>
</table>

Source: Research Findings

who are professional in terms of method and subject matter. At this stage, experts were asked to comment on the selection of the hybrid method for this study, the appropriateness and adequacy of resources, the final resource selection process, and their evaluation and grading of resources based on the CASP form. Fortunately, the quality of resources with a slightly higher index of 30 based on the new CASP farm was confirmed by experts. Regarding the extracted codes and the classification of the codes in the form of categories of results, they were provided to the experts. To evaluate the agreement between the two, the Cohen's kappa coefficient was used. Kappa Cohen measures the agreement between two evaluators, each of whom has arranged N items in a mutually exclusive C class (SMEETON & SMEETON, 1985). What is followed in Cohen's kappa coefficient is the evaluation of the size of the agreement between two people, the phenomenon or source of decision-making, each of which has individually measured our first two quantities. Kappa coefficient and statistical analysis is a numerical measure between -1 to +1. The closer number to +1, the more direct and proportional agreement is. When the kappa value is closer to zero, it indicates less agreement between the evaluators (Viera & Garrett, 2005). Often, Cohen Kappa coefficient values between 0.4 and 0.6 indicate a moderate agreement, values between 0.6 and 0.8 indicate a significant and good agreement, and values above 0.8 indicate an excellent agreement between the two evaluators. The kappa relationship is as follows:

$$\kappa = \frac{p_o - p_e}{1 - p_e}$$

p_o represents the observed agreements and p_e represents the expected agreements. In fact, the Kappa index confirms the reliability of the extracted components of the two transition codes. In order to evaluate the extractive results, the Kappa agreement coefficient was calculated in SPSS software. The value of Kappa index is equal to 0.712. Considering that this coefficient is in the category of values between 0.6 to 0.8, it can be concluded that the agreement coefficient is at a good level and the articles

Figure 4. Conceptual model of sustainable development of cultural tourism
Source: Research findings

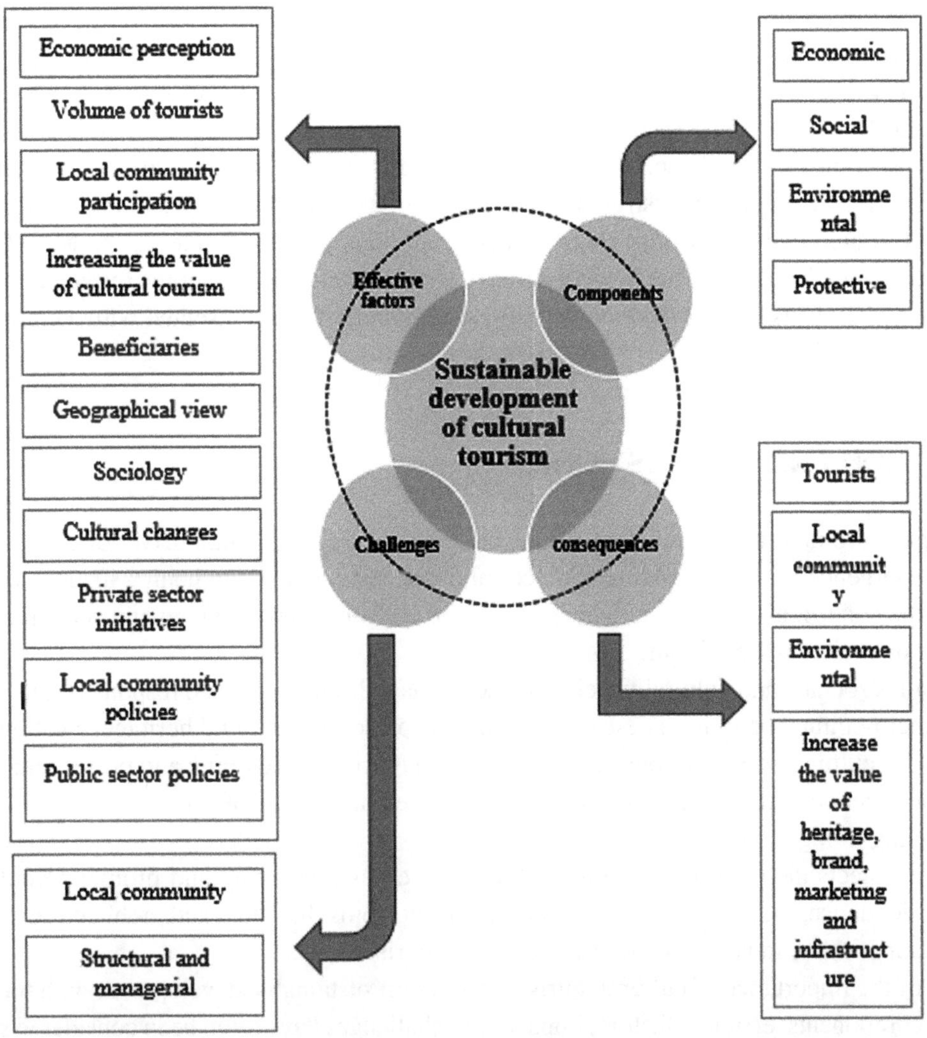

are extracted and analyzed through the meta-method. The composition has a remarkable and good level of reliability and the study is valid.

Step 7: Present the Results and Findings; The subject matter was followed and discussed in detail based on the model of the previous six steps of the meta-combination process in the direction of the research objectives. The results of the research indicate the presentation of 104 primary codes identified based on the studied texts, which in a more abstract step in the form of 29 main and sub-categories are;

Economic, social, environmental and protection (central code: components of sustainable development of cultural tourism), Economic perception, volume of tourists, income from tourism, degree of tourists' satisfaction with cultural product innovation, level of local community participation, number of human resources in tourism industry, local community income from tourism, value of heritage tourism, value added from coefficient invested, Stakeholders, Geographical perspective, Sociology, Cultural change, Private sector initiatives, Local community policies and Public sector development policies (central code: Factors affecting the sustainable development of cultural tourism), Tourists, Environment, Local Community, Cultural Heritage Preservation, Economic Benefits, Socio/Cultural Impacts, Heritage Enhancement, Branding, Marketing, and Infrastructure. The community of 39 final sources and their categorization in the form of 4 central codes were passed to the first question of the research and each of these central codes with local, structural and managerial (central code: Challenges of sustainable development of cultural tourism). By categorizing the primary codes extracted using MAXQDA 2018 software, they were displayed in the form of a tree diagram. Finally, to answer the research question, the conceptual model of sustainable development of cultural tourism was presented with a comprehensive and holistic view (Figure 4).

CONCLUSION AND SUGGESTIONS

The role of culture in terms of social communication and as a fundamental driver of economic development, convergence between economic and cultural discourse and strengthening its transition is very important, the concept of cultural tourism as a factor of economic development and economic resource and welfare of society is significant.

If tourism is socially and culturally well planned, managed, and developed responsibly, it can bring a variety of socio-cultural benefits. These benefits include preserving cultural heritage, creating opportunities for cross-cultural exchanges between residents and tourists, developing and preserving museums, theaters, and other cultural facilities, strengthening and creating a sense of pride among local communities, and raising living standards.

Tourism protects the region's cultural heritage, as well as preserving and promoting cultural patterns in music, dance, theater, clothing, arts and crafts, customs, lifestyle, and architecture - which are important tourist attractions. - Provides by expanding tourism.

In spite of the importance of cultural tourism, studies on sustainable development with the key concepts of its components, efficient factors, impacts and challenges have often been considered separately in the studies. Also, the presentation of a conceptual and comprehensive model in this field was omitted.

In this research, an strive has been made to cope with this problem. Consequently, the goal of this article, further to figuring out the dimensions of the issue, is to offer a comprehensive and holistic conceptual model of sustainable improvement of cultural tourism primarily based on meta-combination method to deal with cultural tourism in tourism science research from the attitude of sustainable development. The science of tourism is concerned with providing a theoretical understanding of tourism. Tourism studies (because the carried-out observation of tourism) are designed to decorate our capacity to successfully control the vacation spot and, in doing so, beautify the nicely-being of the residents of a tourist vacation spot.

In this regard, out of 39 studied articles, 104 codes and 29 main and sub-categories have been identified, which are shown in Tables 1, 2, 3 and 4. Finally, a conceptual model of sustainable development of cultural tourism was presented. The present article in the field of tourism science has increased knowledge in several dimensions. First of all, no research has provided a conceptual model to this comprehensiveness of the subject under discussion so far, which may be due to the choice of meta-composition research method and hidden content analysis. Which has been used to explore and identify this phenomenon, i.e. the development of cultural tourism with a sustainable vision. Therefore, the application of this methodology can be added to the innovations of this research in the field of tourism science studies. Second, some of the categories derived from the concept model are completely new and unique. For example, the first category; The degree of participation of the local community in the tourism process in the form of sub-categories: the number of human resources in the tourism industry, the income of the local community from tourism is not mentioned in any relatively comprehensive research. Third case; This study can be referenced as a comprehensive resource for other researchers in the field of cultural tourism sustainability studies due to the use of meta-combination technique based on the qualitative studies of its predecessors. It can also be used in quantitative studies according to the proposed conceptual model.

Implement the Model on a Case-by-Case Basis at Cultural Heritage Sites

Regarding future studies, it is suggested that researchers develop holistic concepts about sustainable development and tourism development. Also, comprehensive qualitative studies to evaluate the effectiveness of approaches and programs of sustainable development in tourism, systematic studies to establish links between various factors of cultural tourism, provide conceptual models of sustainable tourism development, especially cultural tourism in attractions, destinations, cities and countries and Or presenting a conceptual model for creating a sustainable relationship in tourism are other issues that can be focused on through systematic review to establish this relationship in the future.

Conducting studies in the mentioned fields, in addition to providing a general and comprehensive view to the relevant officials in the field of tourism, policy makers, managers, decision makers, etc., leads to revealing the hidden angles between cultural tourism development and sustainable development. Such studies can serve as a template or roadmap for the sustainable development of cultural destinations and attractions. Conducting systematic studies and preparing related conceptual models will improve the ability to define new indicators and, by its nature, provide methods for evaluating trends.

In order to conduct comprehensive systematic holistic studies, it is suggested that researchers go to find and use previous studies as a reference for their work and by categorizing studies and analyzing their results to conduct research on widespread and important topics.

As mentioned before, one of the strengths of this study is that this research has tried to provide a comprehensive conceptual model for the subject and to provide a specific innovation in the field of tourism science studies.Creating new categories and concepts for future studies can also be considered as one of the strengths of this research. Because this chapter does not provide a quantitative and specific review of quantitative indicators, some of the feedback generated from it points to weaknesses and incompleteness.

In this regard, it is suggested that researchers use a combination of quantitative and qualitative methods, or rather a mixed method, for future studies.

In the end, by citing the results obtained to complete the research process, it is necessary to provide research suggestions for researchers in this field:

1. Research with the aim of implementing the model and developing the generalizability of model findings using quantitative methods.
2. Presenting a model of sustainable development of cultural tourism with a mixed approach.

REFERENCES

Agyeiwaah, E., McKercher, B., & Suntikul, W. (2017). Identifying core indicators of sustainable tourism: A path forward? *Tourism Management Perspectives*, *24*, 26–33. https://doi.org/10.1016/j.tmp.2017.07.005

Amir, S., Osman, M. M., Bachok, S., & Ibrahim, M. (2015). Sustaining Local Community Economy Through Tourism: Melaka UNESCO World Heritage City. *Procedia Environmental Sciences, 28*, 443–452. doi:10.1016/j.proenv.2015.07.054

Awang, K. W. (2010). Tourism policy development: The Malaysian experience. *Arab World Geographer*, *13*(2), 150–158.

Bench, S., & Day, T. (2010). The user experience of critical care discharge: A meta-synthesis of qualitative research. *International Journal of Nursing Studies*, *47*(4), 487–499. https://doi.org/10.1016/j.ijnurstu.2009.11.013

Borges, M. A., Carbone, G., Bushell, R., & Jaeger, T. (2011). *Sustainable tourism and natural World Heritage: Priorities for action. IUCN, Gland, Switzerland.* http://cmsdata.iucn.org/downloads/sustainable_tourism_and_natural_world_heritage_report.pdf

Bowitz, E., & Ibenholt, K. (2009). Economic impacts of cultural heritage – Research and perspectives. *Journal of Cultural Heritage*, *10*, 1–8. https://doi.org/10.1016/j.culher.2008.09.002

Brcar, F., & Colaric-Jakse, L. M. (2017). Evaluation of cultural heritage in Posavje region by tourists and employees. *Challenges of the Future*, *2*(1), 1–15.

Buckley, R. (2012). Sustainable tourism: Research and reality. *Annals of Tourism Research*, *39*(2), 528–546. https://doi.org/10.1016/j.annals.2012.02.003

Camus, S., Hikkerova, L., & Sahut, J. (2012). Systemic Analysis and Model of Sustainable Tourism Sandra Camus Lubica Hikkerova. *International Journal of Business,* 365–378.

Carter, R. W., Thok, S., O'Rourke, V., & Pearce, T. (2015). Sustainable tourism and its use as a development strategy in Cambodia: a systematic literature review. *Journal of Sustainable Tourism, 23*(5), 797–818. doi:10.1080/09669582.2014.978787

Carter, R. W., & Beeton, R. J. S. (2004). A model of cultural change and tourism. *Asia Pacific Journal of Tourism Research, 9*(4), 423–442. doi:10.1080/1094166042000311282

Cerin, P. (2006). Bringing economic opportunity into line with environmental influence: A discussion on the Coase theorem and the Porter and van der Linde hypothesis. *Ecological Economics*, *56*(2), 209–225. https://doi.org/10.1016/j.ecolecon.2005.01.016

Chaplin, I. (2003). Cultural tourism: The partnership between tourism and cultural heritage management, by Bob McKercher and Hilary du Cros. The Haworth Hospitality Press, 2002. xii + 262 pp. $69.95 (hardback); $39.95 (paperback) ISBN 0-7890-1106-9. *International Journal of Tourism Research*, *5*(3), 245. https://doi.org/10.1002/jtr.423

Chenail, R. J. (2009). Communicating Your Qualitative Research Better. *Family Business Review*, *22*(2), 105–108. https://doi.org/10.1177/0894486509334795

Dixon-Woods, M., Cavers, D., Agarwal, S., Annandale, E., Arthur, A., Harvey, J., Hsu, R., Katbamna, S., Olsen, R., Smith, L., Riley, R., & Sutton, A. J. (2006). Conducting a critical interpretive synthesis of the literature on access to healthcare by vulnerable groups. *BMC Medical Research Methodology*, *6*(1), 35. https://doi.org/10.1186/1471-2288-6-35

Dragouni, M. (2017). *Sustainable heritage tourism: Towards a community-led approach*. Academic Press.

du Cros, H. (2001). A new model to assist in planning for sustainable cultural heritage tourism. *International Journal of Tourism Research*, *3*(2), 165–170. https://doi.org/10.1002/jtr.297

Durovic, M., & Lovrentjev, S. (2014). Indicators of sustainability in cultural tourism. *The Macrotheme Review*, *3*(7), 180–189. https://macrotheme.com/yahoo_site_admin/assets/docs/19MR37mj.202135138.pdf

Edwards, J., & Kaimal, G. (2016). Using meta-synthesis to support application of qualitative methods findings in practice: A discussion of meta-ethnography, narrative synthesis, and critical interpretive synthesis: *The Arts in Psychotherapy, 51*. doi:10.1016/j.aip.2016.07.003

Farid, S. M. (2015). Tourism Management in World Heritage Sites and its Impact on Economic Development in Mali and Ethiopia. *Procedia: Social and Behavioral Sciences*, *211*, 595–604. https://doi.org/10.1016/j.sbspro.2015.11.078

FRANC & IŞTOC. (2002). Cultural tourism and sustainable development. *Museum International, 54*(1–2), 126–130. doi:10.1111/1468-0033.00376

Guba, E. G., & Lincoln, Y. S. (1994). Competing paradigms in qualitative research. In *Handbook of qualitative research* (pp. 105–117). Sage Publications, Inc.

Gurira, N. A., & Ngulube, P. (2016). Using Contingency Valuation Approaches to Assess Sustainable Cultural Heritage Tourism Use and Conservation of the Outstanding Universal Values (OUV) at Great Zimbabwe World Heritage Site in Zimbabwe. *Procedia: Social and Behavioral Sciences*, *225*, 291–302. https://doi.org/10.1016/j.sbspro.2016.06.028

Hamimi Omar. (2013). The Development of Sustainable Cultural Heritage Tourism in Malaysia: Implication for Planning and Management. *Newcascle Thesis*. https://theses.ncl.ac.uk/dspace/bitstream/10443/2121/1/Omar,H.13.pdf

Hunt, C. A., Durham, W. H., Driscoll, L., & Honey, M. (2015). Can ecotourism deliver real economic, social, and environmental benefits? A study of the Osa Peninsula, Costa Rica. *Journal of Sustainable Tourism*, *23*(3), 339–357. https://doi.org/10.1080/09669582.2014.965176

Ismagilova, G., Safiullin, L., & Gafurov, I. (2015). Using Historical Heritage as a Factor in Tourism Development. *Procedia: Social and Behavioral Sciences*, *188*(904), 157–162. https://doi.org/10.1016/j.sbspro.2015.03.355

Khlaikaew, K. (2015). The Cultural Tourism Management under Context of World Heritage Sites: Stakeholders' Opinions between Luang Prabang Communities, Laos and Muang-kao Communities, Sukhothai, Thailand. *Procedia Economics and Finance, 23*(October), 1286–1295. doi:10.1016/s2212-5671(15)00563-8

Ko, T. (2005). Development of a tourism sustainability assessment procedure: A conceptual approach. *Tourism Management*, *26*, 431–445. https://doi.org/10.1016/j.tourman.2003.12.003

Liu, G., & Chen, J. (2014). A Dynamic Model for Managing Cultural Tourism. *Asia Pacific Journal of Tourism Research*, *20*, 1–15. https://doi.org/10.1080/10941665.2014.904805

Loulanskia, T., & Loulanski, V. (2011). The sustainable integration of cultural heritage and tourism: A meta-study. *Journal of Sustainable Tourism*, *19*(7), 837–862. https://doi.org/10.1080/09669582.2011.553286

Manrique-de-Lara, P., & Guerra-Baez, R. (2016). Exploring the Influence of Ethical Climate on Employee Compassion in the Hospitality Industry. *Journal of Business Ethics*, *133*(3), 605–617. https://doi.org/10.1007/s10551-014-2415-1

Nocca, F., & Fusco Girard, L. (2018). Towards an integrated evaluation approach for cultural urban landscape conservation/regeneration. *Region, 5*(1), 33–51. doi:10.18335/region.v5i1.160

Padin, C., & Svensson, G. (2013). A multi-layer matrix model of sustainable tourism: Process, measurement areas, gap and reconnection analyses. *European Business Review*, *25*(2), 206–216. https://doi.org/10.1108/09555341311302684

Pereira Roders, A., & Van Oers, R. (2011). Editorial: Bridging cultural heritage and sustainable development. *Journal of Cultural Heritage Management and Sustainable Development*, *1*(1), 5–14. https://doi.org/10.1108/20441261111129898

Popay, J., Roberts, H., Sowden, A., Petticrew, M., Arai, L., Rodgers, M., & Britten, N. (2006). Narrative Synthesis in Systematic Reviews: A Product from the ESRC Methods Programme. *ESRC Methods Programme, 2006*, 93. http://www.ccsr.ac.uk/methods/publications/documents/Popay.pdf

Ring, N., Ritchie, K., Mandava, L., & Jepson, R. (2011). *A guide to synthesising qualitative research for researchers undertaking health technology assessments and systematic reviews*. Academic Press.

Sandelowski, M., & Barroso, J. (2003). Classifying the Findings in Qualitative Studies. *Qualitative Health Research*, *13*, 905–923. https://doi.org/10.1177/1049732303253488

Seyfi, S., Hall, C., & Fagnoni, E. (2018). Managing World Heritage Site stakeholders: A grounded theory paradigm model approach. *Journal of Heritage Tourism*, *14*, 1–17. https://doi.org/10.1080/1743873X.2018.1527340

Smeeton. N. C., & Smeeton, E. A. (1985). The Interface Between English and Statistics. *Teaching Statistics, 7*(2), 34–37. doi:https://doi.org/10.1111/j.1467-9639.1985.tb00574.x

Su, M. M., & Wall, G. (2014). Community participation in tourism at a world heritage site: Mutianyu great wall, Beijing, China. *International Journal of Tourism Research*, *16*(2), 146–156. https://doi.org/10.1002/jtr.1909

Su, M. M., & Wall, G. (2015). Exploring the Shared Use of World Heritage Sites: Residents and Domestic Tourists' Use and Perceptions of the Summer Palace in Beijing. *International Journal of Tourism Research*, *17*(6), 591–601. https://doi.org/10.1002/jtr.2026

Swanson, K. K., & DeVereaux, C. (2012). Culturally Sustainable Entrepreneurship: A Case Study for Hopi Tourism. In K. F. Hyde, C. Ryan, & A. G. Woodside (Eds.), Field Guide to Case Study Research in Tourism, Hospitality and Leisure (Vol. 6, pp. 479–494). Emerald Group Publishing Limited. https://doi.org/10.1108/S1871-3173(2012)0000006029

Thomas, J., & Harden, A. (2008). Methods for the thematic synthesis of qualitative research in systematic reviews. *BMC Medical Research Methodology*, *8*, 1–10. https://doi.org/10.1186/1471-2288-8-45

Torggler, B., & France, C. (2015). *UNESCO's Work on Culture and Sustainable Development Evaluation of a Policy Theme.* https://unesdoc.unesco.org/images/0023/002344/234443e.pdf

Towner, J. (1995). What is tourism's history? *Tourism Management*, *16*(5), 339–343. https://doi.org/https://doi.org/10.1016/0261-5177(95)00032-J

Underberg-Goode, N. (2014). Cultural heritage tourism on Peru's north coast. *Worldwide Hospitality and Tourism Themes, 6.* doi:10.1108/WHATT-03-2014-0013

Vargas-Hernández, J. G. (2012). *Heritage Tourism in Regional Development of.* Academic Press.

Vargas-Sánchez, A., Porras-Bueno, N., & Plaza-Mejía, M. (2011). Explaining residents' attitudes to tourism: Is a universal model possible? *Annals of Tourism Research, 38*, 460–480. https://doi.org/10.1016/j.annals.2010.10.004

Viera, A. J., & Garrett, J. M. (2005). Understanding interobserver agreement: The kappa statistic. *Family Medicine*, *37*(5), 360–363.

Wan, Y. K. P., & Li, X. (2013). Sustainability of Tourism Development in Macao, China. *International Journal of Tourism Research*, *15*(1), 52–65. https://doi.org/10.1002/jtr.873

Xu, J., Lü, Y., Chen, L., & Liu, Y. (2009). Contribution of tourism development to protected area management: Local stakeholder perspectives. *International Journal of Sustainable Development and World Ecology*, *16*(1), 30–36. https://doi.org/10.1080/13504500902757189

Yun, H. J., & Zhang, X. (2017). Cultural Conservation and Tourism Development in the Consolidation Stage of the Tourism Area Life Cycle Model. *Tourism Planning & Development*, *14*(3), 353–368. https://doi.org/10.1080/21568316.2016.1243147

Zimmer, L. (2006). Qualitative meta-synthesis: A question of dialoguing with texts. *Journal of Advanced Nursing*, *53*(3), 311–318. https://doi.org/https://doi.org/10.1111/j.1365-2648.2006.03721.x

APPENDIX

Table 5. Results of evaluation of final articles based on CASP form

No	CASP Criteria (Authors, Year of Publication)	Research Purposes	Method logic	Research Design	Sampling Method	Collecting Data	Reflexivity	Ethical Considerations	Accuracy of Analysis	Clear Expression of Findings	The Value of Research	Total Points
1	(Farid, 2015)	3	3	4	3	3	3	3	3	3	4	32
2	(Khlaikaew, 2015)	4	4	4	3	3	4	4	5	4	4	39
3	(du Cros, 2001)	4	5	4	4	4	3	3	4	4	3	38
4	(Liu & Chen, 2014)	4	5	4	5	5	4	5	4	5	4	45
5	((Bill) Carter * & Beeton, 2004)	3	4	4	5	5	4	4	5	4	5	43
6	(Padin & Svensson, 2013)	5	4	4	5	5	4	5	4	4	4	44
7	(Vargas-hernández, 2012)	4	4	4	4	4	4	4	4	5	5	42
8	(Brcar & Colaric-Jakse, 2017)	5	4	4	5	4	4	5	5	4	5	45
9	(Underberg-Goode, 2014)	4	3	4	4	5	3	4	4	3	4	38
10	(FRANC & IŞTOC, 2002)	4	4	3	3	3	4	4	3	3	4	31
11	(Durovic & Lovrentjev, 2014)	5	4	4	5	5	4	5	4	4	5	45
12	(UNWTO,2004)	5	4	4	4	4	5	4	5	5	5	45
13	(Ismagilova et al., 2015)	4	4	4	3	3	4	5	4	4	5	40
14	(Ko, 2005)	4	4	3	4	3	4	3	4	4	5	38
15	(Su & Wall, 2014)	4	3	4	4	5	4	5	4	4	5	42
16	(Adams, 2010)	3	3	3	3	4	3	3	3	4	3	32
17	(Nocca & Fusco Girard, 2018)	3	4	5	3	4	5	5	5	4	4	42
18	(Hunt et al., 2015)	4	5	5	4	4	4	5	5	5	4	45
19	(Pakdeepinit, 2007)	3	3	4	4	4	3	3	4	5	4	37
20	(Borges et al., 2011)	5	4	5	5	4	3	4	3	4	4	39
21	(Amir et al., 2015)	4	5	4	3	4	4	5	4	4	5	42
22	(Swanson & DeVereaux, 2012)	4	4	3	3	4	3	4	3	3	3	34
23	(Camus et al., 2012)	4	3	4	5	4	5	5	5	5	4	44
24	(Buckley, 2014)	5	4	5	4	5	4	4	4	5	5	45
25	(Manrique-de-Lara & Guerra-Baez, 2016)	4	4	4	4	5	4	5	4	5	5	44
26	(UNESCO, 2015)	5	5	5	4	5	5	5	5	5	5	49
27	(UNWTO,2017)	5	4	5	5	4	5	5	5	5	5	48
28	(UNWTO,2016)	4	5	4	3	4	5	5	5	4	4	43
29	(Wan & Li, 2013)	4	5	5	5	4	3	4	3	3	4	39
30	(Pereira Roders & Van Oers, 2011)	5	5	5	4	5	5	5	5	5	4	48

continued on following page

Table 5. Continued

No	CASP Criteria (Authors, Year of Publication)	Research Purposes	Method logic	Research Design	Sampling Method	Collecting Data	Reflexivity	Ethical Considerations	Accuracy of Analysis	Clear Expression of Findings	The Value of Research	Total Points
31	(Yun & Zhang, 2017)	5	4	5	4	3	4	4	4	5	5	43
32	(Loulanskia & Loulanski, 2011)	5	5	5	5	5	5	4	5	5	5	49
33	(Bowitz & Ibenholt, 2009)	5	4	4	5	4	5	4	4	3	4	42
34	(Dragouni, 2017)	5	5	5	5	4	5	5	5	5	5	49
35	(Seyfi et al., 2018)	5	5	4	5	4	5	5	5	5	4	47
36	(Hamimi Omar, 2013)	5	4	5	5	5	4	5	5	5	5	48
37	(Awang, 2010)	5	4	5	5	5	4	5	5	5	5	48
38	(Gurira & Ngulube, 2016)	5	5	4	4	4	5	4	5	4	5	45
39	(Agyeiwaah et al., 2017)	5	5	5	5	4	5	5	5	5	5	49

Chapter 6
Monitoring Sustainability Along Cultural Routes:
The MED Sustainable Path and Cultural Route Model

Maria Laura Gasparini
https://orcid.org/0000-0001-5562-2577
Center for Advanced Studies in Tourism (CAST), University of Bologna, Italy

Valeria Villalobos
Center for Advanced Studies in Tourism (CAST), University of Bologna, Italy

ABSTRACT

The chapter describes the theoretical and methodological approach followed to develop the MED sustainable path and cultural route (MED S&C Path) model as a practical tool to monitor sustainable tourism development along cultural routes. In the framework of BEST MED (Beyond European Sustainable Tourism MED Path) project, the model aims at improving the management of cultural routes by assessing their sustainability level using a set of criteria and indicators. The implementation process of the model proposes a multi-stakeholder approach which contributes to higher levels of cooperation at local, regional, and transnational level, improving the governance of cultural routes. The MED S&C Path model responds to the concrete need of increasing the tools to measure cultural routes' social, economic, and environmental impact, in order to raise awareness among policy makers about their strategic role in sustainable local development, especially in rural and less-known areas.

INTRODUCTION

Cultural routes (CR) are regarded as new cultural assets, since they connect tangible and intangible cultural and natural heritage elements around a particular theme or a geographical area, promoting an immersive and rich experience. In the case of Europe, the cultural route concept is based on the rela-

DOI: 10.4018/978-1-7998-9217-5.ch006

tionship between tourism and culture, but goes beyond its touristic use to become a real instrument to enhance European shared identity and common heritage.

Since 90% of CR passes through rural territories, they can stimulate cultural heritage preservation and contribute to regional and local development in line with the EU strategies for the four macro-regions (Interreg Europe, 2020), being sustainable tourism a powerful driver. Accordingly, the European Commission perceives CR as tools for economic development and social cohesion, being strategic instruments to implement EU policies through structural and specific programme funds. Interreg Med is one of these programmes developed by the European Commission to strengthen European cohesion and multi-level cooperation across 13 Mediterranean countries.

In this framework, BEST MED (Beyond Sustainable Tourism) is a strategic Interreg Med project being implemented between 2019 and 2022 in eight Mediterranean countries (Spain, Portugal, France, Italy, Croatia, Slovenia, Greece and Montenegro) that aims to achieve integrated planning and management of cultural routes, while contributing to improve the governance of tourism in the Mediterranean.

To this end, the project aimed at developing the MED Sustainable Path and Cultural Route model (MED S&C Path) to improve the management of cultural routes and assess their sustainability level using a set of criteria and indicators. Besides monitoring sustainability, the aim of the model is to achieve greater levels of cooperation between regional authorities, managers of routes and other key stakeholders in the territory, such as businesses' networks, associations, regional parks, research institutions, etc. Additional outcomes of applying the model include favouring the connection between coastal and hinterland destinations, helping to redistribute tourist flows to less crowded destinations, which are highly regarded in the pandemic era.

The lack of tools to measure cultural routes' social, economic, and environmental impact is widely acknowledged (Council of Europe, 2011; Interreg Europe, 2020), and is the reason why cultural routes are not given the importance they deserve as local development tools, especially in rural and less known areas. Precisely, the implementation of the MED S&C Path model aims to fill that gap. Through monitoring sustainability, route's managers and regional policy makers can make evidence-based decisions to better manage the routes and the social, economic and environmental impacts related to them.

Moreover, cultural routes are *"a recent research topic in humanities and social sciences"* (DECRA, 2019, p.47), which provides a further opportunity to advance the body of scientific knowledge in this field, as well as to raise awareness among policy makers about the strategic role cultural routes can have in sustainable local development.

This chapter aims to describe the theoretical and methodological approach followed to develop the MED S&C Path model. In addition, it also addresses the proposed model's implications for monitoring sustainable tourism and improving the governance of cultural routes. To this end, the chapter is organized as follows. After reviewing the main literature on the topics of cultural routes and sustainable tourism monitoring, the MED S&C Path model is presented in detail, including the methodology followed to define it and the toolkit designed to implement it. Finally, the feedback received to improve the model during a number of policy learning seminars within the BEST MED project will be presented and the main challenges and opportunities for further development of the model will be discussed.

THEORETICAL FRAMEWORK OF THE MODEL

Cultural Routes as Cultural Tourism Assets

Cultural routes (CR) are defined as new cultural assets which connect diverse and homogeneous elements of the tangible and intangible cultural heritage in contiguous or geographically dispersed territories, creating a new joint system which enhances their significance (Mariotti, 2012). Pilgrimage is considered the first type of tourism mobility (Timothy & Olsen, 2006), motivated mainly by religious purposes. Nowadays it has evolved to encompass different segments, including spiritual, heritage and secular tourism (Collins-Kreiner, 2019).

The concept of CR evokes the idea of movement, exchange and dialogue between people from different countries and regions. In addition to the physical journey, their significance stems from the stories they tell about the places they pass through (Richards, 2011). Recently, these routes are becoming attractive tourism destinations, since both the path in itself and the cultural and natural heritage along it are part of the cultural experience.

Whether they are pilgrimage routes, natural trails, groups of monuments, commercial roads or paths following major historical figures, a common characteristic shared by all cultural routes is that they are organized as networks of actors (DECRA, 2019). They are complex territorial systems (Berti, 2013), formed by a myriad of existing relationships between a wide range of stakeholders, both at vertical and horizontal level. They are multi-actor networks, with multi-temporal (short term and long term) and multi-scale (local, regional, national and transnational) dimensions, but rooted in stable local systems.

A shared perspective in the literature is that CR are means for preserving and enhancing cultural heritage, while contributing to regional and local development. In this sense, tourism is perceived as a powerful driver to stimulate economic development by clustering activities and attractions, especially in lesser-known areas (Briedenhann & Wickens, 2004).

The Cultural Routes of the Council of Europe Programme

Besides the economic relevance assigned to cultural routes in Europe, they fulfil social and political functions (Severo, 2017). One of the roles temporary mobility and tourism should play is to contribute to the creation of a shared European cultural identity.

According to the Council of Europe (CoE), cultural routes are *"a cultural, educational heritage and tourism co-operation project aiming at the development and promotion of an itinerary or a series of itineraries based on a historic route, a cultural concept, figure or phenomenon with a transnational importance and significance for the understanding and respect of common European values"* (Resolution CM/Res(2013)66).

The cultural routes programme was created in 1987 by the CoE with the basic premise that, by travelling through space and time, European citizens can become aware of their shared common heritage and values. In 2010, the Committee of Ministers of the CoE established an Enlarged Partial Agreement (EPA) to support the development and promotion of cultural routes, recognising them as tools for cultural cooperation, sustainable territorial development and social cohesion.

The European Institute of Cultural Routes (EICR), based in Luxembourg, is in charge of managing the CoE's programme, providing assistance for the certification of routes, as well as capacity building and visibility to existing routes. Currently, there are over 40 cultural routes that are part of the programme, covering a wide range of themes, from history, arts and geography, to religion and agriculture. The routes must fulfil the criteria outlined in the five priority fields of action, namely 1) Cooperation in research and development, 2) Enhancement of memory, history and European heritage, 3) Cultural and educational exchanges for young Europeans, 4) Contemporary cultural and artistic practice, 5) Cultural tourism and sustainable cultural development.

Although the core of the cultural routes programme is on the building process of the European identity based on the values of the CoE (human rights, cultural diversity, intercultural dialogue and mutual exchanges across borders), tourism has a strategic role to play in this process, to position Europe as a unique cultural tourism destination. By creating cross-border sustainable tourism itineraries based on the richness and diversity of European cultures, they contribute to the economic wellbeing, intercultural dialogue and mutual understanding of the regions traversed. The programme offers a model for transnational cultural and tourism management, enabling cooperation between national, regional and local authorities, private sector and civil society.

As stated by Berti (2012), about 90% of the CR cross rural areas and agricultural landscapes, allowing visitors to discover off the beaten path destinations. This is highly relevant in the COVID-19 pandemic era, when limitations to mobility are supporting the rediscovery of less known destinations through slow tourism, as they are perceived as safer for their "isolated" assets.

In the step-by-step guide to the Cultural Routes of the CoE (2015), three main types of cultural routes have been identified:

- *Linear pattern routes:* are mainly those developed along a physical path, such as the Santiago de Compostela Pilgrim Routes. The routes link villages, towns and heritage sites which were connected historically by the presence of the route. These routes are more suitable to create hiking or biking tourism itineraries.
- *Territorial pattern routes:* these are routes that involve territories sharing a common theme or character, such as the Routes of El legado andalusí. A series of paths is generally established to propose an exploration of the different parts of the route.
- *Reticular pattern (archipelago) routes:* These routes are composed of individual items or aggregated elements, presenting different scales and connected by a common theme, such as the Transromanica route. In this case, routes are not characterised by territorial continuity and they may be called "virtual routes", as they connect geographically separated elements under one common theme.

Pilgrimage Trails and Other Types of Itineraries

Pilgrimage routes belong to the linear pattern type of routes and can be considered the forerunner of the modern phenomenon of tourism (Di Giovine & Choe, 2019). A pilgrimage has been defined as *"a journey resulting from religious causes, externally to a holy site, and internally for spiritual purposes and internal understanding"* (Barber, 1993, p. 1) and is a tradition found in all the major religions (Di Giovine & Elsner, 2016).

According to Collins-Kreiner (2016) the phenomenon of pilgrimage is in a phase of "rejuvenation" as it is losing some of its original characteristics (such as the purely religious motivation) and acquiring new ones, such as the personal dimension of spirituality. In effect, the "deregulation" of spirituality as it has been called entails that spirituality is not anymore imposed by religious institutions but instead is "a matter of personal choice" (Kato & Progano, 2017, p.244).

The motivations of current pilgrims might include the search for personal transformation and meaning to one's existence beyond everyday life, but also get to know new territories, learn about local history and culture, meet new people and regain a psychophysical balance, especially after the confinement caused by the pandemic. Therefore, we can speak about secular or contemporary pilgrimage, which is a form of slow tourism, where the journey itself rather than the destination becomes the heart of the experience (Olsen et al., 2018).

Admittedly, pilgrimage has an often-underestimated impact, although it is one of the fastest growing segments of the tourism sector with over 300 million travellers visiting the world's main religious sites each year (UNWTO, 2015). In this regard, besides the CoE's cultural routes programme and historic pilgrimage trails, there are a number of itineraries that have been developed, including new pilgrimage trails, thematic routes (i.e. wine routes), transnational routes linked to natural heritage and even water routes (STAR Cities project).

In Italy, for example, the Ministry of Culture and Tourism launched in 2016 a year-long initiative on walking trails, the so-called "Anno dei Cammini" (year of the routes), which allowed to map over 40 itineraries in the country. With this initiative, the Italian government intended to boost hiking and slow tourism as a way to valorise the historical, natural, cultural and religious heritage of the peninsula. Moreover, a number of criteria were proposed in order to qualify as a "route or itinerary" which included the minimum infrastructure for walkers, services and information available, as well as the presence of a management structure of the route.

For that occasion, the definition adopted by the Italian Ministry considers routes as *"cultural itineraries of particular European and/ or national importance, travelled on foot or by other forms of sustainable mobility, and that represent a way of using the widespread natural and cultural heritage, as well as an opportunity to enhance the natural, cultural and territorial attractions"* (Ministero della Cultura, 2015). Pilgrimage trails are included in this definition because of the valuable contribution they can make in peripheral areas, combining low environmental impact and sometimes unexploited or hidden local development potentials.

This definition, which goes beyond the institutional recognition of the CoE, is the one adopted to design the Med S&C Path model, so as to ensure the transferability of the model from the pilot areas of the project (mainly certified cultural routes of the CoE) to a wider typology of trails and itineraries.

Sustainable Tourism Indicators and Criteria

Sustainable tourism might be regarded as an abstract goal without indicators and other monitoring tools that can make the concept operational (Butler, 1999; Choi & Sirakaya, 2006; Hunter, 1997; Torres-Delgado & López Palomeque, 2014). Since the beginning of the 1990's when the Agenda 21 was introduced in the Earth Summit of Rio de Janeiro, indicators have been recognised for their role in supporting policy makers to see the whole picture of tourism impacts and make evidence-based decisions (Castellani & Sala, 2010; Choi & Sirakaya, 2006; Miller, 2001; Miller & Twinning Ward, 2005; Rametsteiner et al., 2011; Tanguay et al., 2013).

There are two different types of indicators: simple indicators (information collected directly from reality or through simple calculations) and complex, synthetic indicators or indices (which combine simple indicators using a weighting system). A third type of indicators is an indicator system, which groups together simple indicators structured in a particular framework, for instance, the dimensions of sustainability, and the results are interpreted jointly (Torres-Delgado & López Palomeque, 2014).

Several initiatives have been developed worldwide to measure and monitor the impacts of tourism using criteria and indicators. Among those, the Global Sustainable Tourism Council (GSTC) has developed a set of criteria, which is consider the baseline global standard for sustainability in the travel and tourism industry. In turn, the United Nations World Tourism Organization (UNWTO) published in 2004 the guidebook on "Indicators of Sustainable Development for Tourism Destinations", the most comprehensive guide to develop and implement indicators to manage tourist destinations.

More recently, the European Commission launched the European Tourism Indicator System (ETIS) as a common framework to measure and monitor tourism impacts and make informed decisions to manage tourism. The system includes a list of indicators and a toolkit with an implementation guide in order to implement the indicators in a participatory manner with key stakeholders (see European Commission, 2016). The importance of sharing the responsibility for monitoring sustainability is given by the possibility to establish relationships between the several actors working in the destination, to align objectives, resources and work towards common goals.

Certainly, implementing a sustainability monitoring scheme is not free of challenges, related to whether indicators should be standardised or context-specific (Moreno Pires et al., 2014), time and funding constraints, stakeholders' engagement (Agyeiwaah et al., 2017), lack of technical expertise and political commitment (MED Sustainable Tourism Community, 2019; Zabetta et al., 2014).

However, the process of discussing the concept of sustainable tourism helps understanding it and making it a more tangible one. This is known as the "conceptual role" of indicators and it is related to the social learning process that results from bringing a broad range of stakeholders together and facilitating conversations among larger communities (Bell et al., 2011; Farrell & Twining-Ward, 2004; Hezri & Dovers, 2006; Lehtonen, Sébastien, & Bauler, 2016). This social learning process in effect can be considered a pre-condition for more direct uses of the information provided by indicators (Gasparini & Mariotti, 2021).

Monitoring Sustainability Along Cultural Routes

Due to the multi-level governance and transnational and interregional dimensions of cultural routes, developing a model to measure and monitor their impacts by selecting specific criteria and indicators is not a simple task.

Currently there are no mechanisms in place to measure the economic impact cultural routes generate at local level (Council of Europe, 2011). The complexity of gathering information is given by the need to capture data at a much lower level than the national and regional ones which are gathered by Eurostat or national statistics offices. Moreover, most routes lack the technical expertise and financial resources to collect and manage data and there is a lack of coordination and dialogue between institutions that are part of the routes (Interreg Europe, 2020).

As observed in the Routes 4U report on Policies for the Danube region, *"there is a need to improve methodological frameworks to provide better statistics particularly linked to the Cultural Routes performance, including by the development of sets of indicators for monitoring and evaluation of the Cultural Routes and assessing their actual and potential economic and social value in a more systematic manner"* (Council of Europe, 2020, pag.64).

To this end, the model developed in the context of BEST MED project intends to fill that gap, providing a model to gather data at lower levels while improving the governance of cultural routes thanks to a participatory approach to data collection.

MED SUSTAINABLE PATH AND CULTURAL ROUTE MODEL

The MED Sustainable Path and Cultural Route model (MED S&C Path) takes as starting point the criteria developed by the CoE for its Cultural Routes programme, further developing an approach to support the sustainable management of different Mediterranean routes and paths. In order to theoretically define the model, the wider definition of cultural route proposed by the Italian Ministry of Culture and Tourism has been adopted, going beyond the institutional recognition of the CoE, to ensure the transferability of the model.

According to the BEST MED project's consortium, a sustainable cultural route is one that incorporates sustainability principles in its planning and management, achieving sustainable development of socio-cultural and economic activities along the route, while protecting the natural resources. Through monitoring sustainability, cultural route's managers and regional policy makers are able to take evidence-based decisions to better manage the routes and the impacts related to them.

The MED S&C Path model is a voluntary monitoring tool, based on a self-assessment that routes' managers can implement together with key stakeholders, sharing the responsibility to monitor the route's impacts and ensuring tourism acts as a socio-economic driver for the territories along the route. As pointed out by Severo, *"cultural routes management should be based on community participation"* (2017, p.142). Therefore, besides monitoring sustainability, the model contributes to improve the multi-level governance of routes, achieving greater levels of cooperation between regional authorities, the managers of cultural routes and other key stakeholders in the territory (such as businesses' networks, civil society associations, regional parks, research institutions, etc.).

Additional outcomes of applying the model include favouring the connection between coastal and hinterland destinations. In accordance with the recommendations from the Med Sustainable Tourism Community (2019), tourism planning needs to consider the redistribution of tourism flows in space and time, thanks to promoting less-crowded, hinterland destinations and off-season offers, which is made possible thanks to supporting the development of routes and paths. Integrated planning refers not only to geographical integration (like coast and hinterland) but also among different parts of the system that forms the tourism sector, as well as integration of tourism into the overall development policies and plans of a region.

The responsibility of implementing the MED S&C Path Model is directed to two main target audiences, which are also the main users of the information provided by the monitoring exercise:

1. The management structure of the cultural routes: namely the association or legal entity in charge of managing and promoting the route.
2. The Regional Public sector: policy makers at regional level where the section of the route belongs.

The regional public sector and the routes' managers have complementary roles for different aspects, such as promotion and infrastructural interventions. Regions have their own strategies, including sustainable tourism, and cultural routes need support from the regional level in order to align their activities to these strategies and contribute to their implementation. However, usually the organizations managing cultural routes are small in structure which means that they have limited financial and human resources to dedicate to collecting data to monitor sustainability (Università degli Studi di Perugia, 2018).

Therefore, the aim of the model is that regional policy makers can work together with routes' managers to comply with the criteria and indicators, to monitor the sustainability level and improve the governance of the routes (see Figure 1). Moreover, this exercise can help assessing why cultural routes face more difficulties in some regions than in others.

Figure 1. Aim of the MED S&C Path model.
Source: Own elaboration

METHODOLOGY TO DESIGN THE MODEL

The Center for Advanced Studies in Tourism (CAST) from the University of Bologna, external expert for the Lazio Region (BEST MED partner) acted as knowledge provider, developing the theoretical model as well as the methodology and the step-by-step framework to implement it.

The model builds from the results of past projects funded by the European Union regarding cultural routes and pilgrimage trails, as well as international initiatives of sustainable tourism measuring and monitoring, capitalizing these experiences and complementing them with the specific literature on sustainable tourism indicators and cultural routes previously described.

The methodological approach followed is presented in Figure 2 and was developed under Work Package 3 devoted to the study phase of BEST MED. The objective of the study phase was to build the knowledge baseline from which to develop 1) the MED S&C Path model and 2) a network of sustainable tourism observatories at Mediterranean level.

Figure 2. MED S&C Path model methodological framework.
Source: Own elaboration

In order to define the MED S&C Path model, CAST research team firstly performed a structured review of 21 initiatives to select the main criteria, the indicators and draw first conclusions to apply to the MED S&C Path model. Initiatives ranged from international standards, EU funded projects and programmes, study reports and policy briefs dealing with the topics of sustainable tourism monitoring and cultural routes management.

The information about the initiatives was collected through extensive desktop research, including literature review of relevant academic papers and grey literature, as well as consulting the official websites of each project (Deliverables Library section).

The criteria taken to select the initiatives were:

- Initiatives that have developed methodological approaches to data collection and management for sustainable tourism planning
- Initiatives that dealt with sustainable management of cultural routes, pilgrimage trails and other types of paths

As a result of the screening process, a total of 12 initiatives on sustainable tourism monitoring and 9 initiatives on cultural route management were analysed.

Following the full review of existing initiatives, a comparative analysis was performed using a reporting template. The main categories taken into consideration were: 1) the development of criteria and indicators to measure sustainability, 2) whether an implementation framework has been designed to accompany the set of criteria/ indicators, 3) the provision of tools for data collection. The comparative analysis made possible to identify the commonalities across projects, the gaps that could be fulfilled and relevant aspects to take into consideration to design the Med S&C Path toolbox. The lessons learned were used to develop the toolbox which includes a set of criteria and indicators, a step-by-step implementation framework and the tools for data collection (Self-Assessment Audit Sheet, online platform and questionnaires).

Finally, the model has been validated by BEST MED partners as well as local stakeholders that participated in two rounds of policy learning seminars held in each pilot area. Project partners were free to choose how to organize their respective seminars, however they followed specific guidelines according to a seminar format provided by CAST and the results were collected through an evaluation template also provided by CAST, to present the outcomes in a standard form. Sixteen seminars were organized in total and the feedback was integrated in the model.

The results of the comparative analysis, the tools included in the MED S&C Path toolbox and the feedback collected from the policy learning seminars are described in the following sections, as well as their implications for the sustainable management of cultural routes.

REVIEW AND COMPARATIVE ANALYSIS OF EXISTING INITIATIVES

Two matrices summarising the main characteristics of each initiative analysed are provided in Table 1 and Table 2.

When analysing the initiatives on sustainable tourism monitoring it is possible to see that the majority of them (10 out of 12 initiatives) includes criteria, while 7 of them provide also indicators.

The criteria present in sustainable tourism monitoring initiatives were analysed, to include the main dimensions of sustainability and the aspects that must be considered regardless of the type of destination, to understand main positive and negative impacts of tourism. To this end, the division on four main sections adopted by the GSTC (sustainable management, economic, socio-cultural and environmental sustainability) was taken as reference, complementing it with criteria from other international initiatives such as Green Destinations and ETIS.

From those projects suggesting a set of indicators for measuring tourism sustainability, it became evident that most of them build from the ETIS toolkit and adapt the system of indicators to their specific needs. Due to the fact that ETIS is an initiative from the European Commission, it is not surprising that EU funded projects take the system as the suitable base where to start. This is in fact confirmed by Niavis et al. (2019) in a study on 14 projects part of the Med Sustainable Tourism Community, where ten of these projects considered ETIS as the most useful tourism sustainability assessment initiative. As a result, a number of ETIS indicators were integrated as part of MED S&C Path model core and optional indicators.

Monitoring the impacts along cultural routes is certainly different from evaluating impacts at single destination level. Thus, projects and initiatives dealing with cultural and pilgrimage roads were analysed, in order to create a tailor-made model which addresses their specific needs.

It is clear that these initiatives mainly provide recommendations on data collection methods and relevant aspects to consider, but no complete framework to monitor sustainability along cultural routes is provided, which is exactly the aim of the MED S&C Path model. Therefore, the researchers' task

Table 1. Comparative analysis of sustainable tourism monitoring initiatives.

Sustainable Tourism Models and International Standards	Criteria	Indicators	Implementation Framework	Tools for Data Collection
Mitomed Plus – Maritime & Coastal Tourism Management Model	X	X	X	Separate set of indicators and online platform developed by the project
Mitomed Plus - Green Beach Model	X		X	Excel sheet for data collection
Destimed – MEET Ecotourism Standard	X	X	X	Online Monitoring Tool and Ecotourism Footprint Calculator
Co Evolve – Tourism driven strategic plan		X	X	Tourism Sustainability toolkit with an excel sheet for data collection
Emblematic - "Ecojourney Emblematic Mediterranean Mountains" Label	X	X	X	10 ETIS indicators and Action Plan template
MedCycleTour – Transnational Action Plan for development of Euro Velo route	X			Transnational Route Evaluation Report
Sustainable Tourism Programme at World Heritage Sites	X		X	Not provided
GSTC Criteria	X	X		Not provided
Green Destinations	X	X		Online platform
International Network of Sustainable Tourism Observatories - INSTO	X		X	Not provided
European Charter for Sustainable Tourism in Protected Areas	X		X	Not provided
European Tourism Indicator System - ETIS		X	X	Destination data set (excel sheet) and templates for 4 types of survey

Source: Own elaboration

was to select those criteria and indicators relevant for this particular "type of destinations", recognizing that some aspects of sustainability monitoring are out of the control of route's managers and therefore, cooperation with other stakeholders (namely regional policy makers, among others) is needed in order to perform the monitoring exercise.

Particularly relevant for BEST MED project is the INSTO framework (the network of sustainable tourism observatories established by UNWTO) since one of the objectives of BEST MED is advancing the cooperation between tourism observatories at Mediterranean level. Accordingly, when selecting the criteria for the model, the eleven mandatory areas that INSTO requires to monitor tourism sustainability were included, in order to ensure the model's compliance with this framework in the event of future collaboration with this global initiative. The areas are: tourism seasonality, employment, destination economic benefits, governance, local satisfaction, energy management, water management, wastewater (sewage) management, solid waste management, accessibility and climate action.

Several of the projects and initiatives grant a sustainability label and certification, which is generally consider a suitable recognition to highlight the efforts in implementing sustainability practices. However, the multiplication of labels might have the contrary effect in local stakeholders and tourists, confusing them and getting them overwhelmed, instead of persuading them to support the initiative. As far as the

Table 2. Comparative analysis of cultural routes projects and initiatives.

European Cultural Routes and Other Itineraries	Criteria	Indicators	Implementation Framework	Tools for Data Collection
Green Pilgrimage Project		X		Recommendation of different data collection methods
Cultural Routes of the CoE certification programme	X		X	Certification / evaluation form
Routes 4U project	X	X		Recommendations of different data collection methods (including ETIS)
UNESCO Culture for Development Indicators		X	X	Culture for Development Indicators Suite (CDIS)
Impact of European Cultural Routes on SMEs innovation and competitiveness report	X	X		Recommendations of different data collection methods
CERTESS project - "Cultural European Routes. Transferring Experience, Sharing Solutions" Toolkit		X	X	Route Implementation plans
CULT-RING Project - Cultural Routes as Investments for Growth and Jobs			X	Action Plans
University of Perugia Report on monitoring the impacts of visitors of Italian paths	X	X		Questionnaires and surveys for visitors and paths' managers
Policy brief on Cultural Routes of the CoE	N/A	N/A	N/A	Provide recommendations to improve governance and impact monitoring

Source: Own elaboration

MED S&C Path model is concerned, a label is not expected to be provided yet, but instead it aims to be a tool to improve governance and management of the routes.

All of the initiatives analysed stressed the need of adopting a participatory approach to sustainability monitoring, as it is clearly impossible for one actor to be able to access all the needed information to comply with criteria. This aspect is key for BEST MED, as the project intends to contribute to a better governance at regional and transnational level, therefore cooperation between different stakeholders at multiple levels is of particular importance. In order to implement a sustainability monitoring model at destination level, the most common approach is to design a step-by-step methodology, usually starting by forming a working group with key stakeholders, followed by establishing a baseline of the current situation, setting the priorities, collecting data and performing continuous monitoring. This approach was considered suitable for cultural routes and an implementation framework was integrated in the model.

With regards to the tools used for data gathering, online platforms and excel data sheets are the most common tools, since they enable the calculation of quantitative results as well as the comparisons with other destinations. In terms of data collection methods, surveys seem to be the most useful tools, although they can be costly and time-consuming. Consequently, performing this type of analysis in partnership with universities is considered a win-win situation, in order to reduce costs while providing relevant field

Figure 3. Design of the MED S&C Path model and toolbox.
Source: Own elaboration

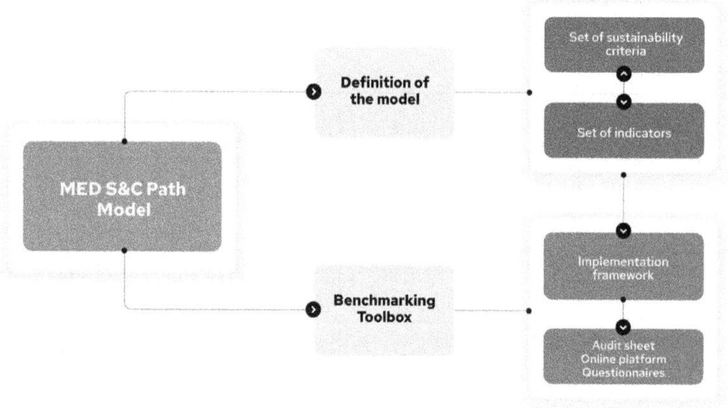

research experience for students. To perform the surveys, a number of the reports and toolkits reviewed present examples and guidelines which have been taken into consideration and adapted to design the questionnaires for the MED S&C Path toolbox.

Finally, several of the reports regarding cultural routes emphasized that one key challenge is to improve the monitoring and evaluation methods to assess the actual and potential impacts routes have on the economy and social tissue of territories, particularly less known and rural areas. This represents an outstanding opportunity for the MED S&C Path model to provide a highly regarded tool to measure the impact that cultural routes, pilgrimage trails and other types of paths generate in Mediterranean regions.

Criteria, Indicators and Toolbox

As a result of the review of existing initiatives, a standard set of criteria appropriate to evaluate the sustainability of a cultural route was developed, including 4 main dimensions: Sustainable Management, Economic sustainability, Socio-cultural sustainability and Environmental sustainability. These criteria are statements which indicate if the destination complies or not with certain requirements to be considered sustainable.

Besides selecting the main relevant criteria for the MED S&C Path model, a number of suggested core and optional indicators were selected and included in the toolbox (see Annex A and Further Reading section). These indicators provide additional guidance to measure the exact level of compliance with the criteria and provide a specific number or percentage that can be benchmarked against established targets. The selected indicators are meant as a guide for stakeholders, they are not an exhaustive list and stakeholders are free to monitor the ones they consider useful.

Both the criteria and indicators are part of the toolbox that has been designed in order to implement the model (see Figure 3). This toolbox contains the following: a step-by-step implementation framework, a self-assessment audit sheet with the sustainability criteria for self-evaluation, a list of core and optional indicators, questionnaires for visitors and cultural routes managers and an online platform where to measure performance and benchmark. In what follows, each of these tools will be briefly presented.

Figure 4. Step-by-step Implementation Framework.
Source: Own elaboration

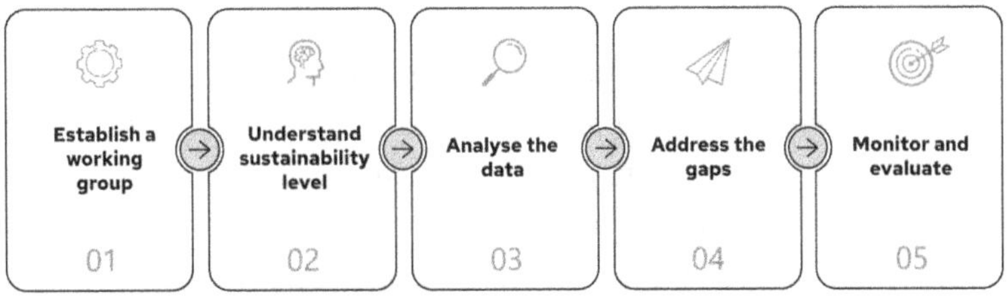

Step-by-Step Implementation Framework

An implementation framework is a process to translate sustainability issues into management actions. It is an essential component of any model since it guarantees that the results stemming from monitoring sustainability will improve tourism management (Miller & Twining-Ward, 2005).

The importance of an implementation framework is given by the possibility of shaping relationships between the several actors working in the destination, sharing the responsibilities for data collection and analysis and benefiting from a social learning process. Moreover, sharing the responsibility for data collection provides a sense of ownership and commitment to the process (European Commission, 2013; European Commission, 2016).

As suggested by Miller & Twining-Ward (2005), the implementation framework was developed based on an adaptive management approach. To begin with, an adaptive management framework, in this sense, includes a variety of monitoring tools and their design, protocols, and fine-tuning. Next, it is followed by data gathering and analysis, which identifies gaps between existing and projected sustainability targets. Finally, it involves the dissemination of results through reports and publications. In the case of the MED S&C Path model, an implementation framework was designed containing the following five steps, as illustrated in Figure 4.

Step 1: Establish a working group. The group should comprise the local managers of the cultural route (who should act as the coordinator of the working group) and regional policy makers (who should provide support for matters outside the competence of routes' managers). Other relevant stakeholders that can be part of the group include the local municipalities, DMOs (Destination Management Organisations), civil society organisations, tourism stakeholders, protected area managers, academic institutions, and national policy makers.

The working group is responsible for putting the model into action, assessing the route's sustainability, and implementing solutions based on the findings.

Step 2: Understand the current sustainability level of the route. Using the "Self-Assessment Audit Sheet" included in the toolbox, stakeholders can evaluate the route's sustainability level. The audit sheet contains the set of sustainability criteria, divided into the model's four main dimensions, that can be compiled simply by responding to yes/no statements. The working group coordinator is responsible for compiling the audit sheet and disseminating the results to the rest of the stakeholders.

The goal is to obtain an initial qualitative evaluation by providing a percentage of compliance with the criteria. This is a preliminary but important result, allowing the manager of the route and regional policy makers to evaluate at which stage actually is with respect to the goal of becoming a sustainable route.

Since cultural routes involve several destinations, it is necessary to define the borders of the area for monitoring purposes. It is possible to implement the model in a short stretch of the cultural route or along the entire length of the route. The difference will be in the number of stakeholders to involve and the overall complexity of the process, if it involves different regions or countries where the route passes. Therefore, according to the chosen length of the monitoring area, it will be important to involve the regional or national authorities, so that administrative borders do not become barriers for cooperation.

Step 3: Address the gaps by using the provided tools. According to the gaps identified in the second step, the working group should discuss the sustainability criteria to agree on a participatory data collection plan, dividing the responsibilities, since certain stakeholders would naturally have easier access to certain information depending on their roles and position.

At this stage, the working group can use the rest of the resources included in the toolbox, specifically, the list of indicators (core and optional), the questionnaires for route's managers and visitors, and the BEST MED Online Platform, to comply with the criteria and improve their level of sustainability, as well as the management of the route.

The information can be sourced as much as possible from sources already available, such as statistical data at regional/ local level, surveys already performed, to reduce the amount of time and resources dedicated to data collection. However, the lack of available data should not be a reason to disregard certain indicators, as one of the advantages of developing indicators is precisely to start collecting valuable data that was previously not being collected. Prioritizing available data will generally mean focusing on economic data only, and the aim of the model is to have an overall assessment of all three dimensions of sustainability.

Step 4: Data analysis. Once the data collection process for the sustainability criteria and indicators has been completed, the findings can be analysed, to see the percentage of compliance with the criteria, as well as the results from the indicators. It is important to note that stakeholders can decide to focus on a limited number of criteria and indicators and increase the monitoring exercise over time.

According to the results, the working group should discuss the most pressing issues, collectively establish priorities, define sustainable tourism development goals along the path, and create an action plan. To this end, the cultural route manager should supervise the action plan's implementation.

Step 5: Regularly monitor and evaluate. Under the leadership of the cultural route manager, the working group is responsible for periodically monitoring the route's sustainability level. For this purpose, the resources included in the toolbox and the information derived from the model can be used to make informed decisions and policy developments. Finally, each destination's specific needs and resources will determine the frequency of monitoring.

Table 3. Self-Assessment Audit Sheet Categories.

1. Sustainable Management	2. Economic Sustainability	3. Socio-Cultural Sustainability	4. Environmental Sustainability
1.1 Cultural Route Planning & Management	2.1 Economic contribution to local economy	3.1 Preservation of Cultural Heritage	4.1 Resource Management
1.2 Quality of Infrastructure	2.2 Seasonality	3.2 Accessibility	4.2 Sustainable mobility
1.3 Health & Safety		3.3 Residents engagement & feedback	4.3 Landscape & Biodiversity protection
1.4 Information & Promotion		3.4 Gender equality	

Source: Own elaboration

Self-Assessment Audit Sheet

The Self-Assessment Audit Sheet is included in the toolbox as a separate spreadsheet, and it is also available in the BEST MED online platform. As previously noted, the audit sheet measures compliance with a set of sustainability criteria divided into the model's four main dimensions: Sustainable Management, Economic Sustainability, Socio-cultural Sustainability, and Environmental Sustainability.

Each dimension is composed of categories with a total of 13 categories and 37 criteria. An overview of the main categories can be seen in Table 3. Please refer to Appendix A for the complete list of criteria included in the Audit Sheet.

Set of Indicators

As recommended by several authors (Coccossis & Koutsopoulou, 2020; Niavis et al., 2019), the initial portion of the indicator selection process involved analysing criteria from scientific literature and the practical experience of past initiatives such as ETIS and other EU funded projects. As a result, it was possible to identify an initial list of indicators that effectively measured sustainable tourism development along cultural routes.

In this case, the first set of indicators was mainly assessed using the World Tourism Organization (UNWTO, 2004) selection criteria:

1. Relevance: This criterion refers to the indicator's bearing on the specific sustainability issue. The ideal indicator will be pertinent to the problem and provide information to help with its resolution, while at the same time making a difference in a choice that affects tourism and the destination's sustainability.
2. Feasibility: An indicator's feasibility concerns the extent to which it is possible to collect and analyse the necessary data. In this regard, it is critical to determine the most effective method to obtain the information, if a data source has been identified, the processes needed to collect the data, and its related financial aspects. In short, the feasibility criteria concern the cost-benefit relation of each indicator.
3. Credibility: Credibility represents the information's reliability and trustworthiness. Data should be obtained from a reliable and scientifically sound source. Furthermore, data should be considered as objective. For instance, data on "seawater cleanliness" will have greater credibility if it is col-

lected and analysed by an independent institute than if it is collected and presented by the beach hotel association" (UNWTO, 2004, p. 41).

4. Clarity: The clarity criterion refers to how easily users may interpret indicators. Users should be able to effortlessly comprehend and act on the information delivered by the indicators. Consequently, indicators should be selected depending on the level of technical knowledge of its audiences.
5. Comparability: This criterion states that the indicator should be capable of reflecting changes over time compared to standards or benchmarks at the same or other locations.

Many studies reiterate similar sets of selection criteria (Agyeiwaah et al., 2017; Lozano-Oyola et al., 2012; Tanguay et al., 2013), and for this reason, the UNWTO's approach was favoured.

Finally, there is no agreed-upon number of indicators to use, so scholars and practitioners employ their own discretion to define the length of the list of indicators. However, these indicators should cover the main sustainability dimensions while remaining pertinent and feasible for the site-specific context (Torres-Delgado & Palomeque, 2014), as indicated in Tanguay et al. (2013). Moreover, validating indicators through participatory procedures involving multiple stakeholders is appropriate in this scenario, and this is precisely the methodology followed by BEST MED through the policy learning seminars.

Following this approach, 26 core indicators were pre-selected to evaluate the common sustainability issues to all destinations/routes. Additionally, a set of 72 optional indicators is also available to accommodate each cultural route's inherent characteristics. Guidelines to use the indicators are also provided in the toolbox, with instructions for data collection. Please refer to Appendix A for the list of core indicators included in the model and the "Additional reading" section for the guidelines to use the indicators (Deliverable 3.4.1).

BEST MED Online Platform

The online platform, developed by the University of Maribor (partner of BEST MED), incorporates destination data from Mitomed+ (Models of Integrated Tourism in the Mediterranean) project, Eurostat and BEST MED's Self-Assessment Audit Sheet. The primary purpose is to provide stakeholders with an intuitive online tool to effortlessly measure compliance with the model's criteria and allow benchmarking among different cultural routes. In addition, the 26 core indicators previously explained are integrated into the platform to further improve comparisons.

The main functionalities of the BEST MED platform are as follows:

- Displaying primary data for cultural routes. For example, the route's name, description, type of route, contacts for its management structure, location, and official website.
- Providing the values of relevant Eurostat indicators for the route's region and country.
- Presenting the values of relevant Mitomed+ indicators for the route's region.
- Compiling the Self-Assessment Audit Sheet with compliance results by sustainability dimension.
- Diagnosis of sustainability issues within the route with the option of monitoring the evolution of the indicators and subsequent benchmarking.
- Interactive map where the route and its corresponding data can be visualised.

As previously indicated, the data included in the Mitomed+ online platform is directly transferred to the BEST MED platform. This allows to improve an existing platform, instead of building one a new one, expanding the data from Mitomed+ project, which was only focused on coastal destinations. Additionally, the platform incorporates Eurostat data at the NUTS2 and national level for all the destinations included in the cultural route. In this sense, the available data comprises: arrivals at tourist accommodations (yearly and monthly), same-day visits, nights spent at tourist accommodations, available number of tourist accommodations, employed persons in tourism, active enterprises in cultures, and generation of waste in tonnes by the service industry (Faculty of Tourism University of Maribor, 2021).

Questionnaires

Two different questionnaires have been designed as part of the toolbox, one for cultural routes' managers and one for visitors. However, the toolbox also includes guidelines to design questionnaires for local businesses and residents if considered necessary (please see the "Additional reading" section for the questionnaires and guidelines, Deliverable 3.4.1)

The design of the questionnaires was carried out based on the ETIS surveys, Green Pilgrimage project surveys (Norfolk County Council & University of East Anglia, 2019) and the "Atlas of Paths" Report (Università degli Studi di Perugia, 2018). In this sense, the questionnaire for cultural routes' managers helps gather general information about the cultural route and its management structure. Additionally, it can be used to collect data related to the services offered to visitors, the average visitor profile, the sustainability practices implemented, and the governance model and cooperation prevalent in the route/path.

Similarly, the questionnaire for visitors is helpful to profile the type of visitors interested in the cultural route and gather their feedback about the sustainability level of the route.

FINE TUNING MED S&C PATH MODEL ACCORDING TO THE RESULTS OF THE POLICY LEARNING SEMINARS

As a final step in the development of the model, a series of Policy Learning Seminars were organised at the local level in each of the eight pilot areas, in order to validate the model theoretically defined during the Study phase, before the Testing phase. As a result, the stakeholder's suggestions were used to enhance the MED S&C Path model and its tools.

In essence, the primary purpose of the policy learning seminars was to compare the most critical needs of each pilot area with the model's criteria and validate the step-by-step implementation methodology. For this reason, each pilot area was asked to report the following outcomes:

1. An assessment of the prioritised positive and negative characteristics, implications, and influences of tourism in the pilot area.
2. A gap analysis between the needs of the pilot area and the criteria included in the MED S&C Path model, divided into "covered", "partially covered", and "not covered" by the model criteria, and classified by sustainability dimension.
3. Evaluation and suggestions from the local stakeholders regarding the step-by-step implementation framework.

Regarding the needs reported as "covered" by the model, the dimension that emerged the most during the Policy Learning Seminars was Economic Sustainability (30%), usually considered a precondition to achieve environmental and socio-cultural sustainability. The other dimensions were given comparable importance, with a slightly higher value given to the Sustainable Management dimension (24%). These percentages vary in each pilot area, as some regions show an even stronger focus on the economic dimension (Andalusia 50%, Algarve 43%, Lazio and Calabria 33% each). In comparison, the socio-cultural dimension is more significant in the Slovenia pilot area (37%), which is considered the leading role of cultural routes.

On the other hand, the needs that were reported as "not covered" by the criteria belong mainly to the Environmental Sustainability dimension (35.3%), followed by Sustainable Management (23.25%), Socio-Cultural Sustainability (23.5%) and Economic Sustainability (17.6%) dimensions. After identifying the aspects that were missing from the criteria, each was analysed on a case-by-case basis for their inclusion in the model.

As part of the aspects reported as "not covered" by the model, carrying capacity was mentioned in several pilot areas as a critical measure to manage sustainable tourism along the routes. However, measuring carry capacity is a complex issue, as it is influenced by several socio-economic and environmental factors (ecological capacity, cultural capacity, social or psychological capacity, infrastructural capacity and management capacity, UNWTO, 2004).

Although the model does not have a single indicator entitled "Carrying capacity", several indicators have addressed the issue transversally. For instance, under the Sustainable Management dimension, the number of tourists on the routes and paths at any given time has been included (criterion 1.1.7 Visitor Management). Under the Environmental Sustainability dimension, the criteria related to the use of water, energy, amount of solid waste and available sewage system can all provide information on the pressure of tourism on the natural resources.

With respect to the step-by-step implementation framework, the findings from the Policy Learning Seminars support the model's proposal of establishing a working group and dividing the responsibilities for data collection between cultural route managers, the network of local actors related to the route and regional policy makers.

As part of this working group, most participants agree on the leading role of the cultural route managing body, as the one who should be in charge of filling out the self-assessment audit sheet, coordinating the data collection responsibilities and leading the implementation of an action plan based on the results. However, it was also recognised that cultural route managers do not normally have the competences and the mandate to perform the protagonist role expected. Moreover, the level of involvement of regional policy makers in cultural routes management changes across regions, some having specific responsibilities by law in terms of infrastructure development and management of trails. Therefore, the degree of responsibility of each actor within the working group is an aspect that remains to be established on a case by case basis. What it is clear though is that splitting the data gathering tasks is imperative, since it is a time- and resource-intensive activity and different actors have access to complementary information.

The suggestions provided by the Policy Learning Seminars' participants were accordingly incorporated into the model's criteria, indicators, and recommendations. Recognising the specificities of each pilot area, only the suggestions relevant to the majority of pilot areas were adopted, allowing particular concerns to be addressed locally, since the model provides a general framework that can be adapted to meet the individual demands of each pilot region.

Additionally, participants agreed that incentives are essential to encourage stakeholders to implement the model. Some of them include a certification or label, an online platform to benchmark results with other destinations, and capacity building to understand the importance of monitoring sustainability. Given the relevance of adequate incentives for the transition towards more sustainable destinations (Blanco et al., 2009; Williams & Ponsford, 2009), these suggestions were incorporated in the model's recommendations as part of the possible further improvement for the final version of the Med S&C Path model, which will be completed following the Testing Phase.

Overall, consulting local stakeholders was a critical phase in the development of the MED S&C Path model since it provided unique insights regarding the model's tools, feasibility, and appraisal. As well as a first-hand glimpse into the local context and the perspectives of stakeholders who would be immediately involved in the model's adoption. These forms of participatory methods have been viewed as useful and beneficial in the context of sustainability monitoring for tackling complicated sustainable development challenges as well as enhancing local development initiatives (Castellani & Sala, 2010).

CONCLUSION

Several conclusions emerged from the design and fine-tuning of the MED S&C Path model.

A toolbox with a step-by-step implementation methodology, such as the one proposed by the model, is an indispensable component of any sustainability monitoring scheme, to ensure the results from criteria and indicators are actually used in tourism management (Miller & Twinning-Ward, 2005). Moreover, by sharing the responsibilities for data collection and analysis, stakeholders benefit from a social learning experience, facilitating the understanding of complex policy problems such as sustainable development. This is known as the "conceptual role" of indicators, which is a precondition for their instrumental use in policy making processes (Gasparini & Mariotti, 2021).

The importance of adopting a multi-stakeholder approach to sustainability monitoring cannot be neglected, as it is impossible for one actor to be able to access all the needed information to comply with criteria. Precisely the intention of implementing the MED S&C Path model in a coordinated way, forming a working group, is to divide the responsibilities for data collection and analysis, so that it does not become a burden for one single actor, while at the same time contributes to improve dialogue and collaboration between public and private sector along the stretch of the route, supporting CR managers to have decision-making power and incidence on the management and impact of these paths.

Clearly, implementing a sustainability monitoring scheme is not free of challenges, related to time and funding constraints, stakeholders' engagement (Agyeiwaah et al., 2017), lack of technical expertise and political commitment (Med Sustainable Tourism Community, 2019; Zabetta et al., 2014). The multi-level governance of cultural routes is a further challenge, with issues related to uncoordinated interventions between local public administrations, poor coordination between tourism operators and lack of consultation and effective involvement of all relevant actors. These are some of the issues the model tackles, as a key aspect of BEST MED project is to contribute to better governance at the regional and transnational level, achieving higher cooperation and coordination between stakeholders at different levels.

In order to improve the management of tourist destinations and, in this case of cultural routes, it is paramount to measure and monitor the positive and negative impacts that tourism generates. Sustainability monitoring is an opportunity for cultural routes to remake themselves as more sustainable destinations, which can also be supported by participatory processes as recommended by the MED S&C Path model.

In the end, these kinds of multi-stakeholder participatory initiatives that bring together industry leaders, community stakeholders, and government agencies are useful to provide long-term directions for actions that are competitive and sustainable. Through the use of sustainability monitoring schemes, government agencies and institutions can therefore help support this transition by enacting greener policies, codes of conduct, and action plans, as well as by convincing stakeholders and tourists of the need to make a commitment to sustainable development (Williams & Ponsford, 2009).

Perhaps with the current COVID-19 pandemic and in the context of cultural routes that traverse mainly rural areas, "over tourism" does not seem a pressing issue, however the crisis has taught that a holistic tourism development model is needed, articulated with other economic activities, making sure destinations are first of all good places to live and therefore to visit.

The MED S&C Path model described in this chapter has been fine-tuned thanks to the feedback provided by BEST MED partners and participants in the Policy Learning Seminars across eight pilot areas. This can be interpreted as a first appraisal and validation of the model and its criteria, but not as an exhaustive evaluation of the model. In fact, a consequent assessment of the model on a more detailed level will be carried out in the following Testing Phase of BEST MED project and a final version of the model will be proposed at the end of the project (expected for June 2022).

Nevertheless, based on the results of the consultation process, we can already affirm that the model constitutes a useful tool to improve the multi-level governance of cultural routes. Thanks to its participatory approach the model responds to the concrete need of improving the monitoring and evaluation methods to assess the impacts cultural routes have on the economy and social tissue of territories, particularly in less known and rural areas.

RECOMMENDATIONS AND FUTURE RESEARCH DIRECTIONS

As a replicable tool, the MED S&C Path model constitutes an outstanding opportunity to measure the impact of cultural routes, pilgrimage trails and other types of paths in Mediterranean regions and beyond, while improving the multi-level governance, both vertically and horizontally, of transnational cultural routes.

The MED S&C Path model can be proposed to the European Institute of Cultural Routes, who is an associated partner of BEST MED, as an instrument for Cultural Routes of the CoE to comply with the 3-year evaluation to renew their certifications. Similarly, the Med Sustainable Tourism Community, as the coordinator of the network of Interreg MED projects concerning sustainable tourism (including BEST MED), can play a key role in the transferability and capitalisation of the model, ensuring the knowledge produced by the network is implemented beyond the life cycle of the projects.

As far as future research opportunities, it will be interesting to incorporate to the model innovative methods for data collection using big data sources, as well as finding synergies with existing monitoring systems and certifications, (i.e. Slovenia Green scheme) as the data already collected by these systems could be integrated in BEST MED platform.

ACKNOWLEDGMENT

The MED S&C Path Model has been developed by the Center for Advanced Studies in Tourism (CAST) research team: Cristina Bernini, Maria Laura Gasparini, Alessia Mariotti, Valeria Villalobos, as external expert of Lazio Region.

This study was conducted with the support of Interreg MED programme, as part of Work Package 3 (Study phase) of BEST MED project, which is co-financed by the European Regional Development Fund.

We would like to thanks all partners of BEST MED project for their contributions to develop the model: El legado andalusí, Andalusian Public Foundation (Lead Partner, Spain), Lazio Region (Italy), University of Algarve School of Management, Hospitality and Tourism (Portugal), Croatian Chamber of Economy, Rijeka County Chamber (Croatia), Ministry of Tourism, General Secretariat for Tourism Policy and Development (Greece), Conference of Peripheral Maritime Regions of Europe (France), National Tourism Organisation of Montenegro (Montenegro), Ministry of Tourism, Regeneration, Justice and Local Administration. Regional Government of Andalusia (Spain), Calabria Region (Italy), University of Maribor, Faculty of Tourism (Slovenia).

REFERENCES

Agyeiwaah, E., McKercher, B., & Suntikul, W. (2017). Identifying core indicators of sustainable tourism: A path forward? *Tourism Management Perspectives*, *24*, 26–33. doi:10.1016/j.tmp.2017.07.005

Bell, S., Eason, K., & Frederiksen, P. (2011). *POINT Policy use and influence of indicators*. Project co-funded by the European Commission within the Seventh Framework (2007-2013).

Berti, E. (2012). *Itinerari culturali del Consiglio d'Europa: tra ricerce di identità e progetto di paesaggio*. Firenze University Press. doi:10.36253/978-88-6655-142-3

Berti, E. (2013). Cultural Routes of the Council of Europe: New Paradigms for the Territorial Project and Landscape. *Alma Tourism*, *7*, 1–10.

Blanco, E., Rey-Maquieira, J., & Lozano, J. (2009). Economic incentives for tourism firms to undertake voluntary environmental management. *Tourism Management*, *30*(1), 112–122. doi:10.1016/j.tourman.2008.04.007

Briedenhann, J., & Wickens, E. (2004). Tourism routes as a tool for the economic development of rural areas-vibrant hope or impossible dream? *Tourism Management*, *25*(1), 71–79. doi:10.1016/S0261-5177(03)00063-3

Butler, R. W. (1999). Sustainable tourism: A state-of-the-art review. *Tourism Geographies*, *1*(1), 7–25. doi:10.1080/14616689908721291

Castellani, V., & Sala, S. (2010). Sustainable performance index for tourism policy development. *Tourism Management*, *31*(6), 871–880. doi:10.1016/j.tourman.2009.10.001

CERTESS. (2020). Retrieved from CERTESS documents: http://certess.culture-routes.lu/documents

Choi, H. C., & Sirakaya, E. (2006). Sustainability indicators for managing community tourism. *Tourism Management*, *27*(6), 1274–1289. doi:10.1016/j.tourman.2005.05.018

Co Evolve Deliverable Library. (2020). Retrieved from Co Evolve: https://co-evolve.interreg-med.eu/what-we-achieve/deliverable-library/

Coccossis, H., & Koutsopoulou, A. (2020). Measuring and monitoring sustainability of coastal tourism destinations in the Mediterranean. *Tourism: An International Interdisciplinary Journal, 68*(4), 482–498. doi:10.37741/t.68.4.8

Collins-Kreiner, N. (2016). The lifecycle of concepts: The case of 'Pilgrimage Tourism'. *Tourism Geographies, 18*(3), 322–334. doi:10.1080/14616688.2016.1155077

Collins-Kreiner, N. (2020). Pilgrimage tourism-past, present and future rejuvenation: A perspective article. *Tourism Review, 75*(1), 145–148. doi:10.1108/TR-04-2019-0130

Council of Europe. (2010). Resolution CM/Res(2010)52 on the rules for the award of the "Cultural Route of the Council of Europe" certification.

Council of Europe. (2011). *Impact of European Cultural Routes on SMEs' innovation and competitiveness*. Council of Europe.

Council of Europe. (2013). *Resolution CM/Res(2013)66*. Retrieved from Council of Europe: https://search.coe.int/cm/Pages/result_details.aspx?ObjectId=09000016805c69ac

Council of Europe. (2015). *Cultural routes management: from theory to practice. Step by step guide to the Council of Europe Cultural Routes*. Council of Europe.

Council of Europe. (2020). *Policies for the Danube region: Transnational cultural policies for the Danube Region (EUSDR)*. Council of Europe.

Council of Europe. (2020). Retrieved from Cultural Routes of the Council of Europe programme: https://www.coe.int/en/web/cultural-routes

Cult-Ring Project. (2020). Retrieved from Interreg Europe: https://www.interregeurope.eu/cult-ring/

DECRA. (2019). *Developing European Cultural Routes for All Erasmus Plus project*. French Federation of European Cultural Routes.

Destimed Deliverable Database. (2020). Retrieved from Destimed: https://destimed.interreg-med.eu/what-we-achieve/deliverable-database/

Di Giovine, M., & Elsner, J. (2016). Pilgrimage tourism. In J. Jafari & H. Xiao (Eds.), *The Encyclopedia of Tourism* (pp. 722–724). Springer., doi:10.1007/978-3-319-01384-8_146

Di Giovine, M. A., & Choe, J. (2019). Geographies of religion and spirituality: Pilgrimage beyond the 'officially' sacred. *Tourism Geographies, 21*(3), 361–383. doi:10.1080/14616688.2019.1625072

Emblematic Deliverable Library. (2020). Retrieved from Emblematic: https://emblematic.interreg-med.eu/what-we-achieve/deliverable-library/

EUROPARC Federation. (2020). Retrieved from European Charter for Sustainable Tourism: https://www.europarc.org/library/europarc-events-and-programmes/european-charter-for-sustainable-tourism/

European Commission. (2013). The European Tourism Indicator System. Toolkit for Sustainable Destinations. *European Union*.

European Commission. (2016). *The European Tourism Indicator System. ETIS Toolkit for sustainable destination management.* Luxembourg: European Union. Retrieved from https://ec.europa.eu/growth/sectors/tourism/offer/sustainable/indicators_en

Farrell, B. H., & Twining-Ward, L. (2004). Reconceptualizing Tourism. *Annals of Tourism Research, 31*(2), 274–295. doi:10.1016/j.annals.2003.12.002

Gasparini, M. L., & Mariotti, A. (2021). Sustainable tourism indicators as policy making tools: Lessons from ETIS implementation at destination level. *Journal of Sustainable Tourism*, 1–19. doi:10.1080/09669582.2021.1968880

Global Sustainable Tourism Council. (2020). Retrieved from Destination Criteria: https://www.gstcouncil.org/wp-content/uploads/GSTC-Destination-Criteria-v2.0.pdf

Green Destinations. (2020). Retrieved from Green Destinations Standard: http://greendestinations.org/wp-content/uploads/2017/11/Green-Destinations-Standard-1.4.2.pdf

Hezri, A. A., & Dovers, S. R. (2006). Sustainability indicators, policy and governance: Issues for ecological economics. *Ecological Economics, 60*(1), 86–99. doi:10.1016/j.ecolecon.2005.11.019

Hunter, C. (1997). Sustainable tourism as an adaptive paradigm. *Annals of Tourism Research, 24*(4), 850–867. doi:10.1016/S0160-7383(97)00036-4

Interreg Europe. (2020). *Policy brief on Cultural Routes in Europe.* Retrieved from https://www.interreg europe.eu/fileadmin/user_upload/plp_uploads/policy_briefs/2019-03-27_PB_Cultural_routes_final.pdf

Kato, K., & Progano, R. N. (2017). Spiritual (walking) tourism as a foundation for sustainable destination development: Kumano-kodo pilgrimage, Wakayama, Japan. *Tourism Management Perspectives, 24*, 243–251. doi:10.1016/j.tmp.2017.07.017

Lehtonen, M., Sébastien, L., & Bauler, T. (2016). The multiple roles of sustainability indicators in informational governance: Between intended use and unanticipated influence. *Current Opinion in Environmental Sustainability, 18*, 1–9. doi:10.1016/j.cosust.2015.05.009

Lozano-Oyola, M., Blancas, F. J., González, M., & Caballero, R. (2012). Sustainable tourism indicators as planning tools in cultural destinations. *Ecological Indicators, 18*, 659–675. doi:10.1016/j.ecolind.2012.01.014

Mariotti, A. (2012). Local System, Networks and International Competitiveness: From Cultural Heritage to Cultural Routes. *Almatourism, 3*(5), 81–95.

Med Cycle Tour Deliverable Library. (2020). Retrieved from Med Cycle Tour: https://medcycletour.interreg-med.eu/what-we-achieve/deliverable-library/

Med Sustainable Tourism Community. (2019). *Ensuring an effective monitoring of tourism sustainability in the Mediterranean Region.* Author.

Miller, G. (2001). The development of indicators for sustainable tourism: Results of a Delphi survey of tourism researchers. *Tourism Management, 22*(4), 351–362. doi:10.1016/S0261-5177(00)00067-4

Miller, G., & Twining-Ward, L. (2005). *Monitoring for a Sustainable Tourism Transition: The Challenge of Developing and Using Indicators*. CABI. doi:10.1079/9780851990514.0000

Ministero della Cultura. (2015). *Direttiva del Ministro 2016 - Anno dei Cammini d'Italia*. Retrieved from Ministero della Cultura: https://www.beniculturali.it/comunicato/direttiva-del-ministro-2016-anno-dei-cammini-d-italia

Mitomed Plus Deliverables Library. (2020). Retrieved from Mitomed Plus: https://mitomed-plus.interreg-med.eu/what-we-achieve/deliverable-library/

Moreno Pires, S., Fidélis, T., & Ramos, T. B. (2014). Measuring and comparing local sustainable development through common indicators: Constrainst and achievements in practice. *Cities (London, England)*, *39*, 1–9. doi:10.1016/j.cities.2014.02.003

Niavis, S., Papatheochari, T., Psycharis, Y., Rodriguez, J., Font, X., & Codina, A. M. (2019). Conceptualising Tourism Sustainability and Operationalising Its Assessment: Evidence from a Mediterranean Community of Projects. *Sustainability*, *11*(15), 1–18. doi:10.3390u11154042

Norfolk County Council & University of East Anglia. (2019). *Green Pilgrimage: A report on methodologies to measure the economic, social and environmental impact of pilgrimage*. Author.

Olsen, D. H., Trono, A., & Fidgeon, P. R. (2018). Pilgrimage Trails and Routes: The Journey from the Past to the Present. In *Religious PIlgrimage Routes and Trails. Sustainable Development and Management*. CABI. doi:10.1079/9781786390271.0001

Rametsteiner, E., Pülzl, H., Alkan-Olsson, J., & Frederiksen, P. (2011). Sustainability indicator development - Science or political negotiation. *Ecological Indicators*, *11*(1), 61–70. doi:10.1016/j.ecolind.2009.06.009

Richards, G. (2011). Cultural tourism trends in Europe: a context for the development of Cultural Routes. In *Impact of European Cultural Routes on SMEs' innovation and competitiveness* (pp. 21–39). Council of Europe.

Severo, M. (2017). European Cultural Routes: Building a Multi-Actor Approach. *Museum International*, *69*(1-2), 136–145. doi:10.1111/muse.12157

Star Cities Interreg Europe. (2021). *Sustainable tourism for attractivity of riverside cities*. Retrieved from Interreg Europe Star Cities: https://www.interregeurope.eu/starcities/

MED Sustainable Tourism Community. (2019). *Ensuring an effective monitoring of tourism sustainability in the mediterranean region*. Interreg Med programme.

MED Sustainable Tourism Community. (2019). *Policy Recommendations by the MED Sustainable Tourism Community*. Interreg MED programme.

Tanguay, G. A., Rajaonson, J., & Therrien, M.-C. (2013). Sustainable tourism indicators: Selection criteria for policy implementation and scientific recognition. *Journal of Sustainable Tourism*, *21*(6), 862–879. doi:10.1080/09669582.2012.742531

Timothy, D. J., & Olsen, D. H. (2006). *Tourism, Religion and Spiritual Journeys*. Routledge. doi:10.4324/9780203001073

Torres-Delgado, A., & López Palomeque, F. (2014). Measuring sustainable tourism at the municipal level. *Annals of Tourism Research*, *49*, 122–137. doi:10.1016/j.annals.2014.09.003

UNESCO. (2020). Retrieved from UNESCO Culture for Development Indicators (CDIS): https://en.unesco.org/creativity/sites/creativity/files/cdis_methodology_manual_0_0.pdf

UNESCO World Heritage Sustainable Tourism Toolkit. (2020). Retrieved from World Heritage UNESCO: http://whc.unesco.org/sustainabletourismtoolkit/

Università degli Studi di Perugia. (2018). Linee guida per l'adozione di strumenti e modelli condivisi di rilevazione dell'impatto economico, sociale ed ambientale dei flussi turistici relativi ai percorsi previsti dal progetto nazionale dei Cammini d'Italia. Perugia.

University of Maribor. (2021). *Deliverable 3.2.3. BEST MED Open Platform*. Interreg Med Programme.

UNWTO. (2004). *Indicators of Sustainable Development for Tourism Destinations: A Guidebook*. World Tourism Organization.

UNWTO. (2015). *Global Report on Cultural Routes and Itineraries*. UNWTO.

UNWTO INSTO. (2020). Retrieved from UNWTO INSTO: http://insto.unwto.org/

Williams, P. W., & Ponsford, I. F. (2009). Confronting tourism's environmental paradox: Transitioning for sustainable tourism. *Futures*, *41*(6), 396–404. doi:10.1016/j.futures.2008.11.019

Zabetta, M. C., Sacerdotti, S. L., & Mauro, S. (2014). Community-based monitoring in tourism sector: An application of the European Tourism Indicators System in the "ATL del Cuneese". *Leadership and Governance for Sustainable Tourism: Proceedings of Summer School, 2014*, 29–36.

APPENDIX: CRITERIA AND CORE INDICATORS

Table 4. Criteria and Core indicators

Dimension	Category	Criteria	Indicator Reference #	Core Indicators
Sustainable Management	1. Cultural Route Planning & Management	Stakeholders involvement	1	Percentage of stakeholders' representation from each community/sector in the route's management structure
		Trained staff	2	Percentage of staff and volunteer adequately trained in tourism, heritage and sustainability
		Route's strategic plan	3	Percentage environmental, social, cultural actions recommended in plan which have been implemented
		Region's sustainable tourism strategy	4	Percentage of the destination with a sustainable tourism strategy/plan with agreed monitoring, development control and evaluation arrangement
		Visitors management	5	Number of tourists on a trail, at one time, in a given time period or season, per year. For example, by measuring the number of credential/ stamps given by local tourist offices
			6	Level of satisfaction by tourists, including perception of cleanliness
	2. Quality of Infrastructure	Infrastructure condition	7	Number of new infrastructure facilities developed related to the management of the Cultural Route
	3. Information & Promotion	Visitor information	8	Number of tourist information offices per tourist
			9	Number of public organisations dedicated to tourism which have included the route in their promotional material
Economic Sustainability	1. Economic contribution to local economy	Tourism flows (volume and value)	10	Number of tourists nights per month
			11	Daily spending per overnight visitor (accommodation, food and drinks, other services)
			12	Average length of stay of tourists (nights)
		Supporting local entrepreneurs	13	Number of SMEs in the tourism, culture and creative industries sector
		Employment and career opportunities	14	Direct tourism employment as percentage of total employment in the destination
	2. Seasonality	Tackling seasonality	15	Percentage of all occupancy in peak quarter or month
Socio-Cultural Sustainability	1. Preservation of Cultural Heritage	Protection of cultural heritage	16	Number of policies, strategies, action plans and tools in place to protect cultural heritage
	2. Accessibility	Accessible facilities	17	Percentage of attractions offering alternative access for those with mobility concerns
	3. Residents engagement & feedback	Residents consultation	18	Residents level of satisfaction with tourism is monitored through surveys and other systematic feedback mechanisms
	4. Gender equality	Gender equality	19	Women/men as a percentage of all formal tourism employment
Environmental Sustainability	1. Resource Management	Solid waste management	20	Percentage of destination area covered by solid waste collection services
		Sewage treatment (wastewater management)	21	Percentage of sewage from the destination treated to at least secondary level prior to discharge
		Water management	22	Water use: (total volume consumed and litres per tourist per day)
		Energy consumption	23	Percentage of businesses participating in energy conservation programs, or applying energy saving policy and techniques
		Sustainability certifications	24	Percentage of tourism enterprises using a voluntary certification/ label for environmental/quality/ sustainability or Corporate social Responsibility along the route
	2. Sustainable mobility	Low impact transportation	25	Percentage of tourists and same-day visitors using soft mobility/public transport services to arrive to and get around the destination
	3. Landscape & Biodiversity protection	Wildlife	26	Number of rules regulating activities such as hunting, fishing, etc.

Chapter 7
A New Framework for Tourism Sustainability and Its Prototyping in Pilot Areas:
Insights From BEST MED Testing Phase

João Carlos Martins
CinTurs, University of the Algarve, Portugal

Claudia Ribeiro de Almeida
https://orcid.org/0000-0002-0790-3152
CinTurs, University of the Algarve, Portugal

Alexandra Gonçalves
https://orcid.org/0000-0003-3796-1801
CinTurs, University of the Algarve, Portugal

ABSTRACT

This chapter highlights the research-action practices developed under the Phase 4 of the Interreg BEST MED Project (Beyond European Sustainable Tourism Med Path, Grant Number: 5987), "Testing." In this project, stage was determinant to prototype and test the best practices towards sustainability, developed by the project's partners, detailing its achievements towards Mediterranean tourism sustainability, resulting from the ground-level operationalization of a theoretical and methodological framework. Here stakeholder participation is determinant to achieve breakthrough advances on more enduring and resilient touristic territories, addressing seasonality and lack of cooperation among tourist actors. This data gathering was developed among public officials, private operators, academics, third sector associations, and citizens in eight Mediterranean areas. Despite the common sustainable and cultural path model, different outputs were found from the participation and auscultation practices.

DOI: 10.4018/978-1-7998-9217-5.ch007

INTRODUCTION

The BEST MED project is a joint initiative of eight Mediterranean areas, developed by academic partners, national public authorities, regional governments, destination management organizations, business associations, public foundations, and a network of other organizations towards the achievement of a shared model of Mediterranean Governance (Bramwell & Lane, 2011; Hall, 2011; Caffyn, & Jobbins, 2003). Its main challenges are to address seasonality and lack of effective cooperation among main tourism actors, promoting a shared and integrated tourism approach, to ensure sustainability principles towards planning and management. The project's main expected result is to achieve tourism sustainable development in the improvement of existing cultural routes and neighbouring territories, entailing the importance of cultural heritage, creativity, and communities' innovation, while preserving the natural environment.

All these processes are developed under the participation and engagement of local stakeholders (Thees et al., 2020; Bichler, 2019, Bichler, & Lösch, 2019), being determinant to analyse their commitment towards the processes of consultation and decision-making around the tourism phenomenon. This awareness is fundamental to address the existence of possible anti tourism discourses and feeling of overtourism by residents (Carreira et al., 2021; Pham et al., 2021; Koens et al., 2018), avoiding the perception that they are taken away from tourism economic and social benefits, and that tourists are more important to policy makers than the local community where tourism occurs.

The team from University of Algarve, as well as other BEST MED pilot areas, present a research-action methodological attitude, being responsible for the compliance of all testing processes in the project. It stands as a common strategy towards the development of all pilot activities, despite the local differences of the partners (which must be considered), as testing grounds for innovation, spaces for experimentation and learning, open to new ideas and approaches. Resulting from a network of local stakeholders, they create open and interactive spaces. The final objective is to Scale up after the piloting, (Hughes et al., 2018; van Popering & van Buuren, 2017; van Buuren & Loorbach, 2009; Vreugdenhil, 2010), to test shared methodologies of operationalization to the future, creating role models for forthcoming similar cases, forms of policy transferability, in fact, test temporary, ground-level achievements, and scientific outputs resulting from prototyping to long term answers. In this particular case, an effective and sustainable tourism governance for Sun and Sea Mediterranean tourism offers, with ground-level achievements in the improvement of existing cultural routes of the project's pilot areas.

The BEST MED project was thought to address the traditional tourism seasonality and lack of cooperation among tourism actors, that influences the coastal areas of the Mediterranean. These dysfunctional elements of the tourism phenomenon in Costal Mediterranean areas, are not new, they are at the foundations of the development of the processes of leisure and space commodification (Karamustafa, & Ulama, 2010; Obrador et al., 2009; Richards, & Wilson, 2006; Rudihartmann, 1986) occurred in these areas. But, despite the presence of several elements of unsustainability, addressed by local actors, academics and as well as several policy makers, before the present COVID-19 pandemic, the tourism phenomenon was being developed under a mass tourism perspective, ending in processes of overtourism on several spaces of the seaside Mediterranean. So, the initial objectives of the BEST MED project, were strongly framed by this mass tourism offer and its main objectives were outlined under an overtourism and seasonality theoretical framework.

The several confinement processes occurring all around the world, as well as a generalized doubts among tourists about the compliance of health measures, has brought not only a reduction on tourism travel and consume, a strong development of national markets, but also a spectrum of uncertainty towards the future of the tourism phenomenon "where the framing of the global tourism system moved from overtourism (…) to nontourism" (Gössling et al., 2020, pp .2). So, it was clear, among the pilot project promotors, but also among local stakeholders, that BEST MED project was a great opportunity not only to address the traditional seasonality and lack of cooperation among tourism actors, but to ensure a better, resilient, and enduring tourism offer for the future. Was seen as a tool to address the future economic, social and cultural recovery that tourism territories and their populations were experiencing. So, the main objective of this chapter is to detail and analyse the procedures held on each pilot area to operationalize the BEST MED Sustainable and Cultural (S&C) Path Model, with the use of participative methodologies, resulting from the local ecosystem of stakeholders: Focus Group session and 1 Day Living lab. More than a simple description of the activities held, we will analyse the partners results in light with the S&C Path Model and its achievements towards Tourism Sustainability in Mediterranean areas. Finally, to create an analytical framework for other territories, where the tourism phenomenon is determinant for the quality of life of local communities. Simultaneously to offer new solutions for the present uncertain context of the tourism phenomenon in the Mediterranean.

BACKGROUND

The main task of the Phase 4 of the BEST MED project, was developed between January and November 2021. Was developed by eight BEST MED partners, under the supervision of the team from University of Algarve, in several territories to perform advances on tourism sustainability, prototyped in the Testing Phase of the BEST MED project, promoting achievements and best practices towards Mediterranean Governance resulting from stakeholders' participation. In total ten routes analysed (Table 1).

S&C Path Model and 4 Dimensions of Sustainability

The BEST MED project developed its own Theoretical Framework, the S&C Path Model. It results from data gathering from Tourism Destination Management Organizations, other Mediterranean Tourism Sustainability projects as MITOMED Plus (2020) and DESTIMED (2020), as well as international boards on tourism as the Global Sustainable Tourism Council (2020), Green Destinations (2016) or the European Tourism Indicator System (2016). The aim of the model, besides monitoring sustainability, is to contribute to better decision making, by route's managers and regional policy makers. Evaluating positive and negative impacts, is desirable to achieve four elements of sustainability: management, social, economic, and environmental sustainability.

Starting as an open theoretical and methodological framework, the S&C Path Model was enriched with the results of the BEST MED testing phase. Learning from ground level participation and past experiences and projects developed in these tourism-oriented sites, "not starting from scratch"; it wishes to highlight its best achievements, as well as to perform innovation answers to the present fragile tourism sector. As an open model, the first task of the BEST MED partners was to discuss its adequacy towards tourism sustainability. These criteria and indicators are accomplished gathering the referred data and by this way achieve a specific percentage of compliance, which will detail the sustainability level of each region. The final model is composed of a self-assessment audit sheet to measure the percentage of compliance with the criteria; a set of indicators, a benchmarking method, and to ensure sustainability in the future and data dissemination, an online platform.

Table 1. Country, Name of the Route, Pilot Area promotor and Nature of the promotor of the BEST MED project pilot areas

Country	Name of the Route	Pilot Area Promotor	Nature of the Promotor
Spain	*La Ruta de las Alpujarras - Routes of El legado andalusí*	*El legado andalusi* (Leader Partner)	Regional Public Foundation
Italy – Lazio	*Vie Francigena del sud*	Lazio Region	Regional government
Portugal	*Rota da Estrada Nacional 2*	University of Algarve	Academic
Croatia	*Iter Vitis* Route – *Kvarner*	*Rijeka* County Chamber	Government
Greece	Route of the olive tree	Ministry of Tourism of Greece	Government
Slovenia	*Iter Vitis* Route – Slovenia	University of Maribor	Academic
Slovenia	European Cemeteries Route – Slovenia	University of Maribor	Academic
Slovenia	Saint Martin of Tours Route	University of Maribor	Academic
Italy – Calabria	Cycle Route of the Parks	Calabria Region	Regional government
Montenegro	*Illyricum Trail* of the Roman Emperors and Danube Wine Route	National Tourism Organization of Montenegro	Government

Source: BEST MED project internal reports

In total five steps were considered to develop the model:

1. ***Establish a working group***. with the objective of establishing relations among the local ecosystem of stakeholders, discussing the existing pilot areas resources, towards its participation and engagement, and setting tangible sustainability goals. It should be composed of regional policy makers, route's managing body and other relevant stakeholders as local municipal officials, local DMO's (Destination Management Organizations), activists and local associations, tourism operators and academics.
2. ***Understand the pilot area current sustainability level***. For that purpose, will be used a Self-Assessment Audit Sheet, with a set of dimensions, variables, and indicators that the working group must respond with Yes/No and assess the existing levels of pilot area sustainability. It is a qualitative evaluation and will provide a quantitative compliance level according to the criteria.
3. ***Analyse the existing data***. After the completion of the Self-Assessment Audit Sheet, is necessary to make an interpretation of the results, and establish a realistic list of priorities and goals in the pilot areas.
4. ***Address the gaps between the model and the local needs of a specific territory.*** Simultaneously to the analysis of the sustainability compliance of the regions with the criteria, is fundamental to enrich the model with local inputs, the contextual indicators that are not contemplated in the model and create a better role model system for the future.
5. ***Monitor and evaluate the sustainability levels***, using an online platform and providing this data for other touristic regions and areas.

So according to the projects' Road Map, the different pilot areas developed a set of different participation processes. Initially, all partners developed a set of interviews and informal contacts with academics, tourism operators and public officials with responsibilities in the areas of tourism and culture. In the testing phase was determinant to develop a series of events with public officials, from cultural, tourism, and municipal organizations, tourism operators, third sector associations and residents. But before discussing the participation practices developed in the last steps of the Phase 4 of the BEST MED project, this chapter will expose some important elements of the State of the Art of this chapter, the Background underlining the research undertaken.

COVID-19 Effects

The tourism approach proposed by Best Med Project is undoubtedly affected by the present health, social and economic crisis resulting from the present pandemic caused by COVID-19. The most prominent research resulting from COVID-19 effects on Tourism (Yeh, 2021; Sigala, 2020; Gössling et al., 2020; Benjamin et al., 2020; Galvani et al.,2020; Romagosa, 2020) has shown unprecedent effects towards on this phenomenon, between a new deal to perform a more resilient, sustainable, socially equitable tourism in the future, or just the continuation in the future of major precedent negative tourism effects, particularly on local communities.

One of the first consequences is the perception that is determinant to develop more local travel experiences. The existing data from Phase 4 Testing Phase of the BEST MED project, that the chapter will develop in the results section, shows that national markets will be determinant on the economic recovery, a stop on the travel democratization of international lower middle classes, a demand for more security and health conditions, a return to certain elements of the past, namely the activity of private Travel Agencies and charter flights. Second, and more profoundly, that the pandemic has brought an unprecedented spectrum of social and economic crisis as an economic cataclysm, related to the decrease on tourism receipts with long term effects, namely on tourist arrivals and receipts, but also in traditional indicators as unemployment rate, overnight stays, hosts, or the number of establishments dedicated to the accommodation of tourists. Third, the COVID-19 pandemic highlighted preexisting touristic elements of low-cost international travel dependency, ecological unsustainability, seasonality, over tourism and precarious workforce, so this can be an opportunity to promote a more sustainable and resilient tourism model to the future, an opportunity to rethink and promote a more local, sustainable, and resilient tourism phenomenon. Finally, a more sceptic vision, that the present uncertain contingency is just a temporary stop on the elements of unsustainability, traditionally associated with the tourism phenomenon, and that in the nearest future the same kind of travel and destinations offers will return to its pre-pandemic times.

The last report from May 2021 from the World Tourism Barometer from UNWTO states important elements of tourism resulting from these unprecedented uncertain times, comparing the years of 2019 and 2020, considering the World, Europe, Southern/Medit. Europe, and the BEST MED countries (Portugal, Spain, Croatia, Slovenia, Italy, Greece, and Montenegro) (Table 2).

The report highlights some interesting assumptions. The year of 2020, according to 2019, show us how the pandemic affected all territories, but was particularly harsh on the Southern Europe-Mediterranean region, with a reduction by 71,1% on arrivals (266,2 million) and less 67,6% on receipts (158465 million USD). It was particularly important in traditional seasonal tourism destinations as Spain with a reduction by 77,3% on arrivals (64,5 million) and less 76,8% on receipts (61231 million USD) or Greece with a reduction by 77% on arrivals (24,1 million) and less 75,8% on receipts (15418 million USD).

Table 2. Changes in International Tourist Arrivals and International Tourism Receipts between 2019 and 2020 (%)

Regions and Countries	International Tourist Arrivals by (Sub)region and Selected Countries and Territories of Destination	International Tourism Receipts by (Sub) region and Selected Countries and Territories of Destination
World	-72,8%	-66,9%
Europe	- 68,5%	-59,1%
Southern/Mediterranean Europe	-71,1%	-67,6%
Portugal	-73,7%	-56,8%
Spain	-77,3%	-76,8%
Croatia	-68,4%	-52,6%
Slovenia	-74,1%	-59,9%
Italy	-60,9%	-60,1%
Greece	-77,0%	-75,8%
Montenegro	-86,0%	-86,6%

Source: UNWTO World Tourism Barometer, Statistical Annex, May 2021

Cultural Tourism and Routes

The contemporary development of Cultural Routes is a significant achievement on the reestablishment of the founding steps of the creation of the tourism phenomenon, recreating tourism offers where cultural heritage, creativity and communities' innovation, ethnic and cultural encounter between visitors and travellers is their main goal. Its main motivation could be related to pilgrimage and religious objectives, reactivating former historical routes as the *Camino de Santiago* a significant cultural route with specific paths in Portugal, France and other areas of Spain towards the North of the Iberian Peninsula in the Spanish region of *Galicia,* sport activities as trekking or cycling, with a strong relation with nature, commercial purposes highlighting local products, and finally cultural reasons, from cultural heritage to artistic practices, promoting a closer connection with local's identity and distinctiveness, all seen as authentic, local and in strong connections with community traditions. Some of them are certified, as the ones accredited by the Cultural Routes of the Council of Europe, or just integrated in regional and national tourism and cultural promotion agendas. They can have a local, regional, national, or even international coverage, re-activating cross-frontier connections.

Under the present COVID-19 pandemic, the authors of this chapter, supported by the auscultation practices held under BEST MED Testing Phase, defend that the development of these forms of Cultural Tourism can be a determinant for economic recovery. This new strategy of cultural tourism is supported by the Word Tourism Organization of the United Nations (UNWTO, 2020), seeing this uncertain moment to revaluate former tourism approaches and to promote a new and resilient tourism view:

1. To support a tourism shift from quantity towards quality, not only changing the ways that tourism has been measured only in statistic forms (number of visitors, income, number of created jobs), but also in a qualitative way, understanding the in-depth effects of tourism in local communities.

At the same time to promote specific communication and marketing practices that highlight the significance and particularities of each tourism area.

2. To diversify the existing cultural tourism products, that, for social distance protocols and heath measures, must ensure safety for travellers, creating alternative tourism products that promote each area culture and identity, but under new safety regulations. These new products, under safety and health protocols can address some of the concerns of the tourists, ensuring that the tourism product is in fact secure for all.
3. To boost community participation and domestic tourism, promoting the engagement between hosts and gests, promoting a more harmonious and community-based experiences in leisure, particularly towards experienced tourists as the national market, which want to know more about their own culture.
4. To customize cultural offers for international visitors, ensuing the same steps taken in the third step, but with an effort to respond to non-national visitors.
5. To promote cultural tourism entrepreneurship and innovation, addressing the economic outcomes resulting from COVID-19 pandemic and supporting small and medium enterprises, cooperatives, and forms of creative and innovative economy, addressing job informality, particularly among women, young people and ethic minorities, the ones more affected by the economic outcomes of the pandemic.
6. To ensure tourism accessibility for all, an issue that was fundamental before this uncertain tourism future.

Cultural routes as structuralized, planned, and integrated itineraries, are innovative tourism products for deprived and depopulated territories, as well as forms of tourism diversification of traditional seasonal touristic destinations. In fact, a significant driver of management, economic, socio-cultural, and environmentally sustainable development.

The development of Cultural Routes is a main feature of the emergence of Cultural Tourism (Richards, 2021, 2018), operating the touristification of former industrial/agricultural areas, but mainly as forms of tourism diversification of long-term areas inflected by leisure travelling. Simultaneously is presented as a process of commodification of local cultural and identities, the reproduction of the tourism offers resulting from globalization, with the creation of new tourism markets and the symbolic competition among them. Initially the idea was to create a ground level, a contextual tourism offers who could differentiate themselves from competitors, but integrated in major tourism circuits, there is a danger of continuous massification of these tourism offers, in fact not differencing totally as supposed to. The response must be developed under a creativity and innovation framework (Richards, & Wilson, 2006). The response to this second level of commodification, after the space commodification of the Sun and Sea tourism offers, and the first cultural tourism wave is to develop models of participation of local communities, creating a creative and innovative tourism model, participated, and engaged by residents, where they have a word to say, avoiding non-authentic tourism offers. Here is fundamental to pass from a Tangible Cultural Heritage appropriation towards more intangible elements of local territories, engaging in narratives, discourses and the community's own vision of their history and identity (Richards, 2011).

Moving away from massified and west European XX century travelling behaviours, based on the attempt to relax in middle class-oriented seaside areas, Cultural Tourism presents itself as a renovated interest towards forms of local and with local's tourism. It asks for a more authentic experience and not recreated by tourism operators based on simplified views of visited communities. It performs cultural heritage reappropriations (Martins, 2020), from tangible inheritance to intangible legacy, where contemporary storytelling, entertainment and consumption are determinant. In fact, performing a cultural realignment to address tourism seasonality (Procopiak et al., 2020; Cabeça et al., 2019).

The promotion of Cultural Routes is a determinant on the development of Scientific Research, which provide an academic certification and validation of the actions proposed. In this case, Bozic & Tomic, 2016, present a set of scientific indicators to evaluate the Cultural Routes, promoting an academic agenda for Tourism areas namely: (i) Its cultural value, particularly towards its connection to local culture of the region, country or wider territories; (ii) the historical value of the route, namely the capacity to promote a significant relation with territorial past; (iii) its artistic value, promoting uniqueness, authenticity; (iv) its social value, indorsing local communities and sense of place attachment, (v) the educational and research values, with its potential to create knowledge useful to other tourism contexts and ensure future sustainability and finally (vi) aesthetic value, highlighting the visual qualities of a specific tourism oriented territory.

Some roads have been seen as potential elements of touristification like de Route 66 in the United States, where several local grassroot associations (Dedek, 2007) tried to highlight its importance. These roads are a role model for other touristic roads, being a branding and iconic historical object (Carden, 2006), an element of authenticity (Wood, 2007), but also a nostalgic and commodified form of heritage tourism (Caton & Santos 2007). Other routes have been used as a tourism product, where centralized management occur as in *Ruta de la Plata* in Spain (Gomez & Pereiro, 2017) or *Ruta Nacional 40* in Argentina (Ferrari et al., 2020) or decentralized as in National Highway 1 in Australia (Kerr, 2018).

These Cultural Routes, being promoted by reduced and temporary funding, must ensure the future sustainability of the actions promoted, engage the local community of institutional stakeholders and local associations, towards a more durable and supported tourism offer.

BEST MED Cultural Routes

As a major fieldwork output of the BEST MED project, towards the development of a sustainable development of Mediterranean touristic areas, the prototyping process in the pilot areas was based in the S&C Path Model towards the monitoring of sustainability indicators, the future consolidation of tourism offers, namely a Route itinerary. To have a more contextual vision about the routes, we present Table 3, where we can have access to specific elements of each BEST MED Route itineraries.

Not being mandatory, the previous working package of the BEST MED project, the study phase, developed between March 2020 and March of 2021 a specific criterion, to be used in the choice of the pilot areas, in order to achieve higher levels of tourism sustainability and comply with the BEST MED S&C Path Model:

1. *To be certified as a Council of Europe cultural route*. Only Cycle Route of the Parks (Calabria) and Rota da Estrada Nacional 2 (Portugal) are not certified under the European program.
2. *To have an international dimension, crossing different countries and establish cross frontier relations*. We will present in the next table the BEST MED cultural routes with presence in several countries, creating relations and synergies among these countries (Table 4).

Table 3. BEST MED Pilot Area, Route itineraries, length, partners, themes, and typology of activities

Pilot Area	Route Itinerary	Length	Main Theme of the Itinerary	Typology of Activities Included in the Route Itinerary
Spain	*La Ruta de las Alpujarras* - Routes of *El legado andalusí*	432,5 km	Cultural Heritage and Natural Resources	Cultural and art events, sport activities; camping and Gastronomy
Italy – Lazio	*Vie Francigena del sud*	198 km	Religious, Pilgrimage and Cultural itinerary	Cultural visits, outdoor activities and gastronomy
Portugal	*Rota da Estrada Nacional 2*	10 km	Cultural Heritage, Gastronomy, Motorcycle and Caravanning Road, Sightseeing, Natural Resources	Cultural and religious activities, gastronomy, sport activities, sightseeing and City tours
Croatia	*Iter Vitis* Route – *Kvarner* Wines Route	432,5 km	Viticulture, Vines and Landscape Sightseeing	Wine tasting and sightseeing
Greece	Route of the olive tree	12 km	Culture of the olive tree Natural; Cultural Resources	Thematic cultural itinerary
Slovenia	*Iter Vitis* Route – Slovenia	300 km	Viticulture, Vines and Landscape Sightseeing	Wine tasting and Sightseeing
Slovenia	European Cemeteries Route – Slovenia	166 km	Cemetery heritage	Education and sightseeing
Slovenia	Saint Martin of Tours Route – Slovenia	470 km	Religious, Pilgrimage and Cultural	Pilgrimagesm hikes and Education
Italy – Calabria	Cycle Route of the Parks	545 km	Cultural and Natural Resources	Cycling and sightseeing
Montenegro	*Illyricum* Trail of the Roman Emperors and Danube Wine Route	1200 km	Roman heritage / Culture and Gastronomy	Visits to Museums, cities, gastronomy and enology tours, active tourism and cycling

Source: BEST MED project internal reports

Table 4. BEST MED Cultural Routes with international dimension

Cultural Route	Countries
Iter Vitis	Slovenia and Croatia, Armenia, Azerbaijan, Cyprus, France, Georgia, Greece, Israel, Italy, Luxembourg, Montenegro, Portugal, Republic of Moldova, Republic of North Macedonia, Romania, Russian Federation, Spain, Tunisia, and United Kingdom
European Cemeteries Route	Slovenia, Austria, Bosnia and Herzegovina, Croatia, Denmark, Estonia, France, Germany, Greece, Ireland, Italy, Netherlands, Norway, Poland, Portugal, Romania, Serbia, Spain, Sweden, and United Kingdom
The Saint Martin of Tours Route	Slovenia, Belgium, Croatia, France, Germany, Hungary, Italy, Luxembourg, Netherlands, and Slovak Republic
Illyricum Trail of the Roman Emperors and Danube Wine Route	Montenegro, Croatia, Bosnia and Herzegovina, Albania, North Macedonia, Serbia, Hungary, Romania, and Bulgaria

Source: Cultural Routes of the Council of Europe (official site)

Figure 1. Vie Francigena del Sud (Italy -Calabria) and Route of the olive tree (Greece).
Source: BEST MED internal report

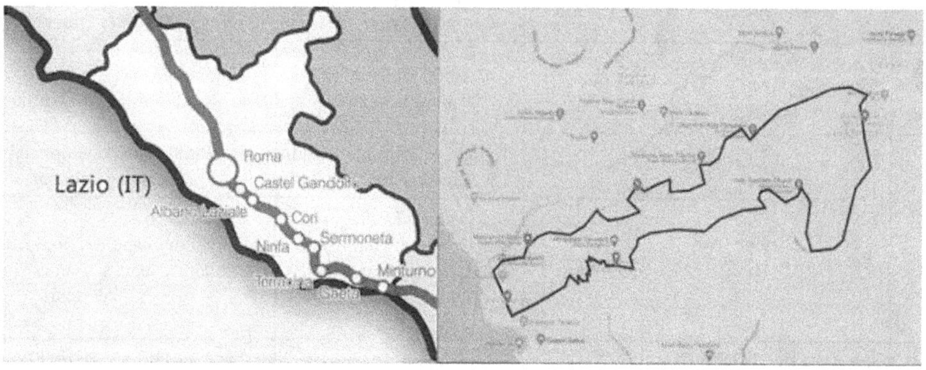

3. *Preferably to be a physical path, establishing a specific itinerary.*

   a. The first one is linear, continuous, resulting from historical paths and infrastructures, as former Pilgrim routes used in the past for travel, creating today, under tourism occupation, relationships among distant territories and connecting cultural assets as castles, buildings, sightseeing spots. That is the case of *Vie Francigena del Sud*, Route of the olive tree and Cycle Route of the Parks (Figures 1 and 2)

b. Others are set as territorial, involving several points in a specific area, aggregated by a common theme, being created a series of paths or itineraries to explore the different parts of the proposed route. That is the case of *La Ruta de las Alpujarras - Routes of El legado andalusí*, *Rota da Estrada Nacional 2*, *Iter Vitis* Route – *Kvarner* Wines Route and Slovenia, and Saint Martin of Tours Route – Slovenia (Figures 3,4,5 and 6).

c. Finally, Reticular pattern (archipelago) or network, composed of distant points, at different scales, but sharing a common theme. This is the case of European Cemeteries Route – Slovenia or *Illyricum* Trail of the Roman Emperors and Danube Wine Route (Figure 7 and 8).

4. *Should be appealing* (on its management, economic, socio-cultural, and environmental dimensions) to a diversity of stakeholders promoting engagement and participation, as well as its future endurance and resilience. The diversity of stakeholders is determinant to see how much participated the discussions around these cultural routes are. For that matter it was created a working group, to develop the future participation arenas, the BEST MED Focus Groups and Living Labs. The members of these participation arenas are mostly composed of Public Regional bodies, on the areas of tourism, planning and culture, as well as managers and officials of these cultural routes, as well as municipal bodies of the cities where these paths are located. It is clear also the presence of academics, with their knowledge in the areas of tourism sustainability from the universities of

Figure 2. Cycle Route of the Parks (Italy - Calabria).
Source: BEST MED internal report

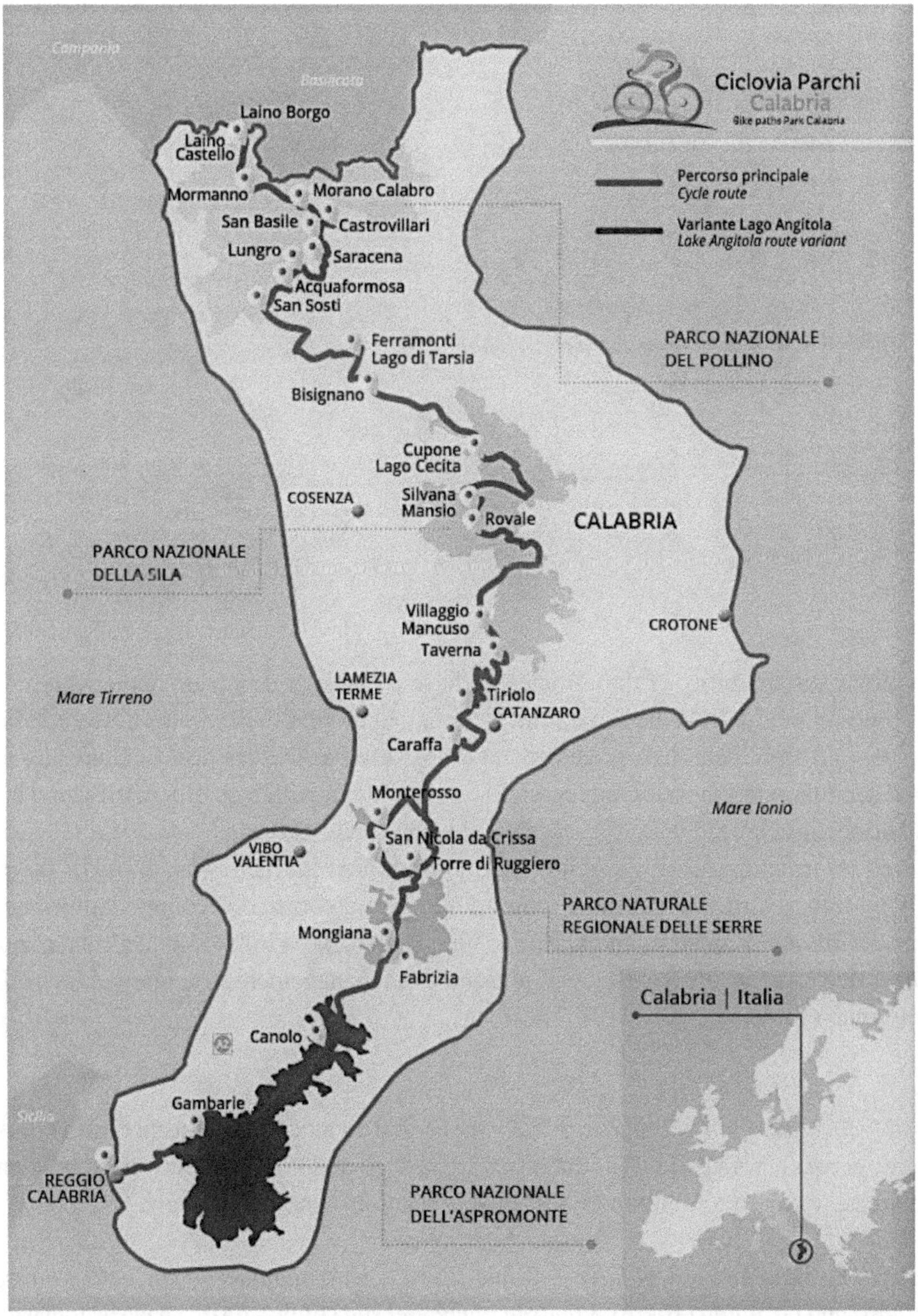

Granada, Bologna, Algarve, Rijeka, and Maribor. It is also clear the lower presence of the private sector, as well as local associations and residents.

5. *To be structured as possible, with a governance board or coordinator with management, dissemination, promotion, and commercialization of the route responsibilities.* That is a fundamental step

Figure 3. La Ruta de las Alpujarras - Routes of El legado andalusí (Spain).
Source: BEST MED internal report

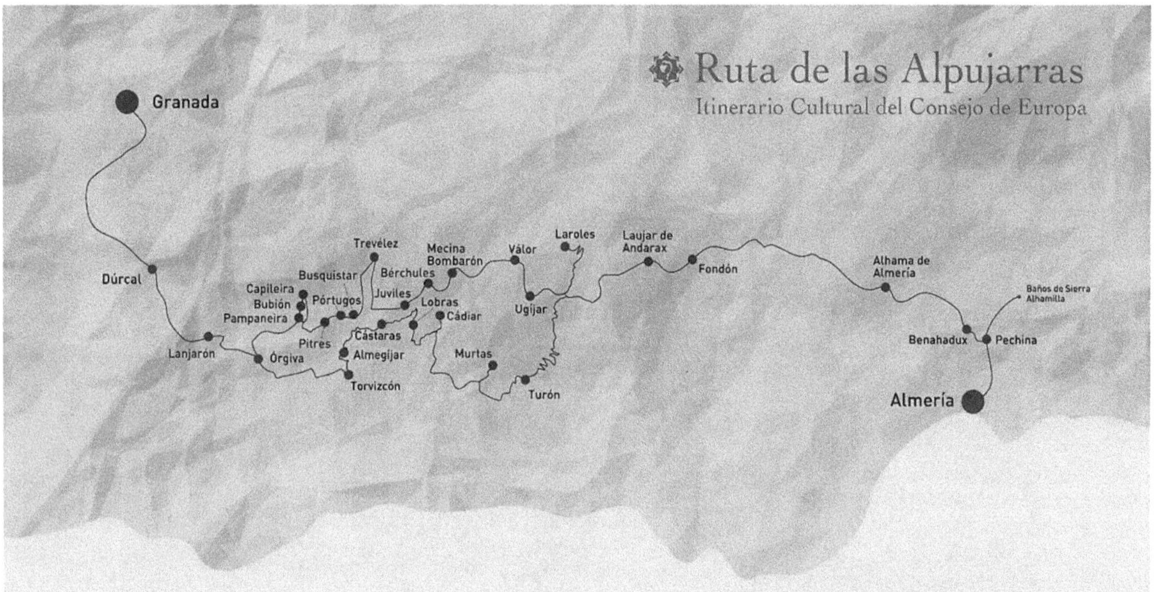

of the future sustainability of these routes, to have a manager or a team dedicated to the success of this tourism product. At the same time is fundamental that these routes have social media and a website, to disseminate their products and tourist experiences. Is also fundamental to have associated accommodation, and other services, most of them with a quality certificated by the route coordination, as occurred with *Rota da Estrada Nacional 2* (Portugal). As well, to have good and structured information about the route and the pilot area tourism offer. Most of the pilot areas comply with these criteria, but three routes don´t have associated accommodation The *Illyricum Trail* of the Roman Emperors and Danube Wine Route, *Iter Vitis* Route in Croatia and Slovenia and Olive tree route in Greece. Below is presented the management structures of each BEST MED cultural route (Table 5).

6. *Preferably located in hinterland areas*, but with a good connection to coastal destinations, creating synergies between both areas. All itineraries are next to major touristic hotspots, but connected to inland areas, trying to develop more sustainable relations among these areas.

In the case of *La Ruta de las Alpujarras* - Routes of *El legado andalusí* (Spain), the nearest spots are *Almeria*, (40km) and *Granada* (16km). *Vie Francigena del sud* (Italy – Lazio) have as tourism hotspots. *Rota da Estrada Nacional 2* (Portugal) in the Algarve, have *Quarteira* (20km); *Albufeira* (35km); *Portimão* (70km) has major seaside tourist spots. Route of the olive tree (Greece) is close to the Southern *Peloponnese* (100km). Cycle Route of the Parks (Italy – Calabria) is located next to *Costa dei Cedri and Costa dei Gelsomini,* (around 30km). *Illyricum* Trail of the Roman Emperors and Danube Wine

Figure 4. Rota da Estrada Nacional 2 (Portugal).
Source: BEST MED internal reports

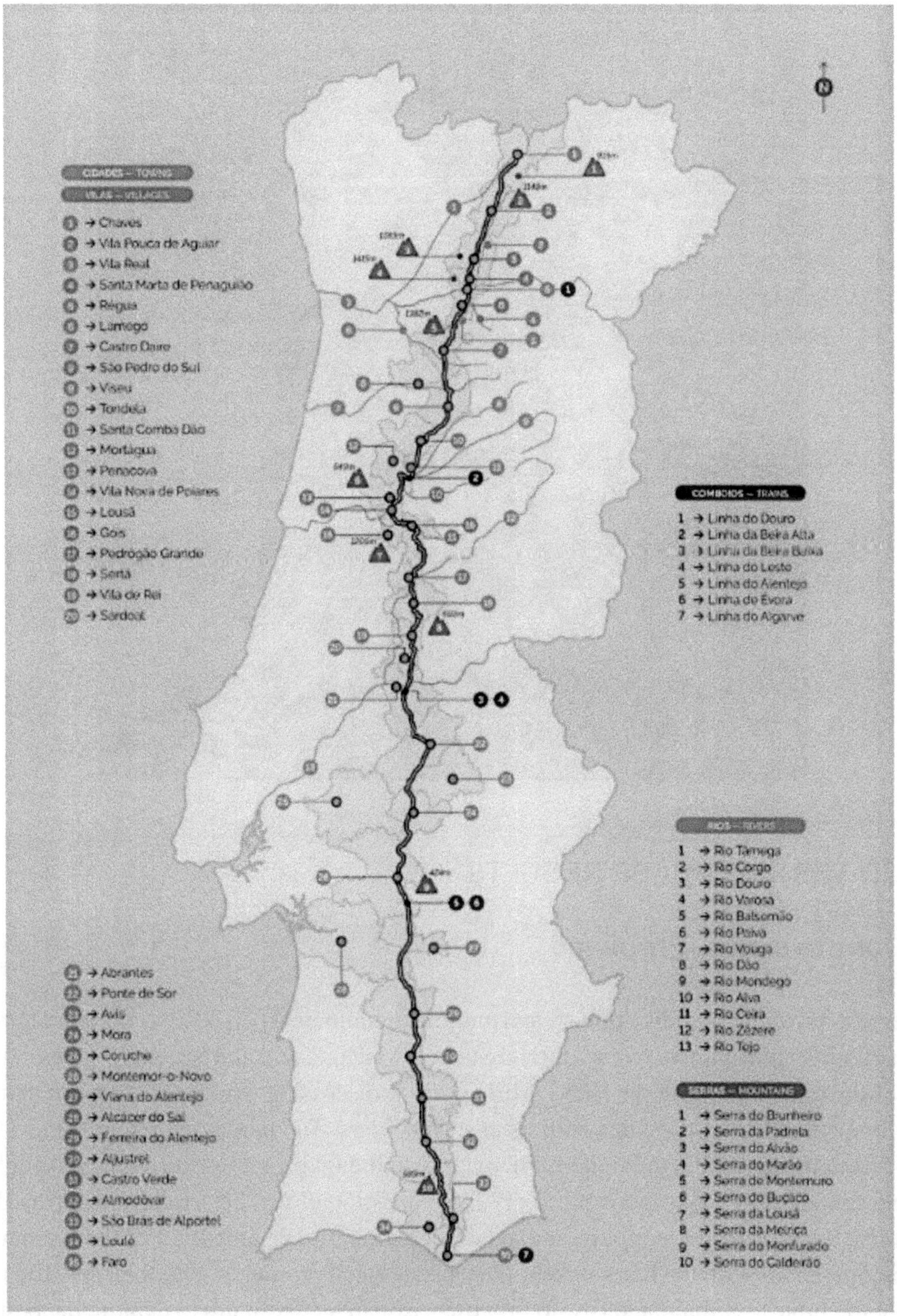

Route (Montenegro) is located next to *Dubrovnik* (90km). All other routes present important touristic hotspots along its path.

Figure 5. Iter Vitis Route – Kvarner Wines Route (Croatia).
Source: BEST MED internal reports

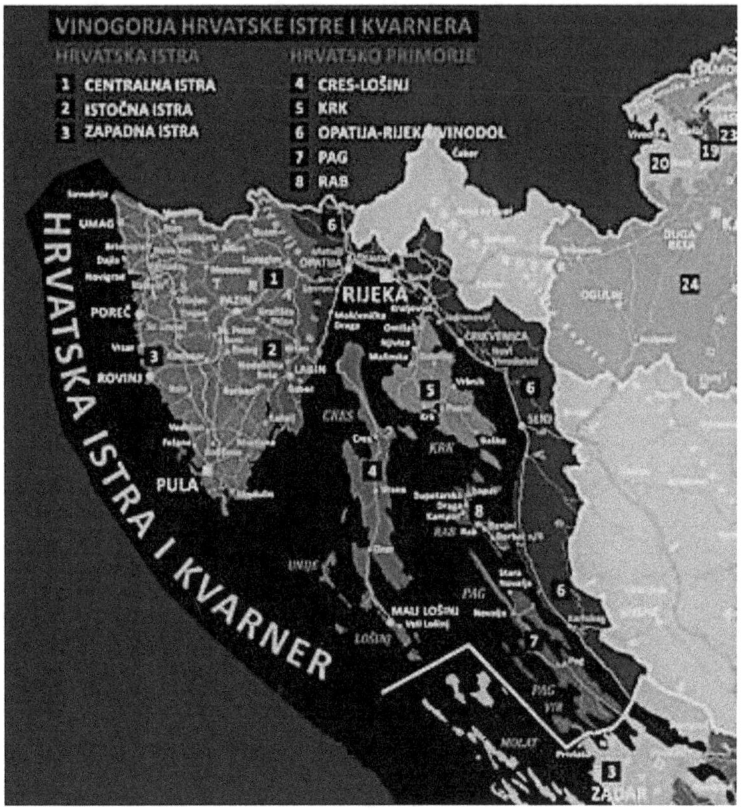

BEST MED TESTING PHASE RESULTS

Issues, Controversies, Problems

This chapter results from a multidisciplinary and multimethodological perspective, disseminating qualitative and quantitative information, from Tourism Studies, Management and Social Sciences, addressed in every step of the testing phase of the BEST MED project, discussing which practices are more suitable to the development of Cultural Routes with local stakeholders. The methodology performed highlights a research-action perspective, where the participation, and engagement of different stakeholders with different roles, intentions, and responsibilities is under question. It was determinant to have a diverse set of stakeholders, to have multiple inputs, and avoid repetition of the same kind of the discourse inputs. So, the main operations with the local stakeholders had a specific order, due to their specific nature and objectives of the project. Simultaneously, to propose a research continuum to explore the best practices to avoid a massified, unsustainable, unauthentic, and standardized tourism offer. The present COVID-19 pandemic has brought significant changes to the touristic, leisure and planning bodies. New demands and ways to perform activities have led to the performance of activities not in presence, but in an online way, so this change was introduced in the performance of development of the project. One of the most determinant elements on community participation, the contribution from single social actors, more or

Figure 6. Saint Martin of Tours Route (Slovenia).
Source: BEST MED internal reports.

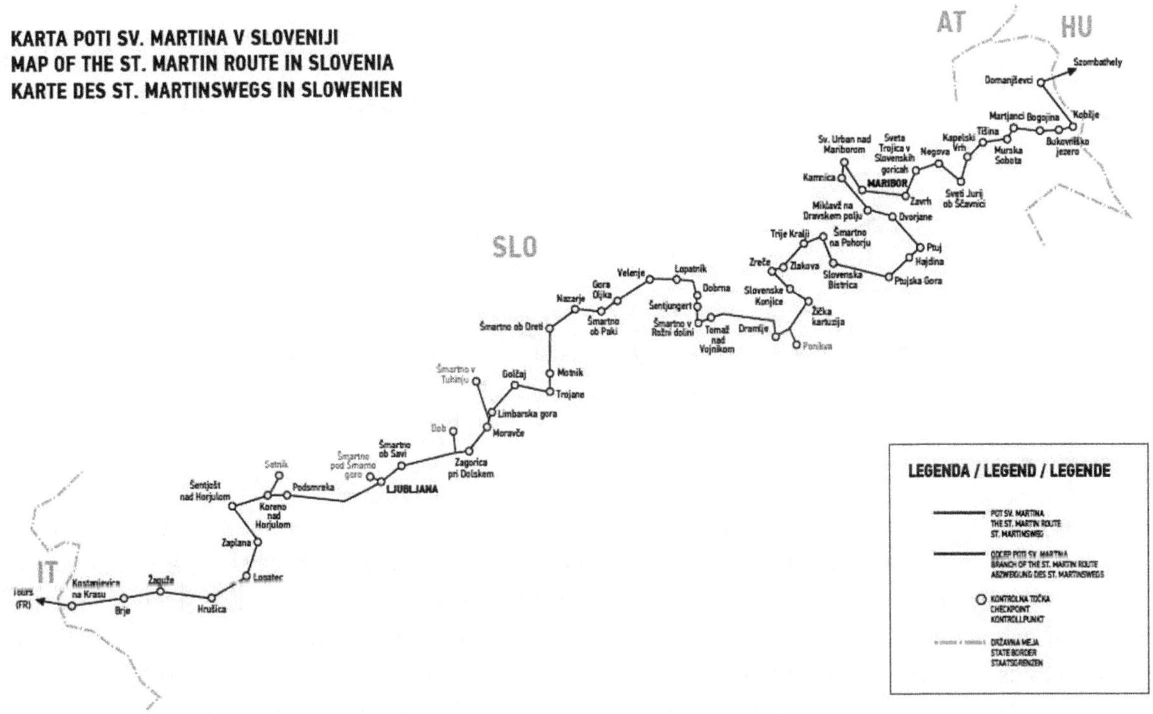

less engaged towards elements of sustainability, have been seriously compromised, for technical reasons (some of these elements have serious informatic difficulties). The traditional ground for community participation, open communitarian assemblies or traditional participation arenas were closed, so it was

Figure 7. European Cemeteries Route (Slovenia).
Source: Official website

Figure 8. Illyricum Trail of the Roman Emperors and Danube Wine Route.
Source: Official website

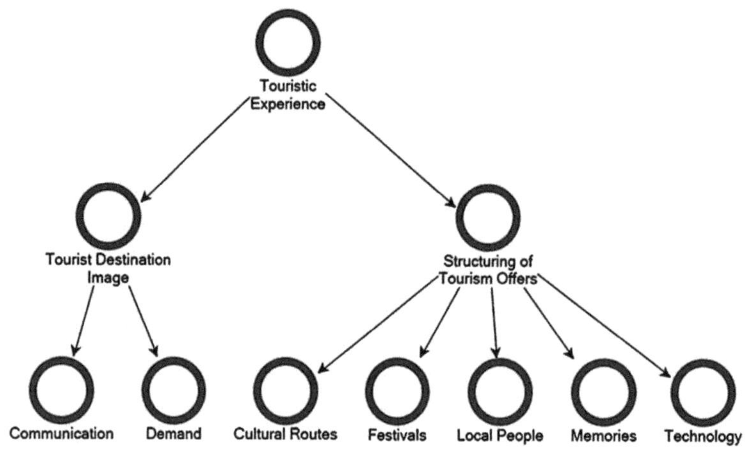

harder to reach these stakeholders. The online methods to achieve participation of local stakeholders have showed, that participation of professional and more engaged stakeholders can be made in faster and smother way, but can be seriously compromised among third sector associations and residents. After the participation practices, most of them made online, it was clear the major participation of public officials and academics, most of them with professional profiles and responsibilities on the tourism sector, a lower participation among private operators, and even lower among third sector associations and citizens. Despite the usual complexities resulting from participation practices, aggravated by the present COVID-19 confinement and consequences, it was possible to test a new Cultural Route in 8 Mediterranean areas towards a more resilient, local, and engaged tourism, where local communities and resources (natural and cultural) are present and highlighted.

Table 5. Route management structure of BEST MED cultural routes

Route Itinerary	Route Management Structure
La Ruta de las Alpujarras - Routes of El legado andalusí (Spain)	Andalusian Public Foundation *El legado andalusí*
Vie Francigena del sud (Italy – Lazio)	*Associazione Europea delle Vie Francigena and Lazio Region*
Rota da Estrada Nacional 2 (Portugal)	*Associação de Municípios da Rota da Estrada Nacional 2*
Iter Vitis Route – Kvarner Wines Route (Croatia)	Croatian Association for tourism and rural development
Route of the olive tree (Greece)	Governing Board of the Routes of the Olive Tree
Iter Vitis Route – Slovenia	*Združenje Konzorcij Turizem v zidanicah*
European Cemeteries Route – Slovenia	Funeral Company Maribor and *Žale* company
Saint Martin of Tours Route – Slovenia	*Kulturni center sv. Martin Tourski – Slovenija*
Cycle Route of the Parks (Italy – Calabria)	Calabrian national and regional park
Illyricum Trail of the Roman Emperors and Danube Wine Route (Montenegro)	Scientific Committee of the RER DWR Route

Source: BEST MED project internal reports

Table 6. Participants on Focus Groups (%)

Partner	Public Sector	Private Sector	Associations	Residents	Academics	Others
Spain	55,6	0	0	0	44,4	0
Italy - Lazio	50	0	10	0	10	30
Portugal	38,9	0	11,1	5,6	44,4	0
Croatia	47,1	11,8	17,6	5,9	17,6	
Greece	52,9	11,8	0	0	11,8	23,5
Slovenia	41,2	0	29,4	11,8	17,6	0
Italy - Calabria	87,5	0	12,5	0	0	0
Montenegro	42,9	0	14,3	28,6	14,3	0
All	**49,5**	**3,9**	**12,6**	**5,8**	**21,4**	**6,8**

Source: BEST MED project internal reports

Focus Group

The BEST MED Focus Group was developed to achieve a strong another level of sectorial stakeholders' participation, a stronger adaptation and acceptance of the S&C Path Model, present the future online platform to monitor sustainability in the future. Secondly, to develop a discussion around 5 themes: Cultural Route Management, Governance, Depopulation in the interior areas & Interior versus Coastline/urban areas, Tourism flows and Seasonality mitigation.

One first important element of these Focus Groups was the nature of origins of the participants, mainly from the public sector and academic field. These are professional bodies strongly related and with responsibilities in the areas of tourism sustainability, management, and cultural routes promotion. We believe that the different confinement moments occurred in the different Mediterranean areas were determinant for the minor participation of local associations and civic moments, residents, and private bodies, because some of them are not so familiarized with online participation platforms, and the traditional spaces for consultation as theatres, public access assemblies, auditoriums were closed (Table 6).

But the most operational outputs were the creation of 5 actions to improve the tourism sustainability in the pilot area, which were presented to local authorities. Considering the 4 dimensions of the S&C Path Model, we developed a content analysis of the 5 actions developed in the BEST MED.

We can observe that most of the actions are oriented towards elements of economic sustainability, namely a major participation of private companies and investors, with the creation of signed agreements with local companies to develop specific kinds of routes, where cultural heritage and gastronomy are highlighted, with experiences of local products consumption, forms of ecotourism and relation with natural resources, more engaged with local communities towards ecological and sociocultural elements of sustainability, as well as cultural paths oriented towards more exclusive and added value. At the same time, a higher presence of the public sector, helping and supporting specific calls to national and European funds. Address seasonality is also an important element, monitoring tourist flows, as well as the development of non-touristic activities, supporting other local economic segments. It was also emphasised the importance of tourism promotion, creating a major route's territorial visibility. Finally, the importance of training of all existing stakeholders from residents to professionals to improve the touristic offer.

Table 7. Participants on Living Lab (%)

Partner	Public Sector	Private Sector	Associations	Residents	Academics	Others
Spain	54,5	18,2	0	27,3	0	0
Italy - Lazio	33	47	13	0	7	0
Portugal	25,	18,8	18,8	18,8	18,8	0
Croatia	31,8	27,3	9,1	13,6	18,2	0
Greece	37,9	27,6	13,8	10,3	6,9	3,4
Slovenia	0	33,3	16,7	0	50	0
Italy - Calabria	42,9	28,6	28,6	0	0	0
Montenegro	33,3	44,4	11,1	11,1	0	
All	**33,9**	**29,6**	**13**	**11,3**	**11,3**	**0,9**

Source: BEST MED project internal reports

Secondly, the most referred actions are related to Management sustainability. It was clear the need for the creation of a more participative, engaged and consultive ecosystem of stakeholders, promoting responsibility sharing as well as the identification of leaders, roles, and responsibilities, creating a system of complementarity among public, private, residents and third sector associations. It was emphasized the need for better coordination among local powers towards management of services, equipment, and infrastructures, sharing strategies to improve the route touristic offer and best practices, promoting at the same time certification and evaluation strategies using digital tools. According to long term results, the participants also recalled that is fundamental to develop these strategies after the BEST MED project ending.

Living Lab

To have a more tourism product orientation, adequate towards local markets, was determinant to develop a more inclusive participation practice, a 1 day Living Lab, asking participants to give us their insight about the best strategies for pilot areas and its cultural routes.

As occurred in the case of Focus Groups, the 115 participants were predominantly from the public sector, but with a higher presence of private bodies. This change can be explained by the main objective of the discussion, to promote innovative ways to improve the regions' tourism offer and to present the different modalities associated with the fam trip (Table 7).

This activity had 3 main objectives. Firstly, to develop a discussion around Cultural tourism; Sustainable development, Tourism and local communities, Tourism and Covid-19 pandemic effects; Rebuilding tourism for the future: tourism recover after Covid-19 and Traditional activities and its tourism use. Secondly, to create a discussion around the governance of the route and the interaction between stakeholders and policy makers. Thirdly, to receive a feedback from the participants, about the future cultural routes of each partner, through a questionnaire.

The first question of the questionnaire was to ask to participants, using a 1-5 scale, how familiar they were the route before the Living Lab. The replies showed an average number of 3.6, which shows that the level of knowledge is not so high, showing that even the local stakeholders need more information about these routes. Afterwards, we asked which are the differencing elements of the proposed route and which elements should be highlighted in a promotional video. For the two categories is clear the importance of Gastronomy experiences, Natural areas, Views, Storytelling around the route and Signs and information spots along the route itinerary. In the bottom of choices are local markets. In tourist oriented cultural routes, the promotion of these itineraries is fundamental for its economic success. In this case, the participants showed preference for social media campaigns with photos and posts (Facebook, Instagram), promotional videos, fam trips with travel agents and tour operators, social media campaigns with photos collected by tourists, the creation mobile phones apps and other devices and digital Brochures. Not surprisingly, the traditional ways of promotion, with leaflets a tv campaigns were the less chosen. Finally, it determinant to know to which tourists these routes are attractive, and we can observe a major interest towards tourists related to gastronomy and culinary, culture, biking and hiking, motorway, adventure, and seniors.

SOLUTIONS AND RECOMMENDATIONS

All these testing processes showed us how challenging is to adapt a theoretical role model for tourism sustainability, asking all partners to be coordinated in the same goals, complying with a short time schedule, under the present COVID19 pandemic, and to achieve collective goals. So after the BEST MED testing phase, will be possible to create an operational and effective framework to understand the relations between studying and testing phases of research-action projects, creating more knowledge about tourism sustainability in the Mediterranean, but also, in a more analytical way, the best practices in projects using prototyping and shared methodologies to achieve common goals, to undergone the challenges of cooperating among different social, economic and cultural realities. It is fundamental that all this effort to promote sustainability in these eight Mediterranean Pilot Areas is continued after the projects ending, particularly by DMO's which have clear responsibilities on the tourism promotion of these routes, by local municipalities with responsibilities in the areas of planning and particularly by public cultural and environmental agencies ensuring socio-cultural and environmental sustainability. Simultaneously, that private tour operators integrate these cultural tours in their touristic offer. The created working group, where private agents, associations and residents have an important role, should continue to monitor the achievement of sustainability levels, claiming for better and resilient cultural routes and tourism territories. Only with a professional cultural route manager, as well as a dedicated team, will be possible to pursue these objectives. The creation of cultural route manager is fundamental to promote a better coordination between municipalities, local entities, and the regional structures, to ensure sustainability in the management of services and facilities as well as to the sharing of the methodology and strategy of promotion-marketing throughout the territory. It will also be determinant for the design, implementation, and operationalization of a strategic policies, mixed management systems (public, associative, private, non-profit, etc.) that enhances the various local resources, human and financial, to ensure services and infrastructures along the routes, as well as sponsorship and communication, identifying the leaders for each activity to be implemented. As governance style of management, should promote bottom-up approaches, throughout the presence of the local community, identifying key people and addressing their responsibilities, the importance of their training, the creation of networks, the creation of signed agreements with local private companies to develop the routes, the search for private investors.

FUTURE RESEARCH DIRECTIONS

Covid-19 pandemic has brought an unprecedented effect on tourism territories, with clear reductions on travel, tourism related income and major increase on unemployment rates, particularly in the areas of accommodations, restaurants, and other tourism sectors. With the increasing reduction of travels bans, as well as the spread of massive vaccination already in the year of 2021, is fundamental to gather data about the economic and social recovery that we are already assisting, and how much time, traditional Mediterranean tourism destinations will take to return to the touristic figures (hosts, overnights, and income) that we assisted in the years of 2019 and 2018. Simultaneously, to know if the pandemic context was really an opportunity to solve some of the unsustainability elements present in Mediterranean tourism, as seasonality, cultural heritage commodification, low economic added value to local communities, high exploration of natural resources, or even the feeling of overtourism expressed by several residents and social movements.

CONCLUSION

Today, under the present consequences resulting from COVID-19 in a global sense, but also according to the performance of these leisure destinations, it is fundamental to search for new answers and approaches to achieve tourism sustainability. Most of the participation processes highlighted the need to develop long-term practices, where the need for professional Management bodies with a shared and participated governance style is determinant to perform the necessary change on Mediterranean tourism offer. Clearly, the most relevant element of all participation practices, was the need for an effective and participated governance of cultural routes, with consequent advances towards tourism sustainability.

The Testing Phase was a determinant moment to promote the S&C Path model, its discussion and dissemination among the project partners, and the improvements made on each pilot area with the support of local sectorial stakeholders. So, it was determinant for the team from University of Algarve to support the other BEST MED partners on participations practices, use the model, understood the existing local conditions of operationalization, its adequacy, its challenges, and its expected results. A major contribution of this testing phase was to create local synergies for the development of the next working packages, but also to ensure the future sustainability of the projects' aims and objectives, to promote a more resilient and sustainable tourism.

The major achievement of the Testing Phase was, first of all, to have a clear picture of each pilot area, its sustainability level, the contextual tourism system of each area. Strongly inflected by the present COVID-19 pandemic, and its catastrophic effects on the tourism system, each pilot area had the chance to create a working group, discuss and propose the best ways to overcome the present crisis. With the S&C Path Model, each pilot area had the chance to monitor its sustainability status, on the management side with the questions around governance and the interconnection among local stakeholders, on the economic side to understand the harsh economic conditions that local economies are passing and how to overcome, on the cultural side to address cultural heritage commodification and lack of participation of the local community, in the environmental side to achieve a better understanding about the consumption of natural resources and address future climate changes. Improving the existing cultural routes, some of them without any sort of monitoring or certification, was fundamental for each pilot area to know more about its own tourism offer, and to achieve a more sustainable, resilient and participated tourism phenomenon. For other touristic areas, BEST MED is a proof that with a consistent tourism model of certification and monitorization, with a shared governance style with the effective participation of local actors, and with the continuity of these practices in the future is possible to achieve a sustainable tourism for the future.

ACKNOWLEDGMENT

This Chapter is financed by National Funds provided by FCT - Foundation for Science and Technology through project UIDB/04020/2020.

REFERENCES

Benjamin, S., Dillette, A., & Alderman, D. H. (2020). "We can't return to normal": Committing to tourism equity in the post-pandemic age. *Tourism Geographies*, *22*(3), 476–483. doi:10.1080/14616688.2020.1759130

Bichler, B. F. (2019). Designing tourism governance: The role of local residents. *Journal of Destination Marketing & Management*, *100389*. Advance online publication. doi:10.1016/j.jdmm.2019.100389

Bichler, B. F., & Lösch, M. (2019). Collaborative governance in tourism: Empirical insights into a community-oriented destination. *Sustainability*, *11*(23), 6673. doi:10.3390u11236673

Božić, S., & Tomić, N. (2016). Developing the cultural route evaluation model (CREM) and its application on the Trail of Roman Emperors, Serbia. *Tourism Management Perspectives*, *17*, 26–35. doi:10.1016/j.tmp.2015.11.002

Bramwell, B., & Lane, B. (2011). Critical research on the governance of tourism and sustainability. *Journal of Sustainable Tourism*, *19*(4-5), 411–421. doi:10.1080/09669582.2011.580586

Cabeça, S. M., Gonçalves, A. R., Marques, J. F., & Tavares, M. (2019). Mapping intangibilities in creative tourism territories through tangible objects: A methodological approach for developing creative tourism offers. *Tourism & Management Studies*, *15*(SI, 1SI), 42–49. doi:10.18089/tms.2019.15SI05

Caffyn, A., & Jobbins, G. (2003). Governance capacity and stakeholder interactions in the development and management of coastal tourism: Examples from Morocco and Tunisia. *Journal of Sustainable Tourism*, *11*(2-3), 224–245. doi:10.1080/09669580308667204

Carden, A. R. (2006). Sign of the times: An examination of the self-concept and emotional motivation of travelers on route 66 (1926-Present). *Journal of Travel & Tourism Marketing*, *20*(2), 51–61. doi:10.1300/J073v20n02_04

Carreira, V., Azeredo, J., González-Rodríguez, M. R., & Díaz-Fernández, M. C. (2021). The Role of Tourism Education in Overtourism and Destination Management. In A. Sharma & A. Hassan (Eds.), *Overtourism as Destination Risk* (pp. 35–48). Emerald Publishing Limited. doi:10.1108/978-1-83909-706-520211003

Caton, K., & Santos, C. A. (2007). Heritage tourism on Route 66: Deconstructing nostalgia. *Journal of Travel Research*, *45*(4), 371–386. doi:10.1177/0047287507299572

Dedek, P. B. (2007). *Hip to the trip: a cultural history of Route 66*. UNM Press.

Destinations, G. (2020). Retrieved from Green Destinations Standard: http://greendestinations.org/wp-content/uploads/2017/11/Green-Destinations-Standard-1.4.2.pdf

European Commission. (2016). *The European Tourism Indicator System. ETIS Toolkit for sustainable destination management*. European Commission.

Ferrari, S., Navarro, V., Mazzoni, E., Albrieu, C., & Biott, J. (2020). Ruta nacional 40 en Santa Cruz (Patagonia, Argentina): Potencialidad turística y propuestas para su planificación responsable. *Revista DELOS*, *13*(37), 409–430.

Galvani, A., Lew, A. A., & Perez, M. S. (2020). COVID-19 is expanding global consciousness and the sustainability of travel and tourism. *Tourism Geographies*, *22*(3), 567–576. doi:10.1080/14616688.2020.1760924

Gössling, S., Scott, D., & Hall, C. M. (2020). Pandemics, tourism and global change: A rapid assessment of COVID-19. *Journal of Sustainable Tourism*, *29*(1), 1–20. doi:10.1080/09669582.2020.1758708

GSTC. (2020). *Global Sustainable Tourism Council*. Retrieved from Destination Criteria: https://www.gstcouncil.org/wp-content/uploads/GSTC-Destination-Criteria-v2.0.pdf

Hall, C. M. (2011). A typology of governance and its implications for tourism policy analysis. *Journal of Sustainable Tourism*, *19*(4-5), 437–457. doi:10.1080/09669582.2011.570346

Hughes, S., Yordi, S., & Besco, L. (2020). The role of pilot projects in urban climate change policy innovation. *Policy Studies Journal: the Journal of the Policy Studies Organization*, *48*(2), 271–297. doi:10.1111/psj.12288

Joijc, S. (2017). Sustainable Tourism in the Albanian Riviera-Cultural Routes Contibution in Sustainable Tourism in the Albanian Riviera. *European Journal of Economics and Business Studies*, *3*(1), 104–113. doi:10.26417/ejes.v7i1.p104-113

Karamustafa, K., & Ulama, S. (2010). Measuring the seasonality in tourism with the comparison of different methods. *EuroMed Journal of Business*, *5*(2), 191–214. doi:10.1108/14502191011065509

Kerr, R. (2018). *Roads, Tourism and Cultural History*. Channel View Publications.

Koens, K., Postma, A., & Papp, B. (2018). Is overtourism overused? Understanding the impact of tourism in a city context. *Sustainability*, *10*(12), 4384. doi:10.3390/su10124384

Martins, J. C. (2020). Tangible Cultural Heritage Re-Appropriation Towards A New Urban Centrality. A Critical Crossroad In Semi-Peripheral Eastern Riverside Lisbon. *Geography, Environment, Sustainability*, *13*(3), 139–146. doi:10.24057/2071-9388-2020-58

Mitomed Plus Deliverables Library. (2020). Retrieved from Mitomed Plus: https://mitomedplus.interreg-med.eu/what-we-achieve/deliverable-library/

Neumann, H. R., Röder, D., & Röder, H. (2020). Diverse and rich fortified cultural heritage of the Iberian Peninsula. Basis for culture tourism with the European Culture Route Fortified Monuments. *Forte Cultura,* 971-976. . doi:10.4995/FORTMED2020.2020.11394

Obrador, P., Crang, M., & Travlou, P. (2009). *Taking Mediterranean tourists seriously*. Ashgate.

Pham, K., Vogt, C., & Andereck, K. (2021). Overtourism: Sustainable Tourism Plan and Stakeholder Involvement. *Travel and Tourism Research Association: Advancing Tourism Research Globally*, 4.

Procopiak, R. K. A., de Almeida, C. R., & Renda, A. I. (2020). Medieval tourism events in the Iberia peninsula. *Ficha Técnica Omnia, 10*(2), 29–40.

Richards, G. (2011). Creativity and tourism: The state of the art. *Annals of Tourism Research, 38*(4), 1225–1253. doi:10.1016/j.annals.2011.07.008

Richards, G. (2018). Cultural tourism: A review of recent research and trends. *Journal of Hospitality and Tourism Management, 36*, 12–21. doi:10.1016/j.jhtm.2018.03.005

Richards, G. (2021). *Rethinking cultural tourism*. Edward Elgar Publishing. doi:10.4337/9781789905441

Richards, G., & Wilson, J. (2006). Developing creativity in tourist experiences: A solution to the serial reproduction of culture? *Tourism Management, 27*(6), 1209–1223. doi:10.1016/j.tourman.2005.06.002

Romagosa, F. (2020). The COVID-19 crisis: Opportunities for sustainable and proximity tourism. *Tourism Geographies, 22*(3), 690–694. doi:10.1080/14616688.2020.1763447

Rúa, C., Sepúlveda, M., Gutiérrez, S., Cárcamo-Vega, J. J., Surco-Luque, J., Campos-Vallette, M., & Pereira, M. (2017). Raman identification of pigments in wall paintings of the Colonial period from Bolivian churches in the Ruta de La Plata. *Conservation Science in Cultural Heritage, 17*, 117–137. doi:10.6092/issn.1973-9494/7945

Rudihartmann. (1986). Tourism, seasonality and social change. *Leisure Studies, 5*(1), 25-33. doi:10.1080/02614368600390021

Sigala, M. (2020). Tourism and COVID-19: Impacts and implications for advancing and resetting industry and research. *Journal of Business Research, 117*, 312–321. doi:10.1016/j.jbusres.2020.06.015 PMID:32546875

Thees, H., Pechlaner, H., Olbrich, N., & Schuhbert, A. (2020). The Living Lab as a tool to promote residents' participation in destination governance. *Sustainability, 12*(3), 1120. doi:10.3390u12031120

UNWTO. (2020). *Cultural Tourism and COVID19*. Available on https://www.unwto.org/cultural-tourism-covid-19

UNWTO. (2021, May)... *World Tourism Barometer, 19*(3). Advance online publication. doi:10.18111/wtobarometereng

Van Buuren, A., & Loorbach, D. (2009). Policy innovation in isolation? Conditions for policy renewal by transition arenas and pilot projects. *Public Management Review, 11*(3), 375–392. doi:10.1080/14719030902798289 van Popering-Verkerk, J., & van Buuren, A. (2017). Developing collaborative capacity in pilot projects: Lessons from three Dutch flood risk management experiments. *Journal of Cleaner Production, 169*, 225–233. doi:10.1016/j.jclepro.2017.04.141

Vlizos, S., & Kosta, E. (2020) Designing Cultural Routes in the region of Sparta, Peloponnese: A Methodological approach. *Proceedings of the 6th UNESCO UNITWIN Conference 2019*, 299-306.

Vreugdenhil, H., Slinger, J., Thissen, W., & Rault, P. K. (2010). Pilot projects in water management. *Ecology and Society, 15*(3), art13. doi:10.5751/ES-03357-150313

Wood, A. F. (2010). Two roads diverge: Route 66, "Route 66," and the mediation of American ruin. *Critical Studies in Media Communication, 27*(1), 67–83. doi:10.1080/15295030903554375

Yeh, S. S. (2021). Tourism recovery strategy against COVID-19 pandemic. *Tourism Recreation Research, 46*(2), 188–194. doi:10.1080/02508281.2020.1805933

Zabbini, E. (2012). Cultural routes and intangible heritage. *Almatourism-Journal of Tourism. Culture and Territorial Development, 3*(5), 59–80. doi:10.6092/issn.2036-5195/3188

ADDITIONAL READING

Du Cros, H., & McKercher, B. (2020). *Cultural tourism*. Routledge. doi:10.4324/9780429277498

IvanovicM. (2008). *Cultural tourism*. . doi:10.13140/RG.2.1.4522.6323

McKercher, B., & Du Cros, H. (2002). *Cultural tourism: The partnership between tourism and cultural heritage management*. Routledge.

Miller, G., & Twining-Ward, L. (2006). Monitoring for a Sustainable Tourism Transition: The Challenge of Developing and Using Indicators. *Tourism and Hospitality Research, 7*. Advance online publication. doi:10.1079/9780851990514.0000

Richards, G. (Ed.). (2007). *Cultural tourism: Global and local perspectives*. Psychology Press.

Richards, G., & Richards, G. B. (Eds.). (1996). *Cultural tourism in Europe*. Cab International.

UNWTO. (2004). *Indicators of Sustainable Development for Tourism Destinations: A Guidebook*.

UNWTO. (2015). *Tourism and the Sustainable Development Goals*. UNWTO.

KEY TERMS AND DEFINITIONS

Cultural Route: Is a tourism product resulting from the integration of identity, community, and heritage aspects towards the creation of specific trails and paths.

Cultural Tourism: Is a type of tourism approach where the main visitor's motivation is to engage, discover, experience, and meet with intangible and tangible attractions of a certain destination.

Focus Group: Is a technique of research, developed as an group interview, where the promotor provides some questions and prerogatives for the participants to discuss.

Living Lab: Is a methodology of research-action where is recreated in physical spaces the testing of new methodologies of social intervention.

Pilot Area: Is a territory of testing new methodologies for social intervention and innovation improvements.

Research Action: Is a type of scientific research where is promoted in a process of engagement with local communities, learning from them and introducing innovation and improvements.

Stakeholders' Participation: Is a source of information to decision-making in social sciences, learning from in-dept engaged social actors.

Chapter 8
Mapping the Cultural Tourism Experience:
Role of Senses, Emotions, and Memories in Cultural and Heritage Experiences and Gastronomic Experiences

Rui Mendonça-Pedro
https://orcid.org/0000-0002-3240-2528
ISMAT, Lusófona University, Portugal & CinTurs, Portugal

Cristina Valente-Pedro
https://orcid.org/0000-0002-3566-6369
Faculty of Economics, University of the Algarve, Portugal

ABSTRACT

The chapter focuses on the role of senses, emotions, and memories model (SEMs) to mapping a cultural tourism experience. Specifically, the results of the survey aimed to identify and assess the role of senses, the emotional states, and the memorable elements within the cultural and heritage tourism experience construction; to identify and assess the role of senses, the emotional states, and the memorable elements within the gastronomic tourism experience creation; and to map and model the relation between the SEMs in cultural and heritage tourism experiences and gastronomic tourism experiences. Results show that SEMs positively impact cultural and heritage tourism experience construction, SEMs positively impact gastronomic tourism experience creation, there are two clear paths to cultural and heritage tourism experience construction (between sight + joy + vividness = MTE and between sight + positive surprise + recollection = MTE), and there is one pattern to gastronomic tourism experience creation (between smell + taste + touch + positive surprise + vividness = MTE).

DOI: 10.4018/978-1-7998-9217-5.ch008

INTRODUCTION

Cultural tourism is a growing area in the tourism industry and has become a crucial tourism industry sector providing economic balance and mitigating seasonality (Sánchez-Sánchez et al., 2020; Vergori & Arima, 2020). According to Yamashita, "Cultural tourism is a program in which tourists appreciate tangible and intangible aspects of culture at the destination, from architecture, visual arts, dance performance, festival, cuisines, to history" (2016, p. 212). Understanding how tourists perceive (senses), create (emotions) and encode (memory) the cultural tourism experience is a critical issue for understanding the stages of experience construction and memorabilia.

According to several authors, the senses play an important role in the perception of stimuli, and the emotions are central to understanding and assessing the meaning of a stimulus, while memory is vital in the process of codifying and storing information (Agapito, 2020; Pedro et al., 2020; Seyfi et al., 2020). Despite the crucial relationships among the senses, emotions and memories (SEMs), to our knowledge, few studies have addressed the role of SEMs in the creation of the cultural tourism experience (Agapito et al., 2017; Dias et al., 2017; Pedro et al., 2020). Given the importance of cultural tourism sector, this is a key gap in the literature, which the present study addresses: mapping the cultural tourism experience through SEMs will clarify how cultural experiences are created and enhance the understanding of how psychological and personal experiences are processed.

This study seeks to explain the role of SEMs on the construction of the cultural tourism experience. First, the role of the senses, emotional states and memorable elements are identified and assessed during the creation of the cultural tourism experience. Second, this research maps and models the relationship between SEMs in the cultural tourism experience; specifically, we consider how the senses activate an emotional state and seek to determine which emotional states trigger the memorable element.

This study adds to our understanding of the construction process for the cultural tourism experience and enriches knowledge to support the design and implementation of this process based on SEMs. The current research will also fill a gap in the literature by investigating the influence of the senses on the emotions and the impact of the emotions on memories to create tourist engagement in the cultural & heritage and gastronomy experiences. The findings of this study can provide insight to destination management organizations (DMOs), destination managers, destination marketers, stakeholders, and others interested to increase destination memories, satisfaction, loyalty, engagement, and cultural contact. All of these results could contribute to the competitive advantage of the destinations and, specifically, to cultural destinations.

BACKGROUND

Cultural Tourism Experience

The cultural tourism experience has become a real trend due to the growing awareness of and interest in culture, heritage, gastronomy, wine/winemaking and other traditions (Sanagustín-Fons et al., 2020; Vergori & Arima, 2020). Cultural tourism promotes multilevel experiential contact, which stimulates the relationship between tourist engagement and memorable tourism experiences (Chen & Rahman, 2018; Guerreiro et al., 2019). According to the World Tourism Organization (UNWTO), there is increasing interest in cultural tourism all over the world (UNWTO, 2018), and related sectors, "especially heritage tourism [...] are amongst the most vital segments of contemporary or modern tourism" (Chen & Rahman, 2018).

The United Nations (UNESCO) defines culture as "the set of distinctive spiritual, material, intellectual and emotional features of a society or a social group, that encompasses [...] art and literature, lifestyles, ways of living together, value systems, traditions and beliefs" (UNESCO, 2009, p. 9). The European Union (EU) adds to this definition the set of attitudes, customs, values and practices which are commonly shared by a group of people (EU, 2010). The UNWTO recently defined cultural tourism as "... a type of tourism activity in which the visitor's essential motivation is to learn, discover, experience and consume the tangible and intangible cultural attractions/products in a tourism destination [...] these attractions/products relate to a set of distinctive material, intellectual, spiritual and emotional features of a society that encompasses arts and architecture, historical and cultural heritage, culinary heritage, literature, music, creative industries and the living cultures with their lifestyles, value systems, beliefs and traditions" (UNWTO, 2018, p. 13).

The focus of the present study is the analysis of the cultural & heritage tourism experience and the gastronomic tourism experience, following in the tracks of the study by Guerreiro et al. (2019), which found that gastronomy (i.e. intangible heritage) and monuments/historical sites/museums and artefacts (i.e. tangible heritage) are the cultural resources with the potential to become tourist products in the Algarve. Cultural tourism has become essential in establishing the attractiveness of tourism destinations (Kim & Jang, 2016; Sukanthasirikul & Trongpanich, 2016). Moreover, gastronomy and food tourism experience is based on the pleasure from food, on taste, on flavour, and sharing meals (Dixit, 2019; Kivela, 2009), in contrast, cultural & heritage tourism experience is based on aesthetic observation, intellectual development, emotional recall, and psychological interpretation (Seyfi et al., 2020; Tolić, 2020). Therefore, given the differences between the cultural & heritage experience and the gastronomic experience perception, construction/creation, and encoding (i.e., senses, emotions, and memories), the researchers implemented two distinctive investigations to improve the in-deep analysis of the cultural tourism experience and a more realistic experience mapping. Study 1, related to the cultural & heritage tourism experience and study 2, connected to the gastronomic tourism experience.

Senses

The senses are vital to human spatial knowledge (Moreira et al., 2017), because they "are the basic means through which humans explore and understand the world" (Lv et al., 2020, p. 106). Tourists communicate with and experience a destination's attractions through the five senses. Sensorial perception is the result of sense stimulation (Krishna, 2012; Krishna & Schwarz, 2014; Lv et al., 2020), and in tourism, sensorial perception/stimulation is the beginning of the experiential process – this is, what tourists sees, tastes, smells, touches, and hears.

Several researchers have investigated the role of the sensory process in the tourism experience (Agapito, 2020; Agapito et al., 2017; Dias et al., 2017; Meacci & Liberatore, 2018) and suggested that the senses build the link between the tourists/subject and the experience/object (e.g. cultural & heritage experience and gastronomic experience). That is, in the tourism experience, the senses provide the principal, physical access to the experiential process, and multisensorial stimulus in the tourism experience is more likely to activate tourists' emotions, memory encoding and reinforcement of positive memories. According to Dias et al. (2017, p. 190), "... the sensory experiences play their role throughout the tourism experience process and transfer more meaning and be aware of tourism experiences and at the same time destination places gained by these means [...] a tourism destination must be memorable and evoke emotions: if this does not occur, its future will end soon". Ultimately, the senses are key in creating the brand fingerprint of the tourism destination definition (Agapito et al., 2017; Meacci & Liberatore, 2018). These arguments resulted in the following hypotheses:

Study 1 – Mapping the Cultural and Heritage Tourism Experience

H1 to H12: Sight, hearing, smell, and touch are positively related to joy in the cultural & heritage tourism experience.

Figure 1 provides a visual representation of the conceptual framework and the related hypotheses regarding the senses and cultural & heritage experience.

Study 2 – Mapping the Gastronomic Tourism Experience

H21 to H35: Sight, hearing, smell, and touch are positively related to joy in the gastronomic tourism experience.

Figure 2 provides a visual illustration of the conceptual framework and related hypotheses considering the senses and gastronomy experience.

Emotions

Emotions are "coordinated reactions to relevant events [...] including cognitive and biological changes, facial and/or bodily expressions, subjective feelings, and action tendencies such as approach or withdrawal" (Fallis, 2013, p. 36). From the perspective of neuroscience, emotions are a set of complex events recruiting distributed cortical and subcortical cerebral structures, in a highly dynamic system, in relation to the nature of the different emotions triggered (Tettamanti et al., 2012). The emotions are central to tourism scholarship, because they are essential for the delivery of a positive, remarkable and memorable experience (Moyle et al., 2019; Shoval et al., 2018), but the emotions are a complex topic to study, particularly in tourism, given their cultural, economic, social, motivational, cognitive and personal behavioural differences (Correia et al., 2017; Tettamanti et al., 2012).

The present research analyses positive basic emotions such as joy, love and positive surprise (Ekman, 1992; Güzel, 2014). Tourism experiences often include positive and pleasant emotions. The presence of positive emotions indicating a positive tourism experience are a major aspect of the holistic experience and contribute to the memorability of trip elements, including cultural, heritage or gastronomic experiences (Sthapit & Coudounaris, 2017). Having an extraordinary cultural tourism experience is the result of positive, pleasurable emotional activation (Sthapit, 2019). According to Correia et al. (2017, p. 163), "tourism is all about places and people". This relationship between places and tourists is established through the tourist experience (Mendes et al., 2010, 2016). The positive emotions elicited during the tourism experience and reinforced through recall enhance the trip's memorability (Williams et al., 2019). Tourists' emotions also influence their evaluation of the experience (Hosany et al., 2015), and experiencing positive emotions during a trip is associated with a memorable tourism experience (Knobloch et al., 2017). Based on these arguments, the following hypotheses will be tested:

Study 1 – Mapping the Cultural and Heritage Tourism Experience

H13 to H18: Joy, love, and positive surprise are positively related to recollection in the cultural & heritage tourism experience.

Figure 1 provides a visual representation of the conceptual framework and the related hypotheses regarding the emotional states and cultural & heritage experience.

Study 2 – Mapping the Gastronomic Tourism Experience

H36 to H41: Joy, love, and positive surprise are positively related to recollection in the gastronomic tourism experience.

Figure 2 provides a visual illustration of the conceptual framework and related hypotheses considering the emotional states and gastronomy experience.

Memories

Creating an exceptional, memorable tourist experience is central to the tourism industry (Knobloch et al., 2017). Memory is among the most important sources of information considered when tourists are deciding to visit or revisit a destination (Chandralal & Valenzuela, 2013). Holiday memories are the product of experiencing positive emotional states during their trip, so a memorable tourism experience is one that is remembered and recalled after the event has occurred (Kim et al., 2012; Sthapit & Coudounaris, 2017).

Some studies have suggested that recollection (Tung et al., 2017; Tung & Ritchie, 2011) and vividness (Gazzaniga et al., 2014; Todd et al., 2012) affect the result of the tourism experience. Kim (2010) and Todd et al. (2012) have observed that highly emotional events and/or tourism experiences are associated with memory vividness, which is therefore, associated with highly arousal emotional events that "evoke a previously unknown heightened subjective richness at the time they are experienced, rendering emotional events in the mind's eye with a special phenomenological character we call emotionally enhanced vividness" (Todd et al., 2012, p. 11201). These highly arousing emotional memories seem to have a persistent vividness that other memories lack (Gazzaniga et al., 2014); Kim (2010) has also suggested that vividness comprehends emotions and perception factors, such as "hearing in my mind", "seeing in my mind", "remembering the spatial layout", "remembering the emotions I lived" and "remembering the setting".

Several authors have reinforced the notion that tourism experiences are valuable when they are stored and consciously remembered through the recollection process (Kim, 2010; Squire, 2004; Tung et al., 2017; Tung & Ritchie, 2011), which is highly influenced by satisfaction, cognition, behavioural elements and physical tasks (Tung & Ritchie, 2011). Recollection may also comprehend constructs such as "reliving", "participating" and "remembering" (Kim, 2010). Thus, the memory of the cultural experience is the result of an experience – that is, what the tourist perceives with the senses as emotionally pleasurable and important enough to be recalled and vivid after it has occurred (Colombo & Gold, 2004; Oh et al., 2007). For those reasons, the following hypotheses were proposed:

Study 1 – Mapping the Cultural and Heritage Tourism Experience

H19 and H20: Recollection and vividness are positively related to memorable tourism experience construction in the cultural & heritage tourism context.

Figure 1 provides a visual representation of the conceptual framework and the related hypotheses regarding the memorable elements and cultural & heritage experience.

Study 2 – Mapping the Gastronomic Tourism Experience

H42 and H43: Recollection and vividness are positively related to memorable tourism experience construction in the gastronomic tourism context.

Figure 2 provides a visual illustration of the conceptual framework and related hypotheses considering the memorable elements and gastronomy experience.

CONCEPTUAL FRAMEWORK AND HYPOTHESES

The conceptual model presented here shows the relationships between the SEMs to create a memorable tourism experience and, specifically, cultural & heritage tourism experience (Figure 1) and a gastronomic tourism experience (Figure 2). In consideration of the different constructs presented in this study, the authors defined the conceptual model as: the senses, which perceive the stimulus from the surrounding environment, triggered/stimulated the emotions, which interpret and understand the meaning of that sensorial stimulus information, and, finally, the highly emotional and satisfactory events construct memories – that is, the memory codifies and stores the information, if it is sufficiently important and extraordinary to the tourist, to be recalled vividly after the tourism experience occurs (Agapito et al., 2014; Krishna, 2012; Pedro et al., 2020; Tung & Ritchie, 2011).

METHODOLOGY

Sample and Data Collection

The population included in the research consisted of national and international tourists aged 18 years and older who arrived at the Algarve through the Algarve International Airport and declared having a memorable cultural tourism experience (Kim et al., 2012). The data were collected via the administration of a structured questionnaire to a convenience sample of tourists at the departure boarding gates in the Algarve International Airport (Matos, 2014; Pedro, 2019).

Measurements

The measurements used in both studies (study 1, cultural & heritage tourism experience, and study 2, gastronomic tourism experience) were identified on the basis of a literature review (Haase et al., 2020;

Figure 1. Cultural & heritage experience, conceptual model, and related hypotheses
Source: Own elaboration from XMind software

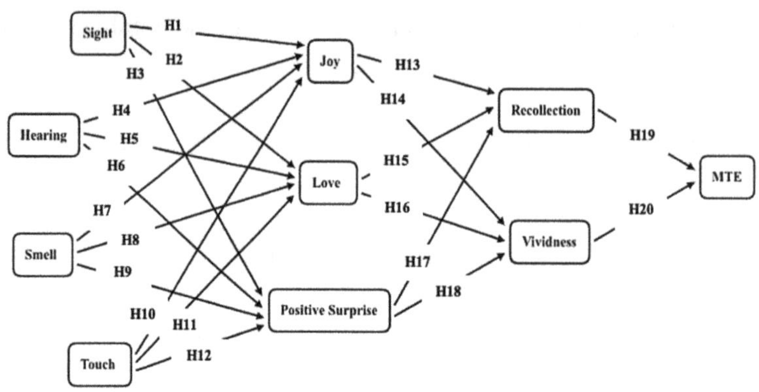

Hosany et al., 2017; Kim & Youn, 2017; Kim & Thapa, 2018), but the taste construct and related items were omitted from study 1, respectively.

PLS-SEM with SmartPLS v.3.0 (Ringle, Wende, & Becker, 2015) was used to synthesize the latent variables hypothesized, to identify the path of relationships between the variables, to estimate the parameters of the structural models and to map the role of senses, emotions and memories in the construction of the cultural tourism experience (Hair et al., 2017, 2019; Usakli & Kucukergin, 2018). The estimated model was evaluated for reflective measurement model fit and structural model fit using PLS-SEM, as shown in Table 1 (Hair et al., 2017, 2019).

Figure 2. gastronomic experience, conceptual model, and related hypotheses
Source: Own elaboration from XMind software

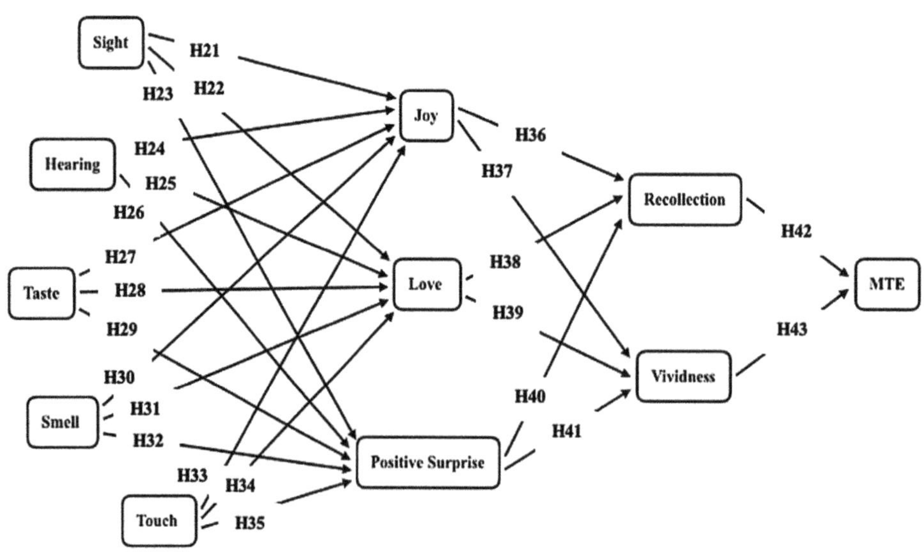

Table 1. Guideline values for PLS-SEM measurement model and structural model

Assessing Reflective Measurement Model		Guideline Values
1st Reliability (indicator loadings)		≥ 0.708
2nd Internal consistency reliability	Cronbach's alpha (Cronbach's α)	Minimum 0.70 (or 0.60 in exploratory research) Recommended 0.70 to 0.90 Maximum 0.95
	Composite reliability (CR)	CR ≥ 0.70
3rd Convergent validity (average variance extracted – AVE)		AVE ≥ 0.50
4th Discriminant validity	Fornell-Larcker criterion	$\sqrt{AVE}$ larger than the biggest correlation with any construct
	Heterotrait-monotrait ratio of the correlation (HTMT)	HTMT < 0.90 (but they are better when lower than 0.85), and must be significantly different from 1
Assessing Structural Model		**Guideline Values**
1st Collinearity (variance inflation factor values – VIF)		VIF ≥ 5 (probable/critical collinearity issues) VIF ≥ 3 to 5 (possible collinearity issues) VIF < 3 (ideal value)
2nd Explanatory power of the model (coefficient of determination – R^2)		$R^2 = 0.75$ (substantial explanatory power of the model) $R^2 = 0.50$ (moderate explanatory power of the model) $R^2 = 0.25$ (weak explanatory power of the model)
3rd Predictive accuracy of the PLS path model (Stone-Geisser test – Q^2) – Blindfolding procedure		$Q^2 > 0$ (small accuracy of the model) $Q^2 > 0.25$ (medium accuracy of the model) $Q^2 > 0.50$ (large accuracy of the model)

Source: Adapted from Hair et al., (2017, 2019)

RESULTS

Sample Profile and Experiential Setting

The data were collected via the administration of a structured questionnaire to a convenience sample of 189 tourists at the departure boarding gates in the Algarve International Airport. Figure 3 summarizes the two different experiential stages/settings at which tourists constructed a memorable cultural tourism experience at the Algarve. The data show that cultural & heritage experiential settings (52.9%) were the slightly more expressive for the construction of a memorable cultural tourism experience than the gastronomic experiential settings (47.1%).

Table 2 summarizes the sample characteristics, including: gender, age, marital status, professional status, Algarve previous visits, education level, and country of origin. The study sample show data females (52.4%) slightly out-number males (47.6%), and the most predominant age groups of this sample are above 51 years old (52.4%). Tourists from UK represents almost more than half of the sample (49.2%), and 68.8% of the respondents are married or living as a couple. Moreover, 46% of the respondents are employed, had mainly the university level (66.1%), and had more than 3 visits to Algarve destination (52.4%).

Figure 3. Experiential setting of the memorable cultural tourism experience construction
Source: Own elaboration from SPSS Statistics v.27.0 software

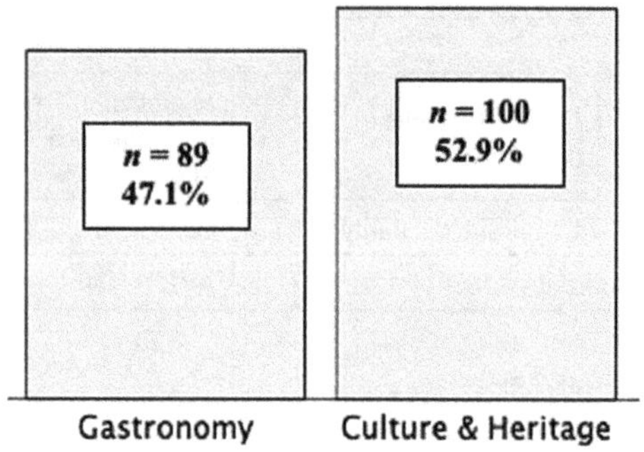

STUDY 1 – MAPPING THE CULTURAL AND HERITAGE TOURISM EXPERIENCE

Descriptive Statistics

The results shown in Table 3 indicate that all the of the variables for the cultural & heritage tourism experience have average neutral (4, Neither agree or disagree/Neutral) to positive values (6, Agree/Very important) on a 7-point Likert scale; the median values, however, indicate that tourists classified all of the constructs with slightly higher values, between 5 (Somewhat agree/Moderately important) and 6 (Agree/Very important).

All of the constructs presented in the study are therefore classified positively for the construction of the cultural & heritage tourism experience based on the perceptions of tourists.

Table 2. Sample profile

Variables	Distribution of Answers
Gender	Female: 52.4%; male: 47.6%
Age	More than 61 years: 29.6%; between 51 to 60 years: 22.8%; between 31 to 40 years: 19.6%; between 41 to 50 years: 15.3%; between 21 to 30 years: 10.6%; under 21 years: 2.2%
Marital status	Married/living as a couple: 68.8%; single: 19.6%; divorced: 8.5%; widowed: 2.6%; other: 0.5%
Professional status	Employed: 46.0%; retired: 29.1%; self-employed: 18.5%; student: 3.7%; unemployed: 2.1%; homeworker: 0.5%
Algarve previous visits	More than 3 visits: 52.4%; 1st visit: 20.1%; 2nd visit: 19.0%; 3rd visit: 7.9%; N/R: 0.5%
Education level	University level: 66.1%; secondary school: 31.2%; N/R: 2.1%; primary school: 0.5%
Country of origin	UK: 49.2%; Ireland: 15.3%; Germany: 12.2%; France: 7.9%; Portugal: 6.3%; Others: 5.8%; Netherlands: 3.2%

Source: Own elaboration

Table 3. Descriptive statistics for the observed variables in Study 1

Constructs	Variables	Mean	SD (±)	Median
Sight	SG1 Aesthetic	5.49	1.50	6.00
	SG2 Attractive	5.94	1.24	6.00
	SG3 Beautiful	5.79	1.56	6.00
	SG4 Pretty	5.68	1.57	6.00
Hearing	HR1 Euphonic	4.40	2.31	5.00
	HR2 Good sounding	4.66	2.18	5.00
	HR3 Melodic	4.40	2.35	5.00
	HR4 Sonorous	4.30	2.39	5.00
Smell	SM1 Fragrant	4.60	2.26	5.00
	SM2 Nice smell	4.71	2.23	5.00
	SM3 Perfumed	4.48	2.26	5.00
	SM4 Scented	4.54	2.24	5.00
Touch	TC1 Comfortable	4.37	2.52	5.00
	TC2 Handy	4.20	2.59	5.00
	TC3 Soothing	4.15	2.52	5.00
	TC4 Well-shaped	4.25	2.57	5.00
Joy	JY1 Cheerful	5.96	1.28	6.00
	JY2 Delight	5.62	1.52	6.00
	JY3 Enthusiasm	5.78	1.36	6.00
	JY4 Joy	5.83	1.28	6.00
	JY5 Pleasure	5.79	1.51	6.00
Love	LV1 Affection	5.17	2.03	6.00
	LV2 Caring	4.97	2.01	5.00
	LV3 Love	4.92	2.18	5.00
	LV4 Tenderness	4.82	2.12	5.00
	LV5 Warm-hearted	5.06	2.01	6.00
Positive surprise	PS1 Astonishment	5.16	1.59	5.00
	PS2 Amazement	5.22	1.70	5.00
	PS3 Fascinated	5.60	1.37	6.00
	PS4 Inspiration	5.37	1.50	6.00
	PS5 Surprise	5.16	1.66	5.00
Recollection	RC1 Reliving	5.31	1.64	6.00
	RC2 Participate in	5.51	1.41	6.00
	RC3 Remember	5.38	1.54	6.00
Vividness	VV1 Hear in my mind	5.05	1.68	5.00
	VV2 See in my mind	5.52	1.34	6.00
	VV3 Feel emotions	5.36	1.50	6.00
	VV4 Revisit setting	5.64	1.36	6.00
MTE	MTE 1 Hedonism	6.01	1.33	6.00
	MTE 2 Refreshment	5.57	1.65	6.00
	MTE 3 Meaningfulness	4.90	1.98	5.00
	MTE 4 Local Culture	5.42	1.72	6.00
	MTE 5 Novelty	5.47	1.75	6.00

Source: Own elaboration from SmartPLS 3.0 software

Analysis of the Proposed Model

Assessing the Measurement Model Fit – Cultural and Heritage Experience

To assess the measurement model, the researchers tested the reliability, internal consistency, and validity of the constructs in the model. Reliability was evaluated using indicator loadings, internal consistency was measured though the Cronbach's alpha and composite reliability (CR), convergent validity was assessed using the average variance extracted (AVE) and discriminant validity was evaluated through the heterotrait-monotrait ratio of the correlations (HTMT).

Before confirming the reliability of the measurement model, the factor loading scores for all of the indicators were analysed according to the threshold values, ≥ 0.708 (Hair et al., 2019), and two indicators were deleted – that is, one indicator at a time with low loading values (item MTE5 = 0.631 and item PS1 = 0.694, less than the threshold values 0.708, see the indicator with * in Table 4); the factor loading scores for all of the other indicators were above the recommended threshold (≥ 0.708), indicating the satisfactory reliability of the constructs. As shown in Table 4, the Cronbach's alpha scores were above the threshold values (minimum 0.70) and ranged from 0.817 to 0.970, the CR ranged from 0.880 to 0.978 (≥ 0.7 threshold values) and the AVE ranged from 0.916 to 0.730 (≥ 0.50 threshold values). Reliability, internal consistency and convergent validity were thus established (Hair et al., 2019; Wong, 2013).

The Fornell-Larcker criterion (1981), which compares the square root of the AVE values with the latent variable correlations, was used to assess discriminant validity. If the square root of the AVE is larger than the biggest correlation with any construct, then discriminant validity is recognized. In this study (Table 5), all constructs met this criterion.

The HTMT ratio of correlations was also applied to assess discriminant validity (Hair et al., 2019). The results indicate that the HTMT ratio of correlations between the constructs VIVIDNESS → MTE (= 0.874) was above the recommended threshold (< 0.85), but under the maximum value accepted (< 0.90) and significantly different from 1 (Carrión et al., 2017; Hair et al., 2019). It therefore indicates satisfactory discriminant validity of the constructs (Table 6) and all of the data used to assess the measurement model fulfilled all of the model fit requirements.

Assessing the Structural Model Fit and Hypotheses Testing – Cultural and Heritage Experience

The bootstrapping technique (5000 re-samples) and path analysis (Ringle et al., 2015) were used to analyse the structural model and test the hypotheses. Collinearity was assessed via the variance inflation factor (VIF) values, the explanatory power of the model was assessed by the coefficient of determination (R^2) and the predictive accuracy of the model was measured using the Stone–Geisser test (Ali et al., 2018; Hair et al., 2019).

Table 7 displays the collinearity results. The data show values ranging from 5.044 to 10.805 in sensorial attributes (HR3-melodic, HR4-Sonorus, SM1-Fragant, SM2-Nice smell, TC1-Confortable, TC2-Handy, TC3-Soothing and TC4-Well-shaped) and emotional states (JY3-Enthusiasm, LV1-Affection, LV2-Caring and LV4-Tenderness), which are above the common cut-off threshold of 5 (Hair et al., 2019). The others VIF values related to sensorial attributes, emotional states, memorable elements, and MTE revealed possible (VIF < 5) to ideal collinearity (VIF < 3).

Table 4. Assessment of measurement model on indicator loadings, Cronbach's α, CR, and AVE in Study 1

Measurements and Items	Loadings	Cronbach's α	CR	AVE
Sight		0.893	0.925	0,758
SG1 Aesthetic	0.884			
SG2 Attractive	0.744			
SG3 Beautiful	0.948			
SG4 Pretty	0.893			
Hearing		0.930	0.950	0.826
HR1 Euphonic	0.902			
HR2 Good sounding	0.824			
HR3 Melodic	0.944			
HR4 Sonorous	0.959			
Smell		0.959	0.970	0.889
SM1 Fragrant	0.951			
SM2 Nice smell	0.970			
SM3 Perfumed	0.926			
SM4 Scented	0.924			
Touch		0.970	0.978	0.916
TC1 Comfortable	0.933			
TC2 Handy	0.966			
TC3 Soothing	0.975			
TC4 Well-shaped	0.955			
Joy		0.909	0.932	0.735
JY1 Cheerful	0.730			
JY2 Delight	0.909			
JY3 Enthusiasm	0.912			
JY4 Joy	0.871			
JY5 Pleasure	0.853			
Love		0.950	0.962	0.834
LV1 Affection	0.930			
LV2 Caring	0.921			
LV3 Love	0.849			
LV4 Tenderness	0.944			
LV5 Warm-hearted	0.920			
Positive surprise		0.878	0.915	0.730
PS1 Astonishment	*			
PS2 Amazement	0.783			
PS3 Fascinated	0.919			
PS4 Inspiration	0.882			
PS5 Surprise	0.827			
Recollection		0.824	0.895	0.740
RC1 Reliving	0.825			
RC2 Participate in	0.887			
RC3 Remember	0.868			
Vividness		0.877	0.916	0.732
VV1 Hear in my mind	0.782			
VV2 See in my mind	0.912			
VV3 Feel emotions	0.880			
VV4 Revisit setting	0.843			
Memorable tourism experience		0.817	0.880	0,649
MTE 1 Hedonism	0.813			
MTE 2 Refreshment	0.886			
MTE 3 Meaningfulness	0.794			
MTE 4 Local Culture	0.719			
MTE 5 Novelty	*			

* Deleted indicators, low loading values under the threshold value of 0.708

Source: Own elaboration from SmartPLS 3.0 software

Table 5. Assessment of discriminant validity according to Fornell-Larker criterion in Study 1

Constructs	HR	JY	LV	MTE	PS	RC	SG	SM	TC	VV
Hearing	0.909									
Joy	0.221	0.858								
Love	0.511	0.480	0.913							
MTE	0.166	0.403	0.423	0.806						
Positive Surprise	0.259	0.483	0.478	0.471	0.854					
Recollection	0.262	0.460	0.349	0.690	0.551	0.860				
Sight	0.276	0.411	0.168	0.360	0.456	0.604	0.870			
Smell	0.507	0.155	0.304	0.290	0.175	0.328	0.295	0.943		
Touch	0.547	0.203	0.462	0.346	0.120	0.322	0.235	0.643	0.957	
Vividness	0.208	0.481	0.332	0.745	0.394	0.722	0.432	0.319	0.338	0.856

Source: Own elaboration from SmartPLS 3.0 software

R^2 was used to assess the explanatory power of the model (Ali et al., 2018; Hair et al., 2019). As shown in Figure 4, the constructs joy ($R^2 = 0.188$), positive surprise ($R^2 = 0.232$) and vividness ($R^2 = 0.269$) had weak explanatory power, but the constructs love ($R^2 = 0.313$) and recollection ($R^2 = 0.353$) demonstrated a weak to moderate explanatory power. The model also indicates that the relationships between the SEMs explain approximately 60% of the construction of the cultural & heritage tourism experience ($R^2 = 0.595$).

The Stone–Geisser (Q^2) value was obtained by applying the blindfolding procedure to all of the endogenous constructs that had reflective measurement models (Hair et al., 2017, 2019). As shown in Table 8, the Q^2 values for these constructs were greater than 0, and the values ranged from 0.122 to 0.366, indicating they had a small (Joy = 0.122, Positive surprise = 0.122, and Vividness = 0.125) to medium (Recollection = 0.231, Love = 0.245, and MTE = 0.366) predictive relevance and validity in the model (Hair et al., 2017, 2019).

Table 6. Assessment of discriminant validity according to heterotrait-monotrait ratio of correlation ($HTMT_{0.85}$) in Study 1

Constructs	HR	JY	LV	MTE	PS	RC	SG	SM	TC
Joy	0.222								
Love	0.532	0.505							
MTE	0.190	0.459	0.477						
Positive Surprise	0.277	0.536	0.552	0.546					
Recollection	0.287	0.520	0.396	0.835	0.632				
Sight	0.279	0.448	0.179	0.413	0.470	0.677			
Smell	0.541	0.159	0.315	0.330	0.185	0.362	0.308		
Touch	0.579	0.217	0.471	0.388	0.134	0.359	0.252	0.660	
Vividness	0.225	0.535	0.358	0.874	0.416	0.836	0.461	0.341	0.363

Source: Own elaboration from SmartPLS 3.0 software

Table 7. Assessment of collinearity through VIF values in Study 1

Measurement Sensorial Indicators	VIF	Measurement Emotional Indicators	VIF	Measurement Memory / MTE Indicators	VIF
SG1	2.969	JY1	2.621	RC1	1.612
SG2	1.721	JY2	4.671	RC2	2.210
SG3	4.531	**JY3**	**5.044**	RC3	2.114
SG4	2.905	JY4	3.113	VV1	1,735
HR1	3.052	JY5	2.906	VV2	3.022
HR2	2.352	**LV1**	**5.349**	VV3	2.499
HR3	**8.579**	**LV2**	**6.974**	VV4	2.256
HR4	**9.748**	LV3	2.900	MTE 1	2.229
SM1	**7.359**	**LV4**	**7.765**	MTE 2	2.870
SM2	**9.725**	LV5	4.422	MTE 3	1.830
SM3	4.622	PS1	------	MTE 4	1.435
SM4	4.299	PS2	------	MTE 5	------
TC1	**5.172**	PS3	2.181		
TC2	**8.791**	PS4	3.385		
TC3	**10.805**	PS5	3.095		
TC4	**6.335**				

Source: Own elaboration from SmartPLS 3.0 software

Figure 4. Coefficient of determination (R^2) in Study 1
Source: Own elaboration from SmartPLS 3.0 software

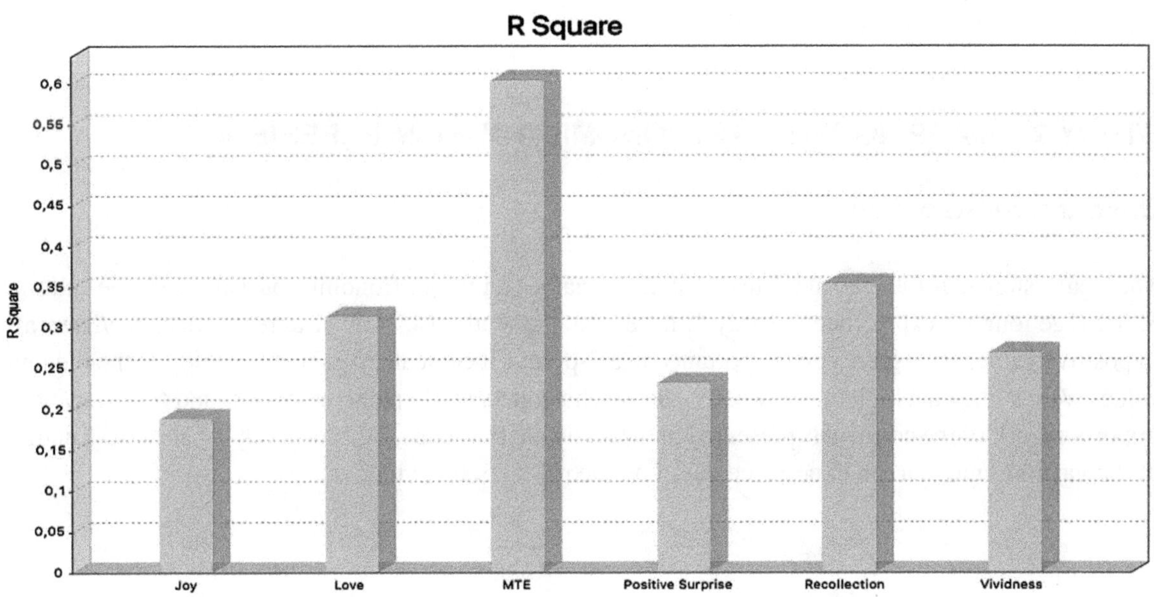

Table 8. Model predictive accuracy (Stone-Geisser Q^2) in Study 1

Constructs	Q^2 (= 1 - SSE / SSO)
Sight	------
Smell	------
Hearing	------
Touch	------
Joy	0.122
Love	0.245
Positive Surprise	0.122
Recollection	0.231
Vividness	0.125
MTE	0.366

Source: Own elaboration from SmartPLS 3.0 software

Table 9 presents the results of the tests of the hypotheses. According to influence of the sensorial stimulus on emotional states, the results supported **Hypothesis 1**, as the corresponding path estimate is positive and statistically significant ($\beta = 0.382$; $t = 3.025$; $p = 0.003$). Similar results were found for **Hypothesis 3**, as the corresponding path estimate is positive and statistically significant ($\beta = 0.442$; $t = 3.802$; $p = 0.000$) and **Hypothesis 5**, as the corresponding path estimate is positive and statistically significant ($\beta = 0.387$; $t = 2.793$; $p = 0.005$).

For the relation between emotional states and the memory process (Table 9), the results supported **Hypothesis 14**, as the corresponding path estimate is positive and statistically significant ($\beta = 0.355$; $t = 3.128$; $p = 0.002$). Similar results were found for **Hypothesis**, as the corresponding path estimate is positive and statistically significant ($\beta = 0.419$; $t = 3.650$; $p = 0.000$), as well as **Hypothesis 19** ($\beta = 0.318$; $t = 2.260$; $p = 0.024$) and **Hypothesis 20** ($\beta = 0.515$; $t = 4.045$; $p = 0.000$). Only 7 of the 20 hypotheses formulated appear to explain the construction of the cultural & heritage tourism experience according to presented research model.

STUDY 2 – MAPPING THE GASTRONOMIC TOURISM EXPERIENCE

Descriptive Statistics

The results shown in Table 10 indicate that all the variables for the gastronomic tourism, as for the cultural & heritage tourism experience in study 1, have average neutral (4, Neither agree or disagree/Neutral) to positive values (6, Agree/Very important) on a 7-point Likert scale. Again, as in study 1, the median values were somewhat higher – between 5 (Somewhat agree/Moderately important) and 6 (Agree/Very important). All of the constructs presented in the study are thus classified positively for the construction of the gastronomic tourism experience based on the perceptions of tourists.

Table 9. Path coefficient analysis of the structural model and hypotheses testing in Study 1

Constructs	Path Coefficient (β)	SD (±)	*t*-Statistics	*p*-Values	Decision
H1 Sight → Joy	0.382	0.126	3.025	0.003**	**Supported**
H2 Sight → Love	0.018	0.125	0.141	0.888	Not supported
H3 Sight → Positive surprise	0.442	0.111	3.802	0.000*	**Supported**
H4 Hearing → Joy	0.094	0.157	0.597	0.551	Not supported
H5 Hearing → Love	0.387	0.138	2.793	0.005**	**Supported**
H6 Hearing → Positive surprise	0.183	0.162	1.132	0.258	Not supported
H7 Smell → Joy	-0.076	0.134	0.568	0.570	Not supported
H8 Smell → Love	-0.094	0.159	0.595	0.552	Not supported
H9 Smell → Positive surprise	0.015	0.141	0.104	0.917	Not supported
H10 Touch → Positive surprise	0.111	0.172	0.644	0.519	Not supported
H11 Touch → Positive surprise	0.306	0.168	1.828	0.068	Not supported
H12 Touch → Positive surprise	-0.089	0.133	0.670	0.503	Not supported
H13 Joy → Recollection	0.243	0.128	1.902	0.057	Not supported
H14 Joy → Vividness	0.355	0.114	3.128	0.002**	**Supported**
H15 Love → Recollection	0.032	0.081	0.396	0.692	Not supported
H16 Love → Vividness	0.072	0.129	0.555	0.579	Not supported
H17 Positive surprise → Recollection	0.419	0.115	3.650	0.000*	**Supported**
H18 Positive surprise → Vividness	0.188	0.118	1.591	0.112	Not supported
H19 Recollection → MTE	0.318	0.141	2.260	0.024***	**Supported**
H20 Vividness → MTE	0.515	0.127	4.045	0.000*	**Supported**

Notes: * $p \leq 0.001$; ** $p \leq 0.01$; *** $p \leq 0.05$
Source: Own elaboration from SmartPLS 3.0 software

Analysis of the Proposed Model

Assessing the Measurement Model Fit – Gastronomic Experience

Before confirming the reliability of the measurement model, the factor loading scores for all of the indicators were analysed according to the threshold values, ≥ 0.708 (Hair et al., 2019) and eight indicators were deleted – that is, one indicator at a time with low loading values according to the following sequence: MTE1 = 0.523; JY2 = 0.604; JY1 = 0.577; SG2 = 0.597; VV4 = 0.605; VV2 = 0.576; MTE3 = 0.669; and PS = 0.684 (Table 11). The loading scores for all of the other indicators factor were above the recommended threshold (≥ 0.708), indicating the satisfactory reliability of the constructs. As shown in Table 11, the Cronbach's alpha scores were above the threshold values (minimum 0.60 for exploratory research) and ranged from 0.685 to 0.958. The CR ranged from 0.823 to 0.970 (≥ 0.7 threshold values), and the AVE ranged from 0.608 to 0.889 (≥ 0.50 threshold values). Reliability, internal consistency and convergent validity were thus established (Hair et al., 2019; Ringle et al., 2014; Wong, 2013).

Table 10. Descriptive statistics for the observed variables in Study 2

Constructs	Variables	Mean	SD (±)	Median
Sight	SG1 Aesthetic	5.55	1.30	6.00
	SG2 Attractive	5.89	1.28	6.00
	SG3 Beautiful	5.70	1.52	6.00
	SG4 Pretty	5.29	1.72	6.00
Hearing	HR1 Euphonic	4.30	2.38	5.00
	HR2 Good sounding	5.10	1.87	6.00
	HR3 Melodic	4.45	2.22	5.00
	HR4 Sonorous	4.48	2.19	5.00
Taste	TT1 Appetizing	6.01	1.43	6.00
	TT2 Flavorful	5.85	1.47	6.00
	TT3 Palatable	5.91	1.58	6.00
	TT4 Tasty	5.80	1.70	6.00
Smell	SM1 Fragrant	4.70	2.13	5.00
	SM2 Nice smell	4.90	2.03	5.00
	SM3 Perfumed	4.52	2.03	5.00
	SM4 Scented	4.67	2.16	5.00
Touch	TC1 Comfortable	4.98	2.08	6.00
	TC2 Handy	4.66	2.23	5.00
	TC3 Soothing	4.74	2.19	5.00
	TC4 Well-shaped	4.46	2.38	5.00
Joy	JY1 Cheerful	5.78	1.68	6.00
	JY2 Delight	5.88	1.24	6.00
	JY3 Enthusiasm	5.88	1.33	6.00
	JY4 Joy	5.84	1.14	6.00
	JY5 Pleasure	6.02	0.89	6.00
Love	LV1 Affection	5.11	1.94	6.00
	LV2 Caring	5.22	1.80	6.00
	LV3 Love	4.83	2.11	5.00
	LV4 Tenderness	4.65	2.08	5.00
	LV5 Warm-hearted	5.12	1.75	6.00
Positive surprise	PS1 Astonishment	4.46	2.08	5.00
	PS2 Amazement	4.55	1.97	5.00
	PS3 Fascinated	5.00	1.73	6.00
	PS4 Inspiration	5.29	1.59	6.00
	PS5 Surprise	4.61	1.87	5.00
Recollection	RC1 Reliving	5.63	1.35	6.00
	RC2 Participate in	5.75	1.32	6.00
	RC3 Remember	5.22	1.68	6.00
Vividness	VV1 Hear in my mind	5.09	1.68	5.00
	VV2 See in my mind	6.01	0.85	6.00
	VV3 Feel emotions	5.35	1.66	6.00
	VV4 Revisit setting	5.93	0.98	6.00
MTE	MTE 1 Hedonism	6.15	1.19	6.00
	MTE 2 Refreshment	5.44	1.74	6.00
	MTE 3 Meaningfulness	4.76	1.93	5.00
	MTE 4 Local Culture	5.28	1.70	5.00
	MTE 5 Novelty	5.48	1.82	6.00

Source: Own elaboration from SmartPLS 3.0 software

The Fornell-Larcker criterion was then used to assess the discriminant validity, as in study 1 (Fornell & Larcker, 1981). In this study (Table 12), all constructs met this criterion.

The HTMT ratio of correlations was also applied to assess the discriminant validity (Hair et al., 2019). Table 13 shows the results, and all of HTMT values are under the recommended threshold (HTMT < 0.85), indicating the satisfactory discriminant validity of the constructs (Carrión et al., 2017; Hair et al., 2019). All the data used to assess the measurement model thus fulfilled all of the model fit requirements.

Assessing the Structural Model Fit and Hypotheses Testing – Gastronomic Experience

As in study 1, the bootstrapping technique (5000 re-samples) and path analysis were performed to analyse the structural model and test the hypotheses (Ringle et al., 2015). Collinearity was assessed with VIF values, the explanatory power of the model was assessed by R^2 and the predictive accuracy of the model was measured using the Stone–Geisser test (Ali et al., 2018; Hair et al., 2019).

Table 14 displays the collinearity results, and the data show same probable collinearity issues because the values ranging from 5.709 to 15.768 in sensorial attributes (HR3-melodic, HR4-Sonorus, TT1- Appetizing, TT2-Flavourful, SM3-Perfumed, SM4-Scented, TC1-Confortable, TC2-Handy, TC3-Soothing, and TC4-Well-shped) and emotional states (LV3-Love, LV4-Tenderness, PS1-Atonishment, and PS2-Amazement) are above the common cut-off threshold of 5 (Hair et al., 2019). The other VIF values related to sensorial attributes, emotional states, memorable elements, and MTE, revealed possible (VIF < 5) to ideal collinearity (VIF < 3).

R^2 was used to assess the explanatory power of the model (Ali et al., 2018; Hair et al., 2019). As shown in Figure 5, the constructs joy ($R^2 = 0.210$), recollection ($R^2 = 0.232$) and positive surprise ($R^2 = 0.392$) had weak explanatory power, while the constructs love ($R^2 = 0.450$), vividness ($R^2 = 0.479$) and MTE ($R^2 = 0.516$) demonstrated moderate explanatory power. The model also indicates that the relationships between the SEMs explain approximately 52% of the construction of the gastronomic tourism experience ($R^2 = 0.516$).

The Q^2 value was obtained by applying the blindfolding procedure to all endogenous constructs that had reflective measurement models (Hair et al., 2017, 2019). As shown in Table 15, the Q^2 values for these constructs were greater than 0, and the values ranged from 0.038 to 0.366, indicating they had a small (Joy = 0.038 and Recollection = 0.119) to medium (Positive surprise = 0.268, Vividness = 0.317, Love = 0.324, and MTE = 0.366) predictive relevance and validity in the model (Hair et al., 2017, 2019).

Table 16 presents the results of the tests of the hypotheses. According to influence of the sensorial stimulus on the emotional states, the results supported **Hypothesis 22**, as the corresponding path estimate is positive and statistically significant ($\beta = 0.428$; $t = 2.895$; $p = 0.004$**). The same is true for **Hypothesis 25** ($\beta = 0.260$; $t = 2.183$; $p = 0.029$***); **Hypothesis 29** ($\beta = -0.229$; $t = 1.993$; $p = 0.046$***); **Hypothesis 30** ($\beta = 0.563$; $t = 2.449$; $p = 0.014$***); **Hypothesis 32** ($\beta = 0.721$; $t = 3.631$; $p = 0.000$*); and **Hypothesis 35** ($\beta = -0.422$; $t = 2.196$; $p = 0.028$***).

For the relation between emotional states and the memory process (Table 16), the results supported **Hypothesis 41**, as the corresponding path estimate is positive and statistically significant ($\beta = 0.373$; $t = 1.908$; $p = 0.050$***), as well as **Hypothesis 43** (Table 16) regarding the relation between memory process and MTE construction ($\beta = 0.570$; $t = 2.773$; $p = 0.006$**).

Table 11. Assessment of measurement model on indicator loadings, Cronbach's α, Composite Reliability, and AVE in Study 2

Measurements and Items	Loadings	Cronbach's α	CR	AVE
Sight		0.860	0.904	0.758
SG1 Aesthetic	0.793			
SG2 Attractive	*			
SG3 Beautiful	0.903			
SG4 Pretty	0.912			
Hearing		0.907	0.936	0.785
HR1 Euphonic	0.901			
HR2 Good sounding	0.765			
HR3 Melodic	0.932			
HR4 Sonorous	0.936			
Taste		0.935	0.953	0.837
TT1 Appetizing	0.973			
TT2 Flavorful	0.953			
TT3 Palatable	0.864			
TT4 Tasty	0.863			
Smell		0.958	0.970	0.889
SM1 Fragrant	0.941			
SM2 Nice smell	0.894			
SM3 Perfumed	0.975			
SM4 Scented	0.960			
Touch		0.953	0.966	0.875
TC1 Comfortable	0.933			
TC2 Handy	0.944			
TC3 Soothing	0.946			
TC4 Well-shaped	0.919			
Joy		0.685	0.823	0.608
JY1 Cheerful	*			
JY2 Delight	*			
JY3 Enthusiasm	0.773			
JY4 Joy	0.816			
JY5 Pleasure	0.748			
Love		0.920	0.941	0.762
LV1 Affection	0.921			
LV2 Caring	0.875			
LV3 Love	0.912			
LV4 Tenderness	0.890			
LV5 Warm-hearted	0.755			
Positive surprise		0.913	0.939	0.795
PS1 Astonishment	0.917			
PS2 Amazement	0.941			
PS3 Fascinated	0.808			
PS4 Inspiration	*			
PS5 Surprise	0.896			
Recollection		0.839	0.903	0.757
RC1 Reliving	0.896			
RC2 Participate in	0.819			
RC3 Remember	0.893			
Vividness		0.764	0.894	0.808
VV1 Hear in my mind	0.876			
VV2 See in my mind	*			
VV3 Feel emotions	0.921			
VV4 Revisit setting	*			
Memorable tourism experience		0.866	0.917	0.787
MTE 1 Hedonism	*			
MTE 2 Refreshment	0.882			
MTE 3 Meaningfulness	*			
MTE 4 Local Culture	0.897			
MTE 5 Novelty	0.882			
* Deleted indicators, low loading values under the threshold value of 0.708				

Source: Own elaboration from SmartPLS 3.0 software

Table 12. Assessment of discriminant validity according to Fornell-Larker criterion in Study 2

Constructs	HR	JY	LV	MTE	PS	RC	SG	SM	TT	TC	VV
Hearing	0.886										
Joy	0.101	0.780									
Love	0.539	0.326	0.873								
MTE	0.363	0.558	0.515	0.887							
P. Surprise	0.429	0.431	0.690	0.481	0.892						
Recollection	0.041	0.438	0.197	0.515	0.355	0.870					
Sight	0.532	0.217	0.575	0.205	0.384	0.197	0.871				
Smell	0.463	0.380	0.414	0.436	0.470	0.123	0.234	0.943			
Taste	0.157	0.188	0.096	0.279	-0.014	0.346	-0.004	0.379	0.915		
Touch	0.608	0.247	0.445	0.326	0.372	0.083	0.434	0.806	0.225	0.936	
Vividness	0.351	0.444	0.584	0.686	0.639	0.477	0.216	0.358	0.065	0.347	0.899

Source: Own elaboration from SmartPLS 3.0 software

DISCUSSION

The TWO proposed models (Figures 6 and 7) explaining the positive relationship between SEMs to establish a memorable cultural tourism experience from a tourist's perspective are supported. The map analysis of the cultural & heritage experience (Figure 6) demonstrated two different patterns for developing a memorable cultural tourism experience according to the emotional states activated and the memorable elements triggered. The first model (Figure 6) starts with the sense of sight, which activates the emotional state of joy and ends with memorable element of vividness. The other model (Figure 6) also starts with the sense of sight, which activates the emotional state of positive surprise and ends with the memorable element of recollection. These results suggest that a memorable cultural & heritage tourism

Table 13. Assessment of discriminant validity according to heterotrait-monotrait ratio of correlation ($HTMT_{0.85}$) in Study 2

Constructs	HR	JY	LV	MTE	PS	RC	SG	SM	TT	TC
Joy	0.230									
Love	0.584	0.370								
MTE	0.413	0.670	0.581							
Positive Surprise	0.466	0.489	0.745	0.537						
Recollection	0.159	0.605	0.224	0.596	0.403					
Sight	0.543	0.276	0.583	0.220	0.359	0.244				
Smell	0.498	0.422	0.442	0.480	0.494	0.139	0.203			
Taste	0.215	0.241	0.127	0.302	0.076	0.377	0.052	0.389		
Touch	0.653	0.275	0.474	0.355	0.397	0.138	0.438	0.848	0.225	
Vividness	0.422	0.561	0.681	0.819	0.760	0.584	0.187	0.411	0.153	0.397

Source: Own elaboration from SmartPLS 3.0 software

Table 14. Assessment of collinearity through VIF values in Study 2

Measurement Indicators	VIF	Measurement Indicators	VIF	Measurement Indicators	VIF
SG1	3.165	JY1	------	RC1	2.565
SG2	------	JY2	------	RC2	1.610
SG3	4.187	JY3	1.179	RC3	2.420
SG4	1.822	JY4	1.720	VV1	1.620
HR1	2.826	JY5	1.576	VV2	------
HR2	1.737	LV1	4.546	VV3	1.620
HR3	**8.733**	LV2	2.910	VV4	------
HR4	**8.684**	**LV3**	**6.179**	MTE 1	------
TT1	**9.988**	**LV4**	**5.709**	MTE 2	1.891
TT2	**8.420**	LV5	1.844	MTE 3	------
TT3	2.638	**PS1**	**7.130**	MTE 4	2.686
TT4	3.430	**PS2**	**8.381**	MTE 5	2.592
SM1	4.832	PS3	1.903		
SM2	3.380	PS4	------		
SM3	**15.768**	PS5	3.068		
SM4	**11.985**				
TC1	**9.828**				
TC2	**7.795**				
TC3	**8.992**				
TC4	**6.163**				

Source: Own elaboration from SmartPLS 3.0 software

experience can be established based on the emotional states of joy or positive surprise. These models explain approximately 60% of the cultural & heritage tourism experiences ($R^2 = 0.595$).

The results of the present study contradict the results of previous studies (e.g. Agapito et al., 2017; Dias et al., 2017; Jelincic & Senkic, 2017; Meacci & Liberatore, 2018) on what multisensorial stimulus makes the tourism experience more remarkable for tourists' emotional activation, memory encoding and reinforcement of positive memories. This can be explained based on the main activity of the cultural & heritage experience – that is, visual contemplation and passive observation of, for example, architecture, visual arts, dance performance, festival, history, traditions, and customs. The resulting emotional states did, however, trigger high levels of memorable elements, namely, recollection and vividness (Gazzaniga et al., 2014; Kim, 2010; Todd et al., 2012; Tung et al., 2017). Moreover, these results are consistent with other studies showing that tourists' positive emotions influence their favourable evaluation of their experience (Hosany et al., 2015), and the experience of positive emotions while engaged in tourism is associated with having a memorable tourism experience (Knobloch et al., 2017). The sense of hearing showed a strong relation with the emotional state of love ($R^2 = 0.313$), but this emotional state was not sufficiently activated to promote memorable elements.

Figure 5. Coefficient of determination (R^2) in Study 2
Source: Own elaboration from SmartPLS 3.0 software

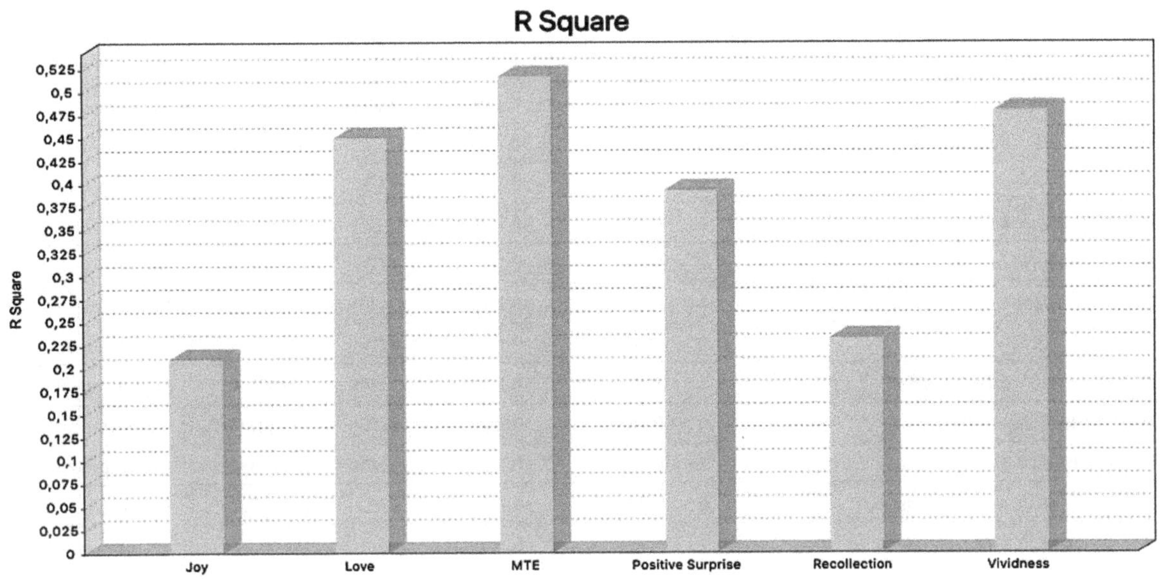

The mapping of the gastronomic experience (Figure 7) demonstrated that developing a memorable gastronomic tourism experience starts with the senses of smell, taste, and touch, which activate the emotional states of positive surprise and ends with the memorable element of vividness. This model explains approximately 52% of the gastronomic tourism experiences ($R^2 = 0.516$).

The results of the present study are consistent with results of previous studies (e.g. Agapito et al., 2017; Dias et al., 2017; Jelincic & Senkic, 2017; Meacci & Liberatore, 2018) based on multisensorial stimulus making the tourism experience more remarkable and likely to activate tourists' emotions,

Table 15. Model predictive accuracy (Stone-Geisser Q^2) in Study 2

Constructs	Q^2 (= 1 – SSE / SSO)
Sight	------
Smell	------
Taste	------
Hearing	------
Touch	------
Joy	0.038
Love	0.324
Positive Surprise	0.268
Recollection	0.119
Vividness	0.317
MTE	0.366

Source: Own elaboration from SmartPLS 3.0 software

Table 16. Path coefficient analysis of the structural model and hypotheses testing in Study 2

Constructs	Path Coefficient (β)	SD (±)	*t*-Statistics	*p*-Values	Decision
H21 Sight → Joy	0.281	0.163	1.721	0.085	Not supported
H22 Sight → Love	**0.428**	**0.148**	**2.895**	**0.004****	**Supported**
H23 Sight → Positive surprise	0.266	0.138	1.932	0.053	Not supported
H24 Hearing → Joy	-0.176	0.164	1.075	0.282	Not supported
H25 Hearing → Love	**0.260**	**0.119**	**2.183**	**0.029*****	**Supported**
H26 Hearing → Positive surprise	0.246	0.130	1.809	0.059	Not supported
H27 Taste → Joy	0.056	0.156	0.358	0.720	Not supported
H28 Taste → Love	-0.033	0.130	0.256	0.798	Not supported
H29 Taste → Positive surprise	**-0.229**	**0.115**	**1.993**	**0.046*****	**Supported**
H30 Smell → Joy	**0.563**	**0.230**	**2.449**	**0.014*****	**Supported**
H31 Smell → Love	0.337	0.185	1.823	0.068	Not supported
H32 Smell → Positive surprise	**0.721**	**0,198**	**3.631**	**0.000***	**Supported**
H33 Touch → Positive surprise	-0.234	0.188	1.246	0.213	Not supported
H34 Touch → Positive surprise	-0.162	0.198	0.822	0.411	Not supported
H35 Touch → Positive surprise	**-0.422**	**0.192**	**2.196**	**0.028*****	**Supported**
H36 Joy → Recollection	0.353	0.216	1.635	0.102	Not supported
H37 Joy → Vividness	0.198	0.102	1.944	0.056	Not Supported
H38 Love → Recollection	-0.111	0.207	0.537	0.592	Not supported
H39 Love → Vividness	0.261	0.186	1.403	0.161	Not supported
H40 Positive surprise → Recollection	0.280	0.210	1.333	0.182	Not supported
H41 Positive surprise → Vividness	**0.373**	**0.196**	**1.908**	**0.050*****	**Supported**
H42 Recollection → MTE	0.243	0.191	1.275	0.202	Not supported
H43 Vividness → MTE	**0.570**	**0.205**	**2.773**	**0.006****	**Supported**

Notes: * $p \leq 0.001$; ** $p \leq 0.01$; *** $p \leq 0.05$
Source: Own elaboration from SmartPLS 3.0 software

memory encoding and reinforcement of positive memories. These results can be explained according to the main activities in the gastronomic experience –taste, smell, and feeling/chewing food. Positive surprise (emotional state) triggered a high level of the memorable element vividness (Gazzaniga et al., 2014; Kim, 2010; Todd et al., 2012). These results are consistent with other studies showing that tourists' positive emotions influence their favourable evaluation of their experience (Hosany et al., 2015), and experiencing positive emotions while engaged in tourism is associated with having a memorable tourism experience (Knobloch et al., 2017). The senses of hearing and sight showed a strong relation with the emotional states of love ($R^2 = 0.450$), but this was not sufficient activated to promote memorable elements. The sense of sight also showed a relation with the emotional state of joy ($R^2 = 0.450$), but, again, this was not sufficiently activated to be memorable.

Figure 6. Hypotheses testing model, coefficient of determination values (R^2), and model predictive accuracy values (Stone-Geisser Q^2) in Study 1
Source: Own elaboration

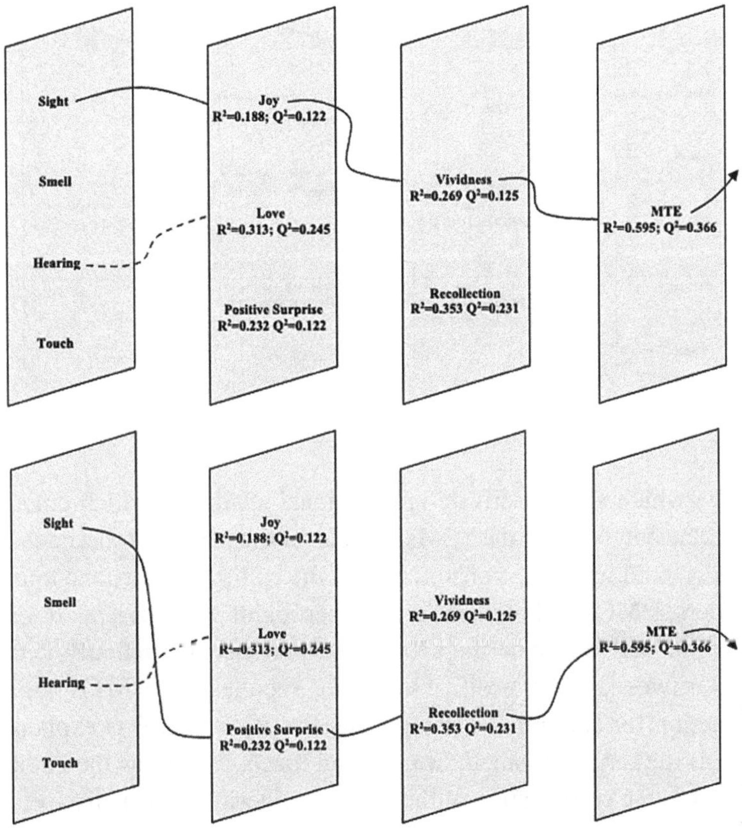

FUTURE RESEARCH DIRECTIONS

Future research should use smaller questionnaires and/or it should be developed on a digital device with a touchscreen (tablet, iPad, etc.), because such devices are more appealing and have user-friendly interfaces. Finally, in the future it seems important to adopt a more detailed posture – that is, mapping more experiences according to the endogenous products at the destination (e.g., 3S, wine, sports and fitness, SPA and wellness and hotel and resort experiences) to create a matrix/map for each tourism experience and increase experience knowledge at the destination.

CONCLUSION

This paper provides an overview of the role of the SEMs on the construction of the cultural tourism experience. Specifically, the purpose of this research was to identify and assess the role of senses, emotional states, and memorable elements within the creation of the cultural & heritage and gastronomic tourism experiences. It also sought to map and model the relationship between the SEMs in these experiences

Figure 7. Hypotheses testing model, coefficient of determination values (R^2), and model predictive accuracy values (Stone-Geisser Q^2) in Study 2
Source: Own elaboration

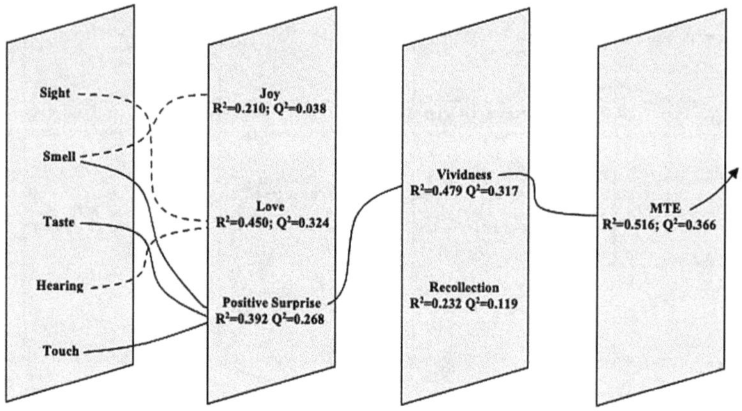

and, more specifically, which senses activate an emotional state, and which emotional states trigger memorable elements. The outcomes of the SEMs models for the cultural & heritage tourism experience are: (a) additional understanding of the construction of the cultural & heritage tourism experience for researchers, stakeholders, DMOs and marketers; (b) assertion of a positive relation between the SEMs and the construction of the cultural & heritage tourism experience, particularly, between sight > joy > vividness > MTE and between sight > positive surprise > recollection > MTE; (c) approximately 60% of experience construction for cultural & heritage tourism ($R^2 = 0.595$) is explained by the variance in the SEMs model; (d) the senses of smell, hearing and touch, as well as the emotional state of love, appear to have no effect in the construction of the cultural & heritage tourism experience; and (e) there appear to be two clear paths for the construction of the cultural & heritage tourism experience according to the SEMs model (Figure 6).

The results of the SEMs model for the gastronomic tourism experience provide: (a) additional understanding of the construction of the gastronomic tourism experience for researchers, stakeholders, DMOs, and marketers; (b) assertion of a positive relation between the SEMs and the gastronomic tourism experience construction, specifically, between smell + taste + touch > positive surprise > vividness > MTE; (c) approximately 52% of the experience construction for gastronomic tourism ($R^2 = 0.516$) is explained by variance in the SEMs model; (d) the senses of sight and hearing, as well as the emotional states of joy and love, appear to have no effect in construction of the gastronomic tourism experience; and (e) there appears to be a distinct path for the construction of the gastronomic tourism experience based on the SEMs model (Figure 7).

This study also improves the understanding of the cultural tourism experience construction process and enriches knowledge for the design and implementation of an experience creation process, based on the SEMs model for cultural & heritage tourism and gastronomic tourism experiences. This research also fills a gap in the literature by investigating the influence of the senses on the emotions and the subsequent impact of the emotions on memories for creating tourist engagement in the cultural & heritage experience and/or in the gastronomic experience. The findings also provide insight to destination stakeholders, DMOs, marketers and others interested in more detailed information about cultural & heritage experience and/or gastronomic tourism experience, as well as strategies for increasing destination memories, satisfaction, loyalty, engagement and cultural contact, as well as mitigating seasonality. All of these results are expected to contribute to the competitive advantage of the Algarve as a cultural destination, because "tourism is all about places and people" (Correia et al., 2017, p. 163).

The present study has several limitations. First, this research used convenience sample, and future should collect more representative data that include proportionately sampling from a destination where the tourist population is known, separately during the low and high seasons. This study also assessed memorable elements before the tourists left the destination and thus measured the short-term memory. To collect more reliable data related to memorable elements, researchers could take another sampling two to four weeks after tourists have arrived at home (Agapito, 2013; Kim et al., 2012). The data were collected via questionnaire and the instrument was rather long (three pages), which was a factor in the withdrawal of participants.

ACKNOWLEDGMENT

This research was supported by National Funds provided by FCT – Foundation for Science and Technology through project UIDB/04020/2020.

REFERENCES

Agapito, D. (2013). *The role of sensescapes in tourist experiences in rural areas*. Faculdade de Economia - Universidade do Algarve. https://sapientia.ualg.pt/handle/10400.1/7776

Agapito, D. (2020). The senses in tourism design: A bibliometric review. *Annals of Tourism Research, 83*(December). doi:10.1016/j.annals.2020.102934

Agapito, D., Pinto, P., & Mendes, J. (2017). Tourists' memories, sensory impressions and loyalty: In loco and post-visit study in Southwest Portugal. *Tourism Management, 58*, 108–118. doi:10.1016/j.tourman.2016.10.015

Agapito, D., Valle, P., & Mendes, J. (2014). The sensory dimension of tourist experiences: Capturing meaningful sensory-informed themes in Southwest Portugal. *Tourism Management, 42*, 224–237. doi:10.1016/j.tourman.2013.11.011

Ali, F., Rasoolimanesh, S. M., Sarstedt, M., Ringle, C. M., & Ryu, K. (2018). An assessment of the use of partial least squares structural equation modeling (PLS-SEM) in hospitality research. *International Journal of Contemporary Hospitality Management, 30*(1), 514–538. doi:10.1108/IJCHM-10-2016-0568

Carrión, G. C., Nitzl, C., & Roldán, J. L. (2017). Mediation analyses in partial least squares structural equation modeling: Guidelines and empirical examples. In H. Latan & R. Noonan (Eds.), *Partial Least Squares Path Modeling: Basic Concepts, Methodological Issues and Applications* (pp. 173–195). Springer. doi:10.1007/978-3-319-64069-3_8

Chandralal, L., & Valenzuela, F.-R. (2013). Exploring Memorable Tourism Experiences: Antecedents and Behavioural Outcomes. *Journal of Economics, Business and Management, 1*(2), 177–181.

Chen, H., & Rahman, I. (2018). Cultural tourism: An analysis of engagement, cultural contact, memorable tourism experience and destination loyalty. *Tourism Management Perspectives*, *26*, 153–163. doi:10.1016/j.tmp.2017.10.006

Colombo, P., & Gold, P. (2004). Multiple memory systems. *Neurobiology of Learning and Memory*, *82*(3), 169–170. doi:10.1016/j.nlm.2004.07.008 PMID:15464401

Correia, A., Oliveira, C., & Pereira, R. (2017). From Emotions to Place Attachment. In A. Correia, M. Kozak, J. Gnoth, & A. Fyall (Eds.), *Co-Creation and Well-Being in Tourism* (pp. 163–177). Springer International Publishing. doi:10.1007/978-3-319-44108-5_13

Dias, J., Correia, A., & Cascais, T. (2017). Traits in Tourists' Experiences: Senses, Emotions and Memories. In A. Correia, M. Kozak, J. Gnoth, & A. Fyall (Eds.), *Co-Creation and Well-Being in Tourism* (pp. 179–194). Springer International Publishing., doi:10.1007/978-3-319-44108-5_14

Dixit, S. K. (2019). The Routledge handbook of gastronomic tourism (S. K. Dixit, Ed.). Taylor & Francis Group. doi:10.1080/13032917.2019.1643641

Ekman, P. (1992). Are there basic emotions? *Psychological Review*, *99*(3), 550–553. doi:10.1037/0033-295X.99.3.550 PMID:1344638

EU. (2010). *Green Paper - Unlocking the potential of cultural and creative industries* (Vol. 148). EU.

Fallis, A. (2013). Emotions and Decision Making. *Annual Review of Psychology*, *53*(9), 1689–1699. doi:10.1017/CBO9781107415324.004

Fornell, C., & Larcker, D. F. (1981). Evaluating Structural Equation Models with Unobservable Variables and Measurement Error. *JMR, Journal of Marketing Research*, *18*(1), 39–50. doi:10.1177/002224378101800104

Gazzaniga, M. S., Ivry, R. B., & Mangun, G. R. (2014). *Cognitive Neuroscience: The Biology of the Mind* (A. Javsicas & S. Snavely, Eds.; 4th ed.). W. W. Norton & Company, Inc.

Guerreiro, M. M., Henriques, C., & Mendes, J. (2019). Cultural and Creative Tourism: The Case of 'Celebrations' in the Algarve Region. *Journal of Spatial and Organizational Dynamics*, *VII*(4), 320–338.

Güzel, F. Ö. (2014). The Dimensions of Tour Experience, Emotional Arousal, and Post-experience Behaviors: A Research on Pamukkale in Turkey. *Procedia: Social and Behavioral Sciences*, *150*, 521–530. doi:10.1016/j.sbspro.2014.09.069

Haase, J., Wiedmann, K. P., & Bettels, J. (2020). Sensory imagery in advertising: How the senses affect perceived product design and consumer attitude. *Journal of Marketing Communications*, *26*(5), 475–487. doi:10.1080/13527266.2018.1518257

Hair, J., Hult, G., Ringle, C., & Sarstedt, M. (2017). *A Primer on Partial Least Squares Structural Equation Modeling (PLS-SEM)* (2nd ed.). Sage Publications., doi:10.1016/j.lrp.2013.01.002

Hair, J., Risher, J., Sarstedt, M., & Ringle, C. (2019). When to use and how to report the results of PLS-SEM. *European Business Review*, *31*(1), 2–24. doi:10.1108/EBR-11-2018-0203

Hosany, S., Prayag, G., Deesilatham, S., Caušević, S., & Odeh, K. (2015). Measuring Tourists' Emotional Experiences: Further Validation of the Destination Emotion Scale. *Journal of Travel Research, 54*(4), 482–495. doi:10.1177/0047287514522878

Hosany, S., Prayag, G., Van Der Veen, R., Huang, S., & Deesilatham, S. (2017). Mediating Effects of Place Attachment and Satisfaction on the Relationship between Tourists' Emotions and Intention to Recommend. *Journal of Travel Research, 56*(8), 1079–1093. doi:10.1177/0047287516678088

Jelincic, D. A., & Senkic, M. (2017). Creating a Heritage Tourism Experience: The Power of the Senses. *Etnološka Tribina, 47*(40), 109–126. doi:10.15378/1848-9540.2017.40.03

Kim, J.-H. (2010). Determining the factors affecting the memorable nature of travel experiences. *Journal of Travel & Tourism Marketing, 27*(8), 780–796. doi:10.1080/10548408.2010.526897

Kim, J.-H., & Jang, S. (2016). Memory Retrieval of Cultural Event Experiences: Examining Internal and External Influences. *Journal of Travel Research, 55*(3), 322–339. doi:10.1177/0047287514553058

Kim, J.-H., Ritchie, J., & McCormick, B. (2012). Development of a scale to measure memorable tourism experiences. *Journal of Travel Research, 51*(1), 12–25. doi:10.1177/0047287510385467

Kim, J.-H., & Youn, H. (2017). How to Design and Deliver Stories about Tourism Destinations. *Journal of Travel Research, 56*(6), 808–820. doi:10.1177/0047287516666720

Kim, M., & Thapa, B. (2018). Perceived value and flow experience: Application in a nature-based tourism context. *Journal of Destination Marketing and Management, 8*(August), 373–384. doi:10.1016/j.jdmm.2017.08.002

Kivela, J., & Crotts, J. C. (2009). Understanding Travelers' Experiences of Gastronomy Through Etymology and Narration. *Journal of Hospitality & Tourism Research (Washington, D.C.), 33*(2), 161–192. doi:10.1177/1096348008329868

Knobloch, U., Robertson, K., & Aitken, R. (2017). Experience, Emotion, and Eudaimonia: A Consideration of Tourist Experiences and Well-being. *Journal of Travel Research, 56*(5), 651–662. doi:10.1177/0047287516650937

Krishna, A. (2012). An integrative review of sensory marketing: Engaging the senses to affect perception, judgment and behavior. In Journal of Consumer Psychology (Vol. 22, Issue 3, pp. 332–351). doi:10.1016/j.jcps.2011.08.003

Krishna, A., & Schwarz, N. (2014). Sensory marketing, embodiment, and grounded cognition: A review and introduction. *Journal of Consumer Psychology, 24*(2), 159–168. doi:10.1016/j.jcps.2013.12.006

Lv, X., Li, C., & McCabe, S. (2020). Expanding theory of tourists' destination loyalty: The role of sensory impressions. *Tourism Management, 77*, 104–126. doi:10.1016/j.tourman.2019.104026

Matos, N. (2014). *The Impacts of Tourism Experiences in the Destination Image. A Marketing Perspective* (PhD Thesis). Faculty of Economics, University of Algarve.

Meacci, L., & Liberatore, G. (2018). A senses-based model for experiential tourism. *Tourism & Management Studies, 14*(4), 7–14. doi:10.18089/tms.2018.14401

Mendes, J., Guerreiro, M., & Matos, N. (2016). Monitoring the Quality of Tourism Experience. *Proceedings of the Asia Tourism Forum 2016 - the 12Th Biennial Conference of Hospitality and Tourism Industry in Asia, 19*(1), 294–304. 10.2991/atf-16.2016.44

Mendes, J., Valle, P., Guerreiro, M., & Silva, J. (2010). The tourist experience - Exploring the relationship between tourist satisfaction and destination loyalty. *Tourism (Zagreb), 58*(2), 111–126. https://sapientia.ualg.pt/handle/10400.1/4517

Moreira, A. C., Fortes, N., & Santiago, R. (2017). Influence of Sensory Stimuli on Brand Experience, Brand Equity and Purchase Intention. *Journal of Business Economics and Management, 18*(1), 68–83. doi:10.3846/16111699.2016.1252793

Moyle, B. D., Moyle, C., Bec, A., & Scott, N. (2019). The next frontier in tourism emotion research. *Current Issues in Tourism, 22*(12), 1393–1399. doi:10.1080/13683500.2017.1388770

Oh, H., Fiore, A., & Jeoung, M. (2007). Measuring Experience Economy Concepts: Tourism Applications. *Journal of Travel Research, 46*(2), 119–132. doi:10.1177/0047287507304039

Pedro, R. M. (2019). *Senses, Emotions and Memories in Tourism* (PhD Thesis). Faculty of Economics, University of Algarve.

Pedro, R. M., Mendes, J., de Matos, N., & Ascenção, M. P. (2020). A Conceptualization of the Senses, Emotions, and Memories in Memorable Tourism Experiences. In P. Pinto & M. Guerreiro (Eds.), *Handbook of Research on Resident and Tourist Perspectives on Travel Destinations* (pp. 208–231). IGI Global. doi:10.4018/978-1-7998-3156-3.ch010

Ringle, C., Silva, D., & Bido, D. (2014). Structural Equation Modeling with the SmartPls. *Revista Brasileira de Marketing, 13*(2), 54–71. doi:10.5585/remark.v13i2.2717

Ringle, C. M., Wende, S., & Becker, J. M. (2015). *SmartPLS 3.0.* Scientific Research Publishing - SmartPLS GmbH, Boenningstedt. https://www.scirp.org/(S(vtj3fa45qm1ean45vvffcz55))/reference/ReferencesPapers.aspx?ReferenceID=2060940

Sanagustín-Fons, M. V., Tobar-Pesántez, L. B., & Ravina-Ripoll, R. (2020). Happiness and cultural tourism: The perspective of civil participation. *Sustainability, 12*(8), 3465–3485. doi:10.3390u12083465

Sánchez-Sánchez, M. D., De-Pablos-Heredero, C., & Montes-Botella, J. L. (2020). A behaviour model for cultural tourism: Loyalty to destination. *Economic Research Journal, 0*(0), 1–18. doi:10.1080/1331677X.2020.1838313

Seyfi, S., Hall, C. M., & Rasoolimanesh, S. M. (2020). Exploring memorable cultural tourism experiences. *Journal of Heritage Tourism, 15*(3), 341–357. doi:10.1080/1743873X.2019.1639717

Shoval, N., Schvimer, Y., & Tamir, M. (2018). Real-Time Measurement of Tourists' Objective and Subjective Emotions in Time and Space. *Journal of Travel Research, 57*(1), 3–16. doi:10.1177/0047287517691155

Squire, L. (2004). Memory systems of the brain: A brief history and current perspective. *Neurobiology of Learning and Memory, 82*(3), 171–177. doi:10.1016/j.nlm.2004.06.005 PMID:15464402

Sthapit, E. (2019). Memories of gastronomic experiences, savoured positive emotions and savouring processes. *Scandinavian Journal of Hospitality and Tourism, 19*(2), 115–139. doi:10.1080/15022250.2017.1402702

Sthapit, E., & Coudounaris, D. (2017). Memorable tourism experiences: Antecedents and outcomes. *Scandinavian Journal of Hospitality and Tourism, 18*(1), 72–94. doi:10.1080/15022250.2017.1287003

Sukanthasirikul, K., & Trongpanich, W. (2016). Cultural Tourism Experience on Customer Satisfaction: Evidence From Thailand. *Journal of Economic and Social Development, 3*(1), 17–25.

Tettamanti, M., Rognoni, E., Cafiero, R., Costa, T., Galati, D., & Perani, D. (2012). Distinct pathways of neural coupling for different basic emotions. *NeuroImage, 59*(2), 1804–1817. doi:10.1016/j.neuroimage.2011.08.018 PMID:21888979

Todd, R., Talmi, D., Schmitz, T., Susskind, J., & Anderson, A. (2012). Psychophysical and neural evidence for emotion enhanced perceptual vividness. *The Journal of Neuroscience: The Official Journal of the Society for Neuroscience, 32*(33), 11201–11212. doi:10.1523/JNEUROSCI.0155-12.2012 PMID:22895705

Tolić, I. (2020). Creating and managing experiences in cultural tourism. In Croatian International Relations Review (Vol. 26, Issue 86). Institute for Development and International Relations. doi:10.1142/10809

Tung, V., Lin, P., Zhang, H., & Zhao, A. (2017). A framework of memory management and tourism experiences. *Journal of Travel & Tourism Marketing, 34*(7), 853–866. doi:10.1080/10548408.2016.1260521

Tung, V., & Ritchie, J. R. B. (2011). Exploring the essence of memorable tourism experiences. *Annals of Tourism Research, 38*(4), 1367–1386. doi:10.1016/j.annals.2011.03.009

UNESCO. (2009). *The 2009 UNESCO Framework for Cultural Statistics (FCS)*. UNESCO Institute for Statistics.

UNWTO. (2018). Report on tourism and culture synergies. In *Tourism and Culture Synergies*. World Tourism Organization (UNWTO). https://www.e-unwto.org/doi/book/10.18111/9789284418978

Usakli, A., & Kucukergin, K. G. (2018). Using partial least squares structural equation modeling in hospitality and tourism: Do researchers follow practical guidelines? *International Journal of Contemporary Hospitality Management, 30*(11), 3462–3512. doi:10.1108/IJCHM-11-2017-0753

Vergori, A. S., & Arima, S. (2020). Cultural and non-cultural tourism: Evidence from Italian experience. *Tourism Management, 78*, 104058. Advance online publication. doi:10.1016/j.tourman.2019.104058

Williams, H. A., Yuan, J., & Williams, R. L. Jr. (2019). Attributes of Memorable Gastro-Tourists' Experiences. *Journal of Hospitality & Tourism Research (Washington, D.C.), 43*(3), 327–348. doi:10.1177/1096348018804621

Wong, K. (2013). Partial Least Squares Structural Equation Modeling (PLS-SEM) Techniques Using SmartPLS. *Marketing Bulletin - Technical Note, 24*, 1–32.

Yamashita, S. (2016). In J. Jafari & H. Xiao (Eds.), *Cultural tourism BT - Encyclopedia of Tourism* (pp. 212–214). Springer International Publishing. doi:10.1007/978-3-319-01384-8_45

ADDITIONAL READING

Bastiaansen, M., Lub, X., Mitas, O., Jung, T., Ascenção, M. P., Han, D., Moilanen, T., Smit, B., & Strijbosch, W. (2019). Emotions as core building blocks of an experience. *International Journal of Contemporary Hospitality Management*, *31*(2), 651–668. doi:10.1108/IJCHM-11-2017-0761

Haase, J., Wiedmann, K., & Bettels, J. (2020). Sensory imagery in advertising: How the senses affect perceived product design and consumer attitude. *Journal of Marketing Communications*, *26*(5), 475–487. doi:10.1080/13527266.2018.1518257

Kapoor, A. (2015). *Consumer Experiences and Emotion Management* (N. K. Malhotra, Ed.). Business Expert Press.

Kastenholz, E., Carneiro, M., Marques, C., & Loureiro, S. (2018). The dimensions of rural tourism experience: Impacts on arousal, memory, and satisfaction. *Journal of Travel & Tourism Marketing*, *35*(2), 189–201. doi:10.1080/10548408.2017.1350617

KEY TERMS AND DEFINITIONS

Cultural Tourism: This is a type of tourism activity in which the visitor's essential motivation is to learn, discover, experience, and consume the tangible and intangible cultural attractions/products of a tourism destination.

Gastronomic Tourism: This type of tourism activity is characterized by a visitor experience linked with food and related products and activities while travelling. Along with authentic, traditional and/or innovative culinary experiences, gastronomic tourism may also involve other related activities, such as visiting local producers, participating in food festivals, and attending cooking classes.

Long-Term Memory: This encompasses information held in mind for a long period of time, ranging from days or months to the entire lifespan. The long-term memory is an outcome of the information from the short-term memory with the addition of a strong emotional connection and a repetition or recall process.

Memorable Tourism Experience: A tourism experience remembered and recalled after the event has occurred.

Senses, Emotions, and Memories (SEMs) Model: This model makes it possible to examine the relation of senses, emotions, and memories in the construction of the tourist experience.

Sensorial Memory: Know as bodily memory, sensorial memory is memory information input from the sensorial stimulus that is held in the mind for 0.2 to 2 seconds.

Short-Term Memory: This encompasses information that is typically held in the mind for a relatively short period of time, between 2 to 60 seconds, before being forgotten. This specifically memory is the resulting from sensorial memory, more an emotional outcome and attention.

Section 2

Stakeholder Analysis

Chapter 9

Moving Towards Sustainable Cultural Tourism Experiences:

A Participatory Journey - The Stakeholders Perspective

Vânia Mendonça
University of the Algarve, Portugal

Ana Isabel Renda
CiTUR, University of the Algarve, Portugal

Nelson M. Silva de Matos
https://orcid.org/0000-0002-6263-5007
CinTurs, University of the Algarve, Portugal & CiTUR, Portugal

ABSTRACT

This research analyses the stakeholders' perception of the development of sustainable cultural tourism experiences. A case study is presented; the chosen place is a council in the Algarve, South Portugal, a mature destination, where the sun and beach are the main attraction. A qualitative approach was used through interviews with key stakeholders, directly and indirectly involved in tourism. The multidimensional analysis shows the relevant findings on how the stakeholders perceive their cultural resources and offer. The concepts of cultural identity, authenticity, cultural preservation, community well-being, pride, and community empowerment emerged, especially through their intangible cultural heritage. It was observed that stakeholders consider local people as their most important cultural resource. Thus, they are also the source of unique, meaningful, and memorable cultural and sustainable experiences.

INTRODUCTION

All tourist activity has a significant impact on economic, social, cultural and environmental levels. The tourism phenomenon has been growing in recent decades, with a broader diversity of tourist destinations

DOI: 10.4018/978-1-7998-9217-5.ch009

also emerging and, consequently, the competitiveness among them, which seek, at all costs, to differentiate themselves from each other. This increase also affects cultural tourism, and the demand for tourist experiences leads the authors to question the sustainability of tourism activity, tourist destinations, and its residents' quality of life. On the one hand, this increase in tourism demand means more economic opportunities that can contribute to the conservation and revitalisation of historic areas, cultural heritage, and the diversification of the cultural offer. On the other hand, the increasing number of tourists also causes a change in the *modus vivendi* of the inhabitants and their cultural values. The increasingly conscious and informed consumer is looking for alternatives to reduce its ecological footprint, with a growing trend due to concern about his travels' impact on the destination (Cohen, Prayag, & Moital, 2014). Thus, when tourist activity is sustainable, it can motivate tourists, generate positive effects on consumer preferences and choices, which, however, depends on the level of knowledge on sustainability (Garg & Pandey, 2021).

Culture is a crucial element in the differentiation of destinations and promotes sustainable development (Dessein, Soini, Fairchlough, & Horlings, 2015), contributing to preserving authenticity, attenuating seasonality, and developing complementary products (Futur & Nedelea, 2018; Henriques, 2014). Based on these assumptions, there is a recognition of gaps in research concerning the synergy between culture and tourism and how to ensure its sustainability in the integration of cultural and tourism industries (Z. Su, Aaron, McDowell, & Lu, 2019).

In this chapter, the authors analyse a council in the Algarve, the district in the south of Portugal; a mature coastal destination, where the sun and beach have been the main attraction, with a high seasonality index (FCT, 2019). Although it continues to be the most representative of the region, nature tourism and cultural tourism (which attract tourists more concerned with sustainability) are expanding and are increasingly promoted and supported by the Algarve Tourism Region. Its growth potential is not only high but also necessary.

The council of Olhão was somewhat apart from the tourism development that has taken place in general in the Algarve region. Partly because it does not have direct access to the beach (islands), so it is not attractive for the same reason that tourists come to the Algarve, but also because its main industrial activity linked to fishing and canning made it an unattractive place (Gonçalves, Oliveira, & Santos, 2015).

To better understand this problematic, this chapter establishes the objective of assessing stakeholders' perception regarding the relevance of the cultural offer and cultural resources to the development of sustainable cultural tourist experiences using a qualitative methodology. Due to the depth of the information that was intended to be obtained, this qualitative approach uses semi-structured interviews (Quivy, 1995). The interview guide was defined and approved by a panel of experts based on the literature review. Key stakeholders were identified, and 30 interviews were undertaken with local and regional government authorities, residents, associations and players of the tourism industry. Interviews took, on average, 50 minutes, were recorded and transcribed.

The content of the interviews was analysed according to the research objectives, for which the software Nvivo was used (Alves, Filho, & Henrique, 2015). After the codification process, a structural model was created, and it allowed an exploratory search of the data collected and framework matrices were generated. This methodology enables assessing stakeholders' perceptions regarding the relevance of the cultural resources and offer and its potential and how it can be structured to develop sustainable cultural tourism experiences.

BACKGROUND

Culture in Tourism

Culture is a complex and dynamic concept to define and much has been written about culture and tourism (Csapo, 2012). The symbiotic relationship between those two results in cultural tourism. According to previous research this definition can vary depending on the perspective of the demand or the supply. For McKercher, Ho, & du Cros (2005) cultural tourism depends on the destination cultural assets to create products for tourists' consumption.

Richards (1996) presents a broader definition where people are motivated to travel by their intention to gather new information and experiences to supply their cultural needs.

In the past museums and monuments were the main reason for cultural tourists to travel, to visit the traditional cultural sites but as the shift occurred culture in tourism has a broader scope and includes performing arts, a proliferation of events and festivals, the introduction of creative activities and recently a shift from tangible cultural heritage to intangible heritage (Artal-Tur, Villena-Navarro, & Alamá-Sabater, 2018).

In this research, culture includes a set of cultural resources (tangible and intangible) including cultural heritage, experiences, knowledge, identity, values, beliefs, traditions, customs, cuisine, and sense of place used in *sustainable development* (Taylor, 2016:32).

Nowadays culture is recognized as key element in the differentiation of destinations and promotes sustainable development (Dessein et al., 2015), contributing to preserving authenticity, attenuating seasonality and providing the development of complementary products (Futur & Nedelea, 2018; Henriques, 2014). The World Tourism Organization (2020) identified several benefits as result of this relationship between culture and tourism, namely: it enriches the visitors' experiences, facilitates intercultural dialogue and cultural appreciation, while enabling the access to culture to a broader audience.

Culture matters in sustainable development (Katriina Soini, Elena Battaglini, Inger Birkeland & Graham Fairclough, 2015:14) as tourism is recognized as a development-promoting activity in which *cultural tourism gains prominence as a force capable of transforming destinations, traditions and lifestyles* (Henriques, 2019: 37). Tourism activity relies on the existence of cultural resources and offers to attract visitors. Recent literature highlights intangible cultural heritage (ICH) as a driving force for tourism (Kim, Whitford, & Arcodia, 2019; Qiu, Zheng, Xiang, & Zhang, 2020; X. Su, Li, Chen, & Zeng, 2020).

Cultural resources are vulnerable to intense tourism activity(Cetin & Bilgihan, 2016), conservation and safeguarding of those resources, tangible and intangible, are a primary measure to achieve cultural sustainable tourism development. The quality of the resources and the cultural offer of a tourist destination determine the attractiveness of the tourism products and, consequently, arouse the cultural tourists' interest. This attractiveness can be achieved when cultural tourism follows the principles of sustainability (Greg Richards & Munsters, 2021). These principles integrate the management of cultural heritage and tourism activities with the local community, generating social, environmental and economic benefits for all stakeholders and considering the quality of life of the host community and the quality of the visitor experience. Holistic planning is necessary through set established strategies for the long-term preservation of resources (UNWTO).

STAKEHOLDERS AND SUSTAINABLE TOURISM

The participation of the stakeholders in sustainable tourism is becoming an emerging theme for academia and industry (Minh, Lam, & Prabhakaran, 2021; Roxas, Rivera, & Gutierrez, 2020; Waligo, Clarke, & Hawkins, 2013). Based on previous research, stakeholders are part of the tourism process and are considered the main support for a successful sustainable tourism implementation (Byrd, 2007).

Goeldner and Ritchie (2003), in an effort to identify the stakeholders in tourism destinations, refer to local residents, local companies, media, employees, government, competitors, tourists, business associations, activists, and tourism developers.

Timur and Getz (2008) refer to the public and private sectors and the local community as the main active players. The literature suggests that stakeholders in tourism are heterogeneous groups with different interests and levels of knowledge about sustainable planning, and shows that the ones participating more in the planning of tourism are also more receptive and supportive towards sustainability and to the involvement of local people (Hatipoglu, Alvarez, & Ertuna, 2016).

However, this joint effort faces a set of challenges such as the absence of a common vision and objectives for long-term planning amongst the stakeholders, *the lack of expertise and training of tourism planning authorities, political traditions that favour centralisation of authority, lack of funding, lack of interest or commitment by stakeholders, competition for the same resources, lack of long-term or strategic planning, and a lack of consensus on specific structures and processes* (Ladkin & Martinez Bertramini, 2002: 75). Moreover, the same authors advocate that factors like the motivation, personality and perceived roles of the stakeholders can also influence the success of this process. Thus, the stakeholders' level of involvement and interest and proper planning, coordination and management of the tourism institutions and tourism destinations contributes effectively to its success (Minh et al., 2021).

Recently, there has been an effort to consider the participation of the various stakeholders to plan in a more collaborative way involving the community.

MAIN FOCUS

This research took place in Olhão, a council located in the Algarve, south Portugal as figure 1 shows.

Olhão is a county with four towns in the central area of the Algarve occupying a territory of 130km^2 (INE, 2021), extending from the Ria Formosa (a protected natural area located on the coast) to the hills in the interior. Olhão is 12 km from Faro International Airport and approximately 50km from the Spanish border.

Olhão is part of the Ria Formosa Natural Park, one of the most important wetlands in Europe, and was considered in 2004 by the International Union for Conservation of Nature as a wetland of worldwide interest (CMO, 2021). Despite Olhão being on the coast, there is no direct access to the beach but a unique experience at any time of the year is provided by the ferry services that serve the islands for the inhabitants, locals and tourists.

The Ria Formosa is made up of a great diversity of ecosystems: islands, beaches, dunes, salt marshes, lagoons and canals. It is also occupied by an enormous variety of flora and fauna, including numerous species of birds. This system has many unique features including long beachs formed of golden sands and crystal clear waters.

Figure 1. Map of Olhão
Source: adapted from d-maps.com

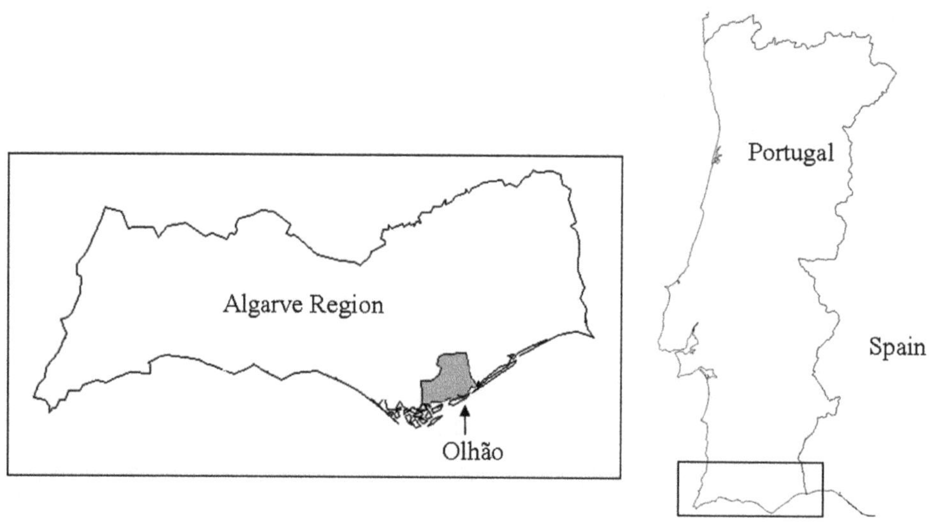

With the largest fishing port in the Algarve, the fishing industry was the main activity (CMO) since the nineteenth century.

In recent years, the increase in tourist demand registered in Olhão has been very significant and unprecedented. The growth of unplanned tourism can jeopardize preserving a territory's cultural resources, traditions and identity. Thus, it is essential to plan to include as many stakeholders as possible involved with tourism activity and their views on this phenomenon. Olhão has a population of 44643 inhabitants of whom 5079 are foreign residents with legal resident status (INE, 2021).

This case study encompasses 30 interviewees in English and in Portuguese language, and the sample includes the public sector (27%), private sector (50%), associations (10%), residents (13%). In figure 2, it is possible to see the sample in detail; the private sector is diverse and includes accommodation, restaurants, local shops, travel agents and tour operators.

Data analysis was carried out using qualitative data analyses software NVivo(framework matrices) to condense a large volume of data from the interviews. Based on the literature, the main dimensions were identified, which are: culture, experience, sustainability and stakeholders, other subcategories emerged during the codification process.

ANALYSIS OF THE CULTURAL DIMENSION

Cultural Tourism Resources

In the analysis of the dimension of culture, the following subcategories were identified: cultural resources, cultural offer and cultural proposals. In figure 3 the private sector stands out for having the largest number of codifications; this can be explained due to the fact that the sample is related to the proportion of sectors. In contrast, the public sector identifies more offers and resources and does not

Figure 2. Interviewed stakeholders

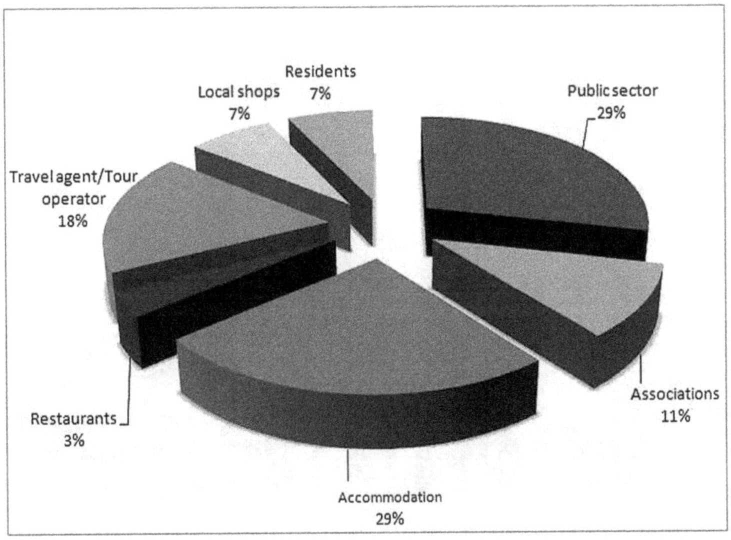

present as many proposals. However, we can observe that the private sector can identify more resources and suggest cultural proposals.

Figure 3. Cultural offer, resources and proposals by sector

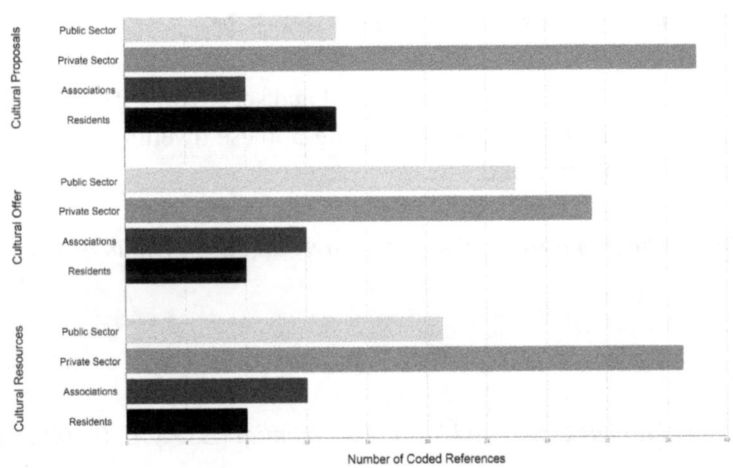

Regarding cultural resources, the interviewees refer, as the main resource, the local people

Its people are also peculiar, it makes us feel positively its locality, its identity, the friendly people, welcoming in general terms (ESA26,05-20-2020)

Figure 4. Olhão, city view with the market buildings on the front
Source: Algarve Drone, 2021

The interviewees representing the private, public and associative sectors perceive the people from Olhão have strong character, with a unique identity, traditions and knowledge embodied and call it *olhanensidade* (ESPR27, 05-20-2020) the greatest asset is to be *olhanense* (ESP4, 05-06-2020)

As part of the main cultural resources, the interviewees also highlighted the town in general, the old quarters due to the uniqueness of the architecture in particular.

What really makes us unique is our people, we say that the historic area of Olhão is nowhere else in the world, it's true, that's what distinguishes us (ESA32, 05-22-2020)

The markets, the architecture can be seen in figure 4 and specifically the cube-shaped, whitewashed houses with large overlapping terraces as shown figure 5, these dwellings form a unique sight of agglomerated cubes in the old quarters of the town.

The roofs, all that life on the roof terraces that is definitely unique in Olhão and very special. (ESPR12, 05-12-2020)

It was found that the interviewees recognize the intangible heritage as an essential asset of their identity

Olhão has a soul (...) Olhão has a soul and has great potential (ESPR10, 05-09-2020)

According to the stakeholders interviewed, the people of Olhão have peculiar characteristics found in their personality, way of speaking and welcoming visitors. Nowadays, it is still present in its traditions and cultural identity. The foreigners who live in the municipality also have a sense of place.

It's the spirit of the place that keeps us here (ESPR27, 05-020-2020)

Figure 5. Olhão, roof terraces view
Source: Photograph by the authors, 2020

Gastronomy is also a cultural resource of great quality and an element of differentiating and memorable experiences.

Gastronomy is an experience that you carry in your belly and you won't forget it (ESPR22, 05-12-2020)

Figure 6 expresses the typology of the cultural tourism resources identified in Olhão. This model shows the typology of the cultural tourism resources at a tourist destination. It can systematize and analyze tourist cultural resources and apply to cities, regions or countries (Richards, 2021). This research found one more main category *Local People*, and one subcategory under the category *Attraction: Intangible cultural heritage.* Those two are strongly associated with the attractiveness and potential for the competitiveness of the place. Regarding community participation, as prior literature has already pinpointed a people-centred approach where they are part of the decision-making process and an important part in developing cultural resources (Hernandez, 2012).

Cultural Tourism Offer

The general opinion of the respondents about the existing cultural offer is that it's limited and very little diversified. However, the various existing festivals were identified with great prominence for the Seafood Festival.

Olhão is the land of the Seafood Festival, basically, it is a poor offer in cultural terms, but economically there is no doubt that it attracts a lot of people, we could always have a better offer (ER20,05-10-2020)

Although the cultural offer is considered limited, respondents believe that there is a lot of potential to grow and innovate. The cultural events attract residents but do not attract tourists, especially the younger generations, and they do not attenuate seasonality.

Figure 6. Model of typology of the cultural tourism resources applied to Olhão
Adapted (Greg Richards & Munsters, 2021)

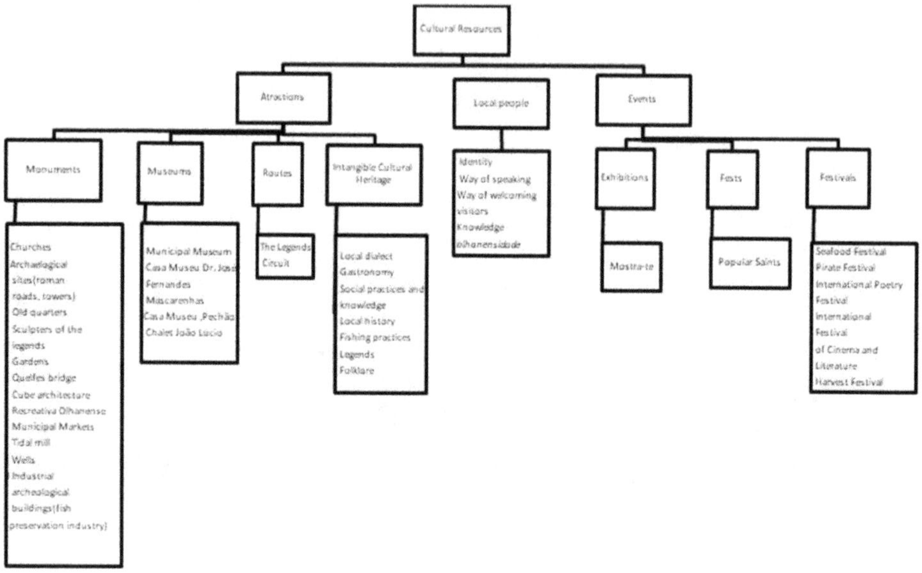

There is not enough offer, especially for the youngest, certainly for the youngest there is not enough offer to attract them in the winter period (ER30, 05-15-2020)

On the other hand, the cultural events that take place are qualified as of great quality.

The concerts at the Conservatório de Música de Olhão are very good with great quality (ESP3, 05-09-2020)

The role of associations in the cultural dynamic and promotion of the municipality providing a cultural programme is also relevant as well as in terms of representation of the community and its traditions.

The clubs and associations in the heritage and cultural part must be reserved a very important role, they are very much a living memory of these things (ESA33, 05-07-2020)

Some stakeholders pointed out that they are unaware of the existing cultural offer because it is of little relevance; it was found that a few stakeholders across all sectors, shared this opinion.

Cultural Proposals

The cultural offer is seen as a way to reduce seasonality when it presents a structured, diversified offer throughout the year, continuously to create a habit. As mentioned in one of the interviews:

To come at another time of the year doing a certain cadence so things don't get done twice and then never happen again (ESP4, 05-09-2020)

Figure 7. Process of creating cultural tourism offers (CTO) according to the stakeholders

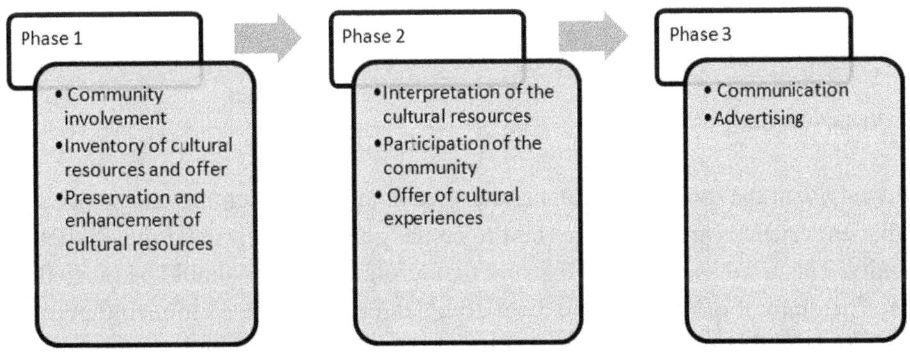

Lack of interpretation of the cultural resources is a recurrent concern. Residents refer to it thus:

It is necessary to interpret the heritage, give meaning and communicate this information (ER38, 05-24- 2020)

Based on the stakeholder's opinion, we can resume how the cultural offer should be structured in figure 7. A three-phase process starts with identifying and inventorying the cultural resources and offers, preserving and protecting the local community. This seems to contribute to the community well-being. There is a second phase of building and implementing meanings, facilitating experiences and the third phase, the communication and advertising those experiences. The community's involvement in the whole process is considered essential, especially in interpreting the heritage to preserve cultural values and avoid excessive commodification of the local resources and local culture in general.

Creative tourism is a way to link tourism with other industries and relationships between people and networks. Experiences based on traditional activities, creation of routes, cultural walks, activities in the Ria Formosa, preserve and rehabilitate the heritage in all local parishes.

We have this vibrant Portuguese community, and we have a very creative and artistic community, and they love it for the combination of architecture and light and all the rich variety of what we have, and people who come to Olhão love it for that (ESPR12, 05-20-2020)

The interviewed stakeholders also mention that cultural spaces should be preserved and musealised. This would value the archaeological legacy of the county, creating cultural spaces that favour the fishing arts, canning industry and cod fishing. Additionally highlighted was the preservation of the natural landscape with symbolic and historical value (Cerro da Cabeça and Cerro de São Miguel). Suggestions included exploring literary tourism and historical recreations involving the population and exploring the urban art's touristic potential that proliferates throughout the city in the form of graffiti. Finally, support the work of local associations, create cultural, gastronomic experiences, and above all, celebrate traditions, as stated by one of the interviewees:

The tourism offer can only be limited by imagination or time (ESPR12,05-12-2020)

EXPERIENCE DIMENSION

Tourism Experience

When questioned about the tourist experience, the stakeholders representing all sectors are unanimous. They agree that experiences are very important to create good memories and that it should involve the local community. The private sector also referred that foreign residents should be given the opportunity to participate. The cultural offer should be diversified, and there is a need to maintain traditions.

In order to structure the tourism offers, as figure 8 shows, the stakeholders suggested holding more events, small but high quality in a variety of places in the county. Various festivals throughout the year are also mentioned. Intangible heritage can be used as a great source to create experiences. As the interviewees state, more emphasis should be placed on experiences. Create circuits and itineraries for independent tourists. Technology can be used to give more autonomy to the visitor and use social networks for advertising.

Destination Image

Despite some issues identified, such as the image being very centred on the urban part of the town and mainly in the Ria Formosa and surrounding islands, thus neglecting the other areas of the county, overall the destination image regarded by the stakeholders is positive and the interviewees highlighted the authenticity of the place though its singular architecture, its community, traditions and gastronomy, but it's very little advertised.

Ria Formosa, seafood, the cubist town, being Olhão is different, the soul of Olhão is different from other towns and cities in the Algarve, a different way of looking at life, happier, more parochial (ESP4,05-06-2020)

The negative aspects identified are the lack of strategy for tourism development, the difficulties of access to housing, the gentrification of the old quarters, especially by foreign residents, but, on the other hand, it is referred that these foreigners could play a more active role if they were invited to collaborate.

SUSTAINABILITY DIMENSION

Impact of Tourism Development on Destinations

The positive impact recognised and most easily identified and referred by the interviewees is the economic one, through revenue generation, job creation, and the economy's development. The stakeholders refer the well-being and psychological satisfaction. They also refer to the possibility of tourists and locals interacting, generating cultural exchange, setting up cultural projects, and creating better access to culture.

Figure 8. Dimension of the tourist experience

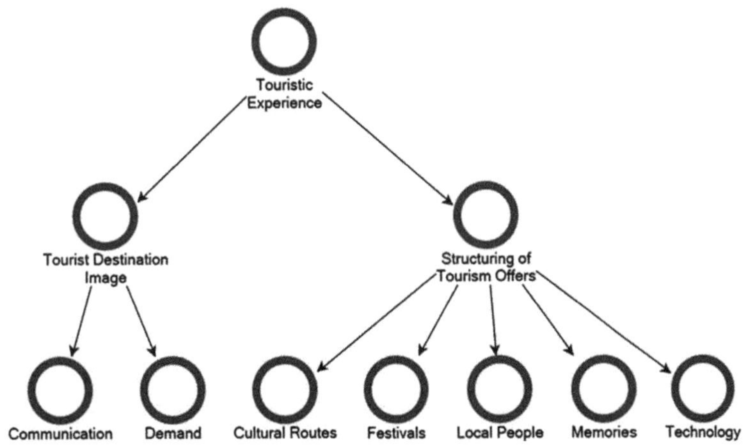

People come here, and this obviously gives us satisfaction (ER28,05-23-2020)

Foreign respondents representing the private sector mention the connection between local people and foreign residents investing in property is very important and consider that the investments are not only financial but also emotional when buying a property to create a better sense of community, although some older local people don't like tourism in Olhão; they see it as a threat (ESPR14, 05-21-2020)

As negative aspects, they refer to the danger of globalization of culture, the excessive dependence on tourism as a monoculture and, in the worst-case scenario, the destruction of resources. They also note the difficulty for the young local people to access housing.

Sustainability of the Tourist Destination Olhão

The majority of the interviewees believe that tourism development in Olhão, so far, has been done in a sustainable way or, at least, in a way that has not damaged the resources. However, they recognize the need for planning for long-term tourism development. Furthermore, according to the stakeholders, to balance host communities and visitors, protecting communities and heritage valuing the authenticity of the place (ESPR11, 2020) is at most importance for the sustainability of tourist destinations. They also state that it has to be sustainable to be good for tourists and viable for the communities. In the sustainability dimension content analysis, several subcategories have emerged (figure 9): the strategies, the challenges, and the impacts of tourism, which generate subcategories. Due to the pandemic context, COVID-19 was also mentioned as a negative factor affecting tourism and showing how vulnerable this activity is.

STAKEHOLDERS DIMENSION

The stakeholders' participation and level of involvement in the tourism planning are seen differently. The debate is mainly about participation, therefore, in general they agree that local community should be heavily involved in the decision-making process, as can be seen in figure 10. Negative opinions towards

Figure 9. Dimension of Sustainability

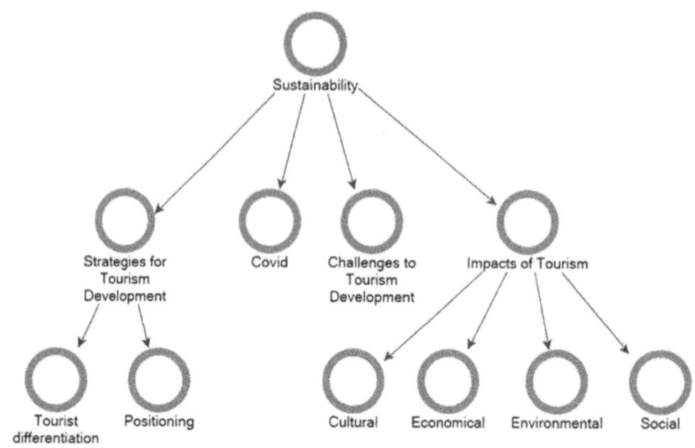

community participation in the planning process are from some interviewees representing the public (8.3%) and private (4.5%) sectors attributed to some residents' low level of education. The public sector (9.3%) also suggested that associations shouldn't actively participate in this process.

Another interviewee from the private sector justifies the participation of all stakeholders in the planning of tourism because culture is about people. The private sector refers that all are important and should follow a democratic approach when planning. The private also underlined if it relies only on the private sector, it's not going to be sustainable, and the public sector should be there to ensure the balance between all participants. The creation of a network was suggested as a way to connect all the entities.

On the other hand the public sector advocates participatory methodologies, in which common objectives and goals should be defined and the public sector should lead the strategy. To be sustainable it must be the result of everyone's participation.

One of the residents' group interviewees associated the importance attributed to the stakeholders' participation with sustainability. It is expected that the public sector should take the initiative, conserve, value and lead by establishing rules. The private sector is essential to generate investment and innovation.

The association sector considers the participation of the public sector, the private sector and the associations themselves important as they represent people and keep memories alive. Planning must reflect the will of all.

When asked about the participation in the operation of the planning in the main, some participants of the private sector do not show a favourable attitude towards the participation of the associations, residents or public. The public and private sectors identify partnerships as a positive form of planning. However, they consider that the public sector should take the lead or create a new entity to coordinate the tourism development at a local level.

According to Tosun (2000) there is a lack of information available for the residents on the tourism development, and this may result on low interest in getting involved.

Among tourism stakeholders, collaboration, coordination, and perceived constraints are some of the difficulties identified in the literature (Ladkin & Martinez Bertramini, 2002) as well as in this research.

Thus, there are differences between the perceptions of the different sectors and how they should collaborate and manage cultural tourism.

Figure 10. Participation of stakeholders in the planning and operationalisation process

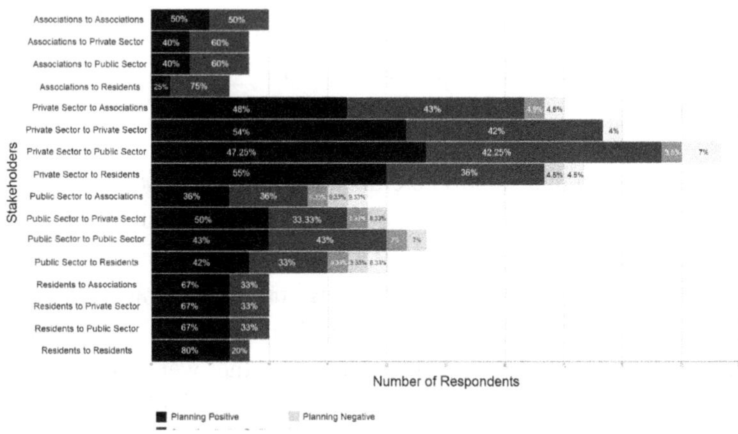

Proposals for Structuring Sustainable Cultural Tourism Experiences

Several respondents expressed the belief that the structuring of tourism experiences should start with the involvement of the local community in the first place and should be informative and raise awareness throughout the process, otherwise it is unlikely to be successful. Experiences that allow tourists to participate in community life and create memories are considered the ideal. Authenticity, one of the main features of the tourist experience must be ensured.

The use of technology such as personal device apps can enhance the experience and also provide more information online for the independent traveller to access.

The main suggestions are to create new products and services, promote more and significant experiences, enhance attractions, and organise cultural festivals (music, theatre).

SOLUTIONS AND RECOMMENDATIONS

The Challenge to Achieve the Balance

This study explores the perception of the stakeholders on the relevance of the cultural resources and cultural offers for the development of cultural, touristic experiences.

As the literature already states, stakeholders are an essential part of the touristic development process, and their involvement is crucial for its success.

In this research, the stakeholders place the local people at the heart of the tourism experience. The local people are not only participants or co-creators of the experience like many authors refer to, but in this case, the local people are perceived as a cultural resource, the main cultural resource.

The biggest challenge is to provide a holistic approach combining all the interests of the various stakeholders to preserve the cultural identity and authenticity of the territory and its communities.

The results of the current study and existing literature have led to developing a model of sustainable cultural tourism experiences. The results of the analysis identified a number of key dimensions with subcategories: cultural tourism resources, cultural tourism offer, touristic experience, destination image and sustainability with the following components associated: authenticity, cultural pride, cultural preservation, community well-being and community empowerment.

According to the model suggested by Richards & Munsters (2021), sustainable cultural tourism bridges different areas and interests. The balance between these parts will allow or result in cultural tourism sustainability.

Authenticity is a characteristic that should be embedded in the tourist experience referred by the interviewed stakeholders and confirmed by the literature (Cetin & Bilgihan, 2016; Domínguez-Quintero, González-Rodríguez, & Roldán, 2019; X. Su et al., 2020). It is also attributed to the local people, traditions, and intangible heritage, as they are peculiar and interesting enough to attract visitors (McKercher, 2002).

The cultural pride was identified through the sense of belonging to the community, pride in its cultural traditions and heritage and especially in the local people that were identified as the main cultural resource. This result is in line with what Besculides, Lee, and McCormick (2002) identified as cultural benefits from tourism, apart from the increase of tolerance and understanding it also increases identity, pride and cohesion.

Joint work with communities is referred to overcome the problem of lack of interpretation of the cultural assets. Their input on cultural interpretation ensures it is relevant and meaningful and authentic to the visitors.

In addition to the various economic benefits that tourism provides easily identified by stakeholders, this research also highlights community well-being, the presence of tourism increases self-esteem and promotes psychological well-being when interaction takes place between locals and visitors. This intercultural exchange also creates more resilient communities. When the level of knowledge about the tourism in the destination is high, it leads to community empowerment through tourism according to Global Sustainable Tourism Council (GSTC).

It can also be considered that in Olhão this positive attitude towards tourism is due to the fact that the sector is in an initial growth phase, which, according to the Tourism Life Cycle Model (Butler, 2006), it's the discovery and adaptation of the tourist destination and the local community to this new activity and also trying to meet the needs of the of tourism growth and meet the expectations of visitors, where the local population perceives tourism as a beneficial activity.

The proposed model in figure 11 shows the balance between the various parties and based on this research, it is suggested that there is direct and indirect interaction between all of them. At the base of the model are the government and the policy support to establish the legal framework and infrastructures, to encourage, facilitate and support bottom-up initiatives with long-term planning.

At the core of cultural tourism are the cultural resources, cultural offer and, according to the interviewees, the local people. Those have an important double role, as an element of attraction and at the same time participating and interacting in the cultural experiences.

This model shows an important relationship between the local community, cultural tourism resources, and stakeholders' participation. Cultural preservation of the cultural resources can be done by public and private sectors, including the local community. The quality of the cultural resources is determinant to attract cultural tourists. The tourism entrepreneurs provide services to tourists, which enriches their experiences and maximises the long-term return from this activity.

The local community here represents the aggregate of people in the old quarters area of Olhão and plays an important role in this process by participating and benefiting from the cultural tourism activ-

Figure 11. Sustainable cultural experiences conceptual model

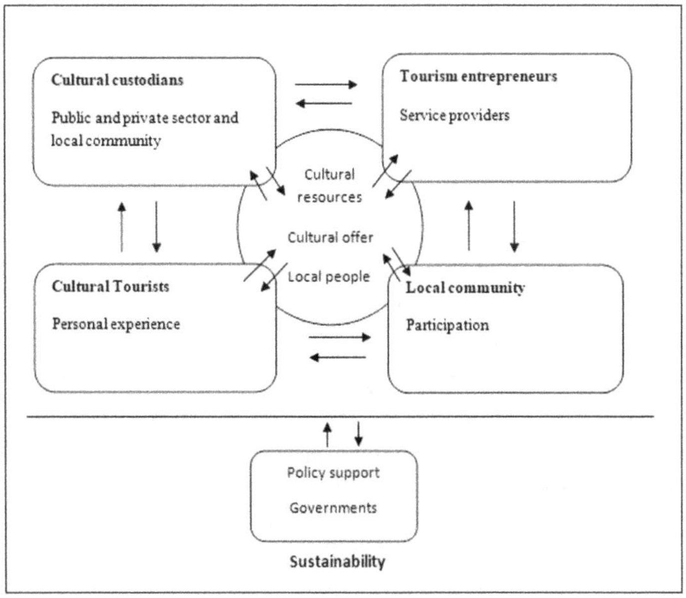

ity by generating income, sharing use of tourism facilities and lifting the population's consciousness regarding the value of the local culture (Greg Richards & Munsters, 2021).

The cultural tourists or cultural seekers are searching for a meaningful experience for which the involvement of all parties in the proposed model is necessary. In this regard Richards (2011) has already identified a tendency where the tourists look for authentic experiences in the daily life of the host communities.

FUTURE RESEARCH DIRECTIONS

For future research, it is suggested to apply the methodology used in this study in other destinations, such as the Algarve region.

To contribute to a better empirical understanding of the multiple factors, including the participation of the stakeholders and the sustainability of the tourism industry and its practical application.

To better understand the extent and nature of the stakeholders' collaboration in tourism planning, more research is needed to identify the barriers and how to overcome these.

Design a development plan in which local people play a central role in the whole process. Investigate what local people want for tourism, products and development.

In addition, more research is needed to better understand the demands of, and analyse the profile of the tourist who visits Olhão. Also, more effort is required to develop digital technology applied to visitor management and interpretation of cultural assets.

CONCLUSION

This research emphasises the relevance of the cultural resources and offer on sustainable cultural tourism experiences. The results show relevant findings on how the stakeholders perceive their cultural resources and offer and how they see themselves and the other stakeholders involved in this process.

Olhão, due to its location and diversity of resources, has characteristics that allow it to provide unique, memorable and sustainable cultural experiences. The cultural tourism experience is still little explored, but it may have great relevance for the municipality in the future.

Results of the present study show that the stakeholders regard culture as a driver for joint participation of host communities and tourists for sustainable development. Cultural resources were identified and should be preserved dynamically with cultural offers that include the participation and cooperation of the local people. Besides, the preservation of tangible and intangible heritage, its interpretation, given meanings and communication are essential for its understanding and enjoyment. It was verified that the stakeholders recognise the local culture as authentic, having the potential to develop the co-creation of diversified tourism products and experiences.

Results state the proposals from the stakeholders and highlight the creation of a cultural offer, well integrated in its context, that means following the dynamics of the place (seasons, festivities, celebrations), and the communities (traditions, intangible heritage) with regular high-quality small-scale events throughout the year, especially festivals in the low season.

According to the stakeholders, the main assets of this destination are the local people and its culture, with a focus on intangible heritage. Gastronomy emerges as the destination's main differentiating product and with high potential to develop. As destinations face challenges to differentiate themselves, it is important to adopt strategic measures that take into account not only the needs of tourists, but also the local community, and the territory's specificities. Thus, it was stated as a priority, by the majority of the stakeholders, the importance of a long-term tourism plan. Also, the diversity of cultural tourism products and innovation in terms of creative tourism and integration of technology solutions, contribute to the improvement of the tourist experience. Thus, cultural tourism has shown to be an excellent opportunity to rethink tourism within a more circular economy, more authentic experiences in a more sustainable way, where the tourism revenue is not the only benefit.

Currently, it is common to find literature about stakeholder participation and more frequent about sustainability in tourism. This study recalls the importance of involving the local communities and the stakeholders in general to plan and develop sustainable tourism. Moreover, the pandemic situation generated by Covid-19 has affected all industries in general, and tourism was no exception, bringing new challenges to this sector. Lessons can be learnt from this situation and it can contribute to tourism management and policy makers' decisions more appropriated and efficient and better address challenges. It is possible to see that there are no better places than others and that valuing what is local (its authenticity involving the local communities and the stakeholders) is a way to ensure its sustainability. This way, it can also make places more resilient and better prepared to face new economic crises or pandemics, thus encouraging responsible tourism.

ACKNOWLEDGMENT

This book chapter is financed by National Funds provided by FCT - Foundation for Science and Technology through project ref. UIDB/04470/2020 CiTUR and UIDB/04020/2020 CinTurs.¨

REFERENCES

Algarve Drone. (2021). https://www.dronealgarve.pt/fotos-aereas-algarve/

Alves, D., Filho, D., & Henrique, A. (2015). O poderoso NVivo: Uma introdução a partir da análise de conteúdo [The powerful NVivo: An introduction from content analysis]. *Revista Política Hoje, 24*(2), 119–134.

Artal-Tur, A., Villena-Navarro, M., & Alamá-Sabater, L. (2018). The relationship between cultural tourist behaviour and destination sustainability. *Anatolia, 29*(2), 237–251. doi:10.1080/13032917.2017.1414444

Besculides, A., Lee, M. E., & McCormick, P. J. (2002). Resident's perceptions of the cultural benefits of tourism. *Annals of Tourism Research, 29*(2), 303–319. doi:10.1016/S0160-7383(01)00066-4

Byrd, E. T. (2007). Stakeholders in sustainable tourism development and their roles: Applying stakeholder theory to sustainable tourism development. *Tourism Review, 62*(2), 6–13. doi:10.1108/16605370780000309

Câmara Municipal de Olhão. (n.d.). https://www2.cm-olhao.pt/en/conhecer-olhao/geografia-do-concelho

Cetin, G., & Bilgihan, A. (2016). Components of cultural tourists' experiences in destinations. *Current Issues in Tourism, 19*(2), 137–154. doi:10.1080/13683500.2014.994595

Cohen, S. A., Prayag, G., & Moital, M. (2014). Current Issues in Tourism Consumer behaviour in tourism : Concepts, influences and opportunities. *Current Issues in Tourism, 0*(0), 1–38. doi:10.1080/13683500.2013.850064

Csapo, J. (2012). The Role and Importance of Cultural Tourism in Modern Tourism Industry. In Strategies for Tourism Industry - Micro and Macro Perspectives. doi:10.5772/38693

D-maps. (2021). https://d-maps.com/pays.php?num_pay=214&lang=pt

Domínguez-Quintero, A. M., González-Rodríguez, M. R., & Roldán, J. L. (2019). The role of authenticity, experience quality, emotions, and satisfaction in a cultural heritage destination. *Journal of Heritage Tourism, 14*(5/6), 491–505. doi:10.1080/1743873X.2018.1554666

Flutur, L., & Nedelea, A. (2018). Cultural Tourism. *Revista De Turism - Studii Si Cercetari In Turism, 0*(25). Retrieved from http://www.revistadeturism.ro/rdt/article/view/382/257

Fundação para a Ciência e a Tecnologia. (2019). *Agenda Temática de Investigação e InovaçãoTurismo, Lazer e Hospitalidade* [Thematic Agenda for Research and Innovation Tourism, Leisure and Hospitality]. FCT. . doi:10.34621/fct.edicoes.agendastematicas-2

Garg, P., & Pandey, A. (2021). Towards sustainable tourism: An empirical investigation. *Foresight, 23*(2), 188–200. doi:10.1108/FS-04-2020-0042

Goeldner, C., & Ritchie, J. R. (2003). *Tourism, principles, practices, philisophies*. Academic Press.

Gonçalves, A. R., Oliveira, A. P. D., & Santos, C. F. (2015). *Fragmentos para a história do turismo no Algarve* [Fragments for the history of tourism in the Algarve]. Retrieved from https://sapientia.ualg.pt/handle/10400.1/8743

Hatipoglu, B., Alvarez, M. D., & Ertuna, B. (2016). Barriers to stakeholder involvement in the planning of sustainable tourism: The case of the Thrace region in Turkey. *Journal of Cleaner Production, 111*, 306–317. doi:10.1016/j.jclepro.2014.11.059

Henriques, C. (2014). Algarve Turismo Cultural. *Revista Turismo Em Análise, 25*(2), 497. doi:10.11606/issn.1984-4867.v25i2p497-499

Henriques.C. (2019). Turismo Cultural como vetor de Sustenabilidade nas Cidades [Cultural Tourism as a vector of Sustainability in Cities]. *Journal of Tourism and Heritage Research, 2,* 327–352.

Instituto Nacional de Estatística. (2021). *Anuário Estatístico da Região do Algarve*. https://www.ine.pt/xurl/pub/410492449

Instituto Nacional de Estatística. (2021). *População residente* [Resident population]. https://www.ine.pt/scripts/db_censos_2021.html

Kim, S., Whitford, M., & Arcodia, C. (2019). Development of intangible cultural heritage as a sustainable tourism resource: The intangible cultural heritage practitioners' perspectives. *Journal of Heritage Tourism, 14*(5/6), 422–435. doi:10.1080/1743873X.2018.1561703

Ladkin, A., & Martinez Bertramini, A. (2002). Collaborative tourism planning: A case study of Cusco, Peru. *Current Issues in Tourism, 5*(2), 71-93. doi:10.1080/13683500208667909

McKercher, B. (2002). Towards a classification of cultural tourists. *International Journal of Tourism Research, 4*(1), 29–38. doi:10.1002/jtr.346

McKercher, B., Ho, P. S. Y., & du Cros, H. (2005). Relationship between tourism and cultural heritage management: Evidence from Hong Kong. *Tourism Management, 26*(4), 539–548. doi:10.1016/j.tourman.2004.02.018

Minh, H. V., Lam, T., & Prabhakaran, S. (2021). Perceptions of Key Stakeholders Towards Sustainable Tourism Development: A Case Study in Mekong Delta, Vietnam. *Journal of Asian Finance, 8*(4), 717–0726. doi:10.13106/jafeb.2021.vol8.no4.0717

Qiu, Q., Zheng, T., Xiang, Z., & Zhang, M. (2020). Visiting intangible cultural heritage tourism sites: From value cognition to attitude and intention. *Sustainability, Switzerland, 12*(1), 1–20. doi:10.3390u12010132

Quivy, R. (1995). *Manual de Investigação em Ciências Sociais* [Handbook of Social Science Research]. Retrieved from https://tecnologiamidiaeinteracao.files.wordpress.com/2018/09/quivy-manual-investigacao-novo.pdf

Richards, G. (1996). The scope and significance of cultural tourism. In *Cultural tourism in Europe* (pp. 19–45). CAB International.

Richards, G. (2011). Creativity and tourism. The state of the art. *Annals of Tourism Research, 38*(4), 1225–1253. doi:10.1016/j.annals.2011.07.008

Richards, G., & Munsters, W. (2021). *The Study of Cultural Tourism in Ten Models*. University of Applied Sciences.

Roxas, F., Rivera, J., & Gutierrez, E. (2020). Mapping stakeholders' roles in governing sustainable tourism destinations. *Journal of Hospitality and Tourism Management, 45*, 387–398. doi:10.1016/j.jhtm.2020.09.005

Soini, K., Battaglini, E., Birkeland, I. N. D., & Fairclough, G. (2015). Culture in, for and as Sustainable Development. Conclusions from the COST Action IS1007 Investigating Cultural Sustainability. In *Culture in, for and as Sustainable Development*. Retrieved from http://www.culturalsustainability.eu/

Su, X., Li, X., Chen, W., & Zeng, T. (2020). Subjective vitality, authenticity experience, and intangible cultural heritage tourism: An empirical study of the puppet show. *Journal of Travel & Tourism Marketing, 37*(2), 258–271. doi:10.1080/10548408.2020.1740141

Su, Z., Aaron, J. R., McDowell, W. C., & Lu, D. D. (2019). Sustainable synergies between the cultural and tourism industries: An eciency evaluation perspective. *Sustainability (Switzerland), 11*(23), 6607. Advance online publication. doi:10.3390u11236607

Taylor, E. (2016). *Culture, tourism and sustainability: an ethnographic study of rural community development in Jamaica* (Unpublished PhD Thesis). Coventry: Coventry University.

Timur, S., & Getz, D. (2008). A network perspective on managing stakeholders for sustainable urban tourism. *International Journal of Contemporary Hospitality Management, 20*(4), 445–461. doi:10.1108/09596110810873543

Timur, S., & Getz, D. (2009). Sustainable tourism development: How do destination stakeholders perceive sustainable urban tourism? *Sustainable Development, 17*(4), 220–232. doi:10.1002d.384

Tosun, C. (2000). Limits to community participation in the tourism development process in developing countries. *Tourism Management, 21*(6), 613–633. doi:10.1016/S0261-5177(00)00009-1

Vargas-Hernández, J. G. (2012). Sustainable cultural and heritage tourism in regional development of Southern Jalisco. *World Journal of Entrepreneurship, Management and Sustainable Development, 8*(2/3), 146–161. doi:10.1108/20425961211247752

Waligo, V. M., Clarke, J., & Hawkins, R. (2013). Implementing sustainable tourism: A multi-stakeholder involvement management framework. *Tourism Management, 36*, 342–353. doi:10.1016/j.tourman.2012.10.008

World Tourism Organization. (n.d.). *Sustainable Development*. https://www.unwto.org/sustainable-development

World Tourism Organization. (n.d.). *Tourism and Culture*. https://www.unwto.org/tourism-and-culture

ADDITIONAL READING

Byrd, E. T. (2007). Stakeholders in sustainable tourism development and their roles: Applying stakeholder theory to sustainable tourism development. *Tourism Review*, *62*(2), 6–13. doi:10.1108/16605370780000309

du Cros, H. (2001). A new model to assist in planning for sustainable cultural heritage tourism. *International Journal of Tourism Research*, *3*(2), 165–170. doi:10.1002/jtr.297

Hatipoglu, B., Alvarez, M. D., & Ertuna, B. (2016). Barriers to stakeholder involvement in the planning of sustainable tourism: The case of the Thrace region in Turkey. *Journal of Cleaner Production*, *111*, 306–317. doi:10.1016/j.jclepro.2014.11.059

Kim, S., Whitford, M., & Arcodia, C. (2019). Development of intangible cultural heritage as a sustainable tourism resource: The intangible cultural heritage practitioners' perspectives. *Journal of Heritage Tourism*, *14*(5–6), 422–435. doi:10.1080/1743873X.2018.1561703

Richards, G. (2018). Cultural tourism: A review of recent research and trends. *Journal of Hospitality and Tourism Management*, *36*, 12–21. doi:10.1016/j.jhtm.2018.03.005

KEY TERMS AND DEFINITIONS

Authenticity: Characteristics that reflects local values.

Cultural Offer: Set of cultural goods or services available for consumption.

Cultural Resources: Cultural assets that exist on a particular site and have particular characteristics capable of attracting visitors.

Cultural Tourism: A type of tourism in which the tourist seeks to learn about and consume the local cultural attractions and products of the tourist destination he or she visits.

Intangible Cultural Heritage: A body of knowledge that is passed down from generation to generation.

Responsible Tourism: A conscious and intentional practice of tourism that makes all participants responsible.

Stakeholders: Person or group of people participating or having an interest in an organisation.

Sustainable Tourism: Takes in account the environmental, social, economic and cultural impacts derived from tourism activity so as not to compromise the resources, and to promote the balance between all parts involved.

Tourism Experience: An individual, subjective and involving interaction at an affective, cognitive, and behavioural level that takes place in a touristic destination.

Chapter 10
Actor Networks and Development of Cultural Tourist Destinations

Oswaldo Ledesma González
University of La Laguna, Spain

Rafael Merinero Rodríguez
University of Pablo de Olavide, Spain

Juan Ignacio Pulido Fernández
University of Jaén, Spain

ABSTRACT

The study analyses the relational structure of the networks of actors of three cultural tourist destinations using social network analysis (SNA). The analytical interest is focused on the degree centrality indicator for socio-centric networks and asymmetrical relationships to obtain the indegree (prestige) and outdegree (influence) of the various actors. The results obtained strengthen three ideas: first, there is a direct relationship between relational dynamics and the development of tourist destinations; second, the relational structure is transformed as tourist destinations evolve; and third, there is a correlation between the number of actors with high centrality and the development of tourist destinations. The study contributes to responding to the demand for literature regarding the need to expand the number of comparative studies that link the relational structure with the development of tourist destinations.

INTRODUCTION

The concept of cultural tourism, although quite widespread, is full of controversies and still elicits not always convergent understandings (Brito, 2009). In this sense, Ashworth (1995), like Richards (1996), point out that the definition of cultural tourism depends on the concept of culture used. On the other hand, the concept of tourist destination, although represents a fundamental aspect of tourism research,

DOI: 10.4018/978-1-7998-9217-5.ch010

it is so common that it is often taken for granted and not defined (Pearce, 2014). With the intention of delimiting the framework of the study, cultural tourism is defined as that which is related to the visit to urban spaces with attractions of heritage interest (monuments, archaeological sites, historical buildings, cultural landscapes, etc.). Regarding the cultural tourist destination, it is conceived from the perspective of the relationships between the tourist actors, that is, individuals or organizations that have an impact on the tourist activity of the destination and that interacts with other actors to identify the needs of the tourists and create an experience for them (Ledesma et al., 2021); from this perspective, a tourist destination is a co-production system (Aarstad et al., 2015), that can be represented by a network composed of actors and the connections between them (Baggio et al., 2010; Ledesma, 2012; Ledesma and Simancas, 2020).

There is a wide and varied set of theoretical contributions that have been used to understand collaboration between actors in tourist destinations, as shown Fyall et al. (2012) and Merinero and Pulido (2016). Network theory is the theoretical framework used in this article and, like other relationship-based theories, such as relational exchange theory, stakeholder theory or collaborative game theory, is driven by the recognition and acceptance of mutual dependence, and the need for reciprocity in the search for shared interests (Fyall et al., 2012). In this way, from this relational perspective, a tourist destination can be analysed using the theoretical and methodological approach provided by Social Network Analysis (from now on, SNA), which allows the structure created by the relationships established between tourism actors to be identified. Using various information gathering techniques, data is obtained regarding the relationships between the actors who make up the various networks in a tourist destination and then analysed using matrix algebra, graphs, and networks, which provide the structural properties of the networks (Merinero, 2015). SNA has been one of the most used methods in tourism studies over the past two decades (Pulido and Merinero, 2018) and it has been implemented to analyse actor networks in tourist areas located on the coast, in the mountains, at heritage sites, etc. For a complete review of the scientific literature that uses SNA in tourism, see Casanueva et al. (2016), Merinero (2015), Merinero and Pulido (2016) and Valeri and Baggio (2021).

In this context, the relational structure of the networks of actors in Ronda, Loja and Osuna is analyzed, three cultural tourist destinations with important attractions of heritage interest that share the particularity that they are "medium cities" in the interior of Andalusia, that is, that they are outside the area of influence of the big cities and have a central role in the environment in which they are located and, therefore, its contribute functionally to the cohesion of the territorial structure of the region. The comparative analysis of the selected case studies is interesting because, despite the fact that they have similar geographical, demographic, heritage, and economic characteristics, which would minimize possible differences in terms of tourist outcomes, they have different tourism developments. The analytical interest is focused on the degree centrality indicator for socio-centric networks and asymmetrical relationships to obtain the indegree (prestige) and outdegree (influence) of the various actors. On the other hand, the analysis of the selected case studies allows to respond to the demands of Baggio et al. (2010) or Merinero (2015), among others, regarding the need to increase the number of comparative studies that link the relational structure with the development of tourist destinations in the research agenda. In addition, this book chapter responds to one of the limitations detected in a previous work of the authors where the convenience of expanding the number of study cases was raised to consolidate the results obtained and strengthen the conclusions (see anonymized).

This work has three objectives: first, to verify if there is a direct connection between cultural tourists destinations development and relational dynamics; second, to check how the relational structure evolves as cultural tourism destinations mature; and, finally, to confirm the premise that supports the existence of a correlation between the number of actors with high centrality and the development of tourist destinations.

SOCIAL NETWORK ANALYSIS AS A THEORETICAL–METHODOLOGICAL FRAMEWORK

SNA is based on the theory that to understand social phenomena we must examine the connections between the elements of those phenomena, and that these connections can be represented as a network of relationships. A network is normally represented by a drawing composed of nodes and connections, which is a mathematical abstraction called a graph. The formal language used to describe the characteristics of graphs is provided by a branch of mathematics known as graph theory. This theoretical-methodological framework highlights the relational nature of social structures and assumes that social phenomena are explained by the relational behaviour of their elements, rather than their attributes (Knoke and Yang, 2008; Wasserman and Faust, 2013). The objective within this framework is to study the structural properties of networks, understood as a set of nodes (actors) and the connections that arise between them (relationships) (Wasserman and Galaskiewicz, 1994), rather than focusing the analysis on the elements that make the connections that sustain the network possible.

Many of the concepts currently used in SNA were introduced in the 1950s and 1960s, which was both a point of departure for this theoretical-methodological framework and a way of expressing social phenomena that were not easy to define using the terminology of that time (Wasserman and Faust, 2013). By the end of the 1980s there was a growing body of academic work that included relational data in their analyses, as well as technical and methodological contributions (Freeman, 2004; Scott, 2013). Today this theoretical-methodological framework is well established in the social sciences (Scott et al., 2008) and has been applied to research in various disciplines, with particular importance over the past few decades due to the an increase in scientific production (Borgatti et al., 2009; Freeman, 2004). Moreover, it has received a great deal of exposure in various disciplines within the social sciences thanks to text books written by Borgatti et al. (2018), Carrington et al. (2005), Hanneman and Riddle (2005), Knoke and Yang (2008), Scott (2013) and Wasserman and Faust (2013). Although analysing relationships is not new in social sciences, SNA provides a relatively recent development in which theoretical hypotheses regarding the nature of relationships and the structural character of networks can be empirically verified (Garrido, 1996).

Over the past few years, the advancement of SNA within the social sciences, along with the development and improvement of software, has made it an attractive strategy for social research, thanks to its flexibility and originality, its capacity to systematize relationships and for its visual representation of social structures (Holgado, 2016), providing opportunities to analyse interactions between elements and examine structural patterns (Scott, 2013). It also allows social phenomena to be examined on various levels simultaneously: ego, dyadic, triadic, sub-structural and structural levels of the entire network (Robins, 2015).

There are two main methodological approaches in SNA: the egocentric or personal network, focused on the individual; and the socio-centric or full network approach, centred on the social group (Laumann et al., 1983; Marsden, 1990). Personal networks are focused on an actor, denominated "ego networks" and on the relationships this actor maintains with other actors, denominated "alters" (individuals) or "alteri" (group). The objective of ego networks is to identify "ego" links without establishing boundaries to the network, as is habitual when analysing complete networks (Wellmann, 2007). The full network approach focuses on analysing the relationships within a defined set of actors, considered as a whole, which involves the creation of boundaries for the target social group.

Unless these limits are already established, it is sometimes difficult for researchers to define a complete network, which is problematic because not covering the entire network can leave behind important information (Knoke and Kuklinski, 1982). To mitigate this issue and obtain the most objectively complete network possible, researchers must decide if they are going to use a nominalist or realist approach (Laumann et al., 1983). In the nominalist approach the limits of the network are defined by the researcher, normally used in studies of formally defined groups (the management board of a company, workers in a hotel, patrons of a foundation, etc.). In the realist approach, the boundaries of the network are defined by the actors themselves, according to mutual relevance, so only actors who are significant to the rest are selected. One of the most used techniques in the realist approach is known as the "snowball" method. In this method, knowledgeable actors (often called "seeds") in the target population are asked to nominate additional actors based on a specific relationship (for example, a collaboration to create a product); the actors named are then asked to nominate other actors, repeating the process until the desired sample size is obtained or the "saturation point" is reached, when no new actors can be identified. This technique is useful for identifying the target population in complete networks, including all important actors beyond the initial list.

As Lozares (1996) points out, the network approach is a formal analysis procedure that evidently is based on relationships, in which information is collected, transformed into data, processed formally, and interpreted substantively, just as in other methods. However, SNA stands apart from the habitual analytical frameworks used in social sciences because it focuses on relational or structural information, rather than attribute data (Verd and Martí, 1999) such as income, age, gender, education, etc. Although such data can also be used, they are not essential to the analysis. Relational data express connections between pairs of actors, which can be personal (friendship, respect, advice, etc.) or based on interactions between individuals (speaking, writing, etc.), affiliation in the same organization, family relationships, transferring resources (goods, money, information, etc.), and so on. In this sense, relational data is essential for SNA, while information based on attributes is secondary.

When discussing relational data, we must differentiate between "content" and "form", although they are inseparable. "Content" refers to the substance of the relationship (information exchange, knowledge sharing, etc.) that constitutes the issue being analysed, while "form" is the abstract expression of the relationship and properties of the relational structure (Lozares, 1996; Verd and Martí, 1999). In any case, in accordance with the numbering system (binary or valued) and the direction of the connection (directed or non-directed), Scott (2013) establishes four types of relational data. The first type is binary and undirected relational data used in the case of symmetrical relationships such as "brother" (A is B's brother and vice versa). The second type is undirected but valued, such as symmetrical relationships in which the value indicates the strength of a relation rather than its mere presence. The third type is binary and directed data, which is the case of asymmetrical relationships such as "lend money", which is represented by an arrow that indicates whether the relationship exists, passing through a 1 if it exists and a 0 if it doesn't. The fourth type is valued and directed data, which are asymmetrical relationships in which the value indicates the strength of the relation and is represented by an arrow.

Regarding the sources used to generate relational data, Nuñez (2008) points out that, in general, researchers can opt for primary (direct) sources and/or secondary (indirect) sources. According to this author, primary sources provide more relevant information on the relationships to be analysed, as they are more precise and specific; methods such as direct observation, focus groups and/or interviews are used to obtain relational data, with the latter being more effective to systematize the knowledge and preferences of the actors regarding the situation being studied. The author also points out that secondary information is obtained by analysing documentation that refers to the social groups that have been defined for the research, including memoirs, other research, publications, annual statistics, among others.

METHODOLOGICAL APPROACH, RESEARCH TECHNIQUES, AND ANALYTICAL MODEL

The methodological approach was to describe the characteristics of tourism development in a tourist destination by examining the relationships between the actors in that destination. Thus, the networks of tourism actors in the case studies were examined from a quantitative perspective of relational analyses, more specifically, from a socio-centric or complete networks perspective, in which all the actors that form part of the networks in the tourist destinations were included, as well as all possible relations between them. Merinero (2011) points out that identifying and describing tourism actors is not overly complex, but the same cannot be said for their relationships. The main contribution of this line of work to tourism studies is precisely the operationalization of these relations. They are treated like production processes, to establish the basic characteristics of the actions that must be carried out in the tourist destination for it to function like a tourism product. In this regard, Merinero (2011) points out that a production process is shaped by four main areas (creation, promotion, marketing, and management), each of which includes a certain number of specific tourism actions that can be carried out in collaboration by various actors.

The methodological scheme of the study has been based on the implementation of four research techniques: (1) first of all, to determine the population under study, unstructured interviews were conducted with key tourist actors of each tourist destination, in order to identify all the actors whose tourism activity had an impact within the territory and prepare the map of actors of each tourist destination; (2) once the target population was determined, semi-structured interviews were conducted with all the tourist actors of the three tourist destinations, in which they were asked what activities they had carried out with the rest of the destination's actors and in which productive areas. In this sense, an actor "A" could name an actor "B" with whom he had established a productive link, without the need for actor "B" to name to actor "A" to carry out the same productive action. In this regard, the productive actions that can be carried out with the involvement of two or more actors do not require relational symmetry; (3) then, all identified actors completed a questionnaire with the aim of obtaining a data panel on the relational dynamics of each tourist destination; (4) and, finally, after obtaining the data from each destination, relational matrices were created to record the information about the relationships between the actors in each destination and production area. The resulting relational matrices are square-shaped adjacency matrices, with the same number of rows and columns. Binary and directed relational data was used to fill in the matrices using a 1 or a 0, depending on if the actor was nominated (1) or not (0) by other actors with whom they had collaborative relationships. The direction of the connection is from the actors who nominated toward the actors who were nominated. Then, using UCINET 6 social network analysis software (Borgatti et al., 2002) the indicators for each matrix were calculated (see Table 1). Finally, VISONE software was used to create and visualize graphs of the relationships between the actors.

The analytical model that was applied is based on the idea that prestige and influence are inherent to a relationship and that both concepts are also closely tied to the idea of centrality. In socio-centric networks

Table 1. SNA indicators used for the analysis

Indicator	Description
Degree centrality	Calculates the number of actors to which a single actor is directly linked. Moreover, the "normalized degree centrality" is the centrality of the actor divided by the maximum possible number of relationships that the actor could have developed, multiplied by 100 in order to express it as a percentage (this is an individual percentage, so the sum of all the actors is not equal to 100%). Asymmetrical relational data provides the "indegree" and the "outdegree" of each actor.
Indegree	Calculates the number of direct relationships received by each actor, thereby identifying actors who are prestigious or prominent for other actors in the network. To interpret the indegree, actors were classified in four categories: very prestigious (0.750-1.000), high prestige (0.500-0.749), average prestige (0.250-0.499) and low prestige (0.000-0.249).
Outdegree	Calculates the number of direct relations initiated by each actor, reflecting the actor's influence and capacity to initiate relations with the other actors in the network. To interpret the outdegree, actors were classified in four categories: very influential (0.750-1.000), high influence (0.500-0.749), medium influence (0.250-0.499) and low influence (0.000-0.249).
Density	Calculates the number of relationships that exist in relation to the total number of possible relationships. It indicates the connectivity of the network. This indicator provides a data point as a proportion and should be multiplied by 100 to convert it to a percentage.

Source: Created by authors

and asymmetrical relationships, degree centrality allows two indicators to be obtained: indegree, which calculates the number of direct relationships received by each actor; and outdegree, which calculates the number of direct relationships initiated by each actor. From this perspective, SNA makes it possible to understand the capacity for action that is present in social relationships, understood as a manifestation of social power, which is a consequence of relationship patterns, that is, that through social relationships there is the development of a capacity to influence the actions of others and have greater opportunities to act (Hanneman and Riddle, 2005). In analytical terms it is possible to understand this manifestation of social power through the degree centrality which, in those cases in which social relationships are the technical manifestation of an asymmetric adjacency matrix, the analysis is possible by differentiating it between indegree and outdegree. So indegree highlights how actors are seen by other actors, allowing prestigious or prominent actors to be identified, so they turn to them to carry out activities, while outdegree shows an actor's capacity to initiate relations with other actors, allowing influential actors to be identified, having the capacity to promote actions by themselves together with the rest of the actors (Hanneman and Riddle, 2005; Knoke, 1990; Wasserman and Faust, 2013).

ANALYSIS OF THE TOURISM MANAGEMENT NETWORKS IN THE MUNICIPALITIES STUDIED

Context and Description of the Case Studies

The municipalities of Osuna, Loja and Ronda were selected as case studies because they have similar geographic, demographic, heritage, and economic characteristics, although their tourism development is different. The municipality, which is the unit of analysis, is considered a tourist destination and defined in relational terms as the group of actors involved in tourism activity. These three tourist destinations are "medium cities"[1]. To understand what is meant by a medium city the setting of the specific area and territorial scale in which it is located must be considered (Baigorri, 2000); in this case, the destinations are non-coastal, medium-sized cities of the Autonomous Community of Andalusia.

The tourist destination of Osuna is in the province of Sevilla, while Loja and Ronda are in the provinces of Granada and Málaga, respectively (see Figure 1). Osuna is the largest municipality in terms of area, with 592 km^2; its main economic activity is agriculture (olive groves, wheat, cotton, etc.), followed by (in order of importance) commerce, construction, industrial manufacturing, and hospitality-tourism. Loja is the second largest tourist destination in terms of area, with 447 km^2; its main economic activity is also agriculture, with a high production of olive groves and a considerable production of the so-called green asparagus. Commerce is also important, followed by construction, hospitality-tourism, and industrial manufacturing. Finally, Ronda has the smallest area, covering almost 398 km^2; its main economic activity is commerce, followed by hospitality-tourism, construction, and professional, scientific and technical activities. Agriculture is less important, although the wine sector is of some interest. All three case studies have similar demographics: Osuna has a population of 17,121 inhabitants, Loja 20,419 inhabitants and Ronda 33,730, according to the Multi-territory Information System of Andalusia (SIMA, 2021).

Regarding the elements of tourist attraction, the three tourist destinations have an important cultural heritage (archaeological sites, monuments, historical buildings, cultural landscapes, etc.). Considering the Ashworth (1995) classification, the interest of the tourist who visits these cultural tourist destinations is based on *Heritage Tourism*, related to visiting urban spaces with heritage appeal. In this sense, Osuna has the archaeological site of "Antigua Urso Romana", museums, historical buildings such as palaces, convents, etc. Loja preserves a good part of "La Alcazaba", a military and administrative centre of an old medieval city, as well as historical buildings in the Renaissance style, churches, etc. Finally, Ronda has the archaeological sites of "La Planilla" (megalithic necropolis from the Bronze Age) and "Acinipo" (Roman city), as well as historical buildings such as palaces, churches, etc.

However, despite the fact that the three case studies have an important cultural heritage, their tourist developments are different. In Osuna and Loja, the hospitality-tourism sector has a minor impact, being the fifth and fourth economic activity, respectively. In Ronda it is of greater importance, being the second economic activity behind commerce (SIMA. 2021). The prominence of tourism in each of these destinations is reflected in their accommodation offer. Although this is not a perfect indicator because it only measures part of overall tourism activity, it does provide an idea of how developed tourism is in a destination, given that the availability of accommodation allows tourists stay longer in the destination and spend more money, adding value to tourism as an economic activity. Osuna has 178 bed spaces, Loja has 447 and Ronda 1,820 (SIMA. 2021). In this sense, there is a correlation between the importance of tourism with the local economy and the number of tourist accommodations and bed spaces.

In short, given that the three destinations have similar geographic locations, relatively similar population sizes and economic activities, an important cultural heritage, and that they compete in the same cultural tourism market, they can be seen as similar tourist destinations, which would allow for possible differences in terms of tourism results to be minimized.

Figure 1. Location of the case studies within the territory of Andalusia
(Source: Institute of Statistics and Cartography of Andalusia, Spanish National Geographic Institute, and GISCO-EUROSTAT. Created by authors)

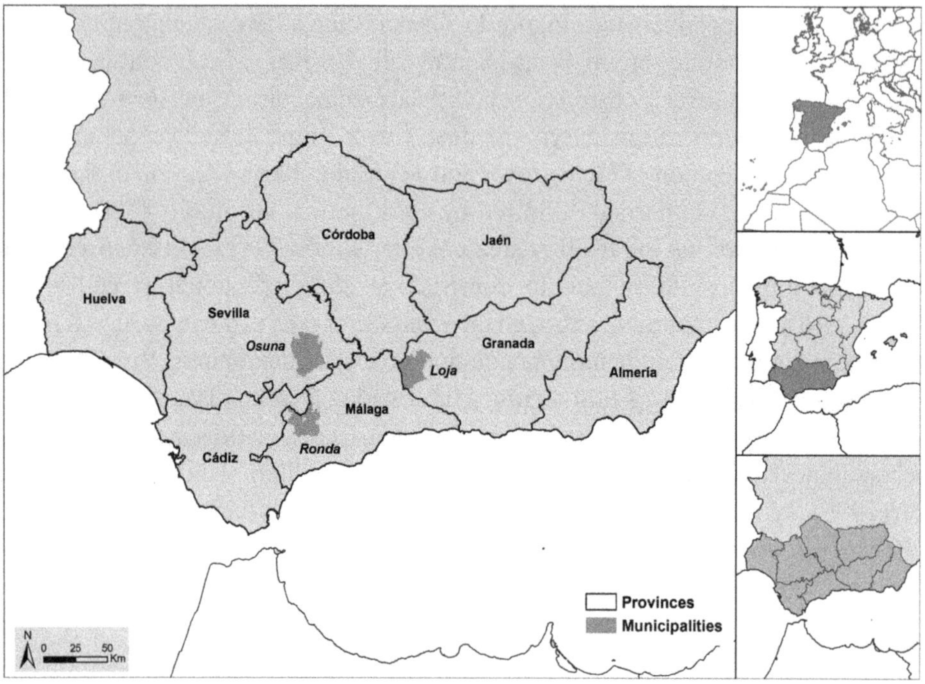

Results of the Analyses of the Networks of Actors Managing the Tourism Product

Description of the Network in Osuna

The Osuna tourism product management network is made up of 30 actors, of which 5 are public (the Council of Tourism, Commerce and Sports, the Provincial Council of Seville, the Osuna City Council, the Local Development Group and the Association of Municipalities) and the rest private. As for the latter, 6 are dedicated to complementary services (cultural and/or tourist associations, intermediation, etc.), 11 are tourist accommodation establishments and 8 are restaurants. The first characteristic that can be highlighted about the network of relationships established between the actors in the management of Osuna's tourism product is that it has a very low density, 6.9%, which means that out of every 100 possible relationships between the tourism actors, there are only 7. Due to this low relational density, the network is not very cohesive.

In terms of indegree, i.e., the prestige of the actors, the fact that all the actors have "low prestige" stands out. Keeping this in this mind, the City Council of Osuna (0.172) and the Patronato de la Colegiata (0.172) have the most prestige for the rest of the actors, the latter being a private actor committed to the defense and enhancement of the cultural heritage (painting, sculpture, architecture, etc.) of the tourist destination. The next group of actors with the most prestige are the Provincial Council of Seville (0.138), the restaurant La Casa del Marqués (0.138) and the complementary services company Viajes Urso (0.138).

Figure 2. Prestige of actors in the Osuna Tourism Management Network
(Source: Created by authors)

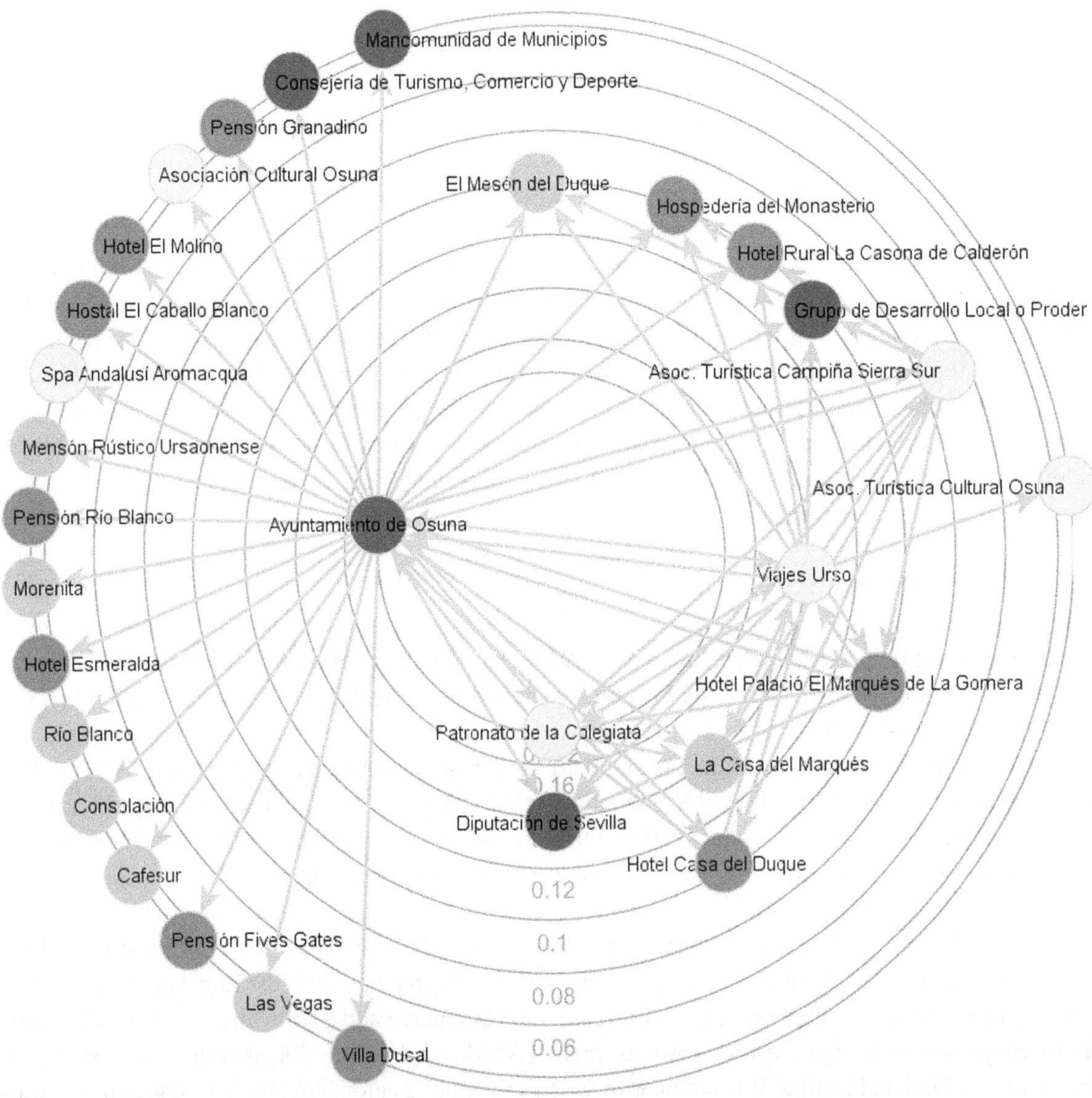

A distinguishing feature of Osuna's prestige network is that the accommodation establishments are not central players. That said, among them, the most prestigious are the hotels Casa del Duque, Palacio El Marqués de La Gomera, La Casona de Calderón and Hospedería del Monasterio which, together with the Local Development Group and the restaurant El Mesón del Duque have an indegree of 0.103. It is also worth noting the low prestige that the Council of Tourism, Commerce and Sports (0.034) and the Association of Municipalities (0.034) have for the rest of the actors in the network. Figure 2 represents the indegree or prestige of the actors in the networks, where actors with higher prestige are located toward the center of the network, and as the prestige of the actors drops, the farther toward the edges of the network they will be located. The graph shows that 17 actors appear in the network when they are mentioned only once by the Osuna City Council, while the Osuna Cultural Tourism Association (0.034) is linked to the network when it is mentioned by Viajes Urso.

On the other hand, if we consider the outdegree (influence), that is, taking into account the number of direct relationships initiated by each actor, the Osuna City Council (0.966) stands out as "very influential", since it has initiated relationships with all the actors in the network except the Osuna Cultural Tourism Association. Meanwhile, the company Viajes Urso (0.414) and the Asociación Turística Campiña Sierra Sur (0.345) have "medium influence". The rest of the actors have "low influence". It is noteworthy that the Ministry of Tourism, Commerce and Sport (0.000), the Seville Provincial Council (0.000), the Association of Municipalities (0.000) and especially the Local Development Group (0.000), given its role as a promoter of the territory, have not initiated any collaborative relationships with other actors and, therefore, have no influence on the management network of the Osuna tourism product. Figure 3, which represents outdegree, shows the Osuna City Council at the center of the network as the actor that has initiated the most collaborative relationships (28 times out of a possible 29). The other actors are located farther from the center and closer to the periphery of the relational structure as they have initiated fewer collaborative relationships. Therefore, the network has a highly centralized, "star-like" structure, highlighting the fact that the Osuna City Council is unquestionably the main driver of the Osuna tourism product management network.

Description of the Network in Loja

The Loja tourism product management network is made up of 29 actors, of which 4 are public (the Council of Tourism, Commerce and Sports, the Provincial Council of Granada, the Loja City Council and the Local Development Group) and the rest are private. Among the private actors, 13 are restaurants, 11 are tourist accommodation establishments and there is an association called Empresarios de Loja. The first characteristic that can be highlighted about the relationships established between the actors in the management of Loja's tourism product is that it has a medium-low density of 28.2%, which means that out of every 100 possible relationships between tourism actors, there are about 30. Due to this density, the network is somewhat cohesive.

What stands out regarding the indegree is that almost 60% of the actors have "medium prestige". Thus, 11 actors share the highest prestige in the network with an indegree of 0.464: the Loja City Council, the Provincial Council of Granada, the Council of Tourism, Commerce and Sports, the Local Development Group, Empresarios de Loja, the hotels Almazara, Los Abades and El Mirador, as well as the restaurants Casa Manolo, Flati and Emilio. It is noteworthy that all the public actors are in this first group of actors with the most prestige and all the private categories are represented, i.e., accommodations, restaurants and complementary services. After these actors there are another 6 that also have "medium prestige", although with lower indegrees: Hotel Las Terrazas (0.429), Hotel La Bobadilla (0.429), Restaurante Casa Anastasio (0.429), Hotel Manzanil (0.393), Restaurante Molino (0.393) and Restaurante Los Molinos (0.357). As for the "low prestige" actors, it stands out that the Hermanos Romero (0.000) and Almería (0.000) pensions are the only actors that have not been named by any of the others for implementing any tourism management action, while at the same time they have not named any other actor, placing them completely outside the relational structure. Figure 4, which represents the indegree, shows that most of the collaborative relationships are established among the 17 "medium prestige" stakeholders, while the "low prestige" stakeholders are located at the edges of the relational structure or outside it. In this sense, and as a characteristic of this network, in the case of the accommodation establishments, it can be seen

Figure 3. Influence of actors in the Osuna Tourism Management Network
(Source: Created by authors)

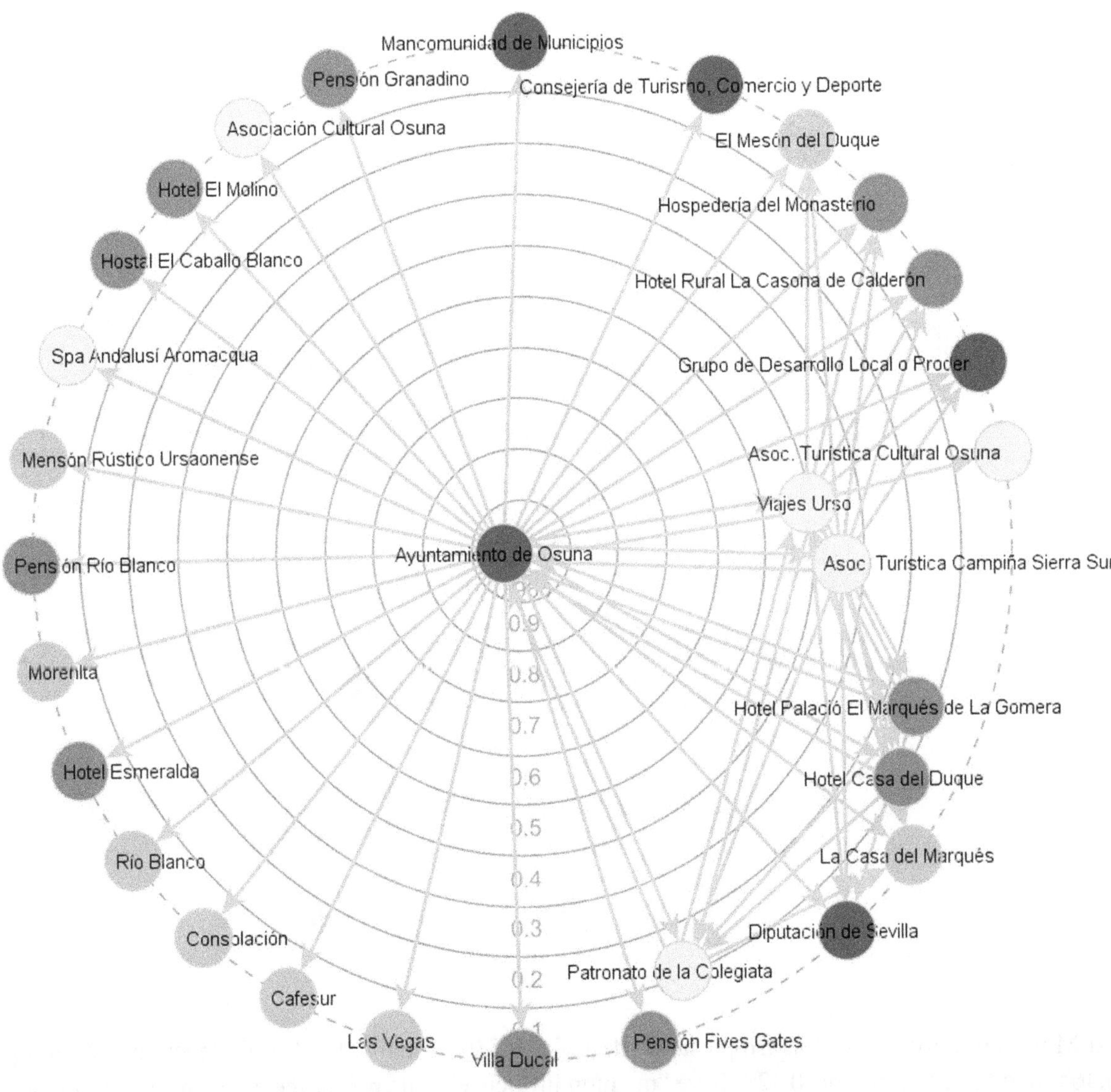

how the hotels are more prestigious in terms of configuring the tourist product of the municipality as they are located towards the center of the network, while the pensions and hostels are on the periphery or outside of it as they are less relevant.

On the other hand, taking into account the outdegree, there are actors in all the categories considered. In this sense, the City Council of Loja (0.893) and the Provincial Council of Granada (0.857) are "very influential" actors, as they have initiated direct collaborative relationships on 25 and 24 occasions, respectively, out of a possible 29. There are 10 actors with "high influence": the Council of Tourism, Commerce and Sports (0.571), the hotels La Bobadilla (0.571), Los Abades (0.571), Manzanil (0.571), Almazara (0.536), El Mirador (0.536) and Las Terrazas (0.536), the restaurants Casa Manolo (0.571)

Figure 4. Prestige of actors in the Loja Tourism Management Network
(Source: Created by authors)

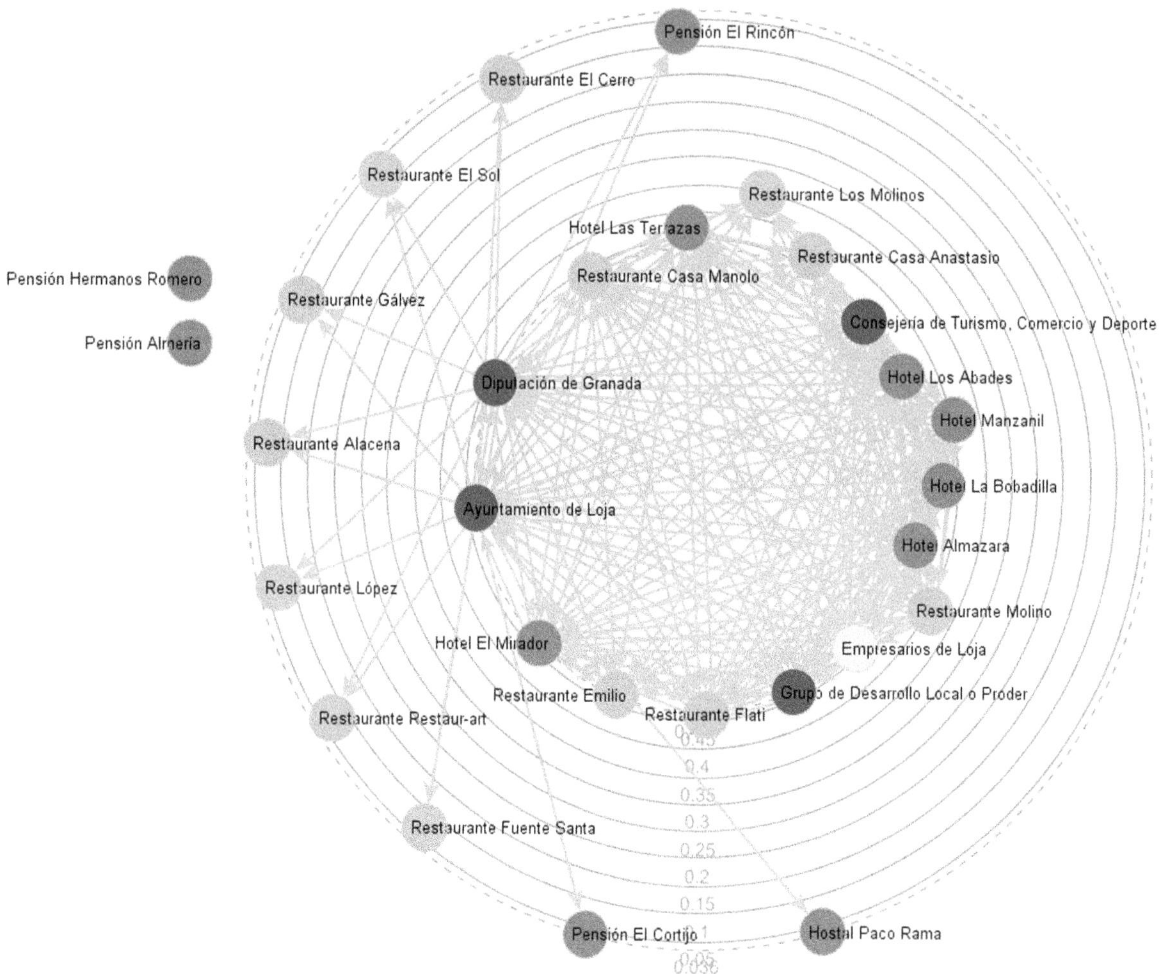

and Molino (0.536), as well as Empresarios de Loja (0.536). Finally, the Local Development Group (0.464) and Flati Restaurant (0.429) have "medium influence". The rest of the actors have not initiated collaborative actions, so they have "low influence". The representation of outdegree in Figure 5 shows the positioning of each tourism actor according to the number of times they have initiated direct collaborative relationships. The "very influential" actors are located at the center of the relational structure, while actors with "high influence" are those closest to them. The graph shows that the influential actors in the management of Loja's tourism product are both public and private. In addition, the graph shows that hotels have a greater outdegree than pensions and hostels, so that the former are not only more prestigious (see Figure 4), but also more influential in the configuration of the tourism product of the municipality.

Figure 5. Influence of actors in the Loja Tourism Management Network
(Source: Created by authors)

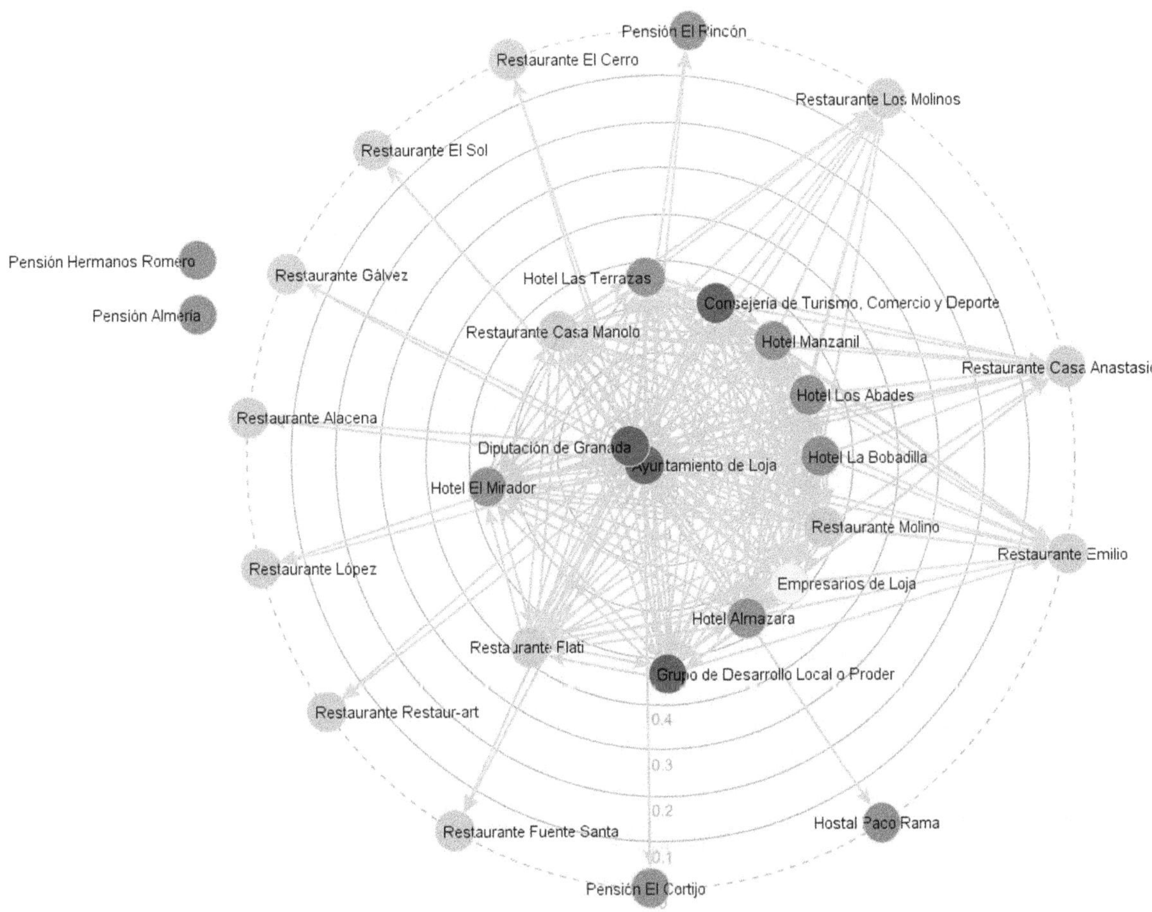

Description of the Network in Ronda

The management network of Ronda's tourism product is made up of 144 actors, of which 8 are public actors (Ministry of Tourism, Commerce and Sports, Malaga Provincial Council, Ronda City Council, Association of Municipalities, Local Development Group, Commercial Office of Tourism, Tourism Office of the Andalusian Government and Ronda Convention Office) and the rest are private actors. As for the latter, 48 are companies providing complementary services, another 48 are tourist accommodation establishments and 40 are restaurants. As in the previous cases, the first characteristic that can be highlighted of the relationships established between the actors in the management of Ronda's tourism product is their density, which in this case is average, 35.8%, which means that out of every 100 possible relationships between the tourism actors, there have been about 36. The result of the density is a relatively cohesive network.

Regarding the indegree, although there are no "very prestigious" actors, it stands out that there are 74 with "high prestige", which is just over 51% of the entire network. Albacara restaurant (0.643) stands out as the most prestigious actor in the whole network, ahead of any public actor. Then there is a large

Figure 6. Prestige of actors in the Ronda Tourism Management Network
(Source: Created by authors)

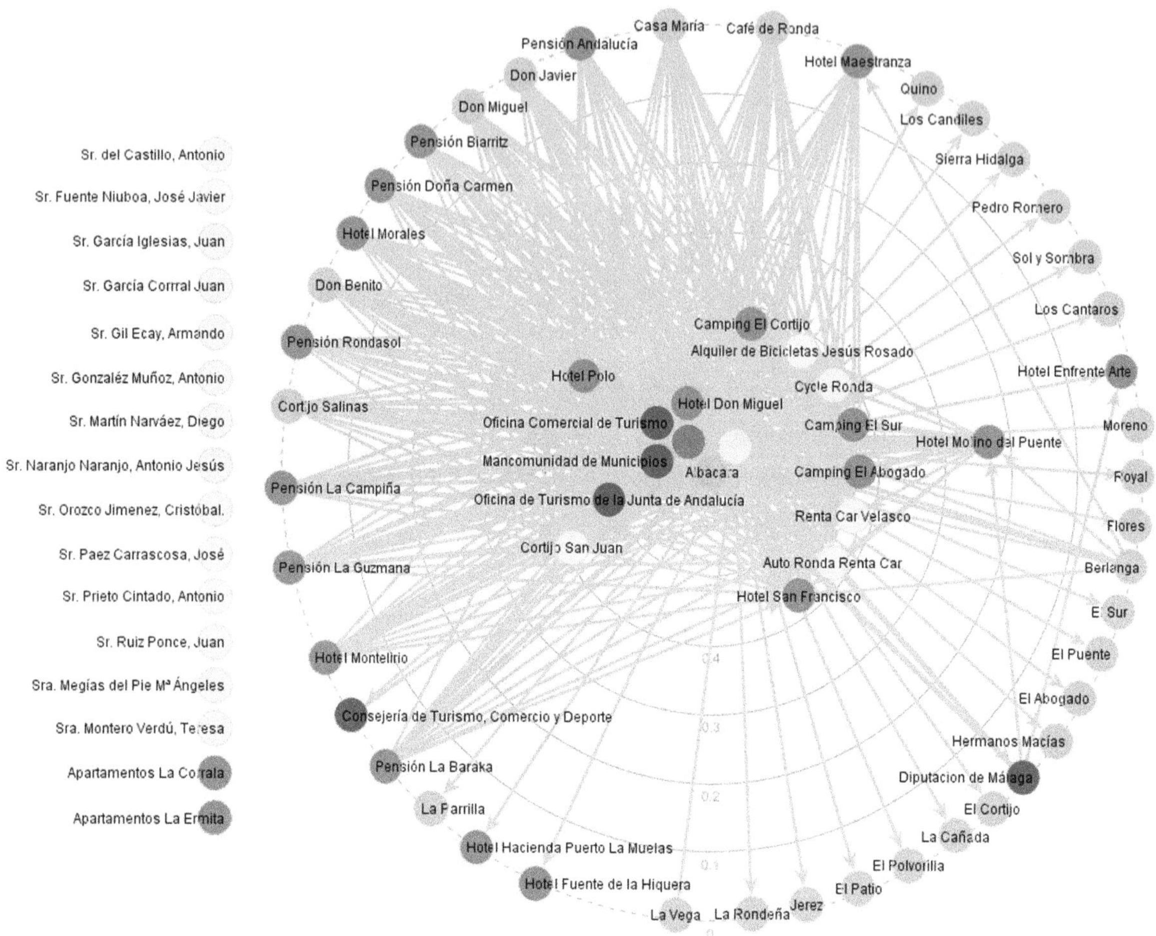

group of 41 actors with an indegree of 0.636 (the City Council of Ronda, the Local Development Group, the Ronda Convention Office, etc.) and after this, another group of 28 actors with an indegree of 0.629 (Viñedos Los Terrones, Aventura Ronda, Taiotur, etc.). The Association of Municipalities (0.622), the Commercial Office of Tourism (0.622), Hotel Don Miguel (0.622) and the Tourism Office of the Andalusian Government (0.615) complete the group of actors with "high prestige". There are 10 actors with "medium prestige" (Hotel San Francisco, Camping El Abogado, Cycle Ronda, etc.) and the rest of the actors have "low prestige". It is noteworthy that both the Ministry of Tourism, Trade and Sport (0.007) and the Provincial Council of Malaga (0.007) are actors of such little relevance for the rest of the actors in the network. Figure 6, which is the representation of the indegree, shows that the relational density is important and that the network is relatively cohesive, with quite a few actors with whom the others have had collaborative relationships, although it also shows that there are a large number of actors who are located at the extremes of the network as they are mentioned on few occasions or even outside the network. In order to represent the network more clearly, the "high prestige" actors with indegrees of 0.636 and 0.629, represented in yellow and purple respectively, have been grouped in the same coordinates.

On the other hand, considering the outdegree, the Ronda City Council (0.825) is the only "very influential" actor. However, 88 actors have "high influence", which represents 61.1% of the total number of actors in the network. Thus, a group of 15 stakeholders made up of accommodation and restaurant companies have an outdegree of 0.587 (Hotel Maestranza, Hotel Molino del Puente, Pensión Andalucía, etc.), making them the most influential stakeholders in the destination behind only the City Council. This is followed by a large group of 42 actors with an outdegree of 0.580, consisting of hotels, restaurants and museums (Café de Ronda, Cantina Estación, Hotel Colón, etc.). Then, a third group of 26 actors have an outdegree of 0.517, formed by public actors, tourist accommodations, wineries and vineyards, etc. (Tourist Office of the Junta de Andalucía, El Juncal, Lunares, etc.). Finally, 4 actors have "medium influence", while the rest have "low influence", highlighting, among the latter, the Local Development Group (0.203) and the Ministry of Tourism, Trade and Sport (0.035), which should have more active roles in terms of managing the tourist destination but are actors with little collaborative initiative. Figure 7, which shows the outdegrees of the actors in the network, has the Ronda City Council at the center of the relational structure and around it the most influential actors. The graph shows that the set of "very influential" or "highly influential" actors is made up of public administration actors and accommodation establishments, restaurants and complementary services companies, so that every category of actors has highly influential representatives. As they lose centrality, the rest of the tourism stakeholders are located closer to the periphery of the network, highlighting the presence of the Regional Ministry of Tourism, Trade and Sport on the edge of the network. In order to represent the network more clearly, the "high influence" actors in the 0.580 and 0.517 outdegree groups are clustered at the same coordinates, represented in yellow and purple, respectively.

CONCLUSION, IMPLICATIONS AND LIMITATIONS

This work approaches the tourist destination as a relational system that can be studied using SNA. The SNA theoretical-methodological framework was employed to analyse the relational structure derived from the interactions that occur between actors in the management of the tourism product in three case studies: Osuna, Loja and Ronda. A socio-centric or complete network approach was used to obtain relational data, as well as to process and analyse the information via the creation of relational matrices using UCINET 6 software and the network density and degree centrality indictors, which allows the indegree (prestige) and outdegree (influence) of socio-centric networks and asymmetrical relationships to be obtained. VISONE software was used to elaborate and visualize the resulting graph of relations between the actors.

To measure the level of tourism development in the destinations selected we considered how important tourism activity was for the local economy, as well as the offer of tourist accommodations in the destinations. Evidently, the volume of accommodations is not a perfect indicator because it only measures part of the tourism sector, but it does give us something to work with since the availability of accommodations is one aspect that marks the level of tourism development and there is a relationship between the importance of tourism for a local economy and the quantity of accommodation establishments and bed spaces available. Despite the fact that Osuna, Loja and Ronda are similar in terms of geography, demographics, heritage and economy, they present differences in terms of tourism development,

A preliminary result obtained from the density analysis of the networks of actors in the management of the tourism product in the selected case studies is that the number of actors is not related to

Figure 7. Influence of actors in the Ronda Tourism Management Network
(Source: Created by authors)

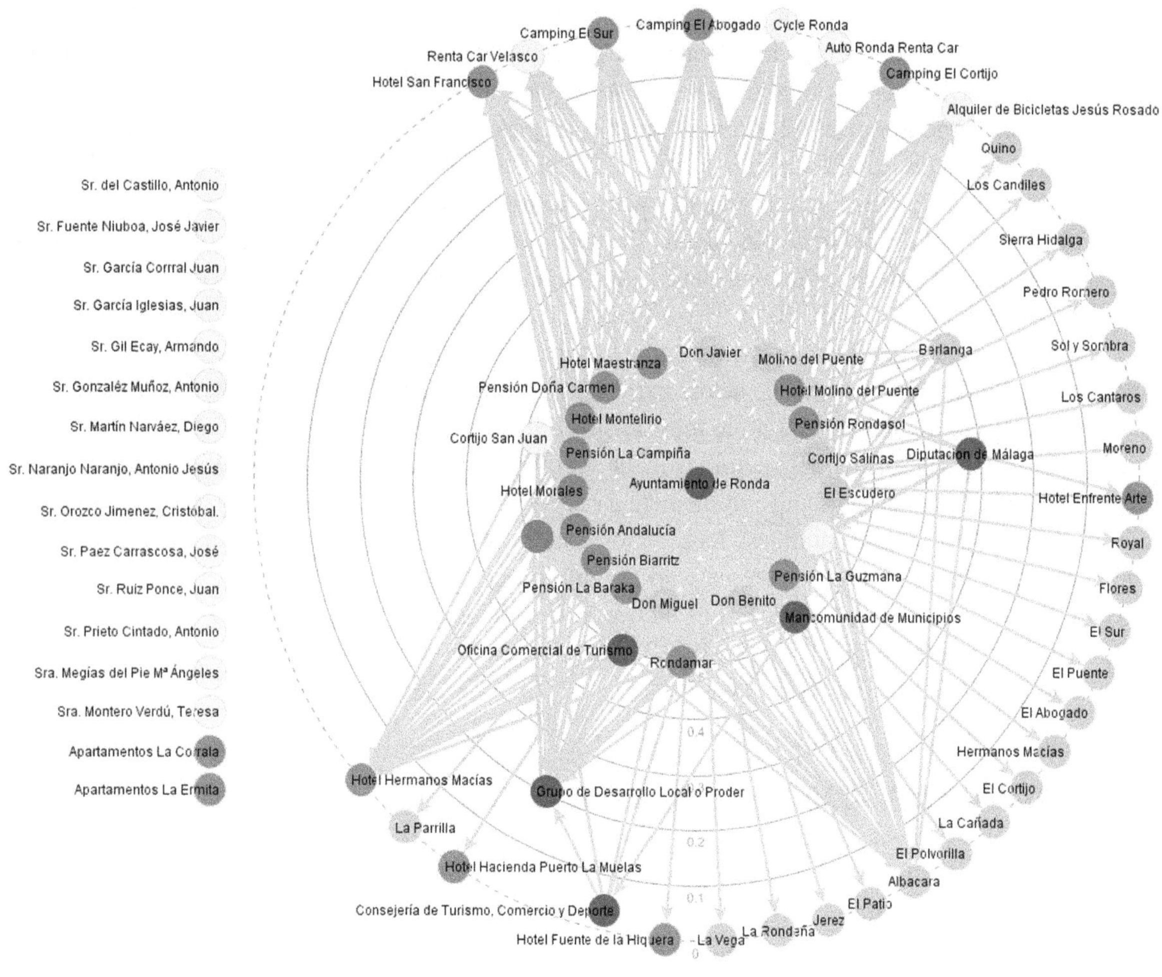

the relational density. In this sense, the networks of Loja and Osuna have practically the same number of actors and yet their relational densities are very different. Thus, while in Osuna there are almost 7 relationships out of a possible 100, in Loja there are almost 30. In the case of Ronda, in addition to the fact that its network of tourism actors has a higher number of actors, it is also has a greater relational density, since there are almost 36 relationships out of a possible 100. In relation to this, the results of the density analysis of the selected cases confirm the existence of a direct relationship between relational dynamics and the development of cultural tourism destinations: the greater the relational intensity, the greater the development, although it is not possible to confirm the causal direction between these two dimensions. Therefore, the first objective of the study is verified. Likewise, the results obtained are coherent with what has been proposed in the scientific literatura (Baggio, 2020; Ledesma et al., 2021; Matías and Pulido, 2012; Merinero, 2011; Merinero and Pulido, 2009; Pavlovich, 2003; Pearce, 2014).

In terms of co-production of the "tourist destination product", the relational dynamics analysed also show that as tourist destinations evolve, more central players appear, in accordance with what is proposed by Aarstad et al. (2015). The results of the analyses indicate that the relational dynamics of the actors influence the evolution of tourist destinations and, in this sense, the structure of the networks of actors of young tourist destinations is characterized by having one or few very central actors, while that, as tourist destinations mature, the structure of the networks of actors is characterized by having a greater number of central actors. Thus, as tourist destinations mature and consequently the number of central players increases, the structure of the network becomes more decentralized, so that the capacity of orchestration (Aarstad et al., 2015) or leadership (Hristov et al., 2018) goes from concentrating on one or a few actors, to being distributed through the network structure. Consequently, it can be said that as tourist destinations mature, the structure of the networks of actors are characterized to have more central actors, so it checks the second objective of the study.

In relation to the above, the degree centrality seems to indicate that as the number of tourism actors in the centre grows, so does the level of tourism development; thus, Osuna, which is the least developed tourist destination among the three case studies, has few central actors, while Loja has more, and there are even more in Ronda. Taking into account the indegree (prestige), the tourist destinations have a different relational structure, since in Osuna all the actors have "low prestige", in Loja almost 60% of the actors have "medium prestige" and the rest have "low prestige", while in Ronda more than 50% have "high prestige", 10 actors have "medium prestige" and the rest have "low prestige". Prestige indicates that all the stakeholders in a tourist destination are not perceived by the other stakeholders as equally important, which is understandable, in line with other studies such as those carried out by Cooper et al. (2009) or Sheehan and Ritchie (2005), which have shown differences in terms of the importance of stakeholders in tourist destinations. In detail, in Osuna one public and one private actor are the most prestigious; in Loja, 4 public and 7 private actors are the most prestigious; finally, in Ronda there are 74 actors who all share being the most prestigious, of which 6 are public actors and the rest are private. These results are consistent with those obtained by Ledesma et al. (2021) and Aarstad et al. (2015) which show that the importance of the public administration increases as tourist destinations grow, both as a regulatory and institutional body, as well as its potential as an agent of development.

As for the outdegree (influence), in Osuna only 6 actors have initiated collaborative relationships (20% of the total), in Loja there are 14 (48% of the total) and in Ronda 94 (65% of the total), which reveals that the capacity of the actors in Ronda to initiate collaborative relationships and, consequently, to influence the rest of the actors is greater than in the other tourist destinations. In any case, the prominent influence of public actors in the three cases analyzed stands out, especially their city councils and, in the case of Loja, also the Provincial Council of Granada. However, as the destinations become more developed, private actors become considerably more influential and they become active in promoting collaborations, with a clear tendency towards a system of intense public-private collaboration. These results are consistent with those obtained by Ledesma et al. (2021) and allow us to confirm the premise that there is a correlation between the number of actors with high centrality and the development of tourist destinations, providing an answer to the third objective posed.

The analysis of the selected case studies allows us to respond to the recommendations of Baggio et al. (2010) or Merinero (2015), among others, regarding the need for more comparative studies linking relational structure with the development of tourist destinations. It also responds to one of the limitations detected in a previous work by the authors, which pointed out the need for more case studies to consolidate the results obtained and strengthen the conclusions. In this sense, the three objectives have been met, and the results obtained confirm the direct association between relational dynamics and the development of tourist destinations. Specifically, as tourist destinations mature, the relational structure has a greater number of central actors, and there is also a direct relationship between the number of actors with high centrality and the development of tourist destinations.

Regarding the limitations of the research, the following stand out: the first has to do with the product offered, since the case studies analyzed are cultural tourist destinations that compete in the same market, so it would be advisable to carry out these analyses in coastal tourist destinations, mountain tourist destinations, etc. The second has to do with the context, since the case studies selected are "medium-sized cities" in the interior of Andalusia. It would be useful to apply the same methodological approach, research techniques and analytical model to other regions of Spain, both on the mainland and the islands, with the aim of minimizing this limitation. Likewise, researchers from other regions of the world are encouraged to carry out similar analyses in order to be able to make comparisons and strengthen the results obtained.

REFERENCES

Aarstad, J., Ness, H., & Haugland, S. (2015). Destination Evolution and Network Dynamics: New Research Agendas. In D. Chambers & T. Rakić (Eds.), *Tourism Research Frontiers: Beyond the Boundaries of Knowledge* (pp. 141–164). Emerald Publishing. doi:10.1108/S1571-504320150000020002

Ashworth, G. (1995). Managing the cultural tourism. In G. Ashworth & A. Dietvorst (Eds.), *Tourism and spatial transformation. Implications for policy and planning* (pp. 265–283). CAB International.

Baggio, R. (2020). Tourism destinations: A universality conjecture based on network science. *Annals of Tourism Research, 82*(102929), 1–13. doi:10.1016/j.annals.2020.102929

Baggio, R., Scott, N., & Cooper, Ch. (2010). Network science. A review focused on tourism. *Annals of Tourism Research, 37*(3), 802–827. doi:10.1016/j.annals.2010.02.008

Baigorri, A. (2000). La red urbana ibérica. In L. de la Machorra & M. Brandâo (Eds.), *La economía ibérica: una fértil apuesta de futuro* (pp. 261–289). Editora Regional de Extremadura.

Borgatti, S., Everett, M., & Freeman, C. (2002). *Ucinet for Windows: Software for Social Networks Analysis*. Harvard University.

Borgatti, S., Everett, M., & Johnson, J. (2018). *Analyzing Social Networks*. SAGE Publications.

Borgatti, S., Mehra, A., Brass, D., & Labianca, G. (2009). Network Analysis in the Social Sciences. *Science, 323*(5916), 892–895. doi:10.1126cience.1165821 PMID:19213908

Brito, M. (2009). *Ciudades históricas como destinos patrimoniales. Una mirada comparada*. Junta de Andalucía, Consejería de Cultura.

Carrington, P., Scott, J., & Wasserman, S. (Eds.). (2005). *Models and methods in social network analysis*. Cambridge University Press. doi:10.1017/CBO9780511811395

Casanueva, C., Gallego, A., & García-Sánchez, M. (2016). Social network analysis in tourism. *Current Issues in Tourism, 19*(12), 1190–1209. doi:10.1080/13683500.2014.990422

Cooper, Ch., Scott, N., & Baggio, R. (2009). Network Position and Perceptions of Destination Stakeholder Importance. *Anatolia, 20*(1), 33–45. doi:10.1080/13032917.2009.10518893

Freeman, L. (2004). *The Development of Social Network Analysis. A Study on the Sociology of Science*. Empirical Press.

Fyall, A., Garrod, B., & Wang, Y. (2012). Destination collaboration: A critical review of theoretical approaches to a multi-dimensional phenomenon. *Journal of Destination Marketing & Management*, *1*(1-2), 10–26. doi:10.1016/j.jdmm.2012.10.002

Garrido, F. (1996). *El análisis de redes en el desarrollo local*. http://pendientedemigracion.ucm.es/info/eurotheo/materiales/hismat/index.html

Government of Andalusia. (2006). *Plan de Ordenación del Territorio de Andalucía. Decreto 206/2006, de 28 de noviembre*. https://www.juntadeandalucia.es/export/drupaljda/pota_completo.pdf

Hanneman, R., & Riddle, M. (2005). *Introduction to social network methods*. University of California.

Holgado, D. (2016). Analyzing Social Networks. *Revista Hispana para el Análisis de Redes Sociales*, *27*(2), 141–145. doi:10.5565/rev/redes.637

Hristov, D., Scott, N., & Michona, S. (2018). Distributed leadership typologies in destination management organisations. *Tourism Management Perspectives*, *28*, 126–143. doi:10.1016/j.tmp.2018.08.003

Knoke, D. (1990). *Political Networks. The Structural Perspective*. Cambridge University Press. doi:10.1017/CBO9780511527548

Knoke, D., & Kuklinski, J. (1982). *Network Analysis*. SAGE Publications.

Knoke, D., & Yang, S. (2008). *Social Network Analysis*. SAGE Publications. doi:10.4135/9781412985864

Laumann, E., Marsden, P., & Prensky, D. (1983). The Boundary specification problem in network analysis. In R. Burt & M. Minor (Eds.), *Applied Network Analysis: A methodological introduction* (pp. 18–34). SAGE Publications.

Ledesma, O. (2012). Actores, gobierno en red y destinos turísticos. El estado del arte. *Turismo. Revista de Estudios de Turismo de Canarias y Macaronesia*, *3-4*, 23–32.

Ledesma, O., Merinero, R., & Pulido, J. I. (2021). Tourist destination development and social network analysis: What does degree centrality contribute? *International Journal of Tourism Research*, *23*(4), 652–666. doi:10.1002/jtr.2432

Ledesma, O., & Simancas, M. (2020). El valor de los datos relacionales en la gestión de destinos turísticos. Una aplicación con Análisis de Redes Sociales (ARS). In M. Simancas and P. Peñarrubia (Coords.), El valor de los datos turísticos (pp. 263-286). Tirant Humanidades.

Lozares, C. (1996). La teoría de redes sociales. *Papers. Revista de Sociologia*, *48*, 103–126. doi:10.5565/rev/papers/v48n0.1814

Marsden, P. (1990). Network data and measurement. *Annual Review of Sociology*, *16*(1), 435–463. doi:10.1146/annurev.so.16.080190.002251

Matías, G., & Pulido, J. I. (2012). Dinámica relacional interorganizacional para el desarrollo turístico. El caso de Villa Gesell y Pinamar (Argentina). *Revista de Estudios Regionales*, *94*, 167–194.

Merinero, R. (2011). Redes de actores y desarrollo turístico. Estudio de casos en Portugal. *Revista de Análisis Turístico*, *11*(1), 9–21.

Merinero, R. (2015). La agenda de investigación del turismo mediante el análisis cuantitativo de redes sociales (QSNA). *Cuadernos de Turismo*, *36*(36), 269–294. doi:10.6018/turismo.36.230991

Merinero, R., & Lara, F. (2011, May). *Las Ciudades Medias del interior de Andalucía. Caracterización y retos para el desarrollo turístico en un nuevo entorno* [Paper presentation]. 15th Congress of the Spanish Association of Scientific Experts in Tourism, Mijas, Málaga, Spain.

Merinero, R., & Pulido, J. I. (2009). Desarrollo turístico y dinámica relacional. Metodología de análisis para la gestión activa de destinos turísticos. *Cuadernos de Turismo*, *23*, 173–193.

Merinero, R., & Pulido, J. I. (2016). Analysing relationships in tourism: A review. *Tourism Management*, *54*, 122–13. doi:10.1016/j.tourman.2015.10.010

Nuñez, J. (2008). *Manual para operaciones básicas con Visone en el análisis de redes sociales para el desarrollo rural*. http://revista-redes.rediris.es/webredes/talleres/MANUAL-VISONE.pdf

Pavlovich, K. (2003). The evolution and transformation of a tourism destination network: The Waitomo Caves, New Zealand. *Tourism Management*, *24*(2), 203–216. doi:10.1016/S0261-5177(02)00056-0

Pearce, D. (2014). Toward an integrative conceptual framework of destinations. *Journal of Travel Research*, *53*(2), 141–153. doi:10.1177/0047287513491334

Pulido, J. I., & Merinero, R. (2018). Destinations relational dynamic and tourism development. *Journal of Destination Marketing & Management*, *7*, 140–152. doi:10.1016/j.jdmm.2016.09.008

Richards, G. (1996). Production and consumption of European Cultural Tourism. *Annals of Tourism Research*, *23*(2), 261–283. doi:10.1016/0160-7383(95)00063-1

Robins, G. (2015). *Doing social network research: Network-based research design for social scientists*. SAGE Publications. doi:10.4135/9781473916753

Scott, J. (2013). *Social Network Analysis*. SAGE Publications.

Scott, N., Baggio, R., & Cooper, Ch. (2008). *Network analysis and tourism: From theory to practice*. Channel View Publications. doi:10.21832/9781845410896

Sheehan, L., & Ritchie, J. (2005). Destination Stakeholders. Exploring Identity and Salience. *Annals of Tourism Research*, *32*(3), 711–734. doi:10.1016/j.annals.2004.10.013

Sistema de Información Multiterritorial de Andalucía. (2021). *Andalucía pueblo a pueblo - Fichas municipales*. https://www.juntadeandalucia.es/institutodeestadisticaycartografia/sima/index.htm

Valeri, M., & Baggio, R. (2021). Social network analysis: Organizational implications in tourism management. *The International Journal of Organizational Analysis*, *9*(2), 342–353. doi:10.1108/IJOA-12-2019-1971

Verd, J. M., & Martí, J. (1999). Muestreo y recogida de datos en el análisis de redes sociales. *Qüestiió. Quaderns d'Estadística, Sistemes, Informàtica i Investigació Operativ*, *23*(3), 507–524.

Wasserman, S., & Faust, K. (2013). *Análisis de Redes Sociales. Métodos y Aplicaciones.* Centro de Investigaciones Sociológicas.

Wasserman, S., & Galaskiewicz, J. (1994). *Advances in Social Networks Analysis.* SAGE Publications. doi:10.1017/CBO9780511815478

Wellman, B. (2007). The network is personal. Introduction to a special issue of Social Networks. *Social Networks, 29*(3), 349–356. doi:10.1016/j.socnet.2007.01.006

ADDITIONAL READING

Romeiro, P., & Costa, C. (2010). The potential of management networks in the innovation and competitiveness of rural tourism: A case study on the Valle del Jerte (Spain). *Current Issues in Tourism, 13*(1), 75–91. doi:10.1080/13683500902730452

Scott, N., Cooper, Ch., & Baggio, R. (2008). Destination networks. Four Australian cases. *Annals of Tourism Research, 55*(1), 169–188. doi:10.1016/j.annals.2007.07.004

Shih, H.-Y. (2006). Network characteristics of drive tourism destinations: An application of network analysis in tourism. *Tourism Management, 27*(5), 1029–1039. doi:10.1016/j.tourman.2005.08.002

Stokowski, P. (1992). Social networks and tourist behavior. *The American Behavioral Scientist, 36*(2), 212–221. doi:10.1177/0002764292036002008

Urry, J. (2003). Social network, travel, and talk. *The British Journal of Sociology, 54*(2), 155–175. doi:10.1080/0007131032000080186 PMID:12945865

Watts, M. (2009). Collaborative implementation network structures: Cultural tourism implementation in an English seaside context. *Systemic Practice and Action Research, 22*(4), 293–311. doi:10.100711213-009-9125-0

Zach, F., & Hill, T. (2017). Network, knowledge, and relationship impacts on innovation in tourism destinations. *Tourism Management, 62*(1), 196–207. doi:10.1016/j.tourman.2017.04.001

Zehrer, A., & Raich, F. (2010). Applying a life cycle perspective to explain tourism network development. *Service Industries Journal, 30*(10), 1683–1705. doi:10.1080/02642060903580698

KEY TERMS AND DEFINITIONS

Actor: Individuals or organizations that have an impact on the tourist activity of the destination and that interacts with other actors to identify the needs of the tourists and create an experience for them.

Cultural Tourism: Tourism typology in which the main motivation of the visitor is to learn, discover, experience, and consume the cultural, material and intangible attractions/products of a tourist destination.

Indegree (Prestige): SNA indicator that calculates the number of direct relationships received by each actor, allowing to know the prestigious or prominent actors for the rest of the actors in the network.

Outdegree (Influence): SNA indicator that calculates the number of direct relationships initiated by each actor, reflecting the influence and ability of an actor to initiate relationships with the rest of the actors.

Relational Dynamics: Guidelines and frameworks that establish the characteristics of the relationships that occur between the actors of a tourist destination.

Relational Structure: Fundamental principles that define the positions that tourism actors occupy in each network of actors.

Social Network Analysis (SNA): Interdisciplinary field of study that aims to analyse the structural properties of networks, understood as a set of nodes (actors) and the connections that occur between them (relationships).

Tourism Development: Set of strategies and plans that aim to increase, develop and/or promote tourism in a destination.

Tourism Management: Multidisciplinary field that includes all activities related to the management of the tourism and hospitality industries.

Tourism Product: Combination of material and intangible elements, tourist attractions, facilities, services and activities around a specific element of interest that generates a comprehensive tourist experience for potential clients.

ENDNOTE

1 The aspects that define medium cities in the interior of Andalusia are: (1) they have a population ranging between 10,000 and 45,000 inhabitants; (2) they have a central role in the area they are located and, therefore, they functionally contribute to the cohesion of the territorial structure of the region; and (3) they are located in the interior of Andalusia and outside the area of influence of its large cities (Government of Andalusia; 2006; Merinero and Lara, 2011).

Chapter 11
The Tourism Value of Terraced Fields Landscapes:
The "Bancales" of the Island of La Gomera (Spain)

Moisés Simancas Cruz
https://orcid.org/0000-0002-0337-0424
University of La Laguna, Spain

José Juan Cano Delgado
https://orcid.org/0000-0001-6735-2033
University of La Laguna, Spain

Jorge Cebrián Ramos
Gesplan, Spain

Carlos Flores Rodríguez
Gesplan, Spain

Desiderio Gutiérrez Taño
University of La Laguna, Spain

Eduardo Martínez Díaz
Gesplan, Spain

María Pilar Peñarrubia Zaragoza
University of Valencia, Spain

Javier Serrano Lara
University of Valencia, Spain

ABSTRACT

Due to the mountainous nature of the Island of La Gomera (Canary Islands, Spain) relatively little flat land is available. This has generated an impressive architecture of terraces with dry stone borders. However, the process of restructuring and readjusting rural areas, mainly due to the emergence of a new model of territorial development and the impossibility of mechanizing tillage, due to the very configuration and morphology of these terraces, have led to the abandonment of agricultural activity on the terraces, with the consequent decay and collapse of the dry-stone walls. This chapter sets out the strategy to develop tourism of the terraces in the north of the Island of La Gomera.

DOI: 10.4018/978-1-7998-9217-5.ch011

INTRODUCTION

The traditional strategy of tourist destinations revolving around the "sun and beach" product is based on tourists spending almost the entirety of their stay in tourist accommodation establishments or in the tourist area where they are located. This has been derived from an erroneous simplification of the tourist area, based in the idea that the main attraction factor is the services and facilities that such areas offer. The consequence of reducing the conception of the tourist area almost exclusively to the urban-tourist space, where tourist accommodation and service facilities are mainly concentrated, is that the territorial setting in which they are located is seen as peripheral and complementary; implying a disconnection with their territorial context.

However, changes in tourist preferences, behaviours, needs, purchasing behaviour, expectations, attitudes, geographical mobility patterns and patterns of consumption of products and services have changed in recent years (Scott et al., 2011; Peñarrubia, 2018). A new tourist profile has emerged, characterized by the combination of services and activities that allow the enjoyment of resources and that shape the tourist experience. This has led to a substantial modification of the above-mentioned centripetal territorial model because tourists are increasingly leaving the coastal tourist areas where they spend the night to carry out various activities in places of tourist interest (Padrón & Hernández, 2017).

This is consolidating what is known as "experience-based or emotional tourism", which promotes getting to know territories through activities and services aimed at generating feelings and sensations, creating experiences that are considered more authentic, genuine and meaningful. This means moving from a transactional model to a relational one, in which travellers are no longer passive elements in the chain, but "co-creators" of the social value generated by the tourist experience (Peñarrubia et al., 2019).

This implies an enhancement of the multiple and diverse tourism resources of the territorial context in which these coastal tourist areas are located. This territorial perspective makes the geographical scale of the destination much smaller. The result is a territorial-tourist system made up of the sum of tourist concentration or departure nodes - the tourist areas that concentrate overnight stays - and the "points of tourist interest", understood as places tourists visit to carry out certain activities or enjoy certain resources. This entails the establishment and/or intensification of complementary, symbiotic or even synergetic relationships between coastal activities and services and those that take place in other areas (Simancas, 2020).

The 85% of the surface has slopes greater than 20%, due to the mountainous nature of the island of La Gomera (Canary Islands, Spain). This explains the relatively little flat land available (see Figure 1).. To counteract this, over the centuries farmers have reshaped the steep slopes, building flat, dry-stone terraced agricultural areas, which were traditionally used for subsistence farming. This has generated an impressive architecture of terraces with dry stone borders. Thus, the terraces, walls, embankments and plains form one of the most unique elements of the landscape and territorial identity of La Gomera.

However, the process of restructuring and readjusting rural areas, mainly due to the emergence of a new model of territorial development and the impossibility of mechanizing tillage, due to the very configuration and morphology of these terraces, have led to the abandonment of agricultural activity on the terraces, with the consequent decay and collapse of the dry-stone walls.

This paper sets out the strategy to develop tourism of the terraces in the north of the island of La Gomera. This is part of the Strategic Plan for the Development of the North of La Gomera, promoted by the Department of Tourism, Industry and Trade of the Canary Islands Government, in collaboration

Figure 1. Map of The Gomera Island
Source: IDECanarias. Grafcan

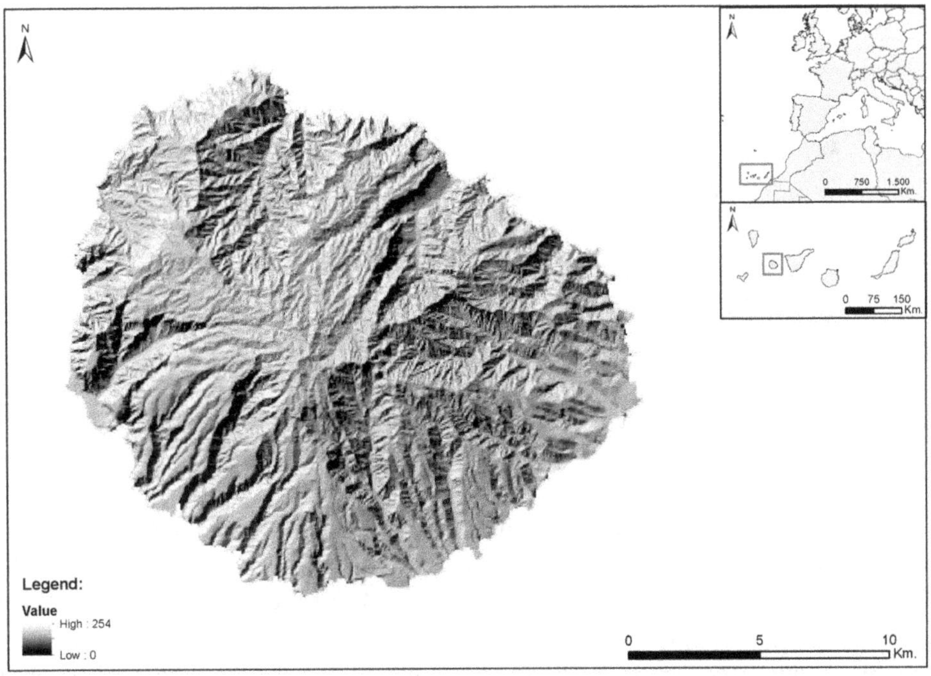

with the Cabildo of La Gomera, and developed by researchers from the ReinvenTUR research team of the University of La Laguna.

PROBLEM STATEMENT

The mountainous nature of La Gomera Islands generates few flat or sub-horizontal areas. For centuries this has forced farmers to build cultivation areas on dry terraces for subsistence agriculture. The result is the generation of "landscapes of hunger" (Gómez, 2007), derived from the "hunger for land" due to the little availability of soils for agriculture (Martín, 2009).This has given rise to an impressive architecture of terraces, with borders (walls) of dry stones, locally called "bancales", "paredones", "andenes" or "llanos" (see Figure 2).

However, the process of territorial restructuring and readjustment of rural areas caused mainly by the emergence of a new model of economic development and the impossibility of employing mechanized ploughing, have led to the abandonment of agricultural activity on terraces, with the consequent decrease in the agricultural area used. Thus, according to the Crop Map of La Gomera (Consejería de Agricultura, Ganadería y Pesca, 2018), the area cultivated on terraces represents an insignificant percentage (0.00155%) with respect to the total. Likewise, 39.5% of the mapped terraces in the municipalities of Agulo, Hermigua and Vallehermoso, in the north of the island, are uncultivated. This abandonment leads to the collapse of the dry stone walls, and consequently erosion linked to runoff, causing soil to be lost and environmental deterioration. This is not only a process of territorial disarticulation, but also

Figure 2. The terraced fields landscapes in Hermigua
Source: Simancas

of deterioration of a heritage element that, in coherence with the European Territorial Strategy (1999), provides identity, singularity and exceptionality at a local, insular and even regional level. In addition, the value of this terraced landscape is enhanced by being a clear reflection of the landscape concept of the European Landscape Convention (Council of Europe, 2000).

Faced with this situation, there are two alternatives: to not intervene and let the terraces totally degrade or to try to maintain some of them; this project is part of the latter. At present, it is impossible to recover the context that gave rise to this type of "heroic" agriculture. Therefore, the maintenance and even the increase of its merit or degree of excellence for the conservation of its structure and the processes it supports require a rethinking of the original functionality through an appraisal of its externalities and integration into the current models of development and land use planning in rural areas. In this context, the challenges for policy development are of great interest capable of overcoming the traditional sectoral and segmented visions, having in consideration of the multifunctionality of these territories (Asiens & Romero, 2010).

The maintenance or recovery of agricultural terraces should be considered as a priority objective, due to its economic, cultural, ecological and landscaping benefits (Lasanta et al., 2013). In this context, the following strategy has been created to develop tourism of the terraces in the north of the island of La Gomera. The aim is to enhance its landscape, environmental and heritage value through tourism. The specific strategies are:

1. To design concrete actions for the transformation of the cultural landscapes in the North of La Gomera into tourist resources. According to 2019 data provided by Promotur Turismo de Canarias, landscape constitutes the second most important factor when choosing the island as a tourist destination (61%). To this we must add that authenticity is the eighth most important factor in choosing the island (31.7% compared to 20.3% of the Canary Islands); the result is that 30% of visitors to La Gomera perform activities in nature, compared to 10% of regional visitors.
2. To design a multi-product offer, which diversifies and differentiates the tourist offer of La Gomera, in general, and of the North of the island, in particular. Although the average stay in La Gomera is 11.47 days, which is higher than that of the Canary Islands (9.09), according to 2019 data from

Promotur Turismo de Canarias, it is necessary to develop places that generate experiences. In this sense, the main motivation for traveling to La Gomera (41.3% compared to l 17.8% of the Canary Islands) is "to get to know and tour the island"; this explains, to some extent, that tourists spend an average of 7.8 hours away from their accommodation. This may contribute to the fact that the rating of the tourist experience in La Gomera (98.9%) continues to be higher than the regional average (97.7%). To achieve this, it is necessary to generate points of tourist interest, understood as those places (landmarks) that have a series of resources that motivate the attraction of visitors to them (Padrón and Hernández, 2017).

3. Contribute to the achievement of the United Nations Sustainable Development Goals 8 (Decent Work and Economic Growth)), 11 (Suntainable Cities and Comumunities), 12 (Responsable Consumption and Production) and 17 (Partnerships to achieve the Goal).

This approach is based on the following premises:

- The need to acknowledge the conception of the cultural landscape generated by the terraces through a single concept: heritage. In other words, that the terraces constitute a heritage landscape. This is an issue that has been the subject of analysis in various studies (Asins and Romero, 2010; Lasanta et al., 2011 and 2013; Tarolli et al., 2014; Arnáez, 2015; Varotto, 2015; Romero et al., 2016; González and Ojeda, 2019).
- The scenery generated by the terraces constitutes a special feature, as a mark left behind by a way of organizing the territory and a certain society. This is the context for understanding the landscape in its full heritage dimension, in line with UNESCO (1972).
- The cultural landscape of the terraces is a constantly evolving heritage (physical and social).
- The need to rethink the ways of making the most of this landscape heritage through the recovery of the concordance between the need for production and the intervention of the territory, as well as the conservation of the "essential features that give it character and personality" (Nogué, 2006).
- It is impossible to recover the original agricultural usage of the terraces, let alone a massive recovery of terraced fields. This generates an unavoidable link between territorial, landscape and heritage management policies (Mata, 2008).
- The terraces of northern La Gomera can be considered as multifunctional tourist-recreational resources (natural, social, cultural, ethnographic, etc.).
- In line with the European Territorial Strategy (European Commission, 1999), a "creative management of cultural landscapes" is necessary. Moreover, their economic use goes beyond mere conservation, since, as stated in the European Landscape Convention (2000), they are an economic resource that creates employment, due to their tourism potential as an important aspect of their sustainable use.

METHODOLOGY OF THE STRATEGY TO PROMOTE THE TERRACES OF THE NORTH OF LA GOMERA FOR TOURISM

The strategy is conceived as a "territorial project", aimed at designing a set of actions to preserve this traditional agricultural landscape of great environmental value and heritage through its enhancement for tourism. The territorial dimension provides a multisectoral and integrated approach, in order to protect rural heritage and diversify activities in rural territories that allow the creation of new sources of income and of employment (Asiens & Romero, 2010). It is proposed as a strategic and operational planning-action process, that will recommend actions - non-binding - articulated around a concept and a methodology based on public participation. Actions will be taken based on concise proposals derived from an initial diagnosis of the situation. The participation of social and economic actors in its elaboration should help to define and prioritize concrete actions.

The strategy is based on the premise that it is impossible to recover the original agricultural operation of the terraces. Therefore, maintaining and even increasing their value or degree of excellence for the conservation of their structure and the processes they support requires reimagining their use. In this sense, the terraces could become structural elements that generate pleasing experiences for tourists/visitors, which are also likely to be enjoyed by residents.

As mentioned, reinventing the terraces as a tourist resource and inserting them as functional elements in the current models of territorial planning and development of rural areas, requires rethinking their functionality and adopting a different dimension to that which they had in their origin. To this end, the aforementioned increase in the interest of tourists in cultural experiences is seen as an opportunity.

Four international experiences have been taken, two national and three regional, in order to have references of recovery of terraced landscapes

The strategy has been developed following the sequential methodological process shown in Figure 3. Several experiences related to the revitalization and enhancement of terraced landscapes and terraces have been analyzed, specifically, four international (The Agricultural Heritage System of World Importance of The Food and Agriculture Organization of the United Nations; Art of dry stone walling, knowledge and techniques on the Representative List of the Intangible Cultural Heritage of Humanity; International Terraced Landscape Alliance; and Associazione Internazionale delle Città della Terra Cruda), two national (Terraces for biodiversity of the Association of Hunting Entities of the Valencian Community; and Recovery of the agricultural and natural landscape in Prado Puente of Manzanares El Real City Council) and three regional experiences (Environmental and Landscape Restoration Plan of the north coast of Fuerteventura; The Canary Islands crop map; and Shared Gardens. Conservation and revaluation of cultivation terraces through community work in Lanzarote). Likewise, the participation of social and economic actors in the process of defining and prioritizing specific actions was an essential task in accordance with the International Charter of Cultural Tourism (ICOMOS, 1999). For this reason, a work session was held with twelve key actors, in which they used an adaptation of the Osterwalder canvas business model; its ultimate aim was to prioritize products and activities, as well as to define some possible business models, providing group knowledge, feedback and consensus.

RESULTS: THE DESIGN OF A STRATEGY TO ENHANCE THE VALUE OF THE TERRACES IN THE NORTH OF LA GOMERA

The result is the proposal of five strategic lines of action (Table 1).

Figure 3. Sequential process for the formulation of the strategy to recover and enhance the tourism value of the terraces in the North of La Gomera
Source: own elaboration

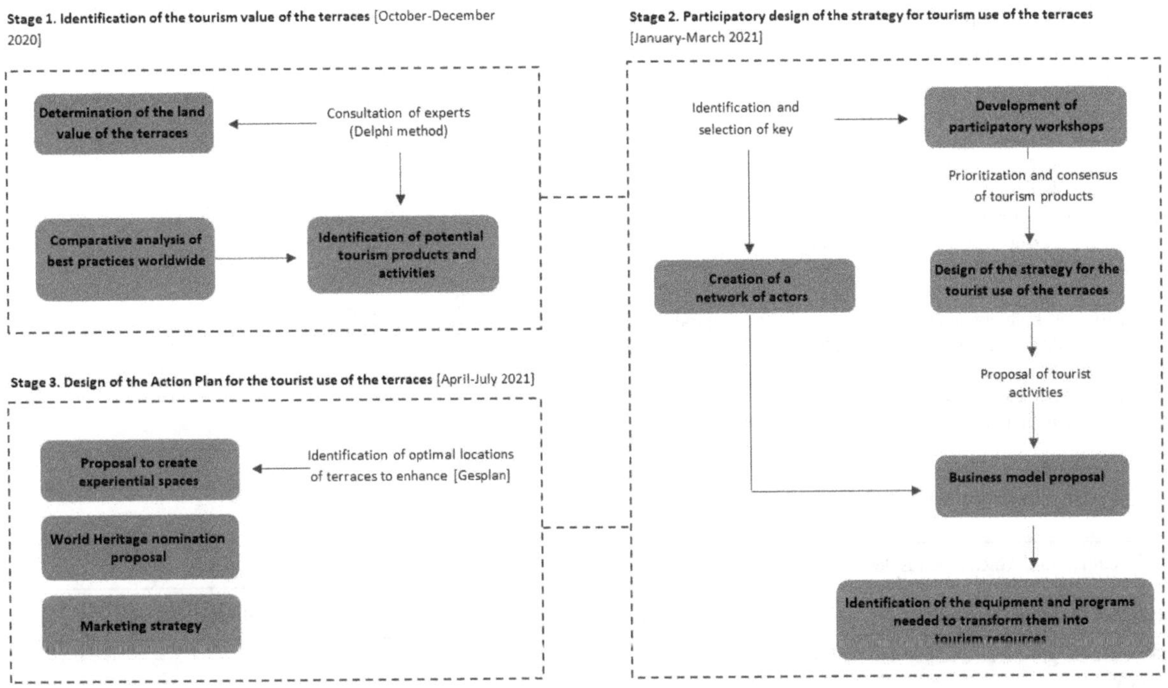

Table 1. Lines of action of the Strategic Plan for the Development of the North of La Gomera

A1	Identification of the potential tourism products to be developed in the terraces	Definition of the tourism potential of the terraces and proposals for products that will enhance their recreational value for tourists
A2	Design of a cluster of local actors involved in the recovery and enhancement of the tourism value of the terraces	Develop a social network -and the different types of social capital- for the actors and organizations that will have a clear influence on the touristic enhancement of the terraces
A3	Design of a territorial or place-based marketing strategy based on cultural landscapes of agricultural terraces	Proposal of a methodology to promote, market and manage a brand based on the terraces of La Gomera
A4	Proposal of a business model linked to the terraces: the creation of experiential spaces	Proposal for the creation of thematic experiential spaces
A5	Proposal to nominate the terraces in La Gomera as a World Heritage Site	Statement of the reasons why they should be declared a World Heritage Site

Identification of Potential Tourism Products to Be Developed in the Agricultural Terraces

The determination of the tourism value of the terraced landscapes of the north of La Gomera must be carried out in a manner far removed from the concepts of classical economics and traditional methodolo-

Table 2. Economic value of the terraces landscape in the North of La Gomera (I)

Description	Possible Actions	Possible Tourism Products
Hernandez et al. (2019) estimate that recovering terraces with better conditions would generate annual revenues of €6,000,000 and the creation of 236 jobs.	Actions to recover the terraces with public and private financing (land stewardship formulas, patronage, corporate social responsibility, etc.)	Site visits
The terraces were built to produce food, mainly for self-consumption. The aim is to recover this function by means of the following solutions: ▪ Use for commercial agriculture ▪ Place for high quality crops, which require meticulous and manual labor ▪ Marginalized areas with low cultivation intensity have high potential for organic farming, as they have low pollution and high biodiversity ▪ They are ideal for housing small vegetable gardens, medicinal plants, mushrooms and fungi, etc. ▪ Recovery of seeds of traditional terrace crops: beans, chickpeas, lentils and peas, etc.	Promotion of zero kilometer or locally grown products Promotion of associated quality brands	Gastronomy tourism (typical and avant-garde)
Vine cultivation, which requires deep soils for its roots to penetrate and a suitable temperature for the grapes to ripen	Development of experiences	Agrotourism Gastronomy and wine tourism
Nurseries to grow fruit and tree and forestry species		Agrotourism
Some of the dry stone walls of the terraces are impressively high, to the point that they are sometimes higher than the width of the area suitable for cultivation	Non-competitive outdoor sports activities that do not require specific infrastructures or group activities. For example, introduction to climbing	Active tourism Sports tourism

Source: own elaboration

gies. Decisions regarding the promotion of this type of agrarian system as a tourism resource should be based on a different conception of value, in which economics is just one of several variables taken into consideration. Likewise, it implies an identification of the multiple and diverse tourism products that can be developed in relation to the terraces. For this purpose, using the Delphi method, a group of experts with the capacity to contribute judgments or opinions was convened. The result is the classification of the value of the terraces into four types (Lasanta et al., 2013): productive (economic), environmental, heritage and intangible (Tables 2, 3, 4, 5 and 6). This involves the creation of a product club.

Design of a Cluster of Local Actors Involved in the Recovery and Enhancement of the Tourism Value of the Agricultural Terraces

The role of stakeholders in the articulation of the territory is essential and strategic in any development process. For this reason, the identification, diagnosis and design of a network of key local stakeholders is an essential line of action.

The importance of social relations is related to the concept of social capital. It generates positive effects or benefits, both for individuals and for territories and/or actions, without which it would be very

Table 3. Economic value of the terraces landscape in the North of La Gomera (II)

<table>
<tr><th colspan="2">Description</th><th>Possible Actions</th><th>Possible Tourism Products</th></tr>
<tr><td>The abandonment of cultivation is an opportunity for the recovery of potential vegetation. This implies a process of natural recovery of semi-natural habitats of high ecological and cultural value linked to human activity (agroecosystems) rich in plant and animal species, a refuge for rare or endangered species and, in general, widely represented within the Natura 2000 and Canary Network of Protected Areas</td><td rowspan="2">According to Lorenzo and Martín (2001), up to 72 different types of birds can be found in Vallehermoso, Agulo and HermiguaFifty-seven percent of the birds are present sporadically or very sporadically in the territory, while 41.5% are common or very common in these three municipalities, and therefore, are easier to see. It is interesting to note that 9.5% of the area of terraces correspond to Special Areas of Conservation for Birds, 12.3% to Special Areas of Conservation and 3.7% to areas protected of the Canary Islands</td><td rowspan="2">Construction of a network of environmentally friendly infrastructures, such as observation trails, thematic signage, delimitation of zones, installation of information booths and, above all, providing the space with the highest possible level of tranquility and silence.</td><td rowspan="2">Birdwatching</td></tr>
<tr><td>Valor de estos sistemas agrarios como hábitat para las aves y otras especies silvestres. Esta contribución es sumamente importante porque aproximadamente el 50% de las especies europeas dependen de los hábitats agrícolas (Grashof-Bockan y Van Langevelde, 2005)</td></tr>
<tr><td colspan="2" rowspan="3">Terraces provide relatively flat surfaces that allow the development of multiple activities</td><td>Development of multiple and diverse outdoor leisure activities.</td><td>Nature tourism
Educational hiking</td></tr>
<tr><td>Creation of routes or tours for stargazing with the accompaniment of trained personnel; possibility of daytime and, of course, nighttime observations through instruments such as telescopes, binoculars, ad hoc; implementation of nighttime routes of trails for stargazing; organization and realization of workshops, talks and activities related to astronomy; promotion of astrophotography, gastronomic activities related to and adapted to the enjoyment of the starry sky; dissemination of how our ancestors interpreted the celestial bodies and what importance they had in their culture; musical events, etc.</td><td>Astrotourism</td></tr>
<tr><td>Camping in tents or glamping</td><td>Camping tourism</td></tr>
</table>

Source: own elaboration

difficult to access or, in some cases, impossible. This is dynamic, where not only the knowledge of a person or institution stands out, but also the recognition, that is, the reciprocity or bi-directionality of relationships. This fact is fundamental for the creation and proliferation of social networks and, therefore, of social capital. Moreover, the possession of social capital tends to accumulate in a few people within a network. Therefore, social capital is not homogeneous, nor is it a fixed and immovable resource

Table 4. Environmental value of the terraces landscape in the North of La Gomera

Description	Possible Actions	Possible Tourism Products
The terraces have contributed -and contribute- to the following environmental services: ▪ CO_2 fixation in soil ▪ Infiltration of runoff water, which would inevitably be lost to the sea, is retained in the ground. ▪ Slowing down runoff, which undoubtedly benefits the water system of the slopes and reduces the dragging of materials. In this way, these terraces promote aquifer recharge. ▪ Soil retention, with the consequent mitigation of the risk of desertification on the island of La Gomera, especially in the areas catalogued in the Strategic Environmental Study of the Hydrological Plans as medium (40.36% of the island's surface area), low (33.4%) and high (25.8%). The terraces affect soil erodibility (stability of the aggregates) and the erosive capacity of precipitation (relationship between the frequency and magnitude of erosion processes and the number of high-energy rainfall events) Prevention and/or mitigation of natural and mixed risks (erosion, river floods, mass movements, forest fires).	Organization of congresses, seminars, meetings and lectures with an informative purpose, involving professionals or amateurs who travel to the site to obtain, basically, knowledge through experiences in the field.	Scientific tourism MICE Tourism (Meetings, incentives, conferences and exhibitions)

Source: own elaboration

(Bourdieu, 1986). Depending on the type of relationships that are taking place between individuals or groups, social capital can manifest itself in several ways.

The creation of the cluster of relevant actors for the conservation and enhancement of the terraced landscapes of northern La Gomera for tourism began with the identification of those people and entities involved directly or indirectly with both actions from the institutional, economic, social and / or technical spheres. The result is the network of relationships according to the typology of the actor presented in Figure 4, where the social capital of cohesion or bonding is identified. This is based on relationships between individuals with similar characteristics or belonging to the same collective, organization or territory (Esparcia et al., 2016). Thus, internal relationships between people of the same status predominate, so horizontal relationships and networks are developed. Closed collectives, organizations or territories with a high density of bonding capital are characterized by their high interconnectedness, trust, their ability to facilitate rules and sanctions, as well as mutual support and great cohesion, which ultimately characterize strong networks (Coleman, 1988).

Social bonding relationships are clear, although it is also possible to identify bridging social capital in supra-municipal relationships. This derives from relationships between individuals or groups that are distant from each other, both physically and because of their different characteristics and/or belonging to other collectives (Woolcock and Narayan, 2000; Esparcia et al., 2016). These (external) relationships are not as strong as bonding, as they have ceased to occur within a collective or territory, with "weak" relationships predominating. On the other hand, they facilitate exchanges of information, innovation, norms, alliances, etc.

Table 5. Heritage value of the terraces landscape in the North of La Gomera

Description	Possible Actions	Possible Tourism Products
Areas of high ethnographic and cultural value. The terraced landscape comprises a multitude and diversity of tangible and intangible heritage assets and cultural identity. They are cultural landscapes that have been shaped over centuries. It implies knowledge about traditional agricultural practices or productive structures of cultural value. Thus, for example, Martín (2009) includes the agricultural space of the terraces of La Gomera among the seven types of traditional agricultural landscapes.		Educational and experiential learning trips
Heritage value. The terraced landscapes are the product of a historical dry stone culture, which: - Constitute a manifestation of popular engineering. - Constitute the survival of an accumulation of traditional knowledge. - Represent the struggle for the subsistence of man in a hostile environment due to the slopes and the lack of fertile soil. - Provide traces of different periods, superimposed and intermingled. - Accumulate a great investment of work for their construction and to guarantee their stability.	World Heritage nomination Development of educational activities	Educational tourism
Territorial value. The terraced landscapes present remarkable qualities that provide representativeness (symbolic character), diversity, stability (permanence in the environment), singularity (distinct or distinguished condition), irreversibility (impossibility that any alteration can be assimilated), fragility (vulnerability), continuity (need for conservation), irreplaceability, significance (importance for the territorial structure) and difficulty of conservation.		Educational tourism

Source: own elaboration

The existence of linked social relationships or *linking* and, therefore, hierarchical with higher public administrations such as, for example, with the Cabildo de La Gomera or the Government of the Canary Islands, can also be observed. This social capital is very similar to bridging, with a predominance of external and weak relationships. Its main feature is that among its relationships or links there is a hierarchical component/link with formal (Woolcock and Narayan, 2000; Esparcia et al., 2016), social and economic institutions, as well as links between people with high levels of prestige or power (Esparcia et al., 2016).

In short, social networks are the key element for capital processes such as the one we intend to generate with the strategy to enhance the tourism value of the terraced landscapes of northern La Gomera. This is due to the fact that each territory will be clearly influenced by certain actors and relationships, as they will tend to have certain types of social capital. In this sense, taking as a reference the various situations derived from the combination of different stocks of bonding and bridging/linking social capital, a social network made up of institutions/entities likely to be involved in the strategy will present high levels of social bonding capital, with solid and cohesive internal structures. In addition, it presents a certain openness to the outside (social capital linking and bridging) that should continue to grow, among all the island groups, in order to gradually evolve towards a scenario where the stock of external social capital continues to increase, in accordance with Esparcia et al (2016).

Table 6. Intangible values of the terraces landscape in the North of La Gomera

Description	Possible Actions	Possible Tourism Products
Aesthetic and perceptual values and elements: the terraced landscape of La Gomera stands out for its visual beauty, as it is a singular attractive landscape with character, charm and majesty.	World Heritage nomination	Cultural tourism
Emotional value: for the local population they generate bonds of identity that are symbolic and experiential, of customs and experiences, of belonging and rootedness to the place. It is about the enhancement of the value of belonging to a vivid landscape as a cultural asset.		Cultural tourism
Artistic-creative value, as scenarios of artistic inspiration		Photography tourism
The terraces are scenarios for the development of recreational activities, sports, with particular relevance to promote leisure among young people in peri-urban areas in different conditions that are healthier than those of urban leisure.	They offer scenarios of quiet enjoyment, for recovery from mental fatigue, emotional stress, depression, anxiety and other disorders.	Health tourism
The contribute to psychological health		Health tourism
The most remote and inaccessible terraces offer people the possibility to experience solitude and quiet, while increasing the capacity for reflection and inspiration.		Wellbeing tourism
Contribution to people's spiritual health, personal enrichment, as well as the practice of relaxation, meditation and internalization techniques.	They provide environments for contemplation, reflection and inspiration; they invoke a sense of belonging; they promote a feeling of connectedness beyond human concerns.	Spiritual tourism
Spaces to promote the social participation of different collectives	Development of activities related to environmental volunteering, recovery of traditional activities, rural practices or transmission of oral tradition, participation in programs for the rescue of flora or fauna and restoration of terraces (master stonemasons), etc.	Educational tourism

Source: own elaboration

Figure 4. Complete social network of institutions/entities on the conservation and enhancement of the terraces of La Gomera
Points: blue (institutional); green (economic), yellow (technical) and red (social).
Black lines (unidirectional relationships) and red lines (bidirectional relationships)
Source: created by authors using UCINET software.

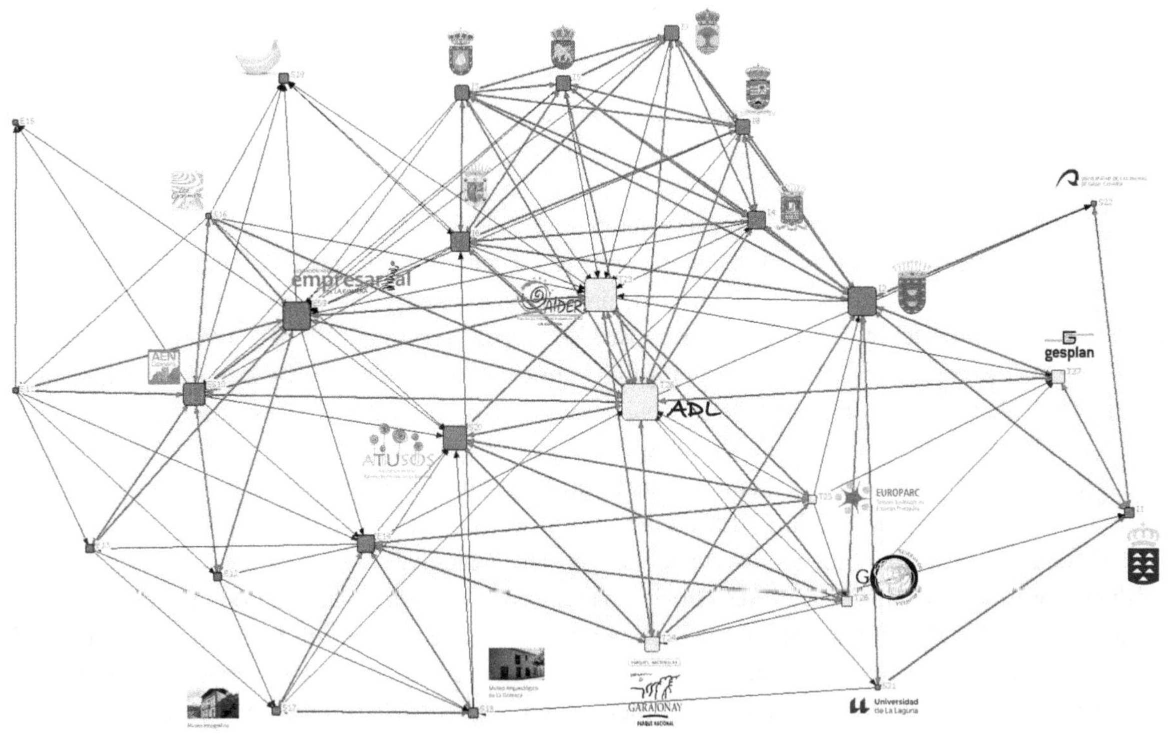

Design of a Marketing Strategy Based on the Cultural Landscapes of Terraced Fields

Tourists demand destinations in which their resources and services are susceptible to the development of differentiated and personalized tourist experiences, which, in turn, generate emotions. In this line, it is proposed to activate in the terraced landscapes of the north of La Gomera products that have always been there but have never been activated because they have not been sufficiently valued. These are in line with current tourism market trends linked to sustainable tourism based on experiential products that generate emotions and orange tourism.

The enhancement of the tourism value of the terraces implies the creation of an experiential product based on its attractions and tourism resources, with joint participation of all the actors, creating what is called a "destination-product". This is where "promomarketing" arises. This is a new form of marketing with which to communicate to the market and sell the product-destination in a single action. It seeks to integrate the promotional and commercial strategies of the public and private actors operating in a local tourism sector into a single actor; in our case it could be a sectoral tourism board. This has two essential requirements: on the one hand, having quality products that guarantee the experiential satisfaction of the tourist, as well as a good level of integration between local operators and the board; and, on the other hand, a commercial proposal that, while offering information about the product, allows it to be easily bookable and purchased.

This form of marketing is a perfect instrument for the strategy to enhance the tourism value of the terraces in the north of La Gomera. This is due to the fact that these can be understood as a product-destination, which allows the construction of collaboration networks between companies of the municipalities where they are located, eliminating competitive features among the actors of the territory. Thus, tourism management acquires a participatory territorial dimension by distributing actions among all the actors. In this way, the creation and management of the product and services correspond to the private sector, which makes them available to the municipalities and the Sectorial Board for their promotion and communication.

An initial step is to identify the tourist and, above all, to convince them to want to travel to La Gomera through communication. This must transmit the differential value of the terraces in relation to other island and even regional resources and attractions. The promotional activity should be empathetic, emotional and innovative. Storytelling is therefore proposed as the most appropriate narrative communication technique, using stories to explain a concept, product or experience. The storyline to be communicated would revolve around the emotion of experiencing sustainability, learning about traditions, customs, magic and legends, the historical construction of the landscape and natural agricultural practices, all linked to the idea of health, physical and spiritual well-being, based on concrete actions, developing the discourse about the destination through the eyes of the tourist.

The focus of the communication would be the terraces themselves, understood as a destination, as a product. The destination introduces these stories in the market with a double function: to create a "hook" to attract tourists and to direct them to the destination's web page, which will be the showcase through which they can learn about the characteristics of the destination and its products. Once the tourist is at the destination, it will be convenient to target them with the direct sale of products and services.

The promomarketing system of the product-destination of the terraces of the North of La Gomera should follow the following procedure:

1. Generation of stories, in the form of tales, that narrate and describe the sensations and emotions of the experiences lived at the destination (storytelling).
2. Definition of an on-line media plan to publish these stories and redirect to the destination's web page created ad hoc.
3. Creation of a tourist web page of the destination that allows, both to inform about the services, proposals and tourist products, satisfying the needs of the tourist, and to commercialize and reserve the service or tourist product. This website would be initially financed and managed by the city councils themselves. Subsequently, after verifying the profitability of the promomarketing system of the product-destination, co-financing between the city council and the agents of the destination could be considered, based on the payment of periodic fees for direct advertising of the products. The website would become a service to tourists that promotes the product-destination through a shopping platform in which all the agents and resources would appear, so that tourists can configure, book and buy, in a modular and customized way, their own product. This platform would be the showcase of the tourist products and services of the northern terraces of La Gomera.

Proposal of a Business Model Linked to the Terraces: The Creation of Experiential Spaces

The current trend is the search for authenticity and identity features of the destination, promoting interaction with society and the consumption of local and "tailor-made" products, services and activities. In this sense, the terraces contribute to the "authenticity" perceived by people who visit or spend the night in the north of La Gomera. They contribute to the added value of a visit to a territory, while integrating original and authentic elements for their enjoyment and products originating from the destination.

This means that we are no longer selling products, but rather experiences. The purpose of the tourist-recreational activity is to have experiences, emotions, feelings, sensations, that fascinate and impact the tourist, creating memories that are more susceptible to be shared through social networks. Therefore, the idea is to generate experiences linked to the terraces in which visitors are the protagonists, rather than mere spectators, by allowing them to participate actively. In other words, the idea is to design unique and exceptional tourism products that are likely to generate memorable and participatory experiences. In this approach the strategy is articulated around experiential tourism. From this point of view, the tourist-recreational experiences linked to the terraces are likely to offer the visitor/tourist the possibility of getting to know the uniqueness of the places, understanding their present and their past and even activating opportunities for identification with the territory, increasing their satisfaction. In this sense, the terraces, turned into tourist-recreational resources, can articulate experiences, feelings, sensations, which fascinate and impact, and precisely for this reason become memorable.

In addition, international charters and recommendations (UNESCO, European Union, Council of Europe, World Tourism Organization, Organization of World Heritage Cities, etc.) have been advocating the need to rethink our models of heritage management and tourism development (Troitiño and Troitiño, 2018). The World Tourism Organization (2016) has set two general objectives in the field of cultural tourism: that it be responsible, inclusive and sustainable, contributing to the socio-economic development of local communities and improving their well-being, and that it contributes to the enrichment and conservation of the cultural identity of the destination, promoting cross-cultural exchanges between visitors and the host community, as well as the promotion and preservation of cultural heritage.

Therefore, the strategy to promote the northern terraces of La Gomera is related to "creative tourism" and "orange tourism". Both types are considered new generation forms of tourism, precisely because of two elements: on the one hand, interaction, so that the tourist is no longer a mere observer to participate in the activity first-hand; and on the other, the relationship with creativity, arts and culture as raw materials, intellectual property rights and copyright, as well as the direct role in a creative value chain (Buitrago and Duque, 2013).

Both encompass those visitors increasingly interested in destination identity, culture and rural heritage, who crave the opportunity to participate in interactive, experiential, recreational, relaxation and educational activities. This generates enormous possibilities for the enhancement (natural, social, landscape, cultural, ethnographic, etc.) of the terraces from a tourist-recreational point of view, introducing new and creative formulas for how they are presented and used.

The existence of cultural and creative spaces is one of the indicators for measuring the creativity of tourist destinations (Hui el al., 2005; Landry, 2008; Pulido et al., 2021). In our case, we propose the creation of what we call "experiential heritage spaces". These are proposed as a projective modality of territorial intervention, where the cultural and natural resources associated with the terraces are intentionally combined, exposed, augmented and promoted to form an agreed upon landscape that tells the story

Figure 5. Location of areas that could be potentially be experiential heritage spaces
Source: own elaboration

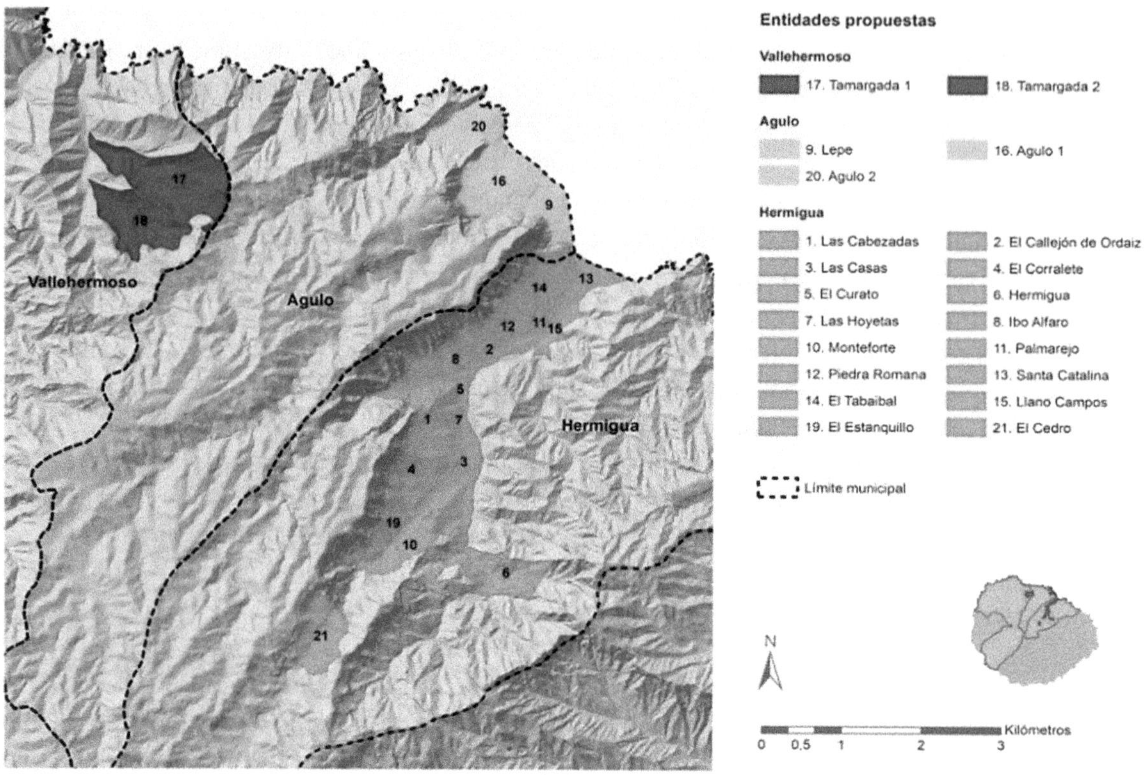

of the territory and its residents (Pérez and Parra, 2004) through its recreational-tourist use. The purpose is for visitors to engage with the terraced landscape, where resources (products) are to be used for the creation of memorable tourism experiences. From this perspective, the conceptual-thematic orientation of the three spaces would be as follows: ethnographic in the municipality of Vallhermoso, agricultural in Agulo and multi-activity in Hermigua.

These experiential heritage spaces are projected as products made up of productive, academic and institutional management components, integrating local economic activities, scientific research, open-air museology, ecomuseums as an expression of living culture, sports, tourism, among other functions that promote territorial income associated with heritage. In this sense, visitors can go from simply viewing the landscape to engaging with it.

A total of 21 territories have been proposed for these three experiential heritage areas (see Figure 5). The common criteria used are the following: terraced surface area, adequacy of the La Gomera Island Management Plan, adequacy of urban planning (municipal scale), property structure and population linked to the area. In turn, two criteria have been used that operate differently: on the one hand, the productive value of the areas located in the municipalities of Agulo and Hermigua; and on the other hand, the value of the ethnographic heritage present in the alternatives located in Vallehermoso.

The idea is for these areas to become places or "points of interest" where visitors can carry out certain activities or enjoy certain resources (Padrón and Hernández, 2017). This will differentiate and diversify the tourist offer of La Gomera, in general, and of the north of the island, in particular. Thus, the whole offer is structured in a combination of products, services, activities or infrastructures for its exploitation, articulated in three areas of terraces that, although distant, form a coordinated unit of action. For this purpose, a business model called "Ethnographic space of the terraces of La Gomera" has been designed with the following main features:

- The basic objective is to increase the value of the destination to enhance its image and attractiveness.
- Public initiative.
- Socioeconomic impact on the entire population and local businesses.
- Exploitation under a public-private approach.
- Scalable, so it can be developed in several phases.
- Geographically distributed in tree

Three types of potential customer segments have been distinguished for the offer, depending on the duration of the visit and the organization of the trip: tourists staying in La Gomera, independent day-trip visitors and day-trip visitors organized by an agency or tour operator (safari or bus).

The value proposition offered to visitors is to get to know, recreate and experience the ways of life, cultural practices and the relationship and interaction with the environment of the island's ancestors, achieving an ethnographic immersion with the local culture and society while enjoying the scenic landscapes. Depending on the level of interaction of the tourists with the different services and infrastructures of the whole offer, three levels are proposed:

1. Scenic. These are scenic trails or routes and self-guided trails through panels and links to interpretive explanations, with mobile access to audio and text in different languages using QR codes. Preferably circular and with different levels of difficulty and travel times (e.g. 30 minutes, one hour and two hours) to make the offer more flexible and viable for visitors with different interests. This product is structured for independent visitors seeking a relatively superficial level of interaction and knowledge of the local area, who are free to choose the time they wish to dedicate to the visit.
2. Guided Experiential. Based on the infrastructures of the previous scenic level, the idea is to add other infrastructures aimed at providing a more experiential service, including, for example, cultivated terraces for demonstration and the possibility for visitors to carry out small and short activities related to the traditional cultivation of the terraces, areas for educational activities, areas for activities of interaction with nature (bird watching, astronomy, silbo gomero[1], etc.).
3. Interpretive. It consists of exhibition areas, video projection room, technology to facilitate the interpretation of the resources of the terraces using virtual reality and augmented reality, etc. It also includes a restaurant, cafeteria and shop-market of local products, preferably operated by local producers.

The main distribution channels, i.e., the circuits or means by which the product reaches the end customer, considered in the proposed model are direct (web / office), tour operators / companies organizing trips to the island, accommodations and tourist information offices. For the direct channel, a double integration can be considered: on the one hand, integrating the products and activities of the area and others "borrowed" in the territory (restaurants, guides, wineries, handicrafts, etc.); and on the other, a backward integration, i.e., integrated into the website and channels already established to market tourism of the island.

In order for the offers to reach the target public, they must be communicated to them. The communication channels that are contemplated to publicize and make attractive the resources linked to the terraces include, advertising at points of interest and contact (port, tourist information offices, tourist accommodation establishments, etc.), fixed billboards, road signs, the island's tourist website and official social networks.

A public investment is proposed to finance the set of actions with the objective of obtaining an indirect social return. The investment should produce an increase in the attractiveness of the destination, which would lead to new clients, increase the loyalty of the tourists who visit and who recommend a trip to the island. The return on the investment comes from the economic development derived from tourist spending and the economic activity generated around it (restaurants, agriculture, local tourist guides, etc.). Another source of income for the maintenance of the developed infrastructures can come from the collection of fees for the administrative concession of the exploitation of the developed ethnographic space. Finally, another potential revenue stream can be provided by companies and organizations who could base their corporate social responsibility strategies on supporting this project.

PROPOSAL TO NOMINATE THE TERRACES IN LA GOMERA AS A WORLD HERITAGE SITE

The issue of landscape is implicitly raised in the UNESCO Convention for the Protection of the World Natural and Cultural Heritage (1972) through the concept of "site". The World Heritage Committee gradually developed the idea of recognition of sites that would be officially classified as Cultural Landscapes, legally linked to the last paragraph of Article 1 of the Convention. The 16th session of the World Heritage Committee in Santa Fe (1992) adopted a new category: Cultural Landscapes.

The National Cultural Landscape Plan, promoted in 2012 by the Spanish Cultural Heritage Institute of the then Ministry of Education, Culture and Sport, was drawn up within the framework of the European Landscape Convention (Council of Europe, 2000). This is defined as "the result of the interaction over time of people and the natural environment (...), product of a process and support of the identity of a community". Terraced and terraced landscapes fall into this category.

The terraced and terraced spaces of northern La Gomera constitute cultural landscapes delimited and defined by their cultural and natural values. It is, therefore, a landscape representative of the essential values of the island's identity, creating its own original essence, selected over the centuries in a territory that has required a great deal of work, adaptation and knowledge of the environment. Therefore, the fifth line of action of the strategy is the attempt to obtain the highest degree of protection both through regional legislation, that is, its declaration as an Asset of Cultural Interest under Law 11/2019, of April 25, Cultural Heritage of the Canary Islands.

This system of protection and management is a necessary condition to apply for inclusion in the World Heritage List as a World Heritage Property, presenting an "outstanding universal value" according to the ten criteria defined in the UNESCO Operational Guidelines for the Implementation of World Heritage (2008).

The terraced landscapes also present the conditions of Integrity and Authenticity. The first refers to the unitary and intact character of the cultural and/or natural heritage and its attributes. The second is linked to the cultural heritage.

CONCLUSION

The revised World Charter for Sustainable Tourism, known as the Charter of the Basque Country of 2015 (replacing the Lanzarote Charter of 1995), includes "cultural and creative tourism as an engine for sustainable tourism development through management responsible for cultural heritage –material and intangible-, arts and cultural and creative industries, from a transversal and global vision of UNESCO's cultural conventions ". In this sense, with proper planning and management, tourism brings economic, social and environmental benefits. These can improve the quality of life of communities, as well as generate employment opportunities, while preserving cultural heritage and contributing to the development of the creative economy.

The terraces are an essential element of the identity of the landscape of La Gomera. The loss of their original function explains why they are currently abandoned. The strategy is based on the premise that it is impossible to recover the original agricultural use of the terraces, and that it is necessary to propose new ones, such as their value for tourism.

From this perspective, the objective of the tourism valorization strategy of the northern terraces of La Gomera Island is to generate leisure experiences linked to them. The objective is to design a holistic strategy based on the determination of its value in the context of the current socioeconomic and territorial model, the participatory identification of tourism products, a holistic approach to the development of sustainable tourism that integrates natural and cultural values, as well as the design of an action plan.

The aim is to promote the recovery of this type of cultural landscape. At the same time, the development of certain tourist-recreational experiences can attract investments, as a consequence of the greater attractiveness and interest aroused in the potential demand and in the actual visitors. It is based on local and personalized management of tourism resources, which requires qualified personnel. It also has a knock-on effect on the island's tourism offer as a whole, with the consequent increase in tourist spending. To this can be added the possibility of increasing income from participation in the enhancement and marketing of products and services, while generating a certain influence on the production of other local products (gastronomic, handicrafts, etc.). Likewise, tourism and recreational activities benefit the local population by favoring the direct and indirect creation of employment, thus increasing residents' job expectations and opportunities.

In short, the strategy is to generate cultural, economic and social development based on responsible and sustainable tourism management and the enhancement of the value of the terraced landscapes. The idea is to create an imaginary, icons and brand around the cultural landscapes of the terraces of northern La Gomera, in conjunction with the local community -the latter participates, identifies with and acts as host- or through its interaction with the tourist/visitor. This perspective can contribute to the North of La Gomera becoming a territory with cultural and creative resources that can attract certain types of travelers/tourists, and whose endogenous development is based on cultural identity.

REFERENCES

Arnáez, J., Lana-Renault, N., Lasanta, T., Ruiz-Flaño, P., & Castroviejo, J. (2015). Effects of farming terraces on hydrological and geomorphological processes. A review. *Catena*, *128*, 122–134. doi:10.1016/j.catena.2015.01.021

Asins, S., & Romero, J. (2010). Gestión, ordenación y protección de las laderas aterrazadas en la Comunidad Valenciana. *Investigaciones Geográficas*, *53*(53), 85–98. doi:10.14198/INGEO2010.53.04

Bourdieu, P. (1986). The Forms of Capital. In J. G. Richards (Ed.), *Handbook of Theory and Research for the Sociology of Education* (pp. 241–258). Greenwood Press.

Buitrago, F., & Duque, I. (2013). *La economía naranja. Una oportunidad infinita*. Banco Interamericano de Desarrollo.

Castillo, J., & Martínez, C. (2014). El Patrimonio agrario: Definición, caracterización y representatividad en el ámbito de la UNESCO. *Boletín de la Asociación de Geógrafos Españoles*, *66*(66), 105–124. doi:10.21138/bage.1782

Coleman, J. (1988). Social Capital in the creation of Human Capital. *American Journal of Sociology*, *94*, 95–120. doi:10.1086/228943

Council of Europe. (2000). *European Landscape Convention, CETS No. 176*. Council of Europe.

Esparcia, J., Escribano, J., & Serrano, J. (2016). Una aproximación al enfoque del capital social y a su contribución al estudio de los procesos de desarrollo local. *Investigaciones Regionales—. Journal of Regional Research*, *34*, 49–71.

European Commission. (1999). *European Territorial Strategy*. Author.

Gómez, D. (2007). *Ordenación territorial*. Mundi-Prensa.

González, A., & Ojeda, A. (2019). Los paisajes de terrazas en las Canarias Orientales (Lanzarote y Fuerteventura): Hacia una tipología actual. *Revista Española de Estudios Agrosociales y Pesqueros*, *253*, 65–92.

González, A., & Ramón, A. (2017). Análisis de la consideración científica e institucional de los paisajes de terrazas agrícolas en Canarias (España). *Revista Española de Estudios Agrosociales y Pesqueros*, *247*, 137–160.

Grashof, C., & Van Langevelde, F. (2005). Green Veining: Landscape Determinants of Biodiversity in European Agricultural Landscapes. *Landscape Ecology*, *20*(4), 417–439. doi:10.100710980-004-5646-1

Hernández, O., Martínez, E., Perdigón, A., Cebrián, J., & Rodríguez, E. (2019). La gestión de los territorios de bancales (caso experimental La Gomera). In *V Congreso Mundial ITLA Re-encantar Bancales* (p. 21). Gobierno de Canarias, Canarias Cultura en Red, Observatorio del Paisaje de Canarias.

Lasanta, T., Arnaéz, J., Ruiz, P., & Lana, N. (2013). Los bancales en las montañas españolas: Un paisaje abandonado y un recurso potencial. *Boletín de la Asociación de Geógrafos Españoles*, *63*, 299–320.

Lasanta, T., Arnáez, J., Ruiz, P., & Ortigosa, L. (2011). Los bancales en la montaña mediterránea. Un paisaje multifuncional en proceso de degradación. In Territorio, paisaje y patrimonio rural (pp. 198-209). Universidad de Extremadura.

Lorenzo, J. & Martín, A. (2001). *Aves del archipiélago canario*. Francisco Lemus Editor.

Martín, V. (2009). Aproximación tipológica a los paisajes agrarios actuales de Canarias. *Papeles de Geografía*, *32*, 97–116.

Mata, R. (2006). Un concepto de paisaje para la gestión sostenible del territorio. In R. Mata & A. Tarroja (Eds.), *El paisaje y la gestión del territorio: criterios paisajísticos en la ordenación del territorio y el urbanismo* (pp. 17–46). Diputación Provincial de Barcelona.

Organización Mundial del Turismo. (2016). Alianza entre turismo y cultura en el Perú – Modelos de colaboración entre turismo, cultura y comunidad. Author.

Padrón, H. & Hernández, R. (2017). Los puntos de interés turístico: relevancia analítica, propuesta metodológica y caso de estudio. *Pasos. Revista de Turismo y Patrimonio Cultural, 15*(4), 979-1000.

Peñarrubia, M. P. (2018). *Los datos estadísticos públicos y su uso en el conocimiento del comportamiento de los turistas en destinos inteligentes* (PhD thesis). Universitat de València.

Peñarrubia, M. P., Simancas, M., & Forgione, G. (2019). Application of geomarketing to coastal tourist areas. *Tourism & Management Studies*, *15*(4), 7–20. doi:10.18089/tms.2019.150401

Pérez, L., & Parra, C. (2004). Paisajes culturales: El parque patrimonial como instrumento de revalorización y revitalización del territorio. *Theoria*, *13*(1), 9–24.

Pulido, J. I., García, J. A., Casado, J., & Rodríguez, B. (2021). Measuring creativity in urban-cultural Spanish destinations. *Boletín de la Asociación de Geógrafos Españoles*, *89*. Advance online publication. doi:10.21138/bage.3074

Romero, L. (2016). Towards a New Valuation of Cultural Terraced Landscapes: The Heritage of Terraces in the Canary Islands (Spain). *Annales, Series Historia et Sociologia*, 26.

Sabaté, J. (2005). Paisajes culturales, el patrimonio como recurso básico para un nuevo modelo de desarrollo. *URBAN. Revista del Departamento de Urbanística y Ordenación del Territorio*, *0*(9), 8–29.

Scott, N., Baggio, R., & Cooper, C. (2011). Network analysis methods for modeling tourism inter-organizational systems. *Advances in Culture. Tourism and Hospitality Research*, *5*, 177–222.

Simancas, M. (2020). Parar para repensar: ¿el fin del turismo como lo conocemos? Fortalezas del modelo turístico canario ante el escenario de incertidumbre y desconfianza de la crisis de la COVID-19. In *Turismo pos-COVID-19: reflexiones, retos y oportunidades* (pp. 689–706). Cátedra de Turismo CajaCanarias-Ashotel de la Universidad de La Laguna. doi:10.25145/b.Turismopos-COVID-19.2020

Tarolli, P., Preti, F., & Romano, N. (2014). Terraced landscapes: From an old best practice to a potential hazard for soil degradation due tol and abandonment. *Anthropocene*, *6*, 10–25. doi:10.1016/j.ancene.2014.03.002

Troitiño, M. A., & Troitiño, L. (2018). Visión territorial del patrimonio y sostenibilidad del turismo. *Boletín de la Asociación de Geógrafos Españoles*, *78*(78), 212–244. doi:10.21138/bage.2716

Varotto, M. (2015). Terraced landscapes of the Alps: Decay, rediscovery, revitalization. In T. Tillmann, M. Bueno de Mezquita. In *II Congreso Internacional de Terrazas. Encuentro de culturas y saberes de terrazas del mundo* (pp. 38-48). Cuzco.

Woolcock, M., & Narayan, D. (2000). Social Capital. Implications for Development Theory, Research and Policy. *The World Bank Research Observer*, *15*(2), 225–249. doi:10.1093/wbro/15.2.225

ENDNOTE

[1] The silbo gomero is a language and communication system that imitates the Spanish language with whistles. This was used by the inhabitants of La Gomera for centuries, due to the rugged terrain (mountains and ravines), which made verbal communication difficult. On September 30, 2009 it was declared Oral, Cultural and Intangible Heritage of Humanity by UNESCO.

Chapter 12
Answers to the Crisis in the Tourism Sector in a COVID Environment:
The Orange Economy Cluster Initiative in the State of Boyacá, Colombia

Danielle Pimentel de Oliveira
https://orcid.org/0000-0003-3572-8002
University of Valencia, Spain

Ramón Hernando Granados
Universidad Pedagógica y Tecnológica de Colombia, Colombia

Sandra Liliana Cely Andrade
Universidad Pedagógica y Tecnológica de de Colombia, Colombia

ABSTRACT

In an unstable environment, amid a global crisis, creativity and innovation still remain as effective tools for local development. Tourism was strongly affected by COVID-19 due to the severe restrictions on travel imposed as a security measure, but it is an actor capable of promoting development in the territories. For this reason, it is beginning to take on a special role in destinations that wish to bet on local idiosyncrasies as a path towards sustainable tourism based on local resources. In this sense, the orange economy is presented as a tool that, together with tourism, offers the necessary instruments to enhance the value of the locals. This chapter presents the Orange Cluster Initiative for the State of Boyacá, Colombia focused on the tourism sector. It is concluded that orange tourism can become a recovery of post-COVID-19 tourism, serving as an element of sustainable union between local agents, tourists, and the environment.

The present chapter is framed within the concession to the Universitat de València of a nominative grant,

DOI: 10.4018/978-1-7998-9217-5.ch012

reference UP2021-044, within the modality foreseen in the Royal Decree289/2021 of 20 April and the concession order: Margarita Salas Aids for the training of young doctors. These grants are funded by the Ministry of Universities and are part of the European Recovery Instrument (Next Generation EU) within the PRTR: Recovery, Transformation and Resilience Plan: 'Modernisation and digitisation of the education system'. Grant reference: MS21-076

INTRODUCTION

Amid a global crisis that started as a health crisis and turned into an economic and social crisis, creativity and innovation have become a new tool for local development.

Tourism, one of the most affected sectors by the crisis caused by SARS-Cov2, which caused COVID-19, due to the severe travel restrictions imposed, is an activity capable of promoting the development of destinations. For this reason, it is beginning to take on a special role in destinations that wish to commit to local idiosyncrasies and the opportunities offered by the territory itself, thus giving rise to a search for the path of sustainable tourism based on local resource. According to some authors, such as Sarantou et al. (2021), Kugapi & Höckert (2020) and Lor et al. (2019), this commitment to localisation can become a stage with unsustainable characteristics, complicated in its implementation and even complex to execute.

The new market trends, together with the already known Crossumer (Barra, 2018) point to a new horizon in the tourism sector, with visitors seeking lesser-known destinations (Vera-Rebollo & Ivars-Baidal, 2020), which offer their culture and way of being a unique and authentic product, with more personalised experiences (Bakas *et* al., 2019) along the lines of sustainable tourism, away from overcrowded destinations, overcrowding and the over-tourism of yesteryear. This idea goes hand in hand with Zehrer's (2009) proposal, which indicates the development of opinions aimed at the design of services and attractions.

All of this opens up a great opportunity for those inland destinations that have a culture marked by their uniqueness, and in the territorial resources that can be made available as new tourist products (Sarantou et al., 2021).

In this sense, the orange economy is presented as a great opportunity, a tool that together with tourism - Creative Tourism or Orange Tourism - offers the necessary instruments for a "valorisation of the local, of the autochthonous" (Pedro Bueno in HOSTELTUR, 2020).

The Orange Economy is configured as a development model where the pillars of social and economic transformation are based on cultural diversity and creativity (SENA, 2017). It is also known as a public-private initiative that opens a new dialogue between agents, seeking to improve the profitability of companies, with a model based on cultural, social and economic development tools, based on the creation, production and distribution of cultural and creative goods and services (Cluster Development, 2021). Within this economy, we can distinguish elemental areas for human beings such as culture and creative industries (architecture, film, design, research, fashion...) sectors that base their proposals on entrepreneurship, sustainability, citizen participation and innovation, being potential successful businesses in the short and medium-term (Cooper *et* al., 2018).

Transforming adversity into opportunity is to act with intelligence. Creative Tourism or Orange Tourism meets this new opportunity and is translated as "a new generation of tourism that involves the tourists themselves and the locals in the creation of the tourism product (co-creation)" (SENA, 2017:1).

- The present chapter is framed within the concession to the Universitat de València of a nominative subsidy, reference UP2021-044, within the modality foreseen in the Royal Decree289/2021 of 20 April and in the order of concession: Margarita Salas aids for the training of young doctors. These grants are funded by the Ministry of Universities and are framed within the European Recovery Instrument (Next Generation EU) within the PRTR: Recovery, Transformation and Resilience Plan: 'Modernisation and digitisation of the education system'.

Following the studies conducted by Fesenmaier and Xiang (2017), creative tourism often follows the line of research that presents a very particular perspective, "from the outside in", i.e. focused on the experience and expectation of the tourist.

As well as being concerned with the tourist experience, Orange Tourism offers visitors to the destination the possibility to actively participate in the daily life of the territory, learning about its culture through local experiences and workshops, where contact with the locals is much closer and more real.

From the population's point of view, the main objective is to (re)discover the territory, focusing on local resources and culture, transforming them into new attractions.

Aspects that Kajan (2014) and Lee and Jan (2019) highlight as levers of local community self-determination, as they incorporate determinants of local culture.

From the visitor's perspective, immersion in the local reality provides a new vision of what the destination and its reality are like, giving them a unique experience. This is known as tourism based on know-how, human relations and creativity as added values. (HOSTELTUR, 2020a).

Within this framework, Orange Tourism is much more than cultural tourism, it is tourism that aims to leave an oversaturated world of offers and enter a new, unique and differentiated sphere. The price element loses its initial strength, imposed on many occasions in mass tourism, giving space for experiences to gain strength, based on a different experience to the offers based on the design of rigid tourist packages and with reduced production costs. It is a kind of tourism that can even repair and restore the idea of commodity tourism (Mesa, 2017).

From the Central East Region of the State of Boyacá, located in Colombia, the Orange Cluster Initiative is being developed, focused on the tourism sector, which is committed to establishing strategies for the future based on actions that add value to the region.

Orange Tourism does not come to displace other types of tourism but is configured as cultural tourism that complements and helps to co-construct the attractions of the destination. An opportunity for any destination where ingenuity, talent and creativity predominate.

Therefore, this research seeks to guide the introduction of Creative Tourism in destinations that wish to develop sustainable tourism, considering the idiosyncrasies and local resources. It, therefore, seeks to describe the process followed to revive the potential of territory with the help of local communities based on the weaknesses found and assumed as challenges by tourism service providers in each of the municipalities covered by this study. Based on the example developed in the State of Boyacá. It describes how the Orange Cluster has been implemented in the region, its development based on the local resources of each municipality involved in the project, adjacent public policy support and management projects, and tourism policies. In addition, the open lines of co-creation of tourism products in the State of Boyacá will be presented, as well as the challenges to be overcome.

From the research carried out for this study, it is concluded that Orange Tourism is timeless, not dependent on the seasons for its development, which makes it an excellent support for the depersonalization of destinations. It is diversified as it optimizes existing resources at the destination, whether tangible or intangible. It is experimental, in the majority of cases it offers visitors unique experiences related to the authenticity of the local idiosyncrasy. It is harmonious as it allows the presence of other tourism segments, such as gastronomy, oenology, ornithology or slow tourism, among others.

In short, Orange Tourism can be a way to recover post-COVID-19 tourism, serving as a link between the local population, the tourist and the environment.

BACKGROUND

Mankind, in general, was not prepared for the magnitude of the impacts of COVID-19, and neither was tourism.

The severe travel restrictions imposed as a security measure and the reduction of tourist movement had a major impact on society and the environment (Baute-Plana & Zamora-Fung, 2021). As a positive result on an environmental scale, it was possible to contemplate the clean Venice Canals (Elcacho, 2020), the skies cleared of pollution in China (Larson, 2020) and in India, the decrease in pollution allowed the Himalayas to be seen again (Justo, 2020). From another, more pragmatic perspective, the epidemiological situation also led to the decline of most small and medium-sized enterprises, forcing them to readjust to the new scenario (Baute-Plana & Zamora-Fung, 2021).

Given the magnitude of the events, the crisis, which began as a health crisis, has taken on other dimensions and has moved on to economic and social levels (Pimentel de Oliveira, 2020).

In this context, tourism is beginning to gain prominence as a tool capable of reactivating the economy of destinations and strengthening local society. (Cerquetti & Cutrini, 2021).

According to Florida (2003), in recent years, creativity has been expanded in several areas, but within tourism it has become a tool that provides advantages to the destination, reducing the negative effects that the activity can and does generate.

The reality of a destination can be told and analysed through different perspectives, according to Thomson (2004) it can be narrated through the actions of its leaders, through the idiosyncrasy of its people, through the imposing beauty of its architecture, through its means of transport and types of accommodation and with the help of its gastronomy. For this reason, the participation of society and the resources of the destination itself in the reactivation of tourism can be an excellent lever for the economic reactivation of tourist areas.

Hasani et al. (2016) reinforces that stakeholders who interact in participatory processes, such as Agenda 21 or in the development and co-creation of products and management, are more likely to perceive the impacts of tourism as positive. For this reason, it will be necessary more than ever before a sustainable development in the post-COVID-19 era in the field of tourism to consider the uniqueness of each destination, as its peculiarity, personality, character and culture that must be reinforced and configured as unique attractions.

In short, it is a matter of considering the strengths of the territory and the local population, enhancing them, integrating different actors capable of achieving the creative regeneration of the destination and at the same time favoring new alliances that manage to reposition the cultural value chain (Richards, 2014).

In this sense, creative tourism strengthens the relationship between the tourist and the destination, creating dynamics that allow a rethinking of the idea of mass tourism (Mesa, 2017). In addition to promoting local revitalization, attracting new businesses, encouraging the administration to explore the creative potential of their localities, creative tourism is also a way to promote the development of local tourism (Álvarez-García et al., 2018).

Creative tourism, or orange tourism, offers the necessary tools to enhance the value of the local and autochthonous. The adaptation of destinations to the needs and demands must be continuous in the evolution of a tourist destination (Mowinckel, 2016), remembering that culture "is a set of resources that can be used for the benefit of our capacity to adapt" (Martín De la Rosa, 2003).

From another perspective, the UN's Agenda 2030 (2016), sets out 17 goals to achieve sustainable development by 2030. When looking at these goals from a more holistic perspective, it is perceived that sustainable tourism would be aligned with several goals linked to these objectives, due to its sustainable characteristics and ability to count on the local population for its implementation.

In this sense, we could affirm that:

- Creative Tourism can create jobs and decent work for the local population and stakeholders involved in the process (SDG8), facilitating the reduction of poverty (SDG1. End Poverty) and hunger (SDG2. Zero Hunger) in the most impoverished destinations.
- Creative Tourism is committed to the professional training of the local population (SDG4. Quality education), concerning craftsmanship or in more precise and technical processes such as reception and services for visitors to the destination.
- It is also capable of empowering vulnerable groups, such as women (SDG5. Gender equality, SDG10. Reducing inequalities), which reduces inequalities and strengthens the local population's sense of belonging to the destination and its culture.
- The creative tourist is a visitor with a sustainable and ecological conscience, and therefore seeks destinations that are aligned with this conviction (SDG6. Clean water and sanitation, SDG7. Affordable and clean energy, SDG12. Responsible production and consumption, SDG13.Climate action, SDG14. Underwater life, SDG15. Life of terrestrial ecosystems);
- Destinations with the implementation of Creative Tourism have certainly chosen to create networks and partnerships to achieve their goals (SDG11, SDG16. Peace, justice and strong institutions, SDG17. Partnerships to achieve the goals).

Why and in What Environment Does Creative Tourism Arise?

After the Second World War, and as a result of the development of transport and paid holidays, mass tourism emerged, initially focusing on sun and beach tourism.

The 1980s brought the popularization of cultural tourism, which, over the years, fell victim to its success with the massification of cities and the standardization of services in the purest Commodity Tourism style.

The explosive growth of tourism in the 1990s, and the need for destinations to stand out from the crowd, has led to the emergence of new, more relaxed tourism perspectives that rely on local people and their culture as unique attractions, thus giving rise to Creative Tourism (Richards et al., 1998; Mossberg, 2007).

The term "Creative Tourism" has its roots in the EUROTEX project initiated in 1998, a programme that aimed to promote cultural and craft tourism, promoting tours and nature visits among craft producers, designers and teachers in countries such as Finland, Switzerland, the Netherlands, Japan, Portugal and Greece (Richards et al., 1998).

Creative tourism began to be researched in greater depth from 2015 onwards when the number of research studies related to the subject began to achieve greater prominence in the academic scientific world (Álvarez-García et al., 2018; Li & Kovacs, 2021).

In 2000, it was theorized by professors Greg Richard and Crispin Raymond, as "Tourism that offers visitors the opportunity to develop their creative potential through active participation in the training courses and experiences that are characteristic of the destinations they choose for their holidays" (Creative Tourism Network, 2020).

Mowinckel (2016:1) describes creative tourism as "where the visitor has an educational, emotional, social and participatory interaction with the place, its culture and its locals", a view also shared by Wattanacharoensil and Sakdiyakorn, (2016). HOSTELTUR (2020:1) reinforces that the "implementation of Creative Tourism facilitates the reduction of the effects of the [COVID-19 health] crisis, creating an environment that stimulates the depersonalization of the destination", as it focuses on the "co-creation of unique experiences from the existing resources of the destination, through the participation of the local population in creative processes related to the autochthony of the tourist territory". Aligned to this idea, the CREATOR team (Duxbury et al., 2020:1) states that creative tourism focuses on: "...creative activity that encourages personal self-expression and interaction between visitors and residents, inspired by local endogenous resources designed and implemented by residents for the benefit of the community".

As observed, local talent and culture, as well as the relationship with the destination, is the mainstay of these activities.

From a more realistic perspective, creative tourism, according to some authors (Li & Kovacs, 2021; Blapp & Mitas, 2018; Duxbury et al., 2019; Duxbury et al., 2020) can also be understood as a consequence of mass tourism, coupled with travellers' desire for a more active and individualized experience with the local host population.

"According to its proponents, creative tourism responds to the needs of tourists dissatisfied with the passive forms of experience production and consumption associated with traditional cultural tourism" (Li & Kovacs, 2021:2).

This trend towards personalization of services responds to the needs of the 'Post-COVID-19 Era' tourist. Given the current pandemic situation, destinations must offer a unique product, which facilitates safety and service delivery.

The Orange Economy in Colombia

In 2017, the Colombian national government issued the "Law 1.834 of 2017", where the main background and elements of the orange economy are identified. Such background and elements are related to the mechanism of protection of intellectual property around the culture offered by the territory. Its objective is the development, promotion, incentive and protection of creative industries, understood as those industries that generate value due to their goods and services, which are based on intellectual property (Congreso de Colombia, 2017).

Likewise, in the "Law 1955 of 2019", the action plan to develop the orange economy in the territory is defined, including this concept in the development plan of the Colombian government 2018-2022 (Congreso de Colombia, 2019:1).

Creative industries, according to the Ministry of Culture et al. (2020b), comprise:

Figure 1. Location of the State of Boyacá
Source: Graphic made by the authors of this article.

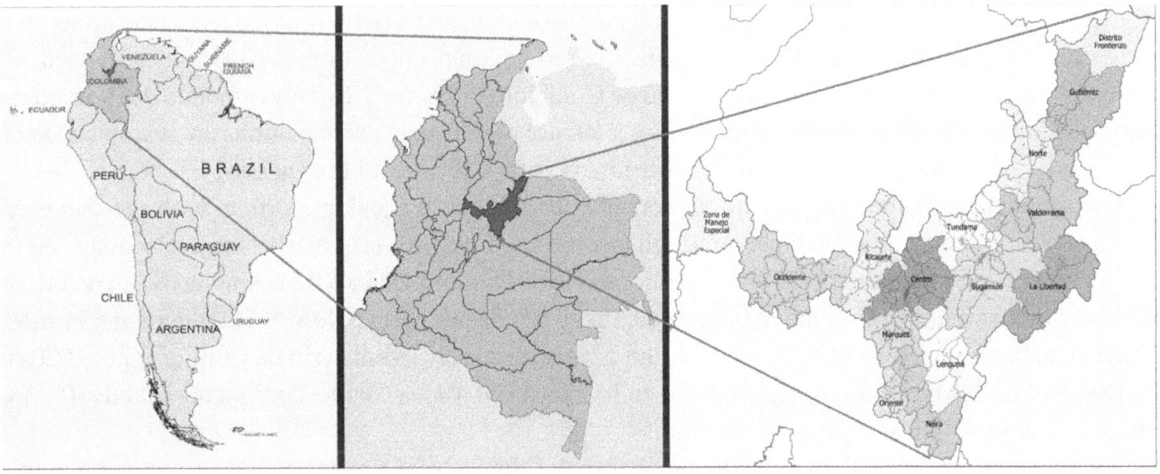

"[...] sectors that combine the creation, production and marketing of goods and services based on intangible cultural content, and/or those which generate copyright protection."

In 2019, the tourist flow in Colombia reached the figure of 4,515,932 visitors, exceeding the target set for this year, reaching 2.7% more than the previous year. According to these data, hotel occupancy reached 57.8%, providing at the same time nominal income to travel agencies with a growth of 3.7% (Ministerio de Cultura et al., 2020a).

Number one tourist attraction in Colombia, is nature tourism, then gastronomy tourism enriched by the ethnic and cultural contributions of indigenous people, Spaniards, Africans and Arabs. (Procolombia, 2012).

Regarding the introduction of the Orange Economy Techniques, these contributed to Colombia nearly 3,2% GDP in 2019, the equivalent to US$8.139 million dollars, generating more than 455.000 jobs. (Departamento Administrativo Nacional de Estadística - DANE, 2021), related to various activities, such as the creation, production and distribution of goods and services with cultural and creative content.

Still, with data from DANE, it is observed that in Colombia 52.7% of the total value added generated by the Orange Economy corresponds to economic activities of partial inclusion, and 47.3% corresponds to activities of total inclusion. As for the participation of the value-added of the Orange Economy in the total national value-added, participation of around 3.0% is observed for 2019 (See Graph 1).

The State of Boyacá

Territorially, Colombia is divided into 32 departments, including the State of Boyacá. Located in the east-central part of the country. It is situated in the eastern mountain range of the Andes, with an area of 23,189 km^2, representing 2.03% of the national territory (Procolombia, 2012) (see figure 1).

The state is divided into 13 provinces and two 'special management zones, totalling 123 municipalities. The State of Boyacá has a population of 1,413,064 inhabitants and a great thermal and geographic variety, where three hundred thousand hectares of "páramos", stand out. This makes the territory an ecological and biodiverse niche (Ministerio de Comercio Industria y Turismo y Fondo de Promoción Turística Colombia, 2013)

Clearly, its greatest asset is nature, however, culture and history are two components that strengthen the destination as a tourist space, being recognized as an important enclave of decisive battles, carried out by Simón Bolívar towards freedom (Ministerio de Comercio Industria y Turismo y Fondo de Promoción Turística Colombia, 2013). In this sense, being a territory of history and tradition, the maintenance of culture and respect for the legacy of their ancestors is part of the local idiosyncrasy.

Tourism demand in the area presents visitors of both genders in equal proportion, with ages between 18 and 60 years, and medium and medium-high income level, according to market research conducted in 2018 in the Province of Gutiérrez (Cámara de Comercio de Duitama et al., 2018). Among the markets that send the most travelers to Colombia, the United States stands out, with a flow of 22%, Peru and France, which in 2019 presented an increase of 19% and 12% respectively (Ministerio de Cultura et al., 2020a).

The reasons for travel are usually related to holidays (90.4%), visits to family and friends (0.7%), business and work (3.6%) and studies (5.3%).

Concerning the Orange Economy, the recognition of the State of Boyacá as a biodiverse and multicultural territory allowed for the approval of a public policy on culture, capable of developing orange economy projects, proposing "a strategic line aimed at reducing the regional concentration of cultural industries, through the exploitation of local potential" (Ministerio de Cultura et al., 2010:1). In this framework, the State of Boyacá has been working in recent years on the state strengthening of the programme of agreements and the programme of incentives for creation, research, training and the circulation of artistic expressions and cultural manifestations (Fondo Mixto de Cultura de Boyacá, 2020).

The implementation of the Orange Cluster Initiative in the state is achieving significant development among the different branches of the economic sector. After the recognition of the existing resources in the Province of Gutiérrez, an average of thirteen cultural, folkloric, business, gastronomic, sports and religious events were observed (Fondo Mixto de Cultura de Boyacá, 2021), which could count on tourism as a sector of union between culture and local development (Li & Kovacs, 2021).

Although the project covers the entire State of Boyacá, in this first stage of implementation and analysis of the project in a COVID period, it was considered prudent that the initial area of analysis should focus on the Province of Gutiérrez, which is made up of the six municipalities: Guacamayas, El Espino, Panqueba, El Cocuy, Chiscas and Güicán de la Sierra. These municipalities will serve as pilot destinations due to their cultural, environmental and economic characteristics, as well as their diversity of attractions that make it possible to offer activities within the scope of creative tourism.

Cluster Initiatives

The concept 'Cluster' was popularized by Michael Porter (1991:207), in the mid 90s, and defined as:

"...the set of companies that operate through networks around a productive activity base of development, concentrated geographically and/or sectorally by concordances and complementarities around the base sector, reach a high degree of specialization, competitiveness and efficiency, generating dynamic processes through which, it is possible to guarantee the success of the Cluster."

Based on the definition of the cluster, Elías y Tuma (2009:69) state that it is possible to understand its usefulness as a "tool for analyzing the productive structure of tourism", being possible to highlight the variability of the chain of activity and links between the participants, whether they are institutions, companies or organizations, giving rise to a wide range of competitive advantages.

According to Cluster Development (2021), their usefulness enables the sophistication of strategies in companies, as they facilitate models of integration and productive consolidation, allow access to markets for greater product innovation and guarantee greater integration between agents, enabling internationalization and impactful ventures.

Furthermore, cluster initiatives facilitate competitiveness, enabling the design of future strategies by creating a dialogue between public and private entities. In short, their implementation improves the quality of the business, allowing investment in human capital, the design of policies to attract investment and the positioning of the region.

To achieve this type of initiatives in the tourism environment, the participation of all stakeholders is of special importance, preceded by a mapping and prioritization of issues, all aimed at an integral tourism governance model.

The development of the Orange Economy Cluster of Central Eastern of the State of Boyacá, which the province of Gutiérrez is a part, is aimed at integrating companies to cooperate rather than compete, aligning themselves under the strategic concept of slow tourism.

MAIN FOCUS OF THE CHAPTER

According to Florida (2003), there are three critical types of creativity: technological creativity, economic creativity and cultural creativity, and only with the union of these three forces, economic growth can be stimulated. For this reason, the project 'Clúster Economía Naranja del Centro Oriente del Departamento de Boyacá' is initially presented to the group of entrepreneurs that are articulated in the value chain of the cluster initiative.

Among these, companies are offering cultural and nature tourism and associated services such as environmental consultancy. Other elements that are also integrated into the value chain of the Orange Economy Cluster are those related to gastronomic production and commercialization, companies with a cultural focus that merge musical tradition, promotion and conservation of folklore and handicrafts. In the field of innovation, companies with ICT offers which include audio-visual production, multimedia, promotion and communications are also presented.

Finally, there is the tourism offer with hotels that integrate Boyacá's culture as the basis of their offer, partners, and travel agencies that offer Boyacá as a destination.

The main objective of the Orange Cluster project is to benefit the business offer, focused on tourism service providers in the region.

Issues, Controversies, Problems

The economic stagnation of the Central East Region of the State of Boyacá is largely due to the pandemic situation strengthened by COVID-19, the low empowerment of the local culture by the locals and the lack of integration of public policies to promote Science, Technology and Innovation (Tc+i), in the service chain. Furthermore, this economic lethargy is also due to the disarticulation between the suppliers of Orange Economy products/services and the institutions of the regional environment.

This is due to a lack of knowledge of the region's economic and cultural potential. Due to its idiosyncrasy, local society does not manage to appropriate its own culture and identity, which generates episodes of mistrust and poor links with the outside world.

Similarly, commercial management is deficient, with no promotion of the offer. No actions have been developed to integrate and benefit the population with its cultural component. An issue of significant interest that extends throughout the country, as the Departamento Administrativo Nacional de Estadística - DANE (2021) states that 82.5% of the 32,141 Colombians interviewed in its annual survey, receive income from handicrafts, and 50.7% claim it to be the main source of income.

In addition, there is a high economic impact as a result of the health crisis, which stagnated business processes, especially tourism and the printing industry, leading to the closure of many companies and job losses, further diluting the economic vocation of the region.

The economic shock suffered in the area had a social impact with the migration of communities to larger regions in search of better living conditions, which widened the gap of poverty and inequality present in the region.

This uprooting, coupled with the severe impact of the pandemic, ends up triggering a vicious cycle of low identity for the roots of the region, in addition to the disappearance of the sense of belonging to the local population, which leads to the loss of socio-cultural wealth and low intention to promote the destination, so that employment opportunities decrease and businesses close, leading the population to seek opportunities in the big cities, resulting once again in the loss of identity and therefore, of cultural heritage, such as figure 2.

In the multi-sectoral survey 'On the effects and expectations of economic recovery in the face of the pandemic in the State of Boyacá 2020' conducted by the Cámara de Comercio de Duitama in 2020 to entrepreneurs, has observed that only 3.9% of entrepreneurs are operating at maximum capacity, 45.1% state that their sales have decreased, 47.1% indicate that they have not sold anything and their business is closed, among other answers (3.9%).

Following the analysis, it was observed that the territory presents the following strategic challenges:

- Recognition of the tourism offer - packaging of products and services.
- Integrating communities.
- Articulate the offer of suppliers.
- Integrating the offer with PROCOLOMBIA.
- Create strategic alliances with a commercial focus.
- Empowerment of ICTs.
- Traceability of the process to integrate the efforts of isolated developments.
- Identification, selection and attraction of the right visitor for the offer.
- Ensure the empowerment and integration of the narrative of culture, tradition and history of the tourism heritage of the territory.
- Strengthen heritage and culture through local gastronomy by integrating the chain.
- Specialized service providers with tailor-made packages.

Given these challenges, it is intended to implement new strategies aimed at the development of more sustainable tourism, integrating and guaranteeing the State of Boyacá's traditions and cultures. For this reason, the proposal for the implementation of the Orange Economy Cluster of Central Eastern of the State of Boyacá is structured.

Figure 2. Vicious circle of rootlessness with the Social and COVID-19 pressure
Source: Graphic made by the authors of this article.

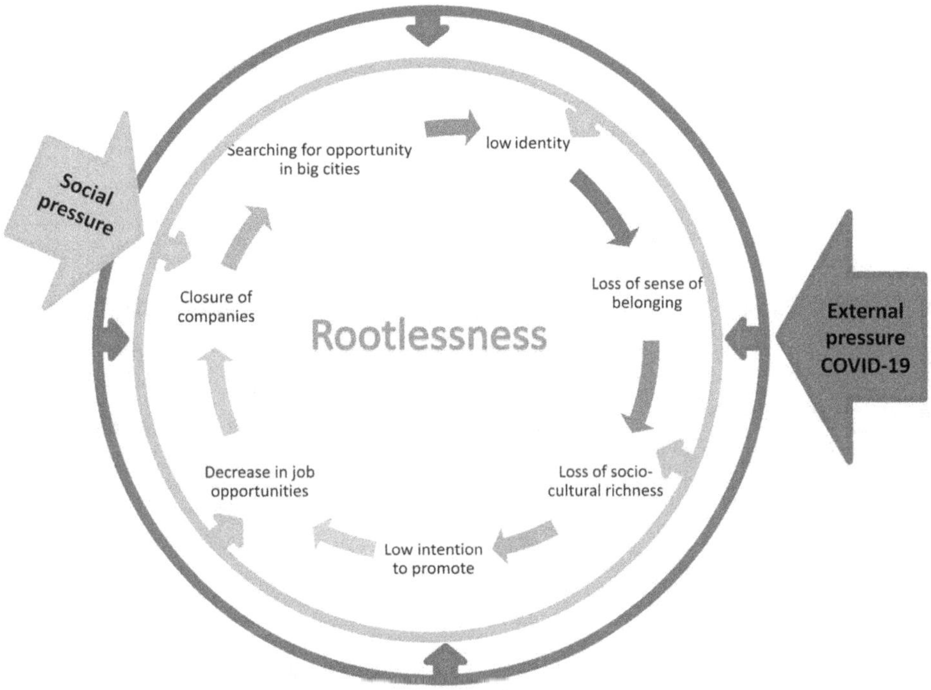

The implementation of the Orange Cluster

Based on the aforementioned challenges, a strategy of variable differentiation is chosen, depending on the market (regional, national and international) and specialization (market, product and sustainability). Bearing in mind that the territory of analysis has the following characteristics, such as Figure 3.

In opting for high differentiation, there is a need to remove barriers to entry, focusing strategic choices on the data shown in table 1.

Based on the aforementioned strategies, aligned with public policies on the orange economy and national, regional and local development plans, the Chamber of Commerce of Duitama, with the support of the Universidad Pedagógica y Tecnológica de Colombia, encouraged the consolidation of the Orange Economy Cluster of Central Eastern State of Boyacá in its area of influence, using the methodology proposed by Michael Porter (1991), for the creation of clusters.

For the development of this methodology and based on the regulations and incentives offered by the government (Ministerio de Hacienda y Crédito Público, 2020), according to Michael Porter's methodology developed by the Spanish consultant in Colombia, Cluster Development, a mapping of the orange supply of the territory was carried out, based on information from the commercial register available in the Chamber of Commerce of Duitama, where the main business potential was identified according to the codification of the International Standard Industrial Classification - ISIC, adopted by the national government (Ministerio de Hacienda y Crédito Público, 2020), finding the types of companies and percentages presented in Table 2.

Figure 3. Strategy analysis
Source: Graphic made by the authors of this article based on data from Cluster Development (2021).

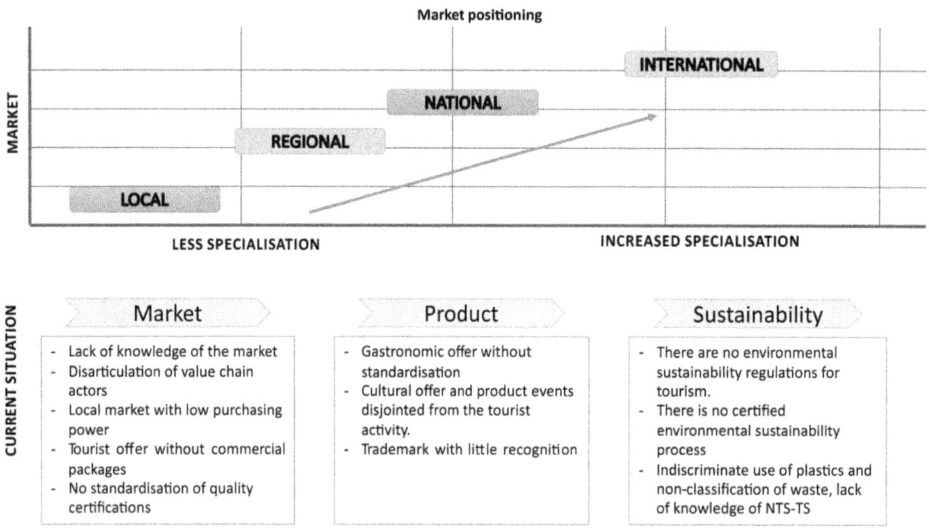

Table 1. Strategic options

	High Differentiation
Product	- Gastronomic offer with 0Km products, certified in clean production and high regional cultural content; - Briefcase of tourist products, with a slow narrative to promote the destination, according to the client, with a calendar of cultural shows; - Recognized and positioned brand that promotes the destination; - Social innovation through identity and economic use of regional culture and tradition.
Market	- Destination positioned and recognized by the certification as a slow destination, with influencers
Sustainability	- Destination with environmental sustainability certification

Source: Self-made elaboration based on data from Cluster Development (2021)

According to figures reported by the Duitama Chamber of Commerce, it is estimated that these 487 companies generate 2,922 jobs and sales of approximately $23,750,000 dollars (Duitama Chamber of Commerce, 2020). These data are extracted from the information of the business register that is kept by

Table 2. Enterprises with access to the government programme

Type of Company	No. of Companies	Representation
Tourism	111	23%
Gastronomy	135	27%
Information and Communication Technologies -ICT	106	22%
Culture-design-fashion	97	20%
Culture-design-fashion	38	8%
Total	487	100%

Source: Cámara de Comercio de Duitama (2020)

the commercial entity, according to the information reported year after year by the companies registered there. From the business register, data is extracted and coded according to economic activity. In this case, the 116 ISICs classified by the Colombian government as belonging to the Orange Economy sector were used. In this analysis it was possible to identify the 12 sectors with the greatest contribution to the economy with a supply of 2,022 companies registered by 2020 in the 12 ISICs of the 116 defined by the Colombian government. Within the aforementioned analysis, it was possible to conclude that there are sectors of greater weight, as they contribute more companies, jobs and sales revenue than others, thus determining that the sectors of Gastronomy, Tourism and Information and Communications Technology-ICT offer made the greatest contribution in the three variables mentioned; with which a first decision was taken to integrate the Cluster initially by these three sectors.

To understand the socio-economic conditions of the orange sector, based on the mapping, a first approach to the local reality was carried out through interviews with entrepreneurs in the region, where the offer, the stakes of each company and the market projection were identified. This first approach gave way to understanding the offer, its advantages and disadvantages, the main weaknesses and the entrepreneurs themselves expressed some opportunities that their daily work had allowed them to identify so far; thus, an analysis of the market and trends was made according to the offer, especially cultural and nature, that the region has.

This information served as a basis for identifying, firstly, the opportunity to take advantage of the integrated nature and cultural offer through tourism; therefore, a benchmarking process was carried out with two destinations with similar and highly competitive offers at a global level, such as Spain and Mexico, respectively. By putting into practice the process of evaluation and comparative analysis indicated above, the possibility of implementing the slow concept was observed as a strategy for integrating the territory's offer into a tourism product that would allow the development of the various orange activities, under the slow umbrella, as a strategy for integrating the cultural, social and business diversity of the area.

After analyzing both the conditions of the local offer, based on benchmarking techniques, for the concepts of nature tourism and cultural tourism in Spain and Mexico respectively, it was possible to identify the most strategic options, where the competitive offer should be prepared towards higher quality tourism. Based on the selection of the market, product and sustainability areas, interventions have been carried out in search of a change in the offer for the domestic market, a national market and especially an international market, seeking to integrate into the global network of Slow Villages.

Finally, according to the strategic options identified as a result of the previous exercises, an action plan was built consisting of projects identified in Table 3 and a governance model for the operation and functioning of the cluster, the latter of which is currently under development among the actors of the territory. Each axis defined projects and actions to be developed among the actors according to the highest priority needs to consolidate the destination, to add value to what is currently being done and, finally, to improve business profitability.

The action plan was agreed between the entrepreneurs and institutions that aimed to consolidate the regional offer, with a time horizon of three years, starting in March 2020 and ending in Feb 2023. The development of this product seeks to benefit the entire service provision chain, made up of four segments, three of which are sectoral and one cross-cutting.

Among the sectoral segments, the first one opts for the nature tourism offer, the second for gastronomy and culture, with products of high tradition and unique added value for the region, and the third segment refers to information and communication technologies. All of the above sectoral segments would

Table 3. Orange Economy Cluster Action Plan

Axis	Project
Market	Consolidate each of the Slow Tourist products currently offered in the territory, integrating the entire chain proposed in the cluster.
	Generate business projects for new businesses or business models in business groups associated with the orange economy, incorporating open innovation processes, collaborative innovation and / or new business development. Thus helping to recover and economically reactivate the sectors associated with cultural and creative activities.
	Train tourism service providers in the central eastern department of Boyacá in good food handling practices, biosafety protocols and food preparation.
	Design the tourism product in the North province of the department of Boyacá as a contribution to socioeconomic dynamization.
Product	Consolidate the regional offer of Slow Tourism of the Eastern Center of Boyacá on the web, through a unique portfolio of the destination.
	Guarantee the adequate promotion of the destination for the positioning of the Slow Tourism regional offer of the eastern center of Boyacá.
	Identify the main gastronomic offer of culture and tradition in the region's most notable for their gastronomic factors.
Sustainability	Consolidate tourist behavior guides and the tourist service provider for the correct development of tourist activity in the Central East of Boyacá

Source: Graphic made by the authors of this article.

be supported by the transversal segment called 'Culture of the Territory', which emphasizes folklore, music and handicrafts.

The next step was the definition of the action plan between entrepreneurs and institutions seeking to consolidate the regional offer, also based on three main axes:

- Consolidation of the offer.
- Market management.
- Integration of the concept of sustainability in the territory.

As a summary, it could be stated that the implementation of the Orange Economy Cluster of Central Eastern State of Boyacá follows the following methodological process, such as figure 4:

RESULTS SO FAR CHALLENGES TO OVERCOME

With the aforementioned strategy, aligned with public policies on the orange economy and national, regional and local development plans, the Chamber of Commerce of Duitama stimulated the consolidation of the Orange Economy Cluster of Central Eastern State of Boyacá. However, with the expansion of the health crisis caused by COVID-19, the government's focus had been on monitoring the priorities of society and restoring the health of the local population.

Meanwhile, the implementation of the Orange Cluster continued, giving way to the generation of a strategic alliance between suppliers and the Chamber of Commerce of Duitama. Based on this, an analy-

Figure 4. Methodological Process of Cluster Implementation
Source: Graphic made by the authors of this article.

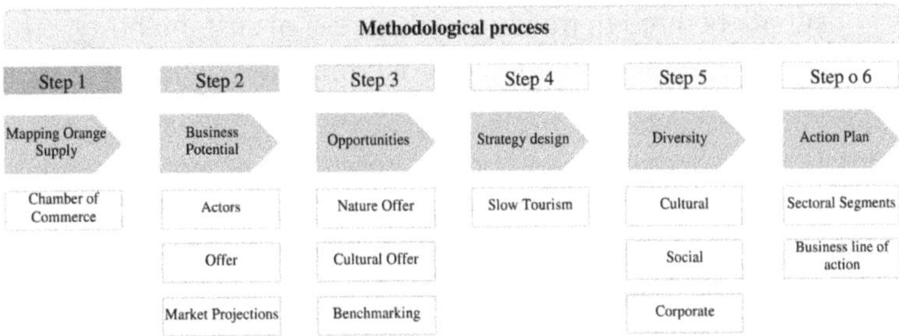

sis of the value chain was carried out, determining the market outlook, and from the suppliers' point of view, the variables that could generate greater added value in the chain were identified.

The convened entrepreneurs understood the importance of working together, in collaboration with third parties, to launch several initiatives ranging from the diversification of the tourism product offered to visitors to gastronomy and complementary services, focused on products and services that add value.

Table 4 shows the proposed challenges for the cluster with a high differentiation value.

Table 4. Proposed Challenges for the Ideal Cluster

Market	Product	Sustainability
Gastronomic offer: products certified in clean production and high cultural content of the region	Technological surveillance, competitive intelligence	Destination with environmental sustainability certification
Biriefcase of tourism products, with a slow narrative to market the destination, according to the client, with a calendar of cultural samples	Internationalization of the offer to high-end segments	Social innovation through the linking of farms and plots of clean agricultural production at the service of tourism activity
Recognized and positioned brand that promotes the destination	Positioned and recognized destination for the certification as a slow destination, with influencers	
Social innovation through identity and the economic use of regional culture and tradition.	Tourism Product Briefcase	
	Recognition for the sustainability of the destination.	

Source: Own elaboration

At present, work is being carried out on the implementation of the most necessary actions, such as:

- The design of the tourism product as an integrator of all the elements of the territory in its axes of nature tourism, gastronomy, ICT from the culture and tradition of the destination, for the promotion of the offer, which still does not have this tool for dissemination and promotion.
- The documentation of the offer in more developed regions for its promotion.

- The consolidation of the offer of the most immediate creative industry, which is intended to show the cultural identity of the territory.
- The preparation of suppliers through training in the areas of sustainability, customer care and service and, in some cases, the implementation of good practices, whether in manufacturing or food production, as appropriate.

All of this is aimed at designing a promotional campaign for the destination, through digital marketing of portals projected with the most consolidated offer in the concept of Slow Tourism in the framework of the national public policy of the Orange Economy.

The municipalities of Gutiérrez, involved in the process, have made progress in the identification and characterization of tourism resources as inputs to define the province's tourism offer.

In this way, the following forty-nine attractions have been defined, as shown in Table 5.

Applying the tourism product design methodology, after reviewing with the local communities the real capacity of each of the resources, those with the greatest recognition and commercial potential were selected, where the 13 resources in table 6 are listed.

Likewise, an inventory of service providers has been carried out to strengthen the offer and integrate the attractions with the alternatives for the provision of services through suppliers. It also seeks to conclude the design of tourism products and integrate the cultural and natural offerings of each of the six municipalities that make up the province of Gutiérrez. All this was configured in the product called "Gutiérrez, paradise in the mountains", such as figure 5.

As a result of the above, and taking into account the configuration of the product, three routes have been proposed for the Gutiérrez region, integrating the cultural, natural and mixed concepts, under the umbrella of Creative Tourism.

Each of the routes includes its activities related to them, but in an integral way and within the territory, forming different routes through various municipalities in the area.

In this way, it has been possible to configure the design of the tourist offer in the following ways:

Route 1: Nature Tourism

This route runs through the west of the territory, and is called "Gutiérrez, a paradise in the mountains".

The route includes walks along ecotourism trails of a high environmental level. The municipalities of Chiscas, El Espino and Panqueba are involved in this route. On these routes, visitors can appreciate the landscape of the region, loaded with frailejones and find quiet spaces that have not yet suffered any kind of anthropic alteration.

In addition, the programme of the route includes the observation of the ecosystems of the páramo and the Andean forests and the recognition of flora and fauna. The route also offers activities related to ornithological tourism where it is possible to observe condors and sightings of animals, where it is possible to spot deer, such as figure 6.

Route 2: Nature Tourism

This route runs through the east of the territory and is called "Nature Tourism Route: Peace, Water and Life".

Table 5. Sites of tourist interest in the Province of Gutiérrez - Boyacá – Colombia

Municipality	Site of Interest
Chiscas	1. The way to the plant - El Arco Bridge
	2. Hacienda el Salado
	3. Pila de los Héroes
	4. Los Menires - "Indigenous Cemetery".
	5. Festivities in honour of the Virgin of Mercedes
	6. Festivities in honour of the Virgen del Carmen
El Cocuy	7. Salt Museum
	8. House of the Matron Genoveva Cújar
	9. Calle del ensayadero
	10. House of Don Simón de Mendoza
	11. House of Doña Carolina Gallo
	12. House of General Baldomero Tarazona
	13. House of General Julián Arango
	14. House of General Santos Gutiérrez
	15. Gregorio Ruiz Charity Hospital
	16. Fairs and Festivities of Carrizalito
	17. Fairs and Festivals of Pachagual of El Cocuy
	18. Aguablanca livestock fair.
	19. Hill of Mahoma
	20. Pachagual Lagoon
El Espino	21. San Ramón Nonato Parish
	22. Main Park
	23. Patron Saint Festivities in honour of San Ramón Nonato
	24. Cintura Lagoon
	25. El Chuscal Municipal Reserve Park
	26. Alto Viento
	27. Alto de Hoyos
	28. El Bathan Waterfalls
	29. Mesa del Salto
Guacamayas	30. Main Park
	31. La Aurora Park
	32. Festivities of the Virgen del Carmen
	33. Maíz Somos
	34. Montoso
	35. El Espadillal
Güicán de la Sierra	36. Chapel of San Antonio de Padua
	37. Chapel of San Ignacio de Loyola or Children's Cemetery.
	38. Monument to the Virgen Morena of Güicán
	39. Monument to the Virgin of Carmen
	40. Bust of Doctor Andrés Girardot
	41. Royal Road Malpaso
	42. Laches Pictograms in Malpaso
	43. Refugio La Casona de San Luis Hot Springs Refuge
	44. Monserrate Hill
Panqueba	45. Nuestra Señora de la Salud Parish
	46. Siberia Farm
	47. Hotel Campestre la Estación
	48. Fairs and festivities in honour of the Virgen de la Salud and Virgen del Carmen.
	49. Mirador Cruz de Mayo

Source: Cámara de Comercio de Duitama (2020)

Table 6. Selected sites of tourist interest. Province of Gutiérrez, Boyacá – Colombia

Municipality	Attractions and Resources	Group
Chiscas	1. Renewable energy trail: Pantanohondo Lagoon;	Natural Sites
	2. Condor and white-tailed deer watching trail: Laguna Las Orozcas, Valle de la cercada, Pico de Rechiniga;	
	3. Las Mercedes sector.	Cultural Sites or Manifestations
El Cocuy	4. Rural rural tour: El Picacho viewpoint, Montebello Bird and Flora Sighting, Finca Agroturística, Mahoma viewpoint. Finca Agroturística Muñoz, Pachagual Lagoon;	Natural Sites
	5. Historic Centre El Cocuy;	Cultural Sites or Cultural Expressions
	6. Demonstrative wool workshops.	
El Espino	7. El Batán waterfall	Natural Sites
	8. Christmas festivities that take place in the town centre.	Cultural Sites or Cultural Expressions
Guacamayas	9. Urban Tour "Getting to know Guacamayas": in the town centre, with a demonstration workshop on basket weaving and espadrille weaving, La Aurora Recycling Park.	Cultural Sites or Cultural Expressions
Güicán de la Sierra	10. Monserrate Hill.	Natural Sites
	11. Cave and waterfall	
	12. Urban tour: Sanctuary of the Virgen Morenita and the city centre, monument to the dignity of the U'wa. Hot springs.	Cultural Sites or Cultural Expressions
Panqueba	13. Cacique de la montaña trail.	Natural Sites

Source: Cámara de Comercio de Duitama (2020)

Like the previous route, the route includes walks along ecotourism trails of a high environmental level. The municipalities of El Cocuy and Güicán de la Sierra are involved in this route. On these routes, visitors have the opportunity to appreciate the scenic beauty and landscape of the region, which also offers the so-called rural community tourism, characterized by direct contact with the local population and their customs.

The programme of this route includes the observation of the typical ecosystems of the páramo, such as the recognition of the flora and fauna of the region. This route presents a tour of turquoise waters in their purest state, associated with these mixed activities in which nature tourism is mixed. This itinerary also allows direct contact with the local community and in this sense offers religious tourism and cultural-historical tourism, such as figure 7.

Route 3: Cultural Tourism

This route runs through the East and West regions of the territory, not touching the central region of Ranqueba and is called "Cultural Tourism: Weaving Gutierrez".

This route is configured as a quick itinerary strategy, which aims to attract weekend tourists or short-stay tourists. What is intended is a partnership between the municipalities, with the purpose of offering attractions and activities designed within the cultural framework of each region. The route not only involves the recognition of attractions, but also experiences with the host community of each destination.

Figure 5. Selected tours and activities for the tourism product
Source: Cámara de Comercio de Duitama (2020)

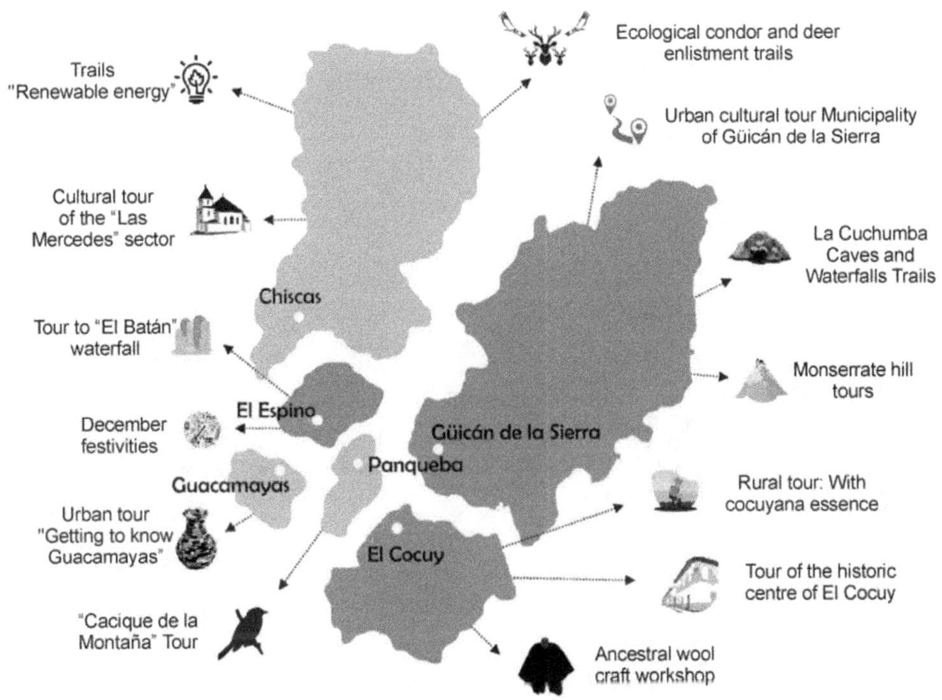

The programming of this route is developed between the municipalities of Chiscas, El Cocuy, El

Figure 6. Nature Tourism Route: Gutiérrez, a paradise among mountains.
Source: Cámara de Comercio de Duitama (2020)

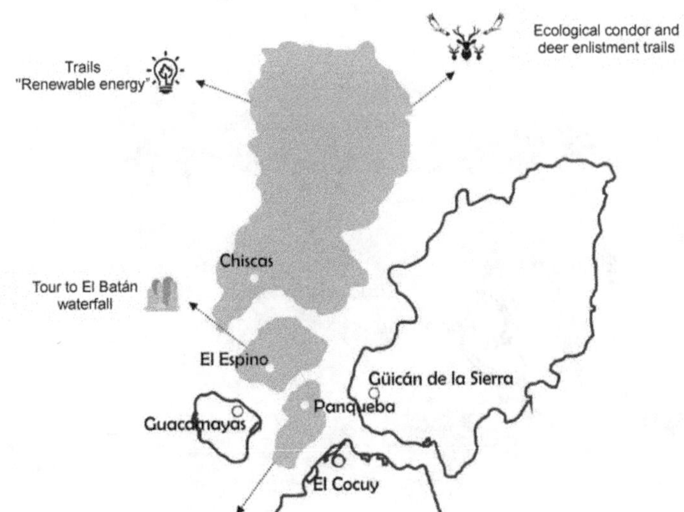

Figure 7. Nature Tourism Route: Peace, Water and Life
Source: Cámara de Comercio de Duitama (2020)

Figure 8. Cultural tourism route: "Weaving Gutiérrez
Source: Cámara de Comercio de Duitama

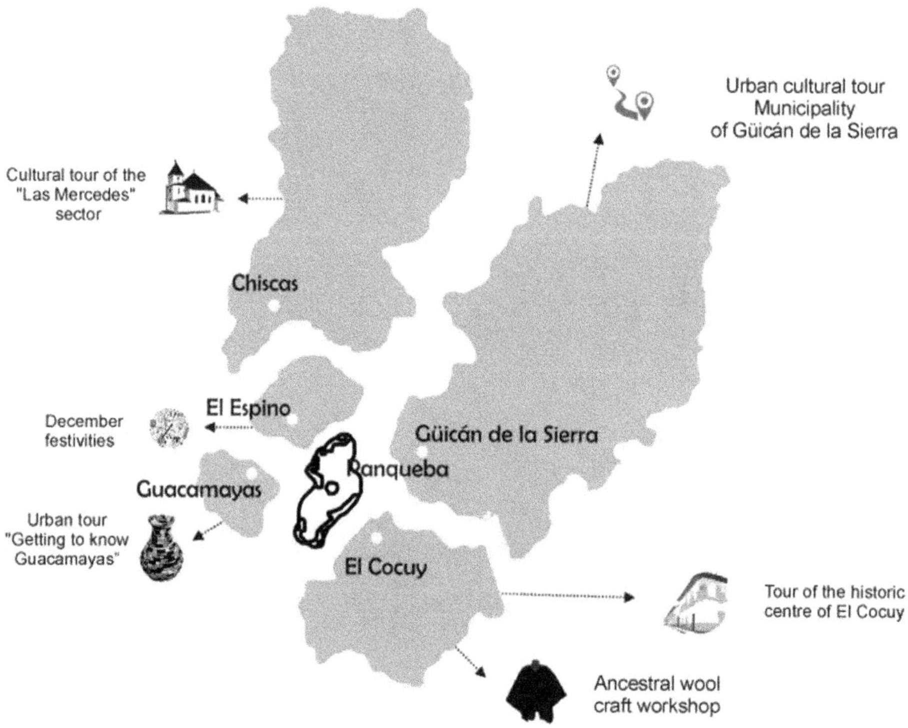

Espino, Guacamayas and Güicán de la Sierra.

The proposal of the itinerary is a total immersion in the local community, in its life and its way of life. It proposes the daily accompaniment of the community, their routines and what they do, as well as the offer of basket weaving workshops and wool craft workshops. All this is linked to the architectural recognition of the area and activities related to the so-called community tourism, such as figure 8.

As regional management projects, some results have been achieved as a result of the collaborative creation exercises between actors. According to the identified strategic options, progress has been made in the following adjacent projects:

- In the consolidation of the product, with the documentation of gastronomic expressions, with components of culture and tradition of the territory, as a result of the joint management of the actors of the region, where the most representative recipes are made with local chefs to promote these cultural icons, through gastro tourism, and the motivation for restaurants with the offer in the region, to integrate the offer from the documented recipes. The memory of three dishes of high regional cultural tradition is built up.
- The design of the tourism product articulated with the slow concept, in the municipalities that precede the province of Gutiérrez, where the main attractions were identified together with the local actors, with co-creation workshops.
- A promotional image was designed for each territory and tourist routes that link the actors, the cultural and natural attractions and the gastronomy of the territory, with the main exponents of the latter being dishes with a high traditional content, such as the "cabrito norteño" (northern goat) and the "chorotas".
- The design of tourist packages integrating and linking the leading offer of greater recognition with suppliers, to articulate the offer, to potentiate investment by consumption and to stimulate the economic reactivation in the whole chain that integrates the offer of Slow Tourism in the Orange Economy Cluster of the Centre East of the State of Boyacá. In this way, a social innovation management is carried out, linking around 500 families among artisans, producers of gastronomic offerings and tourism providers of the territory, to stimulate the economic development of around two thousand local actors who directly and indirectly benefit from the dynamization that Slow Tourism brings to the territory.
- In the area of sustainability, progress has been made in the socialization for the implementation of environmental sustainability standards, through the guides prepared by the Colombian Government, an activity that is developed in partnership with government agencies, seeking to promote sustainable services not only for visitors but also for tourism service providers in the area.
- In environmental matters, the regional autonomous corporation of Boyacá
- CORPOBOYACA (2021) is approached. This organization joined the process of consolidating the cluster with a consultant who has been giving awareness-raising talks to stakeholders on good sustainability practices, which are being carried out along the entire route within the cluster's radius of action. This plan includes guidelines for tourist behaviour during the visit, mitigation of habitat destruction, care for native species, recycling of solid waste at source, replacement of energy sources with renewable energies.
- Collaborative efforts are being made to achieve "Citta Slow" certification in the municipalities of Güicán de la Sierra and El Cocuy, where the Orange Economy Cluster, in association with local governments, is promoting this certification to encourage tourists to enjoy this sensory experi-

Table 7. Public policies and supportive tourism policies

Rule	Scope
Law 2068 of 2020	The law provides for the possibility for local governments to declare areas as tourist attractions. These attractions, once declared as such, must have a conservation programme and a budget for reconstruction, restoration and conservation. https://www.mincit.gov.co/normatividad/leyes/2020/ley-2068-del-31-de-diciembre-de-2020-por-el-cual-s
Decree 1155 of 2020	Regulates the determination, prioritization, approval, financing and execution of the Infrastructure for Special Tourism Projects. https://dapre.presidencia.gov.co/normativa/normativa/DECRETO%201155%20DEL%2020%20DE%20AGOSTO%20DE%202020.pdf
Law 2069 of 2020	Law that establishes a regulatory framework to facilitate entrepreneurship and the growth, consolidation and sustainability of businesses, in order to increase social welfare and generate equity. https://dapre.presidencia.gov.co/normativa/normativa/LEY%202069%20DEL%2031%20DE%20DICIEMBRE%20DE%202020.pdf
Decree 646 of 2021	This policy embraces a long-term vision of the sector, materialized in a strategic plan for the year 2030, which seeks to harmonize the objectives of economic and socio-cultural development of tourism with the need to protect natural capital. https://dapre.presidencia.gov.co/normativa/normativa/DECRETO%20646%20DEL%2016%20DE%20JUNIO%20DE%202021.pdf
Decree 1031 of 2021	"By which numeral 1 of article 53 of Law 2068 of 2020 is regulated, referring to the aids, subsidies or support provided to tourism service providers affected by the declaration of a state of emergency or disaster situation at the national, departmental or municipal level". https://dapre.presidencia.gov.co/normativa/normativa/DECRETO%201031%20DEL%201%20DE%20SEPTIEMBRE%20DE%202021.pdf
"Tourism Sector Plan 2018 - 2022 Tourism: The purpose that unites us".	The Ministry of Industry, Trade and Tourism defines in this document six strategic lines to be developed at the national level, such as: 1. Generation of the tourism sector promotion. 2. Integral management of destinations and strengthening of the country's tourism offer. 3. More investment, better infrastructure and connectivity. 4. Innovation and business development in the tourism sector. 5. Strengthening of human capital for the competitiveness. 6. Promotion of transformative, inclusive and equitable tourism. https://www.mincit.gov.co/CMSPages/GetFile.aspx?guid=2ca4ebd7-1acd-44f9-9978-4c826bab5013
Ministry of Industry, Trade and Tourism.	"By which the departmental development plan of Boyacá - Social Pact for Boyacá: Land that continues to advance - 2020-2023 is adopted". Article 19 of this ordinance determines each of the programmes that allow each of the territories to manage tourism and articulate the institutional framework through tourism councils, committees and technical roundtables with the participation of the private sector. https://asamblea-de-boyaca.micolombiadigital.gov.co/sites/asamblea-de-boyaca/content/files/000488/24351_ordenanza-006---pdd-2020-2023-plan-de-desarrollo-2020-2023_compressed.pdf

Source: Own elaboration. Links consulted by the authors of this article.

> ence. There is a strong leadership of local entrepreneurs and administrators of the territory, with whom we are making progress in identifying actions to achieve the certification of these two municipalities located in the province of Gutiérrez.

In terms of public policies and tourism policies to support the management of the sector, as well as economic reactivation policies, as a framework for the development of the offer of the Orange Economy

Cluster with a focus on slow tourism for the center-east of the State of Boyacá, the following regulations are currently in place to guide and articulate processes, see the table 7.

According to the open-lines of co-creation, it has been identified, after a primary analysis, that the Province of Gutiérrez shows a significant flow of visitors. These visitors usually start their trips from the Province of Tundama, and continue their visit crossing the Northern Province till they finally reach the Province of Gutiérrez.

This route, which begins in Tundama, links the actors and entrepreneurs of the municipalities that make up the road to the distant Province of Gutiérrez.

They are actors who build models of transit tourism, facilitating the route for the tourist, acting in a friendly and welcoming manner.

At present, this route is not developed and this treatment of the locals towards visitors is based on local spontaneity.

It is possible to design a tourist offer related to the culture and idiosyncrasy of the Northern Province, to take advantage of this passage of visitors through the area, taking the opportunity to include elements of cultural interest that add value to the route, such as a route related to the renowned "Queso Paipa" (Paipa Cheese), with the designation of origin, or immersion in this gastronomic art, through demonstrative workshops of its manufacture, offering a high cultural component. It could also include the elaboration of handicrafts that show the tradition and idiosyncrasy of the population of the territory and products related to the flora, fauna and geography of the region.

The design of this offer would encourage the implementation of the bases of creative tourism on the route, empowering the area and promoting the sustainable development of tourism in the area, creating a complete tourist product and not just passing through to the Province of Gutiérrez.

CONCLUSION

The department of Boyacá, a region with a markedly inland character, was already undergoing migration processes linked, to a large extent, to the economic stagnation of the department's Central Eastern region and the decrease in opportunities in the area. This was responsible for the cultural uprooting, which, together with the low integration of public policies to promote TC+i and the local idiosyncrasy, gave way to a vicious cycle of loss of identity and therefore of cultural heritage.

Such processes were strongly accentuated by the pandemic, generating important social and economic effects on a national scale.

The closure of the country and consequently of many departments, such as Boyacá, accentuated the problem and facilitated the establishment of ideas linked to the implementation of processes related to the orange economy, which already existed before the pandemic, taking more and more sense in the framework of territorial regeneration policies.

The new post-COVID-19 market trends set the tone for the design of strategies that could promote the economy from a more sustainable perspective, considering the local population, the environment and, of course, the economy.

Within this framework, Boyacá has been working in recent years on the departmental strengthening of programmes to stimulate creation, research, training, and the circulation of artistic expressions and cultural manifestations.

For this reason, the Chamber of Commerce of Duitama, with the support of the Pedagogical and Technological University of Colombia, stimulated the consolidation of the Orange Economy Cluster of Central Eastern Boyacá, focused on the tourism sector, with the support of the academic community of the region, in the design of the strategy, aligned to the public policies of orange economy, national, regional and local development plans.

The project, whose main objective is to benefit the business offer focused on tourism service providers in the region, relies on local culture as the basis for the reactivation of the economy.

The process was concerned with the articulation of tourism service providers, which has been partially achieved to date, meeting specific goals as the cluster consolidates in the region. The main objective of the Chamber of Commerce and the Universidad Pedagógica y Tecnológica de Colombia is, in addition to carrying out a diagnosis of the reality of the region and its strengths, to promote its social and economic reactivation after the COVID-19 crisis.

Currently, the implementation of the Orange Economy Cluster is achieving significant development among the different branches of the economic sector.

After the study it was observed that the department of Boyacá had significant attractions related to the local idiosyncrasy that could be put in value more beneficially, offering development, growth and better conditions to the society.

The project began with the implementation of the Orange Cluster in the province of Gutiérrez, a region with unique characteristics and expressive attractions, and the action plan for the consolidation of a sustainable offer, within the framework of Creative Tourism, was signed between the entrepreneurs of the region and the institutions involved.

The beneficiaries of the project were identified, as well as the value chain, determining the tourism service providers, the clients, the attractions and products to be offered, as well as the activities that are transversal in the provision of services. Subsequently, and in consensus with the stakeholders, an action plan was agreed upon that includes axes such as the market, the product and sustainability. At present, the aim is to consolidate the governance of the cluster, a process that will last until March 2022.

With the participation of the community, three routes have been designed for the 6 municipalities of Gutiérrez, integrating cultural, nature and mixed concepts, under the umbrella of Creative Tourism. With this, the consolidation of the product articulated with the Slow concept and the promotional image of each territory has been achieved, linking more than 500 families (actors associated with the process) to the cultural and natural attractions.

In terms of sustainability, progress has been made in the implementation of environmental standards, associated with government agencies, seeking to promote sustainable services not only for visitors but also for tourism service providers in the area.

There are still challenges to overcome, considering that the project is a living idea, which is still being implemented. Therefore, expected results have been presented, most of which are linked to the supply and the epistemic culture of the region.

It is concluded that the implementation of the cluster, which focused on addressing the existing socio-economic problems in the region of Boyacá, allowed the articulation of the stakeholders of the tourism sector in the region. In addition, it stimulated economic dynamism, promoted empowerment on the part of local communities and tourism service providers, and became a strategy that favored trust among local communities.

It is expected to be able to achieve an adequate commercialization of the tourism products or packages that are designed and marketed as a result of the implementation of the cluster.

From the research carried out for this study, it is clear that Orange Tourism is timeless, it does not depend on the seasons for its development, which makes it an excellent support for the depersonalization of destinations; it is diversified, as it optimizes existing resources at the destination, whether tangible or intangible; It is experimental, as it offers visitors unique experiences, related to the authenticity of the local idiosyncrasy; it is harmonious, as it allows for the presence of other tourism segments, such as gastronomy, oenology, ornithology or slow tourism, among others.

It can be concluded that Orange Tourism can be a way to recover post-COVID-19 tourism, serving as an element of union between the local population, the tourist and the environment.

ACKNOWLEDGMENT

English revision By Professor Guerly Gutiérrez Barrios.

REFERENCES

Álvarez-García, J., Maldonado-Erazo, C. P., Río-Rama, M. C., & Sánchez-Fernandez, M. D. (2018). Análisis de los Estudios Referentes a los Impactos del Turismo Creativo Indexados en la Base Scopus y WoS. *Revista Portuguesa de Estudos Regionais, 48*, 17-32. https://dialnet.unirioja.es/servlet/articulo?codigo=6438098

Bakas, F. E., Duxbury, N., & Vinagre de Casto, T. (2019). Creative tourism: Catalysing artisan entrepreneur networks in rural Portugal. *International Journal of Entrepreneurial Behaviour & Research, 25*(4), 731–752. doi:10.1108/IJEBR-03-2018-0177

Barra, A. (2018). Crossumer, Prosumer, Fansumer... ¿Qué es todo esto? *Lead Hospitality*. https://www.leadhospitality.es/crossumer-prosumer-fansumer-que-es-todo-esto

Baute-Plana, A., & Zamora-Fung, R. (2021). Una aproximación a la COVID-19 y sus lecciones. *Educación Médica Superior, 35*. http://www.ems.sld.cu/index.php/ems/article/download/2516/1170

Blapp, M., & Mitas, O. (2018). Creative tourism in Balinese rural communities. *Current Issues in Tourism, 21*(11), 1285–1311. doi:10.1080/13683500.2017.1358701

Cámara de Comercio de Duitama. Cámara de Comercio deo Bogotá, & Consejo Provincial de Turismo de Gutiérrez. (2018). *Estudio de Mercados, con enfoque de la demanda turística de producto turístico diseñado para el entorno de la Sierra Nevada del Cocuy, Guican de la Sierra y Chita*. https://ccduitama.org.co/documentos/Observatorio/ESTUDIO%20DE%20MERCADOS%20CON%20ENFOQUE%20DE%20LA%20DEMANDA%20TURSTICA%20DE%20PRODUCTO%20TURSTICO%20DISEADO%20PARA%20EL%20ENTORNO%20DE%20LA%20SIERRA%20NEVADA%20DEL%20COCUY%20GUICAN%20DE%20LA%20SIERRA%20Y%20CHITA.pdf

Cámara de Comercio de Duitama. (2020). *Efectos y Expectativas de Recuperación económica frente a la pandemia COVID-19 en el departamento de Boyacá*. https://ccduitama.org.co/documentos/Observatorio/Efectos%20y%20expectativas%20recuperación%20económica%20Boyacá-%20covid.pdf?_t=1595618870

Cerquetti, M., & Cutrini, E. (2021). The role of social ties for culture-led development in inner areas. The case of the 2016–2017 Central Italy earthquake. *European Planning Studies*, *29*(3), 556–579. doi:10.1080/09654313.2020.1759512

Cluster Development. (2021). Presentación Pública Cluster Economía Naranja Departamento de Boyacá—Colombia. *Clust Development*. https://www.cluster-development.com

Cooper, C., Volo, S., Gartner, W. C., & Scott, N. (2018). *The SAGE Handbook of Tourism Management: Aplications of Theories and Concepts to Tourism*. SAGE Publications Ltd.

Corpoboyaca. (2021). https://www.corpoboyaca.gov.co

Creative Tourism Network. (2020). *The concept: Creative Friendly by Creative Tourism Network*. Creative Tourism Network. http://www.creativetourismnetwork.org/presentation/?lang=es

de Colombia, C. (2017). *Ley nº 1834—23 May 2017. Fomento de la Economía Creativa Ley Naranja*. https://economianaranja.gov.co/media/5ruhnndp/ley-1834-del-23-de-mayo-de-2017.pdf

de Colombia, C. (2019). *Ley nº 1955—25 May 2019. Plan Nacional de Desarrollo 2018-2022 «Pacto por Colombia, pacto por la equidad»*. https://www.funcionpublica.gov.co/eva/gestornormativo/norma.php?i=93970

de Cultura, M., de Comercio, M., Industria y Turismo, & Departamento Nacional de Planeación. (2010). Documento CONPES3659 Política nacional para la promoción de las industrias culturales en Colombia. *Observatorio Cultura y Economía*. https://culturayeconomia.org/blog/documento-conpes-3659-politica-nacional-para-la-promocion-de-las-industrias-culturales-en-colombia/

de Cultura, M., de Comercio, M., Industria y Turismo, & Departamento Nacional de Planeación. (2020a). *¡En 2019 el turismo en Colombia rompió récords!* https://www.mincit.gov.co/prensa/noticias/turismo/en-2019-el-turismo-en-colombia-rompio-records

de Cultura, M., de Comercio, M., Industria y Turismo, & Departamento Nacional de Planeación. (2020b). *Ley Naranja*. https://economianaranja.gov.co/ley-naranja/

Departamento Administrativo Nacional de Estadística - DANE. (2021). Empleo—Economía Naranja. *DANE - INformación para todos*. https://www.dane.gov.co/files/investigaciones/pib/sateli_cultura/economia-naranja/presentacion-rp-4to-reporte-economia-naranja.pdf

Duxbury, N., Bakas, F. E., Vinagre de Castro, T., & Silva, S. (2020). Creative Tourism Development Models towards Sustainable and Regenerative Tourism. *Sustainability*, *13*(1), 2. Advance online publication. doi:10.3390u13010002

Duxbury, N., Silva, S., & Vinagre de Castro, T. (2019). Creative Tourism Development in Small Cities and Rural Areas in Portugal: Insights from Start-Up Activities. In *Creating and Managing Experiences in Cultural Tourism* (pp. 291-304). World Scientific Publishing. https://www.worldscientific.com/doi/abs/10.1142/9789813233683_0018

Elcacho, J. (2020). Efecto secundario de la Pandemia: Canales de Venecia con aguas limpias y peces por el confinamiento del Covid-19. *La Vanguardia - Natural*. https://www.lavanguardia.com/natural/20200317/474234537648/canales-de-venecia-con-aguas-limpias-y-peces-por-el-confinamiento-del-covid-19.html

Elías, S. R., & Tuma, C. M. (2009). Turismo y teoría de clusters, las potencialidades de la localidad de Monte Hermoso. *Aportes y Transferencias, 13*(1), 67-92. http://nulan.mdp.edu.ar/1546/

Fesenmaier, D. R., & Xiang, Z. (2017). Introduction to tourism design and design science in tourism. *Design Science in Tourism*, 3-16. doi:10.1007/978-3-319-42773-7_1

Florida, R. (2003). *The Rise of the Creative Class: And How It's Transforming Work, Leisure, Community and Everyday Life*. University of Toronto Press on behalf of Canadian Public Policy. https://www.jstor.org/stable/3552294

Fondo Mixto de Cultura de Boyacá. (2020). Tunja: Seminario Gobernanza y Patrimonio Cultural: El Papel de la Actividad Turística en el Desarrollo Local. *Fondo Mixto de Cultura de Boyacá*. https://www.fondocultura.org/eventos/?tribe_paged=1&tribe_event_display=list&tribe-bar-date=2020-09-01

Hasani, A., Moghavvemi, S., & Hamzah, A. (2016). The impact of emotional solidarity on residents' attitude and tourism development. *PLoS One, 11*(6), 1–14. doi:10.1371/journal.pone.0157624 PMID:27341569

Hinojosa, V. (2020). El turismo creativo, solución para reactivar el sector durante la pandemia. *Hosteltur*. https://www.hosteltur.com/139045_el-turismo-creativo-solucion-para-reactivar-el-sector-durante-la-pandemia.html?%20shared&utm_source=twitter&utm_medium=social-web&utm_campaign=139045_el-turismo-creativo-solucion-para-reactivar-el-sector-durante-lapandemia&%20utm_content=economia&utm_term=noticia

HOSTELTUR. (2020a). *Creative Tourism Network celebra el día del turismo creativo, naranja y de proximidad*. https://www.hosteltur.com/comunidad/nota/022336_el-creativetourismnetwork-celebrara-el-dia-del-turismo-creativo.html

HOSTELTUR. (2020b). *El turismo poscoronavirus: Hacia un «low cost» de precios intermedio*. HOSTELTUR. https://www.hosteltur.com/lat/136500_el-turismo-poscoronavirus-hacia-un-low-cost-de-precios-intermedio.html

Justo, D. (2020). La drástica caída de la contaminación permite a la India volver a ver el Himalaya después de varias décadas. *SER - Contaminación*. https://cadenaser.com/ser/2020/04/09/internacional/1586428245_618888.html

Kajan, E. (2014). Community perceptions to place attachment and tourism development in Finnish Lapland. *Tourism Geographies, 16*(3), 490–511. doi:10.1080/14616688.2014.941916

Kugapi, O., & Höckert, E. (2020). Affective entanglements with travelling mittens. *Tourism Geographies*, 1–18. doi:10.1080/14616688.2020.1801824

Larson, C. (2020). Cierres por coronavirus dejan cielos más despejados en China. *APNews*. https://apnews.com/article/e09cc102eb67432fa0993c10381a8951

Lee, T. H., & Jan, F. (2019). Can community-based tourism contribute to sustainable development? Evidence from residents' perceptions of the sustainability. *Tourism Management*, *70*, 368–380. doi:10.1016/j.tourman.2018.09.003

Li, P. Q., & Kovacs, J. F. (2021). Creative tourism and creative spaces in China. *Leisure Studies*, 1–18. doi:10.1080/02614367.2021.1948596

Lor, J. J., Kwa, S., & Donaldson, J. A. (2019). Making ethnic tourism good for the poor. *Annals of Tourism Research*, *76*, 140–152. doi:10.1016/j.annals.2019.03.008

Martín De la Rosa, B. (2003). Nuevos turistas en busca de un nuevo producto: El patrimonio cultural. *Pasos. Revista de Turismo y Patrimonio Cultural*, *1*(2), 155–160. doi:10.25145/j.pasos.2003.01.015

Mesa, R. (2017). *Masterclass Turismo Naranja* [Formación; Youtube]. IMF Busness School. https://www.youtube.com/watch?v=g0rAFPzMis4

Ministerio de Comercio Industria y Turismo y Fondo de Promoción Turística Colombia. (2013). *Guía Turística Boyacá—Colombia*. https://www.mincit.gov.co/cmspages/getfile.aspx?guid=91c401c7-2894-4cd6-ada2-2853a2c8922f

Ministerio de Hacienda y Credito Público. (2020). *Decreto Númro 286 de 26 de Feb de 2020*. República de Colombia. https://economianaranja.gov.co/media/hddp0pcx/decreto-286-de-2020.pdf

Mossberg, L. (2007). A marketing approach to the tourist experience. *Scandinavian Journal of Hospitality and Tourism*, *7*(1), 59–74. doi:10.1080/15022250701231915

Mowinckel, K. M. (2016). El turismo experiencial y creativo: El caso de Cantabria. *International Journal of Scientific Management and Tourism*, *2*(3), 195-203. https://dialnet.unirioja.es/servlet/articulo?codigo=5665941

ONU. (2016). *Agenda 2030 y los Objetivos de Desarrollo Sostenible—Una oportunidad para America Latina y el Caribe*. Publicación de las Naciones Unidas. https://www.cepal.org/sites/default/files/publication/files/40155/S1801141_es.pdf

Pimentel de Oliveira, D. (2020). Tarjeta Turística Safety and Security: El pasaporte para turistas y destinos seguros. *Cuadernos de Turismo, 46*. doi:10.6018/turismo

Porter, M. E. (1991). La ventaja competitiva de las naciones. *Harvard Business Review, 1025*. https://d1wqtxts1xzle7.cloudfront.net/53273995/VCN_PORTER-with-cover-page-v2.pdf?Expires=1630409352&Signature=QrDnYFaMwAGNz6VBZ0rbJOGM8VSOmn8XKFMdtGPEm7yhNpXJUOpybrqklWxlZAwRHqi4uSp1-Hoq34~Lg0Kv3RN1ukWzpMnMQnG9C8HFHN2O-NCzx22XQc8YwPw-PltVXM4d4WPk~cKoOVB13PqUe6Hu6kbvuEHejleHLLh8ejKuL3mD~chLEa7l3UHLceiBunaFsNuh3t7upB9schzFbdF~C53DhGOWS3AioyzGbwGPbn1uXDA0Oeb84pkGrZZtZRX~fZgQh9HaNbce-ZK-7kIQFaNAdwNN9aaL84IDJ7OKqZA8LR2MIQBlGy5DnfZ8ZrrUP8zClOCOB2QkXZYk47g__&Key-Pair-Id=APKAJLOHF5GGSLRBV4ZA

Procolombia. (2012). Razones de porqué viajar a Colombia. *Visit Colombia*. https://procolombia.co/especial-colombia-ny-2012/visite-colombia/razones-de-por-qué-viajar-colombia

Richards, G. (2014). Creativity and tourism in the city. *Current Issues in Tourism, 17*(2), 119–144. doi:10.1080/13683500.2013.783794

Richards, G., Van Oss, F., Fernandes, C., Sousa, L., Tzanakaki, K., & Miettenen, S. (1998). *Developing and Marketing Crafts Tourism: The EUROTEX Project*. Atlas. https://d1wqtxts1xzle7.cloudfront.net/37322387/handbook_EUROTEX-with-cover-page-v2.pdf?Expires=1627030970&Signature=gy9-UBaa3YB3CyfIK-5AtU6LkaC1~k-LqqydeSfK273vwfVEuvHHxsqTUk3NL9K0WAIeav-E0FywJ1cjZ-VG-sVtxC3RabpWRBKc3O9LPa3GEcCOx6~e8u1Gkxl74zRGdFmWnYWMO~zRrQv9OBRzicex5eyLdwSVEvr7nh-FfRqpb7N9~o-RUOIeg~3hCm1TnqCDgkVJNPGvN5V5WNnNjeExwwFICdajsnk6PMeiEEfqPL2HGCNSzGW8Viwb1svQQM7Q4jklASLTMADKQIDhMQ3B0Bi07KNWtPxZJXsLJ4DV4Yo-jinAEchI8bTUwHyuWJgqFC3eXJp96Hv~a0fxeh4w__&Key-Pair-Id=APKAJLOHF5GGSLRBV4ZA

Sarantou, M., Kugapi, O., & Huhmarniemi, M. (2021). Context mapping for creative tourism. *Annals of Tourism Research, 86*, 103064. doi:10.1016/j.annals.2020.103064

SENA. (2017). Lo que debes saber sobre la Economía Naranja. *Servicio Nacional de Aprendizaje - SENA*. https://www.sena.edu.co/es-co/formacion/Paginas/economia_naranja.aspx

Thomson, I. (2004). Bonjour blanc: A journey through Haiti (4th ed., Vol. 1). Random House and Vintage Books.

Vera-Rebollo, J. F., & Ivars-Baidal, J. A. (2020). *El impacto del COVID-19 en la actividad turística de la Comunitat Valenciana* (N.º 1; Prospect Comunitat Valenciana, p. 10). https://prospectcv2030.com/wp-content/uploads/2020/04/INFORME-TURISMO-Y-COVID19-CV.pdf

Wattanacharoensil, W., & Sakdiyakorn, M. (2016). The Potential of Floating Markets for Creative Tourism: A Study in Nakhon Pathom Province, Thailand. *Asia Pacific Journal of Tourism Research, 21*(sup1), S3-S29. doi:10.1080/10941665.2014.998250

Zehrer, A. (2009). Service experience and service design: Concepts and application in tourism SMEs. *Managing Service Quality, 19*(3), 332–349. doi:10.1108/09604520910955339

Section 3
Case Studies

Chapter 13
Developing Literary Tourism to Support Local Culture and Tourism:
The Case of Shkodra

Ilda Erkoçi
https://orcid.org/0000-0002-5027-6297
Luigj Gurakuqi University of Shkodra, Albania

ABSTRACT

Literature can provide tourism with new destinations, which can in turn foster cultural and economic development. This chapter contends that Shkodra has potential for the development of literary tourism, and that making better use of literature as part of the cultural heritage and tourism offer in this city would have multiple benefits. Not only would this increase interest in literature as well as create stronger connections with the literary tradition, but it would also offer unique experiences, help connect with local communities, and even support local economy. By using autoethnography as a starting point, this chapter will provide a model on how to use literature as part of cultural tourism by highlighting significant literary locations in Shkodra, Albania, which has been chosen as a case study. That will then be followed by an analytical discussion of the strategies to be used so as to bring literature to the attention of people.

INTRODUCTION

Cultural tourism has now become significant worldwide. It has also become important in a small Balkan country like Albania (which tends to be primarily chosen for its Mediterranean climate and relatively unspoiled natural sites) as visitors, especially foreign ones, are increasingly looking for a more authentic experience such as involvement in the lifestyle of villagers, following the process of artisan work, cooking as well as tasting local food, or even learning more about oral traditions, which are often not part of the usual tourist packages. Such change in the travellers' demands and behaviour calls for changes in

DOI: 10.4018/978-1-7998-9217-5.ch013

what is offered. In the case of Albania, cultural tourism typically focuses on archaeological and religious sites, crafts, museums, music and gastronomy.

Even though literary tourism has grown into an important niche of cultural tourism in many countries in the recent years, this is not the case of Albania. The author of this paper believes that literature should make up an important aspect to be added in the country's cultural offer, for both foreign and domestic visitors. This would not only increase interest in literature and create stronger connections with the literary tradition, but it would also offer unique experiences, help connect with local communities, support local economy as literary tourism is possible in shoulder seasons, and it can also help leverage sustainable tourism.

This paper proposes Shkodra as a site with literary development potential. A city with a rich and diverse cultural heritage, Shkodra is well-known for its literary legacy. It was the birthplace or the adopted home of several great Albanian writers. It has been such a source of inspiration that many of them have turned it into a setting for their works. In addition, it has also attracted foreign authors like Edward Lear or Edith Durham. With its remarkable literary history, the city has a lot of potential for the development of literary tourism as a niche of its cultural tourism. Adding literature to the tourist offer in Shkodra would help redesign and regenerate the city as a creative cultural space using many of its tangible and intangible cultural assets.

This paper starts with a short overview on the relationship between culture, literature and tourism, goes on with an autoethnographic account of the author's literary tour of the city and finally discusses how literature can contribute in the further development of cultural tourism in Shkodra. The main objective of this paper is to explore the literary potential of the city and highlight the benefits of developing literary tourism there and suggest a model of adding literature to the current cultural offer in the city.

METHODOLOGY

The approach used in this paper is qualitative, a combination of primary research through the use of autoethnography, and secondary sources.

The choice of autoethnography was made in order to use personal experience as a starting point for the analysis. The autoethnography is "the sharing of, and reflection upon, subjective, unique and evocative" experiential stories (Harwood and Eaves 2017, p. 146). It is a form of self-narrative that places the self within a social context and gives the researcher the possibility to engage in an analysis of the culture they are a member of and draw broader conclusions about it. It offers an opportunity for "[re]claiming voice" (Holman et al., 2013, p. 32) and can "reveal that which is hidden from conventional approaches" by allowing to observe "self in participation" (Harwood and Eaves, 2017, p. 145). As Ellis (1997, p.117) states, it "adds blood and tissue to the abstract bones of the theoretical discourse."

The autoethnography allows one to focus on telling a personal story which "evokes in readers a feeling that the experience described is authentic and lifelike, believable and possible" (Ellis 1997, p. 133) and encourages others to share their own stories (ibid).

Through observation and reflection, the researcher explores and understands their own culture better as well as help others, both insiders and outsiders gain a different perspective about it. The researcher becomes aware of their position and creates "a space to change the perception becomes aware of their position and creates "a space to change the perception of the resultant reality" leading to "action based on reflection, and reflection based on action" (Blackburn, 2000, as cited in Starrs 2010, p. 2). Starrs believes that for the people involved in autoethnography and those who read autoethnographic accounts, the process of engagement can be transformative: "This kind of writing can inform, awaken, and disturb readers by illustrating their involvement in social processes about which they might not have been consciously aware" (Sparks 2002, as cited in Starr, p. 3).

Even though autoethnography is quite underused and underrated in the academia, Ellis and Bochner (2006, p. 440), believe that this method can contribute in bridging the gap between art and science. It offers an opportunity for new forms of inquiry by using "the empirical evidence of experience to examine the scholastic debate and present a contribution to the debate about the appropriateness of personal experience as a research method" (Harwood and Eaves, 2017, p. 151).

In this paper case, autoethnography allows the researcher to provide an authentic account of her experience as a literary tourist in her own hometown by recreating the spaces visited. The autoethnographic account has been structured as a journal entry in which the author recounts her story as a *flâneur* in the city of Shkodra in Northern Albania. It was used as a way of exploring the literary potential of the city. The account here has been written based on the notes and reflections made by the author of this paper during her literary tours in Shkodra and it reflects all the feelings and memories associated with this experience. Following the account, the author also provides a more detached analysis of the potential of the city in terms of literature, the strategies that can be used by the authorities to better promote and develop the city's literary heritage through tourism, and the benefits associated with such development.

CULTURE, LITERATURE, AND TOURISM

Every now and then, tourism has to reinvent itself so as to meet tourists' demands. With tourists "tired of the same old thing" (Sigala & Leslie, 2005, p.7), tourism authorities and tour operators must look for new ways in providing their services and products to face competition and attract interest. Culture seems to possess that unique quality which plays an essential role in attracting tourists.

Richards (2004) believes that creating a dialogue between tourism and culture has become a necessity as "Tourism and culture no longer inhabit different worlds, but increasingly share the same physical, social and economic spaces." He contends that culture is nothing without tourism: "if the former is a source of values, the latter is crucial in creating such values." Richards adds that a crucial factor which has strengthened the relationship between culture and tourism is sustainability:

In a world in which 'tourists' are often regarded with disdain, because of the supposed negative impacts they have on the environment and culture, cultural tourism is seen as a 'good' form of tourism. Tourists motivated by culture are supposedly more respectful of local culture, they behave themselves better, and most importantly, they leave more money behind.

The beneficial effect of culture on tourism as a way of providing a tourism strategy based on authenticity and sustainability has also been pointed out by other scholars (Costa, 2003; Medeiros, 2005 as cited in Carvalho et al., p. 4).

There is abundant evidence revealing the convergence of culture and tourism as shown by the fact that the major museums and other cultural sites depend on tourists as their main visitors. According to OECD (2008, p. 21), cultural tourists account for 40% of global tourism.

For Ferreira et al. (2009), "The value of the cultural product stands in its "allow[ing] the distinction of territories in an increasingly global world" (p. 506). Among a variety of cultural products, literature has been given growing attention during the last few years as tourism authorities have come to appreciate its distinctive value with relation to local identity and territory. Literature seems to possess that exceptional quality which has the potential to transform the existing cultural tourism into a more creative one with a focus on community involvement. The added value of literature is related to the fact that there is hardly a person who has been left unaffected by a book related to his/her destination, either for the better or worse: "Literature and narratives affect a psychic change in tourists [...] a written text is a key to providing stimulations and the emotive responses to visit and experience a destination" (Jenkins & Lund, 2019, "Preface").

Capdepón (2014) notes that during the last few years, there has been a return of the romantic ideal of travel, reminiscent of the Grand Tour, forerunner of tourism, whose importance stands in its "pav[ing] the way to literary discovery of a landscape: tourism that becomes based on local cultural heritage" (p. 407). Cultural tourism plays a significant role in the appeal of a tourist destination. As an important element of culture, literature can also play a role in promoting the cultural appeal of a place and contributing in enhancing the attractiveness of local tourism. Literary books associated with a place can become the new guidebooks which can open up new paths in the connection between tourism and culture with literature serving as a mediator between fiction and reality, emotion and landscape. For Squire (1993) literary tourism is not just a function of literary influences but rather an instrument through which a variety of cultural meanings and values can be communicated: "All experiences involve meaning and, similarly, form part of wider processes of cultural production and consumption". In the same vein, Carson et al. (2017, p. 380) consider cultural tourism as more than the tourists' economic impact on the city: "it also concerns social and cultural dimensions that occur through tourists' interactions with the city and how intercultural communication takes place" and writers as an effective marketing tool for the city.

The emergence of literary tourism seems to be what Capdepón (2014) calls "the reaction of demand against standardization [...] in the search for the rediscovery of new cultural modalities" (409). There are a number of successful examples of the development of literary tourism worldwide, but especially in English-speaking countries, which could serve as models for other countries in using literature as "an effective tool in branding destinations and creating sustainable marketing strategies" (Potočnik-Topler, 2016, as cited in Vitić-Ćetković et al., 2020, p. 84). Some of these good practices have been mentioned in the Discussion section.

Albania, Shkodra, Tourism and Culture

During the last few years, tourism has become a priority sector in the country (Ministria e Turizmit dhe Mjedisit, [MTM], n.d.) and Albania a rising tourist destination. The tourism sector contributes to the country's GDP and it makes up an important sector in terms of employment. The figures for 2017 were 8.5% and 7.7% respectively, according to Agjencia Kombëtare e Turizmit, AKT [The National Agency for Tourism].

However, even though the National Strategy for Sustainable Development claims to have diversification of the tourist offer as its mission, little has been done so far. This is related to a number of factors such as poor institutional mechanisms specialising in tourism development/ management, poor infrastructure and insufficient investment, lack of qualified staff, lack of a tourist offer which comprises a rich diversity of attractions, unexploited resources, lack of partnership culture among stakeholders, lack of experience, little use of digital technologies, and so on. As Ciro and Toska (2018) put it,

Despite the various approaches adopted, elected governments of the past few years have recognized the potential that the tourism sector has to drive economic development. To this end, numerous strategies have been drafted both on a national and local level, most of which have been unsuccessful in operational terms. Brands such as "Albania: A New Mediterranean Love" and "Albania: Yours to Discover" have, on the one hand, fascinated and lured onetime visitors, as demonstrated by the increasing number of foreigners visiting the country. On the other hand, these branding strategies have further widened the shortcomings deriving from incomplete and unimplemented strategies, structural weaknesses, and sectoral policy failures (p. 84).

The same scholars point out a discrepancy between the government's ambitions for the development of tourism and the latter's minimal contribution to the Gross Value Added. They argue that although the government has been quick to understand the importance of the sector for the country's economy and has taken a few measures to extend the tourism all year round, yet, practically speaking, the application has often failed. Ciro and Toska attribute such failure to a "top-down driven tourism strategy" and the "poorly-skilled and equipped local governments to respond and translate the strategy in operational action plans" (pp. 86-7).

Another problem hindering development is that the tourism sector is not given the necessary financial support; the budget planned for 2018-2022 is less than 0.01% of the GDP and less than 0.1% of the overall government budget (Ministria e Turizmit dhe Mjedisit, 2018, pp. 46).

Ciro's and Toska's analysis of the current tourism situation reveals that the government plays an insignificant role and that this industry mainly depends on the services offered by the private sector. Even though there is legislative framework granting municipalities competences in planning activities which would foster economic development with a focus on tourism, yet, financial problems often make local governments "prone to using opportunistic behaviour to attract investors" (p. 89) which means that sustainability is often overlooked.

Currently, coastal tourism is still the favourite form of tourism in the country, while cultural tourism falls behind despite it having gained some ground as compared to previous years—9% more visitors in 2018 as compared to 2017 (Albanian Institute of Statistics [INSTAT], as cited in MTM). Data shows that cultural sites are higher in demand from foreign tourists (30-35% of them have an interest in cultural tourism, according to the same source) as compared to domestic ones. Research reveals that the cultural tourism offer in Albania is still relatively traditional and it typically depends on tangible heritage.

Shkodra, the Albanian "Cradle of Culture"

Founded in about 400 BC (Akademia e Shkencave e Shqipërisë, 2009), Shkodra has always been a very important area for the country. Under the reign of King Gent, it was the capital city of the Illyrian Kingdom (181-168 BC) before being conquered by the Romans who were well aware of its strategic position. Prior to Ottoman occupation, the city had achieved a high level of development with an excellent institutional and judicial structure, of which the *Statute of Scutari* discovered in Venice a few years ago, is a fine testimony.

The city is rich in water resources as well as flora and fauna. It is just 32 km away from the sea and it is surrounded by the wonderful Alps. Few other cities can boast such natural heritage.

In addition to its natural beauty, Shkodra has also got a rich cultural legacy.

As early as 230 BC, the city had its own currency. The first document in the Albanian language, "Formula e Pagëzimit", by Pal Engjëlli, was written here in 1462 followed one year later by the first printing house. The first photograph was also shot here in 1858, and the first musical band and synoptic station were to be first set up in this city. By the end of the 19th century, museums and archives had been established thanks to the clergy; in addition, there were painting classes, a theatre and the first films started to be shown in 1911 ("Shkodra").

The city preserves a wealth of material and non-material legacy—a castle, an ethnographic museum, the Marubbi photo gallery, churches and mosques, traditional old houses, wonderful folk costumes and songs, and ancient traditions. Over 120 monuments are now considered objects of special cultural importance. In addition to all this, the city is also famous for its artists and personalities in various fields.

A city rich in culture and literature, Shkodra is home to many great figures, especially writers (over 40). Some of the earliest Albanian writers—Marin Barleti, Marin Biçikemi and Dhimitër Frangu were born or studied here. The same is true of the greatest Renaissance and Realism writers—Fishta, Mjeda, Shiroka, Vasa, Migjeni and Koliqi, to mention just a few. Later writers like Martin Camaj and Arshi Pipa who were part of the diaspora, also had their roots in this city. The city also owes a lot to foreign travellers who have often made it known outside the Albanian borders. Prominent among them is Miss Edith Durham.

According to Ndriçim Kulla (2010), "for centuries it was exclusively the only city to have a literary tradition." Scholar Elsie (2004) also recognizes this and highlights the special role of the city in terms of literature:

Even though the first literary writings appeared in the 1600s, it's only in the first two decades of the 20th century, after the country's declaration of independence in 1912 that a real tradition of Albanian prose started. The hub of such a tradition was Shkodra, the cultural capital of Albania (p. 5).

Whether consciously or unconsciously, a writer is often influenced by a place and the people inhabiting it. Writers are not only shaped by place, but the place is also shaped by them, the two becoming inextricably associated with each-other: when we think of Migjeni and Fishta, we think of Shkodra and when we think of Shkodra, from a literary point a view, we can't help linking it with its writers for whom the city was both a source of inspiration and setting for their works.

A literary tourist in my own hometown[1]

There Before the Tribes Arrived

You were

There before the tribes arrived

With your milk in a fissure in the rock

And with your feet in salt water.

They gave you but one name: Shkodra.

And they called you a crowned city

And they cast stones at your head

And ancient iron.

How often did you awaken drenched with blood

And observe yourself in the mirror?

Bearing a woman's name, you bathed in the waters

Of the river and enthroned yourself with fresh garments

Upon the cliff

Your brow shining in the sun over the fields. (Camaj, 1978, p. 31, translated from Albanian by Robert Elsie)

Day 1

I leave home. Feeling as if I were going on some exploring expedition. Anxious, as if I were up to mischief. Can't tell anyone but myself. "You, born and bred in Shkodra, are going to be a tourist in your own town?" – I can hear people laugh at me. But I've promised myself that it'll be something new. Different. As if I have never been there before. See it with the eyes of those who poured down bitter-sweet lines for it. Walk through it with the steps of those who painted wonderful images with their pens.

I decide to start my lit-tour from the lake. I'm not sure why – maybe I need to breathe some fresh air to get inspired for what is to come.

A breeze caresses the linden tree nearby and mixes with the smell of sage coming from the mountains surrounding the lake, in a warm embrace. "Will writers have smelled the same fragrance?" I wonder. "Will their thoughts have soaked up this flavour to then blossom into rhymes which still captivate our senses?" But there is no answer besides the lake gently rocking the boats before it kisses the shore …

And my thoughts float with Harapi's description back in 1918:

It was a fine day. The sun danced on the surface of the lake like a gilded plate. The breeze softened the burning heat of midday. A few waves cradled our little vessel and the rowers cheerfully chattered with the four travellers on board, making sure to be good company …

The villages – Rumija, Kraja and Zogaj were sheltering in the shade of the mountains with their bare edges clothed by sage alone. On the other side, the glowing copper sunrays stretching to Paru and

Veleçik, unveiled breathtaking views: the snow-capped Paru glittered like sparkling diamonds (Harapi 45; my translation).

Like the author, I can't help going back to sweet childhood memories of carelessly stepping over the white pebbles to make my way to the freshness of water:

Oh, dear sweet memories of childhood! Who was happier than we were at the time, looking at life in its full blossom [...] with fantasies like white butterflies unfolding their wings for the first time [...] Oh sweet times that will never come back again! Just the memory of you is strong enough to refresh life that has now been eroded by the waves ... (Harapi 44; my translation).

I wake up from my daydreaming to realize I've been there for too long. Before heading back home, I make sure to get the most of the view and see it as writers saw it in their days:

The sun was going down. Surrounded by a few clouds like puffs of smoke crowning it, the lake was a gilded plate shining in all its grandeur as if the whole of nature had melted in it. In a bidding farewell pose to the king of nature, the clouds were sneaking to steal some of its golden grace ... the sun disappeared leaving behind its burning rays and me ... (Harapi 47, my translation).

I pass by the River Buna and see it embrace the Drin and both enfold the castle. Rozafa is there, vigilant. The water, which, as the legend goes, is the accumulation of her tears, flows under my feet and brings more memories with it ...

Day 2

Shkodra echoes with voices, noise and life.

The newly-risen spring sun pours into people's houses

Illuminating roads and parks in bright colours

Wrapping everything in a slender golden layer. (Koliqi, "Shkodra in the Morning", my translation)

I take up my lit-tour away from the serene lake, in the enjoyable hustle and bustle of the city centre. I want to see what to the eyes unused to its daily routine is the "beautiful, sinful Shkodra, with its streets, gates, mysterious orchards, markets, cafés, bells, minarets [...]" (Kadare, 1991, p. 23). My first stop: Gjergj Fishta's monument – I say "hello" and move on.

I'm now strolling in Gjuhadol, the historic street designed by Idromeno, also known as "little Venice". In addition to its wonderful architecture, this area is also known for its rich heritage of great intellectuals. Gjon Stefa (2011) calls it a "golden oasis" (p. 5) which has produced 123 well-known people, many of which were writers.

Back to Koliqi. I stare at the cobble stones which to him looked like pearls. I reach his house. There it is, locked and lonely. Had it not been for the plaque on the wall, I wouldn't have found it. Looks unbelievable that it's the same place where he returned with a suitcase full of dreams! For a moment, like the one in whose footsteps I'm following, "Shkodra's grandeur, now in sharp decline, caught me by surprise. That house, that room, those silent objects, had once been so full of life …" (Koliqi, 2001, pp. 18-9).

I stop feeling sad for a minute and look up at the first floor from which he would rest his eyes on the wonderful garden and beyond it, "five or six roofs, the façade of a big building with shining window panes, and a bit further the top of a church bell tower" (ibid, pp. 52-3). Even though I cannot enter the home where the writer once lived, I cannot leave without walking through the streets where he once walked, silent streets, which time has not been able to wrinkle.

I learn that quite close there used to be the house of another great man of the city — writer, intellectual and patriot Luigj Gurakuqi, who had penned the sweetest lines to this home:

My mind flies to you, oh my dear home

When my heart longs for you.

It's there I first breathed

It's there I first opened my eyes and had my first joys.

I left my memories with you

In exile I long for you … ("Shpis s'ime" [My Home],1905, my translation)

I leave for my next destination: Pashko Vasa's house. The first impact is that of the door—a grand heavy one as in old times. Behind it: a wonderful garden surrounded by a stone wall. In a corner, a well rises over a mass of hydrangeas. Couldn't wish for a better welcome. I move towards the house, fascinated to walk through an Ottoman-style veranda (Alb. *çardak*), a rare feature survived from the past.

I need to climb a few more steps to come across one of the typical Albanian unpleasant surprises—the house, a rarity of its type in terms of architecture and culture (being a writer's home), has been turned into a hotel! Disappointment gives way to a light burst of joy when I realise that at least it has been maintained quite well and still has authentic style and objects. Such a treat for the nostalgic! I walk over the threshold and get lost in thoughts …

I leave Vasa and head towards another transformed house—that of Filip Shiroka, the poet who begged a swallow to send regards to his home while in exile:

Be off, swallow

Convey my greetings

To the old house where I was born

And greet the lands around it

Where I spent my early years

Be off thither on your flight

And greet my native town! (translated by R. Elsie)

What was once a writer's house is now a restaurant. The only thing to link it with the writer is a plaque on the wall.

Not far from there, I cannot help gazing at another writer's house. All I can do is see it from the street hoping for it to get back the status it deserves, some day. Migjeni's house no longer belongs to the writer…

But I want a happy end for the day, so I try to replace the last image with lines—the only ones that no natural disaster, nor system, nor human activity can destroy:

The sun softly lies down

On the mountain edge at dusk

Leaving all its colours behind

And brightly dressing up

The golden daffodils, violets and rose petals.

Rozafat carves its stones

While the April moon appears like an eyebrow

In that place where deities reside. (Koliqi, "Shkodra in the Evening" in Gjurmët e Stinve, my translation)

Day 3

The last and most intellectual part of this literary tour in my hometown is that of visiting the schools where some of the greatest writers studied. Scholars have often emphasised the role that religious (mainly Roman Catholic) schools and colleges have played in the cultural and social life of the city.

The Jesuits first came to Shkodër in 1841 but had to leave soon as they were driven out by the Muslim fanatics. They would only return a few years later with the support of the Austrian-Hungarian government (Bushati, 1998, p. 602-4). Their arrival soon proved to be very beneficial to the city's life—a musical band was created (1870) as well as an orchestra, dramatic art appeared (1879), several social, cultural and sports societies were established, botanical and ethnographic museums were opened.

Of great importance was the establishment of the so-called Saverian College (now At' Pjetër Mëshkalla High School) in 1877. Most of its first well-selected teachers had studied abroad: Anton Xanoni, Ndre Mjeda, Mati Logoreci, Pjetër Marubi, etc. The school had excellent facilities for the time – different labs and a library with rare books and first editions. It is believed that it shelved over 1000 Albanian titles and 300 foreign ones (ibid, p. 100).

It also boasted some excellent students who would soon become some of the leading intellectuals of the city: Lazër Mjeda, Faik Konica, Luigj Gurakuqi, Lazër Shantoja, Anton Harapi and many more. Ndre Mjeda is often considered as the most significant figure of this school (Koliqi, 2003).

The second school is that of the Franciscans. The Franciscan Order can be traced back to the 13th century, but their influence became greater in the late 19th – early 20th century. Like the Jesuits, the Franciscans contributed a lot to the social and cultural life of the time. They collected, studied and published oral literature. Also, they translated into Albanian some of the best-known world masterpieces and published literary and cultural magazines such as "Hylli i Dritës". Their library was considered to be the best in the whole city with 30.000 works, plus magazines and newspapers. In addition, both schools also had their own museums (Bushati, 1998, pp. 64-5).

Gjergj Fishta is regarded as the best representative of the Franciscan literary school, but other well-known writers also studied there, such as Vinçens Prendushi and Zef Pllumi, as well as a number of religious leaders, scholars and linguists like Shtjefën Gjeçovi, Bernardin Palaj, Donat Kurti, Marin Sirdani, Justin Rrota, etc.

I end my literary tour with mixed feelings—I am excited, happy, surprised, but at the same time I cannot help feeling disappointed and sorry for what has been lost, mainly due to negligence. Yet, I keep dreaming …

A voice says:

Live! Don't just close your eyes and dream.

The voice says: Sing to life

Sing to truth!

Yet, Shkodra cannot help but dream.

And a hope kindles in me

As in your heart.

This is how in the midst of despair dreams appear.

Because I and you, Shkoder, cannot help but dream. (Gaspër Pali, as cited by Stefa, p. 100, my translation)

DISCUSSION

The above account provided a simple model literary itinerary in Shkodra. It highlighted a few places with literary associations as well as pointed out problems such as the inaccessible houses of a few writers. The analysis below explains in more detail why Shkodra has potential for the development of literary tourism, suggests a few strategies that could be developed, discusses the benefits associated with such development, and also tackles the problems and challenges related to such development,

So far, no place in Albania has an official status associated to literature, even though some indirect use of literature is made as is the case of Kadare's house or Byron's trail (see Erkoçi, 2016; Erkoçi, 2013). Research is little or lacking and no records are kept concerning the influence of literature in shaping people's perceptions of place and motivating them to visit or the extent to which a literary piece affects one's decision concerning the choice of a tourist destination. Similarly, no on-site research has ever been conducted to explore if the visit changes the reader-tourist's initial perceptions.

Shkodra is the third literary trail the author of this paper suggests after Kadare's house and hometown (Gjirokaster) and Byron's trail in Southern Albania. The choice of this city was made because the researcher believes that it possesses what Herbert (2001) calls general and exceptional qualities. The city is an important urban and cultural hub in the country (general quality). It is the home or adopted home to several major writers and it is the setting of several important literary pieces (exceptional qualities). In addition, places associated with literature happen to be in significant city areas such as near the historic centre (writers' houses and the schools where they studied) or landscapes described in the works such as neighbourhoods, the castle, the lake, and so on, which are often typical tourist stops. Hoppen et al. (2014) believe that the more exceptional qualities a literary place possesses, the easier it is to develop and use it for tourism purposes.

Shkodra's literary heritage dating back centuries ago, could be of interest to both residents who "read the book through images" which they see every day, as well as those who don't know the city, who "read" it through their imagination Reijnders (2015). As Capdepón (2014) puts it, "whereas visitors observe, locals experience" (408).

However, the role that literature as a cultural component could play has never been seriously considered. In Shkodra (and in the whole country) this aspect is under-researched and underexploited. As the autoethnographic account revealed, not much is done to promote the literary heritage of the city. Even though a few literary activities are organised as part of the cultural calendar in the city, they lack visibility and the necessary distinctiveness to attract more people. In addition, the writers' houses and their personal belongings are not accessible to the public. Another drawback is that very little has been translated into other languages, which means that there is no real drive for foreign visitors. If literature were exploited in more diversified and interesting ways, the benefits would be multiple. Below are a few strategies suggested by the author and the benefits of their implementation.

Developing Literary Tourism in Shkodra: A Few Suggestions

It has already been mentioned that the potential of literature in Albania and especially in Shkodra has been overlooked. One of the benefits of developing literary tourism here could be its contribution in facilitating the shift from mass to niche tourism for those no longer attracted by overcrowded places, "looking for tailor-made products that provide a unique experience in an authentic environment" (Munster, 2021, p. 56) and information "to satisfy their cultural needs" (Richards, 1996) as shown by the above autoethnographic account.

In Shkodra, the current situation could be improved by redesigning existing tourism offerings. To begin with, what is needed is *a new perspective* of the relationship between culture and tourism, especially one which involves intangible culture more actively. A *diversification* of cultural offerings is necessary to meet the demands of an increasing class of emancipated tourists in search of new, authentic experiences, but also arising more interest among those who are not directly interested in a cultural experience. In each case, a variety of forms need to be provided by the local authorities and tour operators so as to make destinations more appealing. In order to achieve this, Delgrado (2001) believes that "the new cultural tourism has to base itself, above all, on the offering of experiences with three basic elements: diversity, interactivity and context" (as cited in Richards, 2014, p. 3).

In order to meet such criteria, *promotion* is of vital importance. Cultural tourism scholars like Caca et al. (2016), list cultural heritage as an important aspect to be considered with concern to Albanian tourism. Similarly, Marku (2014) states that most cultural resources in the country are "of very high quality and would likely welcome high numbers of tourists if they were more widely publicized within target populations." As such, it is essential to make literature a permanent component of the tourist offer in the city and disseminate the city's literary heritage as a way of boosting interest in culture and literature. This would be the first step in increasing its visibility.

Next, *distinctiveness*. If it is to be competitive, cultural tourism in Shkodra needs to offer exceptional, authentic and creative experiences and literature is a great tool to help achieve this. It could be used in a number of ways besides the traditional ones. Events are indispensable as they "add value to the attractions because they bring them to life" (Munsters, 2021). For example, the responsible people could organise activities like "Choose a place and write" which would encourage the community to get involved and see the city with different eyes. Another strategy could be an open invitation for a commissioned book set in Shkodra which could then turn into a literary guide. Literary puzzles and quizzes could also be used to involve and raise interest among schoolchildren.

Involving both domestic and foreign visitors in creative workshops, for instance, would provide for what is lacking in tangible heritage (such as writers' houses in the case of Shkodra). Literature could also be used to enable the development of creative activities of a performative kind which would actively engage visitors and provide ground for the display or development of (new) individual skills (Ray, 1998 as cited in Al-Ababneh, 2019, p. 112). This could in addition, lead to more creative activities "following the advantages gained from knowledge spillover and networking."

As an element of the intangible culture, *storytelling* would be a great way to introduce a glimpse of a variety of cultural aspects as "stories have always played significant roles in prescribing behavior, establishing meaning and constituting identity", in every culture (Godsil & Goodale, 2013, as cited in Vitić–Ćetković et al., 2020, p. 86). In the case of Albania, there is interest in this aspect as shown by the multiple views received by the "Myths and Legends" section in this paper author's personal website "Mapping Literary Albania." As such, storytelling, could play a role as a strategy for branding, innovation, sustainability as well as visitor engagement ("Heritage, Tourism and Hospitality", 2018, as cited in Vitić–Ćetković et al., 2020, p. 88).

Storytelling also contributes in enhancing the visitor's experience by involving them through inviting them to actively participate. This, in turn, would add to the uniqueness of the experience by having the visitor perform, not just consume a service and as such develop their own creative potential, leading to memorable experiences. Guides and actors would also help make the experience even more unique (Edinburgh or Dublin would be great models to follow in this case).

Another positive aspect of employing this specific tool is that visitors are likely to pass these stories on to other people as "our desire for unique experiences is tempered by our need to tell others about them" (Richards, 2004), hence encouraging more place-related visits.

In addition to involving tourists creatively, Richards (2011) highlights the fact that the destination itself needs *transformation* by becoming "more creative in designing characteristic experiences linked to place which give tourists a specific motivation to visit." Literature is a great tool to use for redeveloping culture by diversifying it and as such, giving a fresh breath to a destination associated with it while also contributing to local economic growth. Óbidos in Portugal would be an excellent example to follow. A small medieval town (often called 'village') of only 3800 inhabitants, this once wedding present for queens, managed to transform into one of the most appealing destinations thanks to a literary project started in 2011 aiming at reviving life and revitalizing the economy there. In 2012, the largest Portuguese bookstore was opened in a church in Óbidos, followed by 11 more as well as a literary hotel. This transformation from a historical village into a major cultural and literary centre, gained Óbidos the title of UNESCO City of Literature in 2015. Such a successful case is just a model Shkodra could follow.

Another way of providing some diversification and transformation in the city would be, for instance, by adding signs with quotes which could be placed in key city spots with literary associations, ideally translated into English. A great example of this is, for instance, "Stars and Stories" in Edinburgh. Also, for increased effectiveness, more books should be translated into other languages to attract foreign visitors in addition to the domestic ones.

Currently, cultural projects that merge Shkodra's literary context with its natural landscapes or built environments are lacking. As such, it could be a good idea to project passages of stories on important buildings (following Edinburgh's model of "Message from the Sky", for instance), which would not only keep visitors in suspense and motivate them to go to the next destination following the hints, but it would also help showcase less known, but important areas in the city. Several sites could be connected through a literary trail following a certain itinerary, which would further enhance the existing cultural attractions as well as add new ones with a literary focus. And this is related to the next point of this paper, employment of technology.

Digital technology is a great instrument to attract visitors and bring literature to their attention. Some Australian readers, for example, who traced a literary trail developed by an iPhone application reported that they enjoyed feeling part of the story (Carson et al., 2017, p. 388). Technology can be used from showcasing literature on the social media to creating literary maps, podcasts, and so much more. It could be used playfully for example by creating holograms which enable having imaginary conversations with writers (Martino Alba as cited in Quinteiro and Baleiro, 2017). Another way would be creating e-novels which encourage readers to visit a city by giving them clues. An example is that of Japanese writer Higashi Moriyama who has partnered with the city of Kyoto, Japan to write a novel which will draw visitors to the city (Wisegeek).

Integration of the *local library* could also be a good idea. A great example of this is Hefei Children's Library in China where a special tourism shelf called "Poems and Futures" has been set up. There, one can find guides for both domestic and foreign travel destinations as well as travel notes and literary works associated with certain destinations (see Li & Liu, 2019). We could do the same in Shkodra by creating a themed section in the library to meet people's needs for specialized knowledge and information related to literary tourism. It could also turn into a shared space, a meeting spot for tourists and locals. The existing library in Shkodra can hardly turn into an attraction per se due to its boring architecture (unlike many other counterparts worldwide), but at least it can serve as a cultural tourism hub, expanding the social impact of libraries.

Literary tourism is also a great form of creative tourism as "travel directed towards an engaged and authentic experience, with participative learning in the arts, heritage, or special character of a place"

(UNESCO 2006). Thus, the image of a place includes many aspects crucial to its identity and character, and literature has always played a great role in *reflecting the identity* of Shkodra and its people. As such, it is essential to intertwine local identity with the local heritage also as a way of *connecting* visitors with residents. This is why the author claims that better use of the cultural component with a focus on literature should be made as it could be a factor influencing future visitors in choosing Shkodra as a destination. Literature offers a chance to immerse oneself into the culture of the other, and as such foster intercultural understanding.

Involving the community and *capitalising on local resources* could have a number of benefits. Ciro and Toska (2018, pp. 85-6) believe that if endogenous development potential is used efficiently, it can help foster local identity and self-identification of the locals with their own territory and local resources. They suggest that this can be achieved by introducing alternative forms of tourism. This in turn would contribute to local economic development, by "lifting the development pressure off of the coastal area and shifting the focus towards inland, less explored and promoted destinations". Another benefit would be more attention and investments towards local resources, which could help reduce migration and poverty, proving a self-sustainable sector for the local economy. The researchers believe that the important thing is to get started as small activities with short term effects, which can later trigger more important, long-term development processes. The local government has a responsibility in encouraging such practices by providing the necessary strategic framework. However, in order to achieve literary tourism goals and to foster sustainable and responsible tourism development, management and marketing, it is crucial to connect the academia to the tourism industry. For a better development of the literary offer, several stakeholders should be involved (not just tourism authorities but also small businesses and ordinary people).

LIMITATIONS AND FUTURE RESEARCH DIRECTIONS

The current study is anything but exhaustive. It marks just the first step in exploring the potential of literary Shkodra and suggesting ways of making literature an integral part of the city's cultural offer. Further research is needed with regard to literature as related to concepts like authenticity and sustainability as well as to better understand the market demand for literary tourism.

While autoethnography was used as an innovative approach, it is limited in that it may be somehow subjective as the researcher's experience may be may vary from other people's perceptions and understanding.

CONCLUSION

This paper made use of autoethnography as a way of collecting data regarding the literary potential of the city of Shkodra in Northern Albania. Research showed that the cultural tourism offer in Albania is still relatively traditional and it typically depends on tangible heritage. As such, there is a need for innovation. Intangible culture needs a push so as not to be overshadowed by the mass coastal tourism. This paper suggested that heritage tourism including literature can help support current tourism activities as well as enhance the development of new ones. This would also be in line with the 2019-2023 Albanian National Strategy for Sustainable Development mission for the diversification of the tourist offer and the development of new tourist products and services with a focus on the promotion of local products and fighting seasonality by proving a year-round tourist offer.

Literary tourism as an alternative form of tourism broadens destination choice beyond the mainstream ones. While its natural attractions are the main drive for tourists in Shkoder, literature could act as a pull factor to promote the city's rich intangible culture. Constructing special literary trails and literature spaces could further enhance the tourism appeal of Shkodra as a destination. It would attract not just visitors interested in culture and more specifically literature, but most probably also 'convert' some or arise curiosity in those who were not in a specific place for its literary appeal.

The intangible forms of culture are often the ones most closely linked with the concept of identity. As such, highlighting literature would not only market a somehow marginalised element of the Albanian culture but also keep local culture and identity alive, and revive aspects which have not been given due importance during the last few years, such as storytelling or public reading.

Shkodra can capitalise on the use of literature to offer innovative and creative ways of animating tourists at the cultural tourism destination. This would foster the creation of unique visitor experiences through participation and performance in events such as themed writing or storytelling, and enhance a better connection between readers and writers, increasing interest in literature. In addition, special stories would facilitate the distinctiveness of the branding of the city as a destination.

The emergence of a variety of cultural forms including literature could help increase interest in local literature and reading in general. This would particularly have educational benefits for school-aged visitors, who should be turned into a target group. Literary events, for example could be used because of their informative and educational value as well as the potential to provide entertainment. As such, engagement with literature would enable learning as well as a fuller experience of place.

Studies have shown that literature could also help showcase less visited city areas/architecture. The use of literature could be used to valorise the city and highlight places which would be otherwise inaccessible.

Furthermore, considering literature as a normal practice within cultural tourism would add value to tourist experience by making it more meaningful and authentic. The more dynamic and interactive it is, the more impression it will make and live on in the visitor's memory.

Giving more priority to literature as an important element of culture would also contribute to economic development as showed by a few studies mentioned above. A positive result related to this would be increased job opportunities, especially for small businesses like those involving artisans.

Through careful planning and management, Shkodra should be able to create a distinctive literary image among other cultural forms. Making literature an integral component of the city's cultural map would lead to sustainable development and better conservation, as well as help revive the literary identity of Shkodra, and position the city in the list of must-see destinations in the country and the whole Balkan region.

REFERENCES

Akademia e Shkencave e Shqipërisë. (2009). *Fjalor Enciklopedik Shqiptar* [Albanian Encyclopedic Dictionary]. III. Author.

Al-Ababneh, M., & Masadeh, M. (2019). Creative cultural tourism as a new model for cultural tourism. *Journal of Tourism Management Research*, *6*(2), 109–118. doi:10.18488/journal.31.2019.62.109.118

Bushati, H. (1998). *Shkodra dhe Motet*. Rozafat.

Caca, E., Ladias, Ch., & Polo, A. (2016). The Development of tourism in Albania and the importance of cultural tourism. *Tourism and Culture in the Age of Innovation*, 109-119.

Camaj, M. (1978). *There before the tribes arrived.* http://www.albanianliterature.net/authors/modern/camaj/camaj_poetry.html

Capdepón, F. P. (2014). The literary landscape and its connection with cultural tourism. *Cuadernos de Turismo*, *33*, 407–409.

Carson, S., Hawkes, L., Gislason, K., & Cantrell, K. (2016). Literature, tourism and the city: Writing and cultural change. *Journal of Tourism and Cultural Change*, *15*(4), 380–392. doi:10.1080/14766825.2016.1165237

Carvalho, I., Baptista, M. M., & Costa, C. (2012). Binding culture and tourism through literature: The potential of literary tourism. *International Journal of Hospitality and Tourism*, *2*(2), 1–27.

Ciro, A., & Toska, M. (2018). Sustainable tourism development in Albania. *Annual Review of Territorial Governance in Albania*, *I*, 84–93. doi:10.32034/CP-TGAR-I01-06

Ellis, C. (1997). Evocative autoethnography: Writing emotionally about our lives. In W. G. Tierney & Y. S. Lincoln (Eds.), *Representation and the Text: Re-Framing the Narrative Voice* (pp. 116–139). State University of New York Press.

Ellis, C., & Bochner, A. (2006). Analyzing analytic autoethnography: An autopsy. *Journal of Contemporary Ethnography*, *35*(4), 429–449. doi:10.1177/0891241606286979

Elsie, R. (2004). *Tales from Old Shkodra*. Dukagjini.

Elsie, R. (2005). *Albanian Literature: A Short History*. Palgrave.

Erkoçi. I. (2013). A modern-day journey in Byron's footsteps in Albania. *Academic Journal of Interdisciplinary Studies, 2*(4), 97-106.

Erkoçi, I. (2016). Ismail Kadare as tourist attraction. *DosAlgarves. A Multidisciplinary e-Journal, 28*, 1-12.

Erkoçi, I. (2017). *Stats and Insights*. Mapping Literary Albania. https://wordpress.com/stats/day/mappingliteraryalbania.wordpress.com?startDate=2021-09-19

Ferreira, A., Alén, E., Liberato, P., & Liberato, D. (2019). Literary tourism: A cultural trip? *Advances in Tourism. Technology and Smart Systems*, *171*, 505–515. doi:10.1007/978-981-15-2024-2_44

Ghetau, L. & Esanu, L.V. (2011). Literary tourism as a promoter of cultural heritage. *Sustainable Development and Planning, 150.*

Harapi, A. (1995). *Valë mbi valë* [Wave after Wave]. Academic Press.

Harwood, S., & Eaves, S. (2017). In search of an autoethnographic method. *16th European Conference on Research Methodology for Business and Management Studies*, 145-153, https://www.pure.ed.ac.uk/ws/portalfiles/portal/33004807/ECRM_FINAL_REVdoc x.pdf

Herbert, D. (2001). Literary places, tourism and the heritage experience. *Annals of Tourism Research, 28*(2), 312–333. doi:10.1016/S0160-7383(00)00048-7

Holman Jones, S. (2013). Introduction: Coming to know autoethnography as more than a method. In Handbook of Autoethnography. Left Coast Press.

Hoppen, A., Brown, L., & Fyall, A. (2014). Literary tourism: Opportunities and challenges for the marketing and branding of destinations? *Journal of Destination Marketing & Management, 3*(1), 37–47. doi:10.1016/j.jdmm.2013.12.009

Jenkins, I., & Lund, K. A. (2019). *Literary Tourism: Theories, Practice and Case Studies*. CABI. doi:10.1079/9781786394590.0000

Kadare, I. (1991). *Ardhja e Migjenit në Letërsinë Shqipe* [The Emergence of Migjeni in the Albanian Literature]. 8 Nëntori.

Koliqi, E. (1933). *Gjurmat e Stinve* [The Traces Left by the Seasons]. Academic Press.

Koliqi, E. (2001). *Pasqyrat e Narcizit*. Camaj-Pipa.

Koliqi, E. (2003). *Vepra, V*. Faik Konica.

Kombëtare e Turizmit, A. (n.d.). *Turizmi në shifra për vitin 2018* [Tourism in Albania in 2018]. https://akt.gov.al/wp-content/uploads/2019/06/STATISTIKA-TURIZMI-2018.pdf

Kulla, N. (2010). Shkodra, Mjeda dhe Koliqi. *Gazeta Shekulli*.

Li, Y., & Liu, X. (2019). "Library + Tourism" A new direction for the sustainable development of libraries. *Recruiting and Managing the New Generation of Employees to Attract New Markets and Create New Services*. http://library.ifla.org/id/eprint/2703/1/S09-2019-li-en.pdf

Message from the Skies. (n.d.). *The Guardian*. https://www.theguardian.com/stage/2020/jan/03/message-from-the-skies-shorelines-review-edinburgh-venues

Ministria e Turizmit dhe Mjedisit. (2018). *Strategjia Kombëtare për Zhvillimin e Qëndrueshëm të Turizmit 2018-2022* [National Strategy for Sustainable Tourism Development 2018-2022]. https://www.turizmi.gov.al/wp-content/uploads/2018/02/Strategjia-e-Turizmit-18-22-draft-1.pdf

Ministria e Turizmit dhe Mjedisit. (n.d.). *Turizmi në një vështrim* [Tourism at a Glance]. https://turizmi.gov.al/turizmi-at-a-glance/

Munsters, W. (2021). *The Study of Cultural Tourism in Ten Models*. https://www.academia.edu /48862966/The_Study_of_Cultural_Tourism_in_Ten_Models_Wil_Munsters

Obidos. (n.d.). https://www.cityofliterature.nl/en/cities/obidos/

OECD. (2008). Roles and impact of culture and tourism on attractiveness. In The Impact of Culture on Tourism. OECD.

Quinteiro, S., & Baleiro, R. (2017). *Literature and Tourism: Fundamental Concepts* (2nd ed.). University of Lisbon.

Reijnders, S. (2015). Stories that move: Fiction, imagination, tourism. *European Journal of Cultural Studies, 19*(6), 672–689. doi:10.1177/1367549415597922

Richards, G. (1996). *Cultural Tourism in Europe*. CABI.

Richards, G. (2004). *Towards a sustainable form of cultural tourism?* https://www.academia.edu /2006332/ Towards_a_sustainable_form_of_cultural_tourism

Richards, G. (2011). Creativity and tourism: The state of the art. *Annals of Tourism Research, 38*(4), 1225–1253. doi:10.1016/j.annals.2011.07.008

Richards, G. (2014). *Tourism trends: The convergence of culture and tourism*. https://www. academia. edu/9491857

Shkodra. (n.d.). *Qendra e Studimeve Shqiptare "Ernest Koliqi" Shkodër & Bashkimi Katolik i Publicistëve Shqiptarë*. http://kulturserver-hamburg.de/home/shkodra/shkodra/labeat.html

Squire, S. J. (1994). The Cultural Values of Literary Tourism. *Annals of Tourism Research, 21*(1), 103–120. doi:10.1016/0160-7383(94)90007-8

Starr, L. J. (2010). The use of autoethnography in educational research: Locating who we are in what we do. *Canadian Journal for New Scholars in Education, 3*(1), 1–9.

Stars and Stories. (n.d.). *List*. https://www.list.co.uk/event/657195-canongate-stars-and-stories/

Stefa, G. (2011). *Oazi i Artë i Gjuhadolit* [Gjuhadol's Golden Oasis]. Gjergj Fishta. Wisegeek. What is Literary Tourism? https://www.wise-geek.com/what-is-literary-tourism.htm

Vitić–Ćetković, A., Jovanović, I., & Potočnik Topler, J. (2020). Literary tourism: The role of Russian 19th century travel literature in the positioning of the smallest European capital – Cetinje. *Annales. Series Historia et Sociologia, 30*(1), 81–98.

ADDITIONAL READING

Arcos-Pumarola. (2018). Literary urban landscape in a sustainable tourism context. *Human Geographies, 12*(2), 175–189. doi:10.5719/hgeo.2018.122.3

Gentile, R., & Brown, L. (2015). A life as a work of art: Literary tourists' motivations and experiences at Il Vittoriale degli Italiani. *European Journal of Tourism, Hospitality and Recreation, 6*(2), 25–47.

Mansfield, C. (2015). *Researching Literary Tourism*. TKT.

Robinson, M., & Andersen, H. C. (Eds.). (2003). *Literature and Tourism: Essays in the Reading and Writing of Tourism*. Thomson Learning.

Smith, Y. (2012). *Literary Tourism as a Developing Genre: South Africa's Potential*. University of Pretoria.

Sparkes, A. C. (2000). Autoethnography and narratives of self: Reflections on criteria in action. *Sociology of Sport Journal, 17*(1), 21–43. doi:10.1123sj.17.1.21

KEY TERMS AND DEFINITIONS

AKT: Agjencia Kombëtare e Turizmit - Albanian National Agency of Tourism.

Autoethnography: A research instrument for collecting primary data based on personal experience ("auto") which is used as a starting point to reflect, describe and interpret ("graphy") social and cultural practices, experiences, and beliefs ("ethno").

INSTAT: Instituti i Statistikave të Shqipërisë - The Albanian Institute of Statistics.

Literary Tourism: A type of cultural tourism which has developed into an important tourism niche during the last years, literary tourism is travel related to literature, typically, either to places linked with a specific author (hometown, house, grave) or a book (setting, a trail in the footsteps of characters, etc), but it can also take a number of other forms.

MTM: Ministria e Mjedisit dhe Turizmit - The Albanian Ministry of Tourism and Environment.

OECD: The Organisation for Economic Co-operation and Development. The main focus of this international organisation is to develop policies aiming at well-being, equality, and opportunity for everyone.

Shkodra: A city in Northern Albania, often referred to as the cultural capital of the country 1A literary map of the city, made by the author of this paper, can be accessed here: https://www.google.com/maps/d/u/0/viewer?mid=1urDuR0SCKOFG2Ha6MfK1opCj8SE&ll=42.06657408949859%2C19.51263043371592&z=15.

Chapter 14
Promoting the Development of Local Communities Through "Touripilgrimage": The Portuguese Inner Way of Santiago de Compostela

Raphael Campana Marinho
University of Trás-os-Montes and Alto Douro, Portugal

Pedro Azevedo
University of Trás-os-Montes and Alto Douro, Portugal

Xerardo Pereiro
University of Trás-os-Montes and Alto Douro, Portugal

ABSTRACT

This chapter offers a "touripilgrimage" that takes place on the Portuguese Way of the Interior of Santiago de Compostela–PIWSC and how such a phenomenon can add new and different experiences to the "turipilgrims" while helping to promote the development of the local communities inserted along the way. In accordance, firstly, there is the discussion of a theoretical framework associated to the role of local identity in the sustainable development of tourism. Additionally, there is debate about the transformations in the act of pilgrimage. Secondly, it follows a case study approach based on the PIWSC. The results of this exploratory analysis highlight elements that enhance the possibilities of increasing exchanges between locals and visitors in order to insert the visitor into the local context and the community to develop through the convergence of its endogenous attributes. The dialogue between the different local actors is essential to establish guiding strategies that will compose the dynamics of the communities.

DOI: 10.4018/978-1-7998-9217-5.ch014

INTRODUCTION

This chapter aims at investigating the "touripilgrimage" that occurs in the Portuguese Inner Way of Santiago de Compostela – PIWSC, and how this phenomenon can add new and different experiences to the "touripilgrims" while helping to promote the development of local communities inserted along the way.

It is noted that the development referred to here, is not centered on economic development or in function of it, but comprises a broader form. Great emphasis should be placed on improving the quality and intensity of social exchanges both between resident-residents and resident-visitors.

In this way, a great opportunity arises for both sides. The community can structure itself to offer the "touripilgrims" gastronomic, cultural, landscape experiences, among others, adding its diverse competitive advantages in order to promote the increase of interactions between locals and visitors and, thus, enhance the development of their endogenous capacities. In turn, the touripilgrim has the opportunity to choose new experiences and make a more immersive and genuine visit.

The specific objectives of this research are to characterize the profile of the PIWSC touripilgrims, present the endogenous attributes of this route, and highlight the potential of touripilgrimage in contributing to the development of local communities. To fulfil these objectives, the structure of the chapter includes the following sections: theoretical framework, methodology, results and conclusion.

In the first section, the theoretical framework there is debate about the transformations in the act of pilgrimage, then, discusses the role of local identity in the sustainable development of tourism. Subsequently, PIWSC is presented describing some of its main characteristics and attributes capable of contributing to promote the transformation of the local community along the way.

The second section presents the methodological elements. Chapter is based on certain empirical materials, various research-studies documents, semi-structured interviews and ethnographic field diary.

Regarding the discussion of the results, in the third section, the concern is to characterize the profile of the PIWSC "touripilgrims" and analyze the potential that exists within the PIWSC to raise the level of social exchanges between locals and visitors, in order to insert the visitor into the context local, and the community to develop through the convergence of its endogenous attributes.

Finally, the conclusion shows that the great challenge is to carry out a management that, among the various local actors, shares the feeling of valuing local attributes in favor of the residents themselves. In addition, for the desired development to reach the PIWSC, it is necessary to promote and develop it as a whole, in an integrative way, thus surpassing isolated or punctual actions in a given location.

THEORETICAL FRAMEWORK

The theoretical framework begins with a review of the transformation in the connotation of the act of pilgrimage over time. This situation points to an approximation of what was previously an essentially religious act towards a slightly more "profane" side of current society, being used in different contexts and approaches, such as in tourism.

Then, the emphasis is on the question of the perception of local values and identities in a community. How to value and encourage the development of these virtues? How do such actions provide the possibility of developing sustainable tourism?

Finally, we present a small review of the importance of the Saint James Way as a tourist-cultural route and detail some of the characteristics that make the PIWSC a unique Jacobean route.

Transformations in the Act of Pilgrimage

The act of pilgrimage is intrinsic to human nature because pilgrimage is associated with human evolution itself (Cardita, 2012). However, it was from monotheistic religions that pilgrimages had a great impulse, especially with Christianity through its thousands of saints (Santos, 2000). Added to this, the fact that the Catholic Church relates pilgrimages to penances, making pilgrimages a way for the faithful to "pay for their sins" and be led to a new life, cleansed and purified.

However, the disciplinary diversity through which pilgrimages are studied helps to demonstrate the evolution of the act itself and also the way in which the pilgrimage relates to current society. For example, with recent studies in different areas: anthropology (Belhassen, Caton, & Stewart, 2008), geography (Collins-Kreiner & Wall, 2015), and religious studies (Vilaça, 2010). The pilgrims' motivations stand out as the focus of the work (Amaro, Antunes & Henriques, 2018; Damari & Mansfeld, 2014; Liutikas, 2017), as well as the pilgrim's profile (Guilarte & González, 2016; Gusmán, Lopes, Lois González & Santos, 2017; Parga-Dans, 2012) and the touristification of modern pilgrimage (Blom, Nilsson & Santos, 2016; Jensen & Lovrod, 2017; Nilsson, 2018).

The current motivations that have led to the growing number of modern pilgrims differ from traditional pilgrimage practices (Kim, Kim & King, 2016). The motivation of pilgrims linked to a religious purpose continues to exist, but they are not limited to devotional visits to sacred shrines (Courtney, 2013). However, studies show that the number of pilgrims who walk the paths for other reasons is increasing (Oviedo, Courcier & Farias, 2014). Among these various reasons, the following stand out: cultural immersion, a spiritual adventure or to get away from the hustle of modern life and get to know oneself (Carbone, Corinto & Malek, 2016; Pereiro, 2017). This difference in motivation can be seen through the data from The Pilgrim's Reception Office in Santiago de Compostela. In Data for the years 2011 and 2019 point to an increase in the number of pilgrims with non-religious motivations, as show in Table 1.

The decrease in the number of pilgrims in 2020 was mainly due to COVID-19. The months of May and June were not considered due to the fact that the Pilgrim's Reception Office does not provide data for these months.

In this sense, it can be said that when starting their journey, each pilgrim is looking for a personal expression of their own spirituality. For this reason, both the motivation to make the pilgrimage and the impact it has on the pilgrim's personal life can be very heterogeneous (Ivakhiv's, 2003).

Given the appropriation of pilgrimage mainly for the tourist aspect, religious tourism can be considered as an element for the promotion of a region or a country contributing to a positive transformative impact on the community (Kouchi, Nezhad, & Kiani, 2016; Sánchez, Fernández, & Lara, 2016).

The modern pilgrimage is made through the contact with nature, the appreciation of the landscapes, the preference for typical gastronomy and the emphasis on cultural elements. Thus, it is necessary to manage these circuits, in order to promote the social, cultural and economic benefits that they can leverage for the affected territories, highlighting the creation of a coherent, authentic and sustainable cultural tourism experience (Carbone et al., 2016; Dimitrovski, Ioannides, & Nikolaou, 2020).

It is noted the proximity between tourism and pilgrimage in the current scenario, therefore, this research adopts the term "touripilgrims" to represent the individual who performs the pilgrimage (Pereiro, 2018). At the same time that one is a pilgrim, one is also a tourist. And many times, he alternates between pilgrim/tourist during his own walk.

Table 1. Motivation for the pilgrimage – Saint James Way

Year	Motivations	Pilgrims	%
2011	Religious and others	93,147	50.80
	Religious	78,969	43.07
	Non-religious	11,250	6.14
	Total	183,366	100.00
2019	Religious and others	169,314	48.71
	Religious	140,110	40.31
	Non-religious	38,154	10.98
	Total	347,578	100.00
2020	Religious and others	24,788	45.78
	Religious	16,976	31.35
	Non-religious	12,379	22.86
	Total	54,143	100.00

(Data Source: The Pilgrim's Reception Office in Santiago de Compostela, 2011-2020)

Local Identity and Sustainable Tourism

A community is neither monolithic nor unique from the standpoint of identity. It finds itself in a constant construction and deconstruction through a group of people with different interests, narratives and practices. The community has social mechanisms of internal and external differentiation and limits that can produce a certain sufficiency and community sovereignty, being composed of several subgroups that dynamically interrelate and represent the different focuses of power. The community has social relationships and hierarchies that can lead to conflicts and disaffection, thus causing changes. Changes can be developed in different ways: with or without tensions, threats and exclusions (Torres, 2019).

A community can be defined according to different perspectives, namely in the fields of anthropology and social sciences, as can be analyzed in Table 2.

At the base of a community are its individuals. The choice of the most important values for the individual comes from the individual, attributing a meaning of values and a hierarchy of values through the individual's subjective attitudes. Such values directly affect an individual's behavior because people tend to find actions more attractive and consistent with their own values. The main source of values formation is the social and cultural environment, thus including family, friends, professional colleagues, the community and society as a whole (Feather, 1995; Roccas & Elster, 2014).

Identification with a group and the emotional value of being a fraternity member in a group is an integral part of the individual and his or her sense of identity. Some affiliations are especially important for self-concept and can provide the senses of self-esteem, stability, or security. The exchange of experiences, the feeling of belonging and acceptance, solidarity in sharing common values and goals in life are part of the construction of the individual, and therefore of the community. Socialization is the process of acquiring a social identity. People learn specific social and cultural norms and religious beliefs. Rituals, parties and festivals, as well as lectures and sharing of experiences, help in the process of socialization

Table 2. Perspectives and theoretical concepts of communities

Perspectives	Theoretical Concepts and Authors
a) Essentialist;	Meaningful and affective places (Lewicka, et al., 2019);
b) Primordialist;	Local communities and identities are fixed, natural and ancient (Viera,1998);
c) Psychological;	How communities relate and interact in society (Graves, 1967);
d) Dichotomic;	Distinctions between society / community (Giddens, & Giddens, 1973: 65).
e) Symbolic - representational;	The social construction of the community (Cohen, 2013);
f) Institutional,	Normative, where communities are guided by organizations (Bates, 1970), rules and values;
g) Political Ecology;	According to Eric Wolf, community is ecological, struggle for resources, inequalities, differences in social structures (1972);
h) Political Economy;	Spaces and power, conflicts over power and identity conflicts: Eric Wolf (1956) and Claude Meillasoux (1981).

Source: self-elaboration

as they are able to provide a strong orientation towards values (Tajfel, 1981). In this context, pilgrimage can serve as an instrument for socialization and he formation of a social identity.

The more developed a person's social ties, the more diverse their personal and social identity becomes. The more these relationships are developed with people who occupy a different position in the structure of society, the more different attitudes, beliefs and forms of behavior the person can develop. This difference in relationships also helps to determine the versatility of personal values, as well as the number of symbolic places that mark these values in space (Tajfel, 1981).

Cultural identities are defined from points of identification through culture and history where it is formed through the time lived (past and present) by a given society and are under constant construction and reconstruction within the temporal environment and influenced by individual memory and that community (Hall, 2002; Canclini, 2006). In this way, attributes and values contemplated today can be left aside by the community itself if it understands that the bases that support it have been modified.

As a community synthesizes itself around a custom or a tradition and produces an identity, it can be said that one of the aspects of local identity would be the sharing of the feeling of belonging and the collectivity in which everything that is shared by the that inhabit this place becomes interchangeable. In these territories, identity is celebrated at parties and dramatized also in daily rituals (Canclini, 2006).

In the long run, a locality will develop by virtue of its capacity for social and political organization to shape its own future. Therefore, it is necessary to directly relate to the availability of different forms of intangible capital in the region or locality (Boisier, 2007). Social capital is one of the main factors that explain regional differences and the level of development of regions. This is because social capital is composed of a set of factors of a cultural nature that increase the propensity of social actors to collaborate and to undertake collective actions (Bandeira, 1999). In this way, the high level of social capital streamlines situations and provides opportunities for finding solutions to community problems. In turn, the low level of social capital leads to a weak regional identity process.

It can be said that the historical evolution of tourism is centered on three phases. The first of it until the 1960s, in which tourism was limited to leisure activities with a small industrial dimension. Then, between the 1960s and 1990s, the movement of people began to be present in everyday life, causing mass tourism. At this time, it evolved into a commercial and industrial dimension of great geographic and economic scope. Finally, there is the current phase, in which tourism emerges as an opportunity to enhance the sustainable development of territories, adding cultural, experiential and creative tourism (Costa, 2014).

The difference between tourist destinations is beyond the services and leisure facilities they provide to the visitor. They often take place in the projection of the place as a unique place, whose experimentation will transform the visitor (Lopes & Pereira 2017).

The involvement of local people is critical for tourism to be successful in promoting development (Jarvis, Stoeckl & Liu, 2016; Sinclair-Maragh & Gursoy, 2015; Petrović, Blešić, Ivolga & Vujko, 2016). It starts with the understanding that the support of the local community will be determined by their perception of the positive and negative impacts of the activity (Rasoolimanesh & Jaafar, 2017). Therefore, involving the community from the early stages of the project allows for greater receptivity and support from residents (Sinclair-Maragh & Gursoy, 2015).

In this sense, sustainable tourism seeks to promote, based on an open and flexible strategy, adapted to local territorial singularities, the development and planning of tourist activities. The central focus is the respect and long-term preservation of natural, cultural and social resources, in order to contribute positively and balanced both to the economic, social, cultural and environmental development of the locality (Careto & Lima, 2006).

It is noticed that there is a need to transform the visitor into an ally, in order to contribute to the increase of territorial cohesion and consequently to the strengthening of local identity, as the loss of identity cohesion can lead to the disappearance of the sustainability of the identity itself. Such a situation could change the identity as a whole and compromise ongoing initiatives. Therefore, this transformation is only possible with the coordinated and effective participation of the different social actors that are impacted by local tourism (Torres, 2019).

The Saint James Way

In 1987, the Saint James Way was declared the first Cultural Itinerary in Europe by the Council of Europe and bears the title granted by UNESCO as a World Heritage Site since 1993. However, its history dates back to the ninth century and the first pilgrimages to the sepulcher of the Apostle St. James (Pardellas & Padín, 2014).

For centuries it was coursed by kings, nobles and many other people who set out to walk. Still, the Way lost popularity at the end of the Middle Ages (Devereux, 2003). Interest in the Way resurfaced only in the second half of the 20th century, being driven by several illustrious pilgrims, such as Pope John Paul II in 1982, who made a pilgrimage for the IV World Youth Day (Alves, 2011).

Currently, the Saint James Way is a tourist-religious attraction and identification for the European people. In Santiago de Compostela, the development of tourist activities is concentrated in the historic city center and in the area around the cathedral. Tourism transformed the meaning of the city, which started to regain a magical component as a place of arrival and meeting for all types of visitors (Fernández, González & Lopez, 2016).

The Way is made up of a rich cultural heritage, both tangible (churches, hostels, monasteries, bridges) and intangibles, represented by natural landscapes. In this sense, they reinforce unique rituals and traditions, giving the walking experience a symbolic and emotional meaning by connecting the pilgrim to

Table 3. Number of the touripilgrims – Saint James Way

Year	Touripilgrims	Annual Increase	Increase in the Period
2011	183,366		
2012	192,488	4,97%	
2013	215,880	12,15%	
2014	237,983	10,24%	
2015	262,516	10,31%	89,55%
2016	277,854	5,84%	
2017	301,036	8,34%	
2018	327,378	8,75%	
2019	347,578	6,17%	
2020	54,143	-84,42%	-70,47%

(Data Source: The Pilgrim's Reception Office in Santiago de Compostela, 2011-2019)

the route. Thus, the Way exemplifies values, social practices and rituals and typical of that community. In addition, it also provides pilgrims with the opportunity to access renovated places while at the same time adding another element to their experience: enjoying socializing with other tourists, pilgrims and local people (Gómez, Lopez, Antelo & González, 2021).

Sitting at the table, among the pilgrims, for community meals in the hostels is one of the rituals that still take place along the Way and that allow both an exchange of experiences and values and an assimilation of traditions and feelings of identity, both by pilgrims and by residents (Thomson, 2021).

The increase in the number of "touripilgrims" arriving in Santiago de Compostela is notorious. This fact, in addition to being visually verified by the number of people at Obradoiro Square and the increase in the offer of accommodation around the city's historic center, is confirmed by the figures released by The Pilgrim's Reception Office in Santiago de Compostela. In Table 3, data for the period from 2011 to 2019 point to an increase of almost 90% in the number of "touripilgrims".

More than that, it is also important to emphasize that these are the official numbers of people who, when they finish their pilgrimage, complete the ritual by registering at the Office. In fact, the number of people who circulate in Santiago de Compostela tends to be much higher when taking into account the number of those who do not register and also those who visit the Sanctuary without making a pilgrimage.

For the year 2020, this successive increase in the number of touripilgrims was not repeated due to the restrictions imposed globally due to the COVID-19 pandemic. However, as soon as the barriers to the movement of people begin to be removed, the flow of pilgrims will naturally be able to resume.

METHODOLOGY

The methodology used in this qualitative research was the ethnographic fieldwork developed through a combination of documentary research, field diary and semi-structured interviews in which the characteristics of the path, its landscapes, points of interest were collected, as well as testimonies from residents, touripilgrims, hospitallers and local merchants and agents.

Data collection took place between 2019 and 2021 during two distinct opportunities to walk through the PIWSC. Data collection was somewhat hampered due to the restrictions imposed by the COVID-19 pandemic, making the collection period longer. Despite the time difference in data collection, the results obtained did not vary. At the same time, results of previous surveys within the PIWSC were also used as a source of secondary data.

The stages taken in the PIWSC sought to represent the official stages recommended by the PIWSC organization. The entire route followed the indications recommendations of the yellow signs and arrows, as well as was accompanied by GPS tracking previously made available by the PIWSC organization.

As they went through the stages, the researchers made their notes in the field diary and interviewed local actors from the most diverse areas. Among them, "touripilgrims", residents, members of local governments, church members, members of associations linked to the Saint James Way, merchants and hospitallers. The diversity of the interviewed actors is important in the search to contemplate the widest possible spectrum of opinions within the community. Non-probabilistic samples were used, being selected by the researcher's subjective criteria, sometimes being samples for convenience, sometimes by judgment, such as through the "snowball sample" approach.

In this way, to explore the critical reading of the phenomenon observed with a greater amplitude, the combination of investigation methods and techniques was fundamental (GIL, 2002; Richards & Munsters, 2010).

The collected data were analyzed using Content Analysis (Bardin, 1977), in order to enable an aggregation into topics that were later presented and discussed in the Results Analysis section.

THE PORTUGUESE INNER WAY OF SANTIAGO DE COMPOSTELA - PIWSC

The territory crossed by the Portuguese Inner Way of Santiago de Compostela – PIWSC corresponds to the center and north of Portugal, a geographical area characterized as a markedly rural territory with low population density, as shown in Table 4, considering the resident population data.

Specifically, the PIWSC starts in Viseu and travels through the interior of Portugal to Chaves (214 Kms), when it arrives in Spain via Verín and continues to Santiago de Compostela (for another 182 Kms), as illustrated in Figure 1.

It was recently revitalized and is still a Way of St. James with a low number of pilgrims. Data collected by associations and tourist offices point to around 400 pilgrims per year on this route. This is one of the factors that attract new people, the possibility of experiencing the path in a more isolated and intense way (Pereiro, 2017). As well, it stands out as an opportunity to encourage project initiatives that can leverage the endogenous virtues found in communities throughout the PIWSC.

In October 2021, the PIWSC had its path certified by the Portuguese government through Ordinance No. 457/2021 under the terms of Decree-Law No. 51/2019. The application was created by the Portuguese Federation of the St. James Way in partnership with the municipalities through which the PIWSC crosses and was submitted to the Certifying Entity, of Turismo de Portugal, I.P. and the Directorate General for Cultural Heritage, which legitimized an integrated work of recognition and preservation of the cultural and natural heritage associated with the St. James Way.

This milestone reached by PIWSC evidences a joint effort of several local agents in the elaboration of a certification dossier that, in addition to the detailed description of the original PIWSC route, devises a

Table 4. Population of the municipalities crossed by the PIWSC in 2019

Municipality	Resident Population in the Year 2019
Chaves	39,343
Vila Pouca de Aguiar	11,961
Vila Real	49,962
Santa Marta de Penaguião	6,599
Peso da Régua	15,773
Lamego	24,895
Castro Daire	13,823
Viseu	97,249
Total	259,605

(Data Source: PORDATA, https://www.pordata.pt/)

plan to enhance this itinerary through the definition and implementation of a set of support tools to the pilgrim and friend of the St. James Way.

The Portuguese part is full of attributes that have the ability to provide development for communities. Geographically, the PIWSC crosses the area of eight Portuguese municipalities (Viseu – Castro D'Aire – Lamego – Peso da Régua – Santa Marta de Penaguião – Vila Real – Vila Pouca de Aguiar – Chaves)

Figure 1. The Portuguese Inner Way of Santiago de Compostela (PIWSC)
Source: own elaboration based on Google Earth

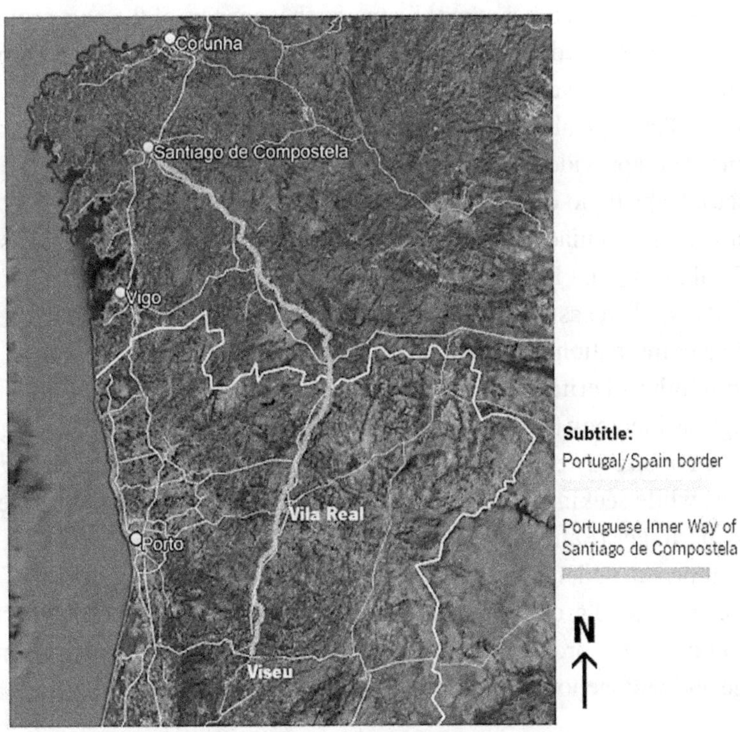

each with its own characteristics, differences in social and cultural perspectives, as well as points of interest on the PIWSC route (CPIS, 2021).

Previous studies were concerned with tracing the profile of "touripilgrim" in the PIWSC. This is predominantly covered by Portuguese adults, who make the pilgrimage in groups, focusing on the months of April and September, due to the weather. In addition, they declare themselves mostly Catholics and that the PIWSC was not the first Way of St. James to travel. The motivations range from religiosity, to being a less traveled path and, therefore, which can lead to greater introspection, as well as historical issues: to know the history of the region where you were born (or live) and increase your local community identification (Pereiro & Gomes, 2019).

One of the factors that makes PIWSC unique is that it is marked by constant transitions. We can highlight the different landscapes along the journeys, from the beautiful vineyards in the Douro region to the impressive mountain regions and thermal areas. The rural environment shares space with the urban environment and small towns with larger cities. The Jacobean heritage is present all along the way and in different forms. Gastronomy is a separate category: the variety is so great that, for each location, the "touripilgrim" finds a different tasty dish.

Along the PIWSC route, there are communities' heirs to ancestral traditions and customs characteristic of the mountain in the municipalities of Viseu, Castro Daire, Vila Pouca de Aguiar, which created a cultural mountain landscape, namely, in the mountains of Montemuro, Alvão and Padrela. In turn, in the municipalities of Lamego, Peso da Régua and Santa Marta de Penaguião, we come across communities established in a wine-growing cultural landscape, in the renowned Alto Douro Wine Region. The municipality of Vila Real has some settlements located in Alto Douro Vinhateiro and others in the Alvão Mountains, that is, in both types of cultural landscapes. Finally, the municipality of Chaves has other characteristics, such as the thermal areas and marks the transition to Spain. In addition to the protected religious heritage, such as churches, hermitages, crosses and crosses, the sacredness along the PIWSC is demonstrated through numerous other symbolic marks that are part of the non-protected heritage. The hundreds of cultural elements listed in the unprotected heritage sometimes have a popular, community and identity origin (Santana & Alves, 2019).

Eat well at PIWSC! The typical dishes of each location join with the various festivities celebrated throughout the calendar and provide the "touripilgrim" with a variety of flavors. The highlights are Vitela de Lafões (Viseu), Cabritinho do Montemuro (Castro D'Aire), Coelho Bravo (Lamego), Arroz de Cabidela (Santa Marta de Penaguião), Falacha which is a typical sweet from the parish of Moura Morta (Peso da Régua), Covilhetes (Vila Real), Arroz de Cogumelos com Vitela (Vila Pouca de Aguiar) and Pastel de Chaves (Chaves). The gastronomic richness, coupled with the welcome offered in the slightly smaller parishes, are a combination that enriches the "touripilgrimage" experience.

This cultural and religious heritage is part of its identity, constantly evoking the historical origin of the Way and the religious ideology that inspired it (Gómez *et al.*, 2021). These heritage resources, many on the way to extinction or already extinct in many places, can contribute to increasing the attractiveness of small communities, while seeking, through local authorities, to strengthen the defense and preservation of these identity heritages (Santana & Alves, 2019).

In this sense, soft tourism, understood as tourism based on valuing the potential of the local territory (Krippendorf, 1982), has gained ground in local administrations, especially in more rural areas. In fact, soft tourism makes it possible to promote the local economy, in addition to contributing to the preservation of local heritage and endogenous potential.

In recent years, soft tourism has also reached tourists and pilgrims who travel the different Saint James Way. By demanding a closer relationship with residents, and consequently a greater number of social exchanges and interactions, the creation and development of various activities and endogenous opportunities in the territories through which these routes pass, especially cities and rural communities, is enhanced (Azevedo, 2021).

ANALYSIS OF RESULTS

The revitalization of the PIWSC in the 21st century started with the book by Professor Arlindo de Magalhães Ribeiro da Cunha (1995), on the history of the Portuguese ways to Santiago de Compostela. The research is based on historical accounts of pilgrims, oral social memories and extensive fieldwork. In which it is stated that the PIWSC is, of all the ways to Santiago, the one with the greatest concentration of Jacobean heritage in Europe (Cunha, 1995).

Based on that work, an attempt was made to develop an inter-municipal and transnational cooperation project between the municipalities involved in the PIWSC. At first, it was to identify, with historical bases, the route traced. With this, it was possible to proceed with the signaling of the route, with the collaboration of tourism technicians from the chambers, volunteers and the community itself. The creation of hostels to host pilgrims was the next stage of the project. Thus, through the cooperation of several local agents, on April 24, 2012, the first official journey of this new stage of revitalization of the PIWSC was celebrated.

The revitalization is recent, however, it is already able to demonstrate some positive results, others negative, as well as opportunities that should be analyzed by local managers and the community.

The PIWSC is visited by a number of around 400 touring peregrines a year, a new reinvention of tradition that presents itself as a potential alternative to other pilgrimage routes to Santiago de Compostela, some of which are already massive, such as the French route. This is one of the factors that make pilgrims choose PIWSC: to escape the massification of the Paths. Therefore, those who choose to do the PIWSC, most of them have already done another St. James Way before and, therefore, have different expectations. The PIWSC, due to the fact that it is at an early stage of development, is able to provide the touripilgrim with an experience closer to the local community, at the same time that it provokes a series of reflections along the various landscapes that make it.

Another profile that stands out in the PIWSC is that of "touripilgrims" who live in the municipalities (or nearby) where the PIWSC passes and who carry out the steps in a non-sequenced and linear manner. It is common for a pilgrim to start walking on the first stage, sleep at the point of arrival, and on the following day leave for another stage of the Way. And so, it does, day after day, until it reaches the Cathedral of St. James the Apostle in Santiago de Compostela. However, the data show that a good part of the PIWSC "touripilgrims" perform a stage (being able to enjoy the attractions of the arrival location) and return to their homes. At another time, they return from where they left off and continue walking. In other words, the PIWSC is often not covered in a single long journey of consecutive days of pilgrimage.

Some of the explanations found for this phenomenon, when analyzing the data, are that many times the "touripilgrim" wants to: a) know a little more deeply about their own history, the places that built their cultural identity base; b) an experience more "slow" where it becomes possible to reflect and contemplate over the unique aspects of each landscape without having to think about step out while thinking with the manning of hostels; c) benefit from a double sense of pilgrimage due to the fact that the PIWSC largely shares a route that leads to the Sanctuary of Fátima.

The multi-motivational aspect of the "touripilgrim" is evident in the PIWSC. Religiosity, worship and preservation of the Jacobean heritage are found from the large churches where St. James is patron to the hundreds of small chapels. In fact, even to a lesser extent territorial, the PIWSC is the Portuguese path that most concentrates parishes, hermitages and Jacobean brotherhoods (Cunha, 2019). In the same way, the "touripilgrims" present other motivations for choosing the pilgrimage at PIWSC, such as: sporting, cultural, influenced by groups of friends who have traveled before and described the beautiful landscapes.

When analyzing the profile of the PIWSC "touripilgrim" combined with its multi-motivations, and considering that it tends to return to the PIWSC at different times, the endogenous potential of communities gains great prominence as factors that can transform the reality of these small localities. Such opportunities must be aligned so that in a coordinated and sustainable way they can generate economic, social, cultural results (and overflowing their influence to even more complex issues such as the fixation and maintenance of people in rural areas, which is one of the current challenges for the northern interior of Portugal). All of this, at the same time that it promotes an increase in social exchanges, causing an increase in territorial cohesion, in addition to preserving local material and intangible heritage, fostering new small businesses such as those linked to the offer of accommodation and local gastronomy, for example.

Opportunities knock at the door of PIWSC however, challenges are also present. There is a constant problem with signage along the way, whether due to the lack of markings, or the vandalism suffered by some or even the residents' lack of knowledge that this marking is there to guide the pilgrim. In addition, some stages demand a great physical effort from the pilgrim and are not adequately covered by sufficient infrastructure support. A crucial downside is the lack of official PIWSC information online. Despite the PIWSC organization providing tracking, the inexistence of an operational official website (which consolidates information, relevant points and tourist attributes of the eight municipalities) and the low insertion of the PIWSC on social platforms distance and create barriers for the pilgrim.

At the same time, managing an inter-municipal and international cooperation project is not one of the easiest tasks. There are several social actors involved, in different types of organizations, with different visions, opinions and objectives for the PIWSC. The management of this combination of forces is in charge of the leadership of the municipality of Vila Pouca de Aguiar, whose work is to align these expectations in order to translate a harmonious return to the councils. It is also noteworthy that since 2016 the PIWSC has been part of the European Federation of the Ways to Santiago and, more recently, the Portuguese Federation of the Ways to Santiago created in 2019, headquartered in Vila Pouca de Aguiar, with the objective of promoting, disseminating, organizing and managing of the Ways of Santiago in Portuguese territory.

There is still a lack of harmony and homogeneity between the councils when dealing with the PIWSC. Nothing to stop it from developing and helping to produce great returns for the localities. But at the same time, is necessary provide the "touripilgrim" with a "minimum standard" in delivering this cultural-tourist route.

PIWSC is a reinvented historical path that creates a new representation of the rural spaces of northwest Iberia. Through a political, technical and economic process, new rurality's have been shaped through the creation of narratives about rural landscapes, their heritage and gastronomic potential. On this route, you can experience a unique St. James Way as a model of cultural and touristic route.

CONCLUSION

For this research, the first point is to understand that the act of pilgrimage has been transformed over time. What was previously a ritual exclusively linked to religion and faith, nowadays divides space with several other aspects. Today, the pilgrim is motivated to walk for different reasons, some more sacred: religious and spiritual; others more profane: as a sport, opportunity to escape from stress and a more hectic life. At the same time, the pilgrim can have multi-motivations or even alternate them during his journey.

The second point is to consider the change in the profile of the pilgrim and how they make the pilgrimage. The profile started to be more diversified, ranging from young people to the elderly, linked or not to religion and with access to a larger cultural capital. In addition, technology was also incorporated into the pilgrimage, either with the "trackings" of the routes to be followed, or with the recording and sharing of experiences on various social networks. It can be said that the pilgrimage today begins even before the first step is taken, and ends long after reaching the sacred site.

For this reason, communities must update the way they look at pilgrims. They stopped looking for just accommodation and assistance and started wanting to find many other experiences along the way. Secular religious practices remain, but their combination with tourism gives it new practices, approaches and meanings. Thus, the term "touripilgrim" becomes appropriate to designate the person who also incorporates elements of a tourist when making a pilgrimage. In this way, a great opportunity arises for both sides. The community can structure itself to offer the "touripilgrim" gastronomic, cultural, landscape experiences, among others, adding its diverse competitive advantages in order to promote the increase of interactions between locals and visitors and, thus, enhance the development of their endogenous capacities. In turn, the "touripilgrim" has the opportunity to choose new experiences and make a more immersive and genuine visit.

PIWSC is a unique path due to several factors. The constant transitions in the landscape characterize the Way. Climbing elevations in rural areas can contemplate nature, then return to a lower point and enter a small parish. In addition to visiting the beautiful vineyards in the Douro region, further on the "touripilgrim" meets the thermal areas. The dichotomy of the rural environment contrasts with the urban environment, placing small towns and larger cities in the same route. However, the presence of the Jacobean heritage is not reserved only for the great monuments in the biggest cities, it is present all the Way through in different ways, including in small parishes and in the identity of the local community itself. The gastronomy of a region can be appropriated as a strong element of local identity. Along the PIWSC it is vast and allows the "touripilgrim" to take a "gastronomic tour" within a pilgrimage route.

Two other points of PIWSC can be highlighted. The first of these is the profile of pilgrims. Most of those who visit PIWSC are Portuguese (residents of the region), which can add an element of search for knowledge and appropriation of the identity and values base itself. The second point is that there is a significant portion of visitors who make the journey in non-sequential stages. That is, they do not start their journey in Viseu and go through a stage each day and continue the next day. The usual thing is to perform a stage on a weekend and return home. Then, they return to the stage where they stopped and do one more.

Such elements enhance the possibilities of increasing exchanges between locals and visitors, in order to insert the visitor into the local context, and the community to develop through the convergence of its endogenous attributes. The dialogue between the different local actors is essential to establish guiding strategies that will compose the dynamics of the communities. It must be considered that today's pilgrim is a potential tourist of tomorrow, just as today's tourist is a potential pilgrim of tomorrow.

In addition, the conservation, preservation and promotion of a set of local identities and values, material and immaterial heritage, makes the communities involved develop social capital that will generate new wealth for them, in addition to enhancing the way they manage them. Also transforming the social exchanges that take place in the daily lives of communities. In this way, the levels of cohesion and trust between the locals tend to be higher, which tends to provoke new opportunities for intensification in social exchanges, generating a cycle of potentiation in the development of local communities.

It is noted that the development referred to here, is not centered on economic development or in function of it, but comprises a broader form. Great emphasis should be placed on improving the quality and intensity of social exchanges both between resident-residents and resident-visitors.

It should be noted that the revitalization of the PIWSC is recent, and, therefore, it is a project that is being consolidated and developed within the community itself. Thus, it is necessary to capture and develop in individuals the identity elements and values that rescue their own history and reinforce the feeling of belonging and the cultural appropriation of such a wide range of heritage throughout the PIWSC.

The challenge is to make a management that, among the different local actors, shares the feeling of appreciation of the local attributes for the benefit of the residents themselves. In the case of PIWSC, which consists of eight municipalities, this becomes even more challenging. Isolated or isolated actions in a given location are not enough. For the desired development to reach PIWSC, it is necessary to promote and develop it as a whole, in an integrative way and with shared management between the municipalities. But there are already some actions for this purpose, and municipalities are seeking to align themselves more and more and launch joint actions. The PIWSC Certification achieved in 2021 is an example of this integration. Such recognition demonstrates the potential of PIWSC and the capacity of local agents to jointly develop and implement actions with the objective of strengthening and valuing the communities that make up the PIWSC.

As limitations of the research, the researchers understand that although the results presented manage to demonstrate a horizon with which local agents can work, discuss and develop actions, they cannot be extrapolated nor considered as a complete "picture" of reality. The interviewees were selected in order to guarantee the representativeness of the communities, without the concern to establish a statistical validation, being intentionally selected with the objective of exploring the different views and opinions about the PIWSC.

ACKNOWLEDGMENT

The acknowledgments section has been removed in order to allow the blinded peer review of the chapter, but will be added in the final version of the chapter.

REFERENCES

Alves, S. M. P. P. (2011). *A Antevisão do Peregrino na Iconografia de São Tiago no Caminho Português de Santiago entre Viseu e Chaves. Subsídios para a criação de uma rota turística* (Master's dissertation). Universidade Católica Portuguesa.

Amaro, S., Antunes, A., & Henriques, C. (2018). A closer look at Santiago de Compostela's pilgrims through the lens of motivations. *Tourism Management, 64*, 271–280. doi:10.1016/j.tourman.2017.09.007

Azevedo, P. (2021). The Ways of Saint James in Trás-os-Montes and Alto Douro as an example of soft tourism in rural areas. *European Countryside, 13*(2), 314–329. doi:10.2478/euco-2021-0020

Bandeira, P. (1999). *Participação, Articulação de Atores Sociais e Desenvolvimento Regional.* Texto para discussão, Instituto de Pesquisa Econômica Aplicada (IPEA), nº 630.

Bardin. (1977). Análise de Conteúdo. Academic Press.

Bates, F. L. (1960). Institutions, organizations, and communities: A general theory of complex structures. *Pacific Sociological Review, 3*(2), 59–70. doi:10.2307/1388201

Belhassen, Y., Caton, K., & Stewart, W. P. (2008). The search for authenticity in the pilgrim experience. *Annals of Tourism Research, 35*(3), 668–689. doi:10.1016/j.annals.2008.03.007

Blom, T., Nisson, M., & Santos, X. (2016). The way to Santiago beyond Santiago. Fisterra and the pilgrimage's post-secular meaning. *European Journal of Tourism Research, 12*, 133–146.

Boisier, S. (2007). Territorio, estado y sociedad en chile. La dialéctica de La descentralización: entre la geografía y la gobernabilidad (Thesis). Universidad De Alcalá, Alcalá De Henares.

Canclini, N. G. (2006). *Culturas Híbridas: estratégias para entrar e sair da modernidade.* Editora da Universidade de São Paulo.

Carbone, F., Corinto, G., & Malek, A. (2016). New Trends of Pilgrimage: Religion and Tourism, Authenticity and Innovation, Development and Intercultural Dialogue: Notes from the Diary of a Pilgrim of Santiago. *AIMS Geosciences, 2*(2), 152–165. doi:10.3934/geosci.2016.2.152

Cardita, A. (2012). Peregrinação: Possibilidades de compreensão crítica de uma experiência. (Sociologia). *Revista da Faculdade de Letras da Universidade do Porto, 24*, 195–213.

Careto, H., & Lima, S. (2006). *Turismo e Desenvolvimento Sustentável.* GEOTA - Grupo de Estudos de Ordenamento do Território e Ambiente.

Cohen, A. P. (2013). *Symbolic construction of community.* Routledge. doi:10.4324/9780203131688

Collins-Kreiner, N., & Wall, G. (2015). Tourism and religion: spiritual journeys and their consequences. In The Changing World Religion Map (pp. 689-707). Springer. doi:10.1007/978-94-017-9376-6_34

Costa, C. (2014). Gestão Estratégica do Turismo: Evolução Epistemológica dos Modelos e Paradigmas, e Tendências para o Futuro. In Turismo nos Países Lusófonos: Conhecimento, Estratégia e Territórios (pp. 19-40). Escolar Editora.

Courtney, M. (2013). A review of the literature exploring the occupation of contemporary pilgrimage. *Journal of Occupational Science, 22*(2), 170–182. doi:10.1080/14427591.2013.764816

CPIS. (2021). Available at: https://cpis.utad.pt/

Cunha, A. de M. R. da. (1995). *Caminhos Portugueses de Peregrinação a Compostela. Itinerários portugueses* (Identificação dos caminhos e coordenação da investigação). Santiago de Compostela/Porto, Xunta de Galicia/Centro Regional de Artes Tradicionais.

Cunha, A. de M. R. da. (2019). Viseu > Vila real > Chaves. Da devoção ao turismo: O Caminho Português Interior de Santiago de Compostela Caminho Português Interior de Santiago de Compostela. In Património cultural jacobeu, turismo e peregrinação: O Caminho Português Interior de Santiago de Compostela (CPIS) (pp. 79-112). PASOS.

Damari, C., & Mansfeld, Y. (2014). Reflections on pilgrims' identity, role and interplay with the pilgrimage environment. *Current Issues in Tourism*, *19*(3), 1–24.

Devereux, C. (2003). Spirituality, pilgrimage and the road to Santiago: questions for cultural tourism. In *Religious Tourism and Pilgrimage*. Tourism Board Of Leiria/Fátima.

Dimitrovski, D., Ioannides, D., & Nikolaou, K. (2020). The tourist-pilgrim continuum in consumer behaviour: The case of international visitors to Kykkos Monastery, Cyprus. *Journal of Tourism and Cultural Change*, 1–19.

Feather, N. T. (1995). Values, valences, and choice: The influences of values on the perceived attractiveness and choice of alternatives. *Journal of Personality and Social Psychology*, *68*(6), 1135–1151. doi:10.1037/0022-3514.68.6.1135

Fernández, B. M. C., González, R. C. L., & Lopez, L. (2016). Historic city, tourism performance and development: The balance of social behaviours in the city of Santiago de Compostela (Spain). *Tourism and Hospitality Research*, *16*(3), 282–293. doi:10.1177/1467358415578473

Giddens, A., & Giddens, A. (1973). *The class structure of the advanced societies*. Hutchinson.

Gil, A. C. (2002). *Como elaborar projetos de pesquisa*. Atlas.

Gómez, L. A. E., Lopez, L., Antelo, M. A. P., & González, R. M. L. (2021). The Timeline of The Way St James: Motivations and Impacts of the Pilgrimage on Personal Lives. In D. Liutikas (Ed.), *Pilgrims: Values and Identities* (pp. 72–82). CABI. doi:10.1079/9781789245653.0072

Graves, T. D. (1967). Psychological acculturation in a tri-ethnic community. *Southwestern Journal of Anthropology*, *23*(4), 337–350. doi:10.1086outjanth.23.4.3629450

Guilarte, Y. P., & González, R. C. L. (2016). El Perfil Del Visitante En Santiago De Compostela: Tradición Y Actualidad. *Cuadernos de Turismo*, *37*(37), 305–322. doi:10.6018/turismo.37.256251

Gusmán Correia, M., Lopez, L., Lois González, R., & Santos, X. (2017). The Challenges of the First European Cultural Itinerary: The Way to St. James. *Almatourism - Journal of Tourism, Culture and Territorial Development*, *8*(6), 1–19.

Hall, S. (2002). *A identidade cultural na pós-modernidade*. Editora DP&A.

Ivakhiv, A. (2003). Nature and self in new age peregrimage. *Culture and Religion*, *4*(1), 93–118. doi:10.1080/01438300302812

Jarvis, D., Stoeckl, N., & Liu, H. (2016). The impact of economic, social and environmental factors on trip satisfaction and the likelihood of visitors returning. *Tourism Management*, *52*, 1–18. doi:10.1016/j.tourman.2015.06.003

Jensen, R., & Lovrod, H. M. (2017). Pilgrimage, past and present. Norwegian perspectives with emphasis on Santiago de Compostela. *Cuadernos de Estudios Gallegos*, *64*(130), 467–489. doi:10.3989/ceg.2017.130.13

Kim, B., Kin, S. S., & King, B. (2016). The sacred and the profane: Identifying pilgrim traveler value orientations using means-end theory. *Tourism Management*, *56*, 142–155. doi:10.1016/j.tourman.2016.04.003

Kouchi, A. N., Nezhad, M. Z., & Kiani, P. (2016). A study of the relationship between the growth in the number of Hajj pilgrims and economic growth in Saudi Arabia. *Journal of Hospitality and Tourism Management*, *36*, 103–107. doi:10.1016/j.jhtm.2016.01.008

Krippendorf, J. (1982). Towards new tourism policies: The importance of environmental and sociocultural factors. *Tourism Management*, *3*(3), 135–148. doi:10.1016/0261-5177(82)90063-2

Lewicka, M., Rowiński, K., Iwańczak, B., Bałaj, B., Kula, A. M., Oleksy, T., Prusik, M., Toruńczyk-Ruiz, S., & Wnuk, A. (2019). On the essentialism of places: Between conservative and progressive meanings. *Journal of Environmental Psychology*, *65*, 101318. doi:10.1016/j.jenvp.2019.101318

Liutikas, D. (2017). The manifestation of values and identity in travelling: The social engagement of pilgrimage. *Tourism Management Perspectives*, *24*, 217–224. doi:10.1016/j.tmp.2017.07.014

Lopes, J. R., & Pereira, A. M. (2017). Patrimônio cultural, turismo e desenvolvimento local: Estudo de caso da Cidade Velha, ilha de Santiago, Cabo Verde. Sociabilidades Urbanas –. *Revista de Antropología Social*, *1*(2), 45–60.

Meillassoux, C. (1981). *Maidens, Meal and Money - Capitalism and the Domestic Community*. Cambridge University Press.

Nilsson, M. (2018). Wanderers in the shadow of the sacred myth: Pilgrims in the 21st century. *Social & Cultural Geography*, *19*(1), 21–38. doi:10.1080/14649365.2016.1249398

Oviedo, L., de Courcier, S., & Farias, M. (2014). Rise of pilgrims on the Camino to Santiago: Sign of change or religious revival? *Review of Religious Research*, *56*(3), 433–442. doi:10.100713644-013-0131-4

Pardellas, X. X., & Padín, C. (2014). El Camino De Santiago Portugués En Galicia Y Las Estrategias Turísticas Locales. *Revista de Análisis Turístico*, *17*, 51–59.

Parga Dans, E. (2012). *El perfil de la peregrinación en el Camiño Santiago-Fisterra*. https://digital.csic.es/handle/10261/45922

Pereiro, X. (2017). Turiperegrinos portugueses no Caminho Português Interior de Santiago de Compostela. *Revista Turismo e Desenvolvimento*, *27*, 413–423.

Pereiro, X. (2018). Ancient ways, new cultural tourism routes: The Inner Portuguese Way to Santiago de Compostela. In E. G. Pellón (Ed.), *Rural Worlds, Social Sustainability and Local Landscapes in Globalisation Era* (pp. 289–315). Case Studies in Southern Europe.

Pereiro, X., & Gomes, C. (2019). Perfil, motivações e experiências dos peregrinos do Caminho Português Interior de Santiago de Compostela. In X. Pereiro (Ed.), *Património cultural jacobeu, turismo e peregrinação: O Caminho Português Interior de Santiago de Compostela (CPIS)* (pp. 79–112). PASOS.

Petrović, M. D., Blešić, I., Ivolga, A., & Vujko, A. (2016). Agritourism impact toward locals' attitudes-an evidence from Vojvodina Province (Serbia). *Journal of the Geographical Institute Jovan Cvijic*, *66*(1), 105–123. doi:10.2298/IJGI1601105P

PORDATA. (2021). Available at: https://www.pordata.pt/

Rasoolimanesh, S. M., & Jaafar, M. (2017). Sustainable tourism development and residents' perceptions in World Heritage Site destinations. *Asia Pacific Journal of Tourism Research*, *22*(1), 34–48. doi:10.1080/10941665.2016.1175491

Richards, G., & Munsters, W. (2010). *Cultural Tourism Research Methods*. CABI. doi:10.1079/9781845935184.0000

Roccas, S., & Elster, A. (2014). Values and religiosity. In V. Saroglou (Ed.), *Religion, Personality, and Social Behavior* (pp. 193–212). Taylor & Francis.

Sánchez, V. L., Fernández, M. D., & Lara, J. Á. S. (2016). Economic impact of a religious and tourist event: A holy week celebration. *Tourism Economics*, *23*(6), 1255–1274. doi:10.1177/1354816616675996

Santana, M. O. R., & Alves, A. (2019). A paisagem cultural sagrada no Caminho Português Interior de Santiago. In X. Pereiro (Ed.), *Património cultural jacobeu, turismo e peregrinação: O Caminho Português Interior de Santiago de Compostela (CPIS)* (pp. 59–77). PASOS.

Santos, G. E. de O. (2000). Importância das peregrinações para o turismo mundial. *Revista Turismo Em Análise*, *11*(2), 38–44. doi:10.11606/issn.1984-4867.v11i2p38-44

Sinclair-Maragh, G. S., & Gursoy, D. (2015). Imperialism and tourism: The case of developing island countries. *Annals of Tourism Research*, *50*, 143–158. doi:10.1016/j.annals.2014.12.001

Tajfel, H. (1981). *Human Groups and Social Categories: Studies in Social Psychology*. Cambridge University Press.

Thomson, A. X. (2021). Commensality, Communitas and Contestation on the Camino Francés: Making Pilgrims through Meals. In D. Liutikas (Ed.), *Pilgrims: Values and Identities* (pp. 94–104). CABI. doi:10.1079/9781789245653.0094

Torres, F. E. J. (2019). *Bem-estar comunitário e visitantes através do Caminho em Santiago. Grandes narrativas, ideias e práticas culturais na cidade*. ANDAVIRA.

Vieira, B. (1998). The Construction of National Identity – On Primordialism and Instrumentalism. *Human Affairs*, *8*(1), 29–43.

Vilaça, H. (2010). Pilgrims and pilgrimages: Fátima, Santiago de Compostela and Taizé. *Nordic Journal of Religion and Society*, *23*(2), 137–155. doi:10.18261/ISSN1890-7008-2010-02-03

Wolf, E. (1956). Aspects of Group Relations in a Complex Society: Mexico. *American Anthropologist*, *58*(6), 1065–1078. doi:10.1525/aa.1956.58.6.02a00070

Wolf, E. (1972). Ownership and Political Ecology. *Anthropological Quarterly*, *45*(3), 201–205. doi:10.2307/3316532

ADDITIONAL READING

Bond, N., & Falk, J. (2013). Tourism and identity-related motivations: Why am I here (and not there)? *International Journal of Tourism Research*, *15*(5), 430–442. doi:10.1002/jtr.1886

Cohen, E. (2006). Religious tourism as an educational experience. In J. D. Timothy & D. H. Olsen (Eds.), *Tourism Religion and Spiritual Journeys* (pp. 78–93). Routledge.

Collins-Kreiner, N. (2018). Pilgrimage-tourism: Common themes in different religions. *International Journal of Religious Tourism and Pilgrimage*, *6*(1), 8–17.

Kurrat, C. (2018). Biographical motivations of pilgrims on the Camino de Santiago. In: *Book of Abstracts. 10th Annual International Religious Tourism and Pilgrimage (IRTP) Conference*. University of Santiago de Compostela.

Liutikas, D. (2015). Indulgence feasts: manifestation of religious and communal identity. In A. Jepson & A. Clarke (Eds.), *Managing and Developing Communities, Festivals and Events* (pp. 148–164). Palgrave Macmillan.

Navruz-zoda, B., & Navruz-zoda, Z. (2018). The destination marketing development of religious tourism in Uzbekistan: a case study. In S. Yasuda, R. Raj, & K. Griffin (Eds.), *Religious Tourism in Asia: Tradition and Change trough Case Studies and Narratives* (pp. 149–157). CAB International. doi:10.1079/9781786392343.0149

Prat, J. (2011) ¿Por qué caminan? Uma mirada antropológica sobre el Camino de Santiago. In La cultura sentida. Homenaje al professor Salvador Rodriguez Vecerra. Signatura ediciones.

Schnell, T., & Pali, S. (2013). Pilgrimage today: The meaning-making potential of ritual. *Mental Health, Religion & Culture*, *16*(9), 887–902. doi:10.1080/13674676.2013.766449

KEY TERMS AND DEFINITIONS

Community Development: Development promoted by local actors that result in identity, cultural, social, heritage and economic benefits for the locality.

Endogenous Capacities: Virtues, local potential, and locality elements of capacity that can provide community development.

Local Actors: People from different sectors of a locality inserted in the dynamics of power and decision within the community.

Local Identities: The set of elements, values and traditions of a locality that differentiates it from another.

Multi-Motivations for Pilgrimage: Characteristic of the modern pilgrim who does not have only a single motivational factor for carrying out the pilgrimage, such as the religious factor.

Sustainable Tourism: A type of non-predatory tourism that is concerned with the preservation of a locality's natural, social and cultural resources as well as local development.

Touripilgrims: Individuals endowed with multi-motivations who do a pilgrimage.

Chapter 15
Local Perspectives on Cultural Tourism and Cultural Sustainability:
The Case of the Cyclades, Greece

Theano S. Terkenli
https://orcid.org/0000-0001-6982-6132
University of the Aegean, Greece

Vasiliki Georgoula
https://orcid.org/0000-0001-6144-4471
University of the Aegean, Greece

ABSTRACT

The objective of the chapter is to explore cultural tourism perceptions, practices, concerns, and prospects among local residents and business representatives in the Cyclades, specifically three sites (Andros, Syros, and Santorini), and to draw their implications for local cultural sustainability. For this purpose, it relies methodologically on an on-site intensive questionnaire survey, effectuated in the context of the SPOT HORIZON2020 EU project (2020-22) on cultural tourism. The findings show that the role of culture as an actual tourism attraction and the potential for further growth in cultural tourism, and consequently in local development, are broadly recognized from all sides involved in the survey. Conversely, the role of tourism in these destinations' cultural development, management, and appropriation is viewed with some trepidation and ambivalence, while, in the relative absence of the state, local residents and tourism- or culture-related businesses appear to have taken most aspects of cultural tourism planning and management in their hands.

INTRODUCTION

This case-study analysis is placed in the broader context of analytical efforts towards understanding

DOI: 10.4018/978-1-7998-9217-5.ch015

the role that cultural tourism plays in small-scale destinations and its implications for local cultural sustainability. Our research focuses on the Cycladic Islands, among the most world-renowned tourism destinations in Greece, and highly competitive vis-à-vis other top global destinations (Berg & Edelheim, 2012; INSETE, 2020; World Population Review, 2021), but also extremely dependent on tourism for their economic survival. The objective of the chapter is to explore cultural tourism perceptions, practices, concerns and prospects, among local residents and business representatives in the Cyclades—and specifically the three islands of Andros, Syros, and Santorini—and to link these findings with prospects and challenges in achieving local cultural sustainability. The archipelago of the Cyclades was selected as our case study also on the basis of its especially rich present and past cultural heritage.

In the current context of extremely tentative, fluid and ambivalent tourism mobilities (Knezevic et al., 2021), it is crucial to develop an understanding of distinctive trends and patterns from the bottom up, and especially so in tourism-dependent economies and societies, such as the Greek islands. The concept and framework of cultural sustainability were employed in order to discuss in more depth the variable interrelationships between culture and tourism in the development of cultural tourism and in overall local sustainability. The cultural dimension of sustainability issues, resource uses, practices, demands and interests etc., including tourism prospects and repercussions, adds a further level of analysis and operationalization to the three pillars of 'sustainability', the basis on which tourism values, processes and choices ought to be negotiated and effectuated.

THEORETICAL BACKGROUND

According to its original instatement (UN, 1987), sustainability is defined as meeting our own needs without compromising the ability of future generations to meet their own needs, and, at a basic level, has been broadly associated with addressing environmental, social and economic concerns and pursuits, the 'three pillars of sustainability' (UN, 2015; Adams, 1995). The simplest definition of sustainability is the 'capability of being maintained at a certain rate or level' (Gruen et al., 2008). Although not explicitly mentioned as such, the cultural dimension of sustainability runs through all basic UN and scientific documents referring to sustainability, including the UN 2030 Agenda for Sustainable Development (UN, 2015), in terms both of underlying principle and scope, but also in terms of specific goals, such as decency, equality, inclusivity, literacy, well-being, respect for cultural diversity, human rights and fundamental freedoms, etc. Relevant scientific inroads and institutional efforts have multiplied in recent years, aiming at increasing and improving 'sustainability's' practical adaptability and ways of implementation (Janoušková et al., 2019).

During the past decade, culture has been valorized as a fourth pillar of sustainability, and has been used theoretically and empirically to illustrate the foundational role it plays in understanding the workings of sustainability itself and in meeting its goals, in various scientific and non-scientific areas and (Duxbury 2021; Helgadóttir et al. 2019; INSETE 2020; Empacher 2002; Terkenli 2000). Culture represents the highest-level, most deeply ingrained, holistic, integrative, and stable system of reference in human life, and has been increasingly acknowledged as basic and pivotal to spatial analysis. As such, it engulfs all tangible and intangible manifestations of human life and obviously permeates all principles and precepts of sustainability. Defined as 'a whole way of life' (Williams, 1958; Throsby, 2008), culture can be understood as a necessary foundation for meeting the aims of sustainable development research and practices, in many realms, among which none the least is tourism (Tsartas 2003; Saarinen 2006; Coccossis et al 2011; Hughes & Scheyvens, 2016)—a broader definition of culture, also adopted by our study.

'Sustainability' as a concept offers opportunities for bringing together environmental, social and economic considerations and goals (UN, 2015; Helgadóttir, 2019), especially when the cultural variable is added to the equation (Duxbury, 2021; Thimm, 2019; Auclair & Fairclough, 2015; Terkenli 2000; Oikonomou 2000). As a term, 'cultural sustainability' was introduced in 1995 by the World Commission on Culture and Development (WCCD), defined at the time as 'inter-and intra-generational access to cultural resources' (WCCD, 1995 as cited in Thimm 2019). In the past, this concept has been considered as overlapping with social sustainability (Murphy, 2012; Thimm, 2019), but also broadened (Throsby, 2003) to incorporate the following principles: material and non-material well-being, intergenerational equity, intra-generational equity, maintenance of diversity, precautionary principle, maintenance of cultural systems, and recognition of independence. The importance of establishing a fourth pillar of sustainability (Culture 21, 2011; Hawkes, 2001) notwithstanding, such an enriched understanding of sustainability with the addition of the cultural dimension, unless specified, operationalized and applied in empirical examples and situations, is often susceptible to abstract theorization. As a result, during the past few decades, the use of the concept of sustainability in a very broad range of contexts continues to be often inefficient, chaotic and, to a great extent, fraught by uncritical over-simplification, favoring established practices, interests and/ or conceptualizations of human–environment relationships (James et al., 2015; Saunders at al., 2020; Oikonomou, 2000). Specifically, as widely acknowledged, the deployment of the term 'sustainability' in a variety of real-world situations has been used as a catch-all concept, devoid of consideration regarding questions of power, politics, but also of its cultural underpinnings and repercussions (Katz, 1998; Sneddon 2000; Sheppard et al., 2009).

Among recent efforts towards deciphering the various perspectives and prospects on cultural sustainability in relevant scientific endeavors (Pavlis & Terkenli, 2017; INSETE, 2020; Boström, 2012; UNWTO, 2018), the framework of cultural sustainability analysis by Soini and Birkeland (2014) stands out as perhaps the most comprehensive and systematic analytical scheme proposed so far. Spanning various political, cultural and normative positions, and cautioning that its use ought to be highly contextualized, Soini & Birkeland (2014) organized and classified the uses of the concept of 'cultural sustainability', into seven categories: cultural heritage, cultural vitality, economic viability, cultural diversity, locality, eco-cultural resilience and eco-cultural civilization. The category most pertinent to our study is obviously 'economic viability', since tourism, in all of its forms and versions, is first and foremost an economic activity (Statista, 2021; Su, 2018). As such, tourism often conflicts with other destination aspects and activities, such as the social, the cultural and environmental, as represented by the three other pillars of sustainability (Shaw & Williams, 1994; Terkenli et al., 2007; Tsartas et al., 2020). Perspectives of 'economic viability' view "culture, and particularly heritage… as a resource for economic viability, and local and regional development, such as in the form of tourism" (Soini & Birkeland, 2000, p. 217). Furthermore, in order to decipher such interrelations between tourism and culture better, our analysis and discussion of the relationship culture-tourism was applied specifically to cultural tourism, namely the type of tourism that links culture and tourism, par excellence. Simply put, 'cultural tourism' refers to the compound set of activities of tourism planning, effectuating and experiencing a destination, with the—broadly defined—motive of culture. For the purposes of our study, the UNWTO definition was adopted of cultural tourism as "a type of tourism activity in which the visitor's essential motivation is to learn, discover, experience and consume the tangible and intangible cultural attractions/products in a tourism destination" (Tsartas et al., 2000, p.18).

Figure 1. Model of cultural tourism, showing the relationship of culture, tourism, and cultural tourism at the destination level and its implications for sustainability.

The whole problematic on cultural sustainability not only applies to but also compounds the concept's application and operationalization in the case of tourism, a synthetic and multifaceted phenomenon, by definition. Furthermore, the predominance of economic pursuits in tourism practices often comes in contrast or conflict with cultural priorities, values and practices at the destination. On the other hand, according to Liu, "most sociocultural changes brought about by tourism development are beneficial and the unique role of tourism in promoting modern values, social progress and cultural evolution should be greatly appreciated" (2003, p. 467). This necessitates more in-depth and integrative scientific understanding of how such often contested and conflictual cultural tourism dynamics (i.e., economics versus culture) play out in local (cultural) sustainability and its future prospects for the destination. As laid out in the ICOMOS International Charter for Cultural Heritage Tourism 2021, certain "This calls for … responsible and diversified cultural tourism development and management contributing to cultural heritage preservation; community empowerment, social resilience and wellbeing; and a healthy global environment" (p. 1).

The factor of tentativeness and contestation in the applicability of the cultural sustainability concept to tourism, and specifically cultural tourism, has not yet been adequately addressed by social scientific and tourism research—precisely the research area which this chapter aspires to contribute to. For this purpose, it is essential to bring together all involved sides to such analysis, by surveying understandings, perceptions, opinions and attitudes regarding cultural tourism and its repercussions for local (cultural) sustainability among local residents and relevant local businesses. The theoretical/ conceptual model proposed below (Figure 1) purports to aid in the clarification and understanding of the basis on which these bottom-up processes and relationships are effectuated.

The conceptual model in Figure 1 graphically represents the relationship of culture, tourism, and cultural tourism at the destination level and its implications for local cultural sustainability. Cultural tourism appears as the common ground between culture and tourism (the two circles coming together). Both the realms of culture and tourism, as well as the interface between them (cultural tourism), generate and instigate a series of processes which take place at the destination level, which (may) lead to local cultural sustainability.

Figure 2. The case study in the Archipelago of the Cyclades: Andros, Santorini, and Syros Islands.
Source: Cartography and Geoinformatics Laboratory, Department of Geography, University of the Aegean.

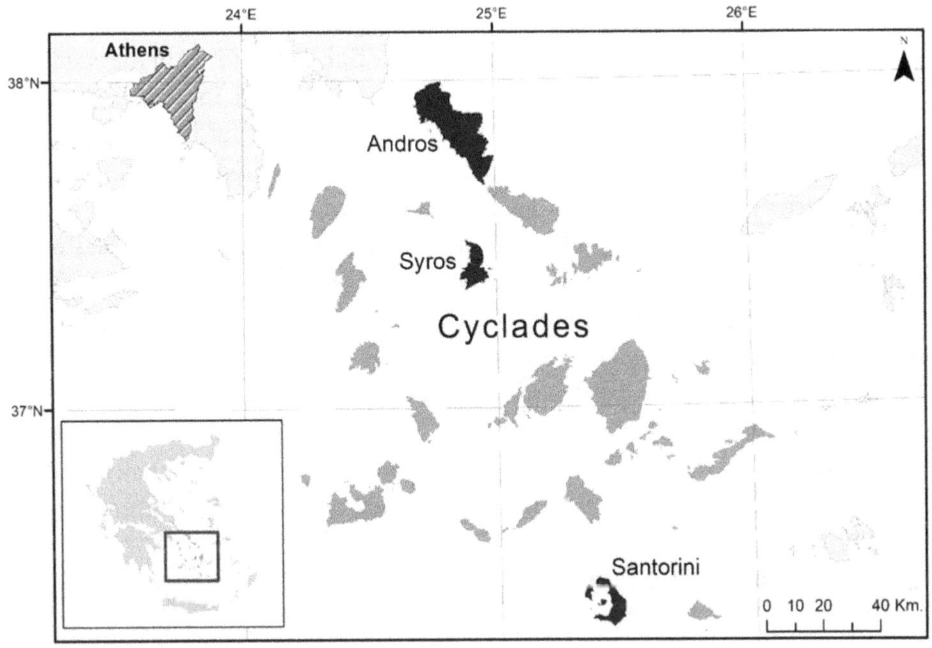

THE CASE STUDY AND METHODOLOGY

Cyclades: The Research Setting

The selection of our specific case study stemmed from the fact that the Cyclades (Map 1), and especially the islands of Mykonos and Santorini, are among the most world-renowned and highly-competitive global-tourism destinations in Greece (WTTC, 2020; INSETE, 2020; Coccossis, 2001). Besides their strong 3Ss (sea-sand-sun) allure, they also boast striking natural/environmental assets, great landscape diversity, and rich cultural traditions and heritage, dating back to the antiquities (Prokopiou et al., 2018; Berg, 2012). Indicatively, international tourist air arrivals in the Cycladic islands in 2019 reached 994,000 according to official statistical data (INSETE, 2020). However, it is their cultural attractions, taken all together, that render them a most significant pole of both local and international tourism attraction, constituting a competitive advantage vis-à-vis other similar destinations. This fact, as will be highlighted through our research findings in the next section of the chapter, has significant implications for these islands' economic survival and development. Syros Island was selected for our study purposes as the main administrative center of the Cycladic Archipelago and an island with a fairly vibrant economy that does not rely solely on tourism, as well as for its cultural tourism attractions (festivals, architecture, etc.). Santorini was selected on the basis of its worldwide reputation as a top tourist destination. Finally, Andros was selected as a mainly domestic tourism and second-home destination, as well as for its alternative (ecotourism and art tourism) attractions.

As regards the cultural activities and sites in this area, Andros is known nationally and beyond as a Cycladic island featuring a relatively high level of arts and ecological awareness activities (i.e. hiking tourism/ recreation paths). Unsurprisingly, the majority of the received responses from respondents of Andros Island were positively associated with the visual and the performing arts, like music, theatre, and visual arts exhibitions. Traditional feasts and celebrations with local dance and music were mentioned a lot, as were also art exhibitions. Similarly, in the case of Santorini, the majority of the acquired responses were positively associated with the performing arts, such as music, theatre, local dance and music fiestas, and other cultural events. Santorini's image, however, is very different for both insiders and outsiders alike to that of Andros and the majority of other Cycladic Islands. It is renowned for its tourism appeal and amenities/ infrastructures, its breath-taking caldera landscape and its traditional settlements. Accordingly, residents also brought up 'nightlife', as the island hosts a great number and wide variety of bars, restaurants and clubs, as well as activities pertaining to sports and ecological tourism which have recently been added to wine tourism and cultural tourism. Lastly, Syros is the capital and administrative center of our case study area, the Cycladic Islands and known both nationally and internationally for its cultural activity, most notably film and music festivals. Generally speaking, the residents of Syros who participated in our survey replied very positively with regard to its cultural features and seemed to appreciate the wide variety of local and international events taking place annually on their island, in the fields of both visual and performing arts (i.e. theatre, film, dance events, concerts, festivals, etc.).

Research Design and Methodology

The objective of the chapter is to explore cultural tourism perceptions, practices, attitudes and concerns, among local residents and business representatives in the Cyclades and to draw their implications (prospects and challenges) for local cultural sustainability. For the purposes of our study and our primary data collection, we employed the questionnaire survey method and applied it to three selected Cycladic islands, with different features as regards their cultural tourism profile. More specifically, the study aimed to a) to sketch those basic parameters of cultural tourism in these islands that relate to local (cultural) sustainability, and b) to use these findings to discuss future prospects of local (cultural) sustainability, in the discussion that follows. In order to capture all principal bottom-up perspectives to cultural tourism and sustainability, three extensive and in-depth questionnaires (one for local residents and one for business representatives) were constructed and implemented by the authors through on-site visits to the islands of Andros, Syros and Santorini, during the period of July 2020 to September 2020.

Specifically, the questionnaires combined both quantitative and qualitative variables, and were organized for the field survey assumed in the context of the SPOT HORIZON2020 EU project, which focuses more generally on cultural tourism and Europeanization, aiming towards sustainable development benefitting host communities and regional economies [SPOT Proposal Part A, p.3, http://www.spotprojecth2020.eu]. Informed consent was obtained for such work with the human subjects of our study, according to EU ethics requirements, in the context of the SPOT project, from the Mendel University Committee of Ethics (coordinating institution of SPOT). After obtaining and analyzing the research survey outcomes, the concept of 'cultural sustainability' was employed as a context of discussion of the variable interrelationships between culture and tourism, with an emphasis on the role of these interrelationships in the development of cultural tourism and overall local sustainability, from a bottom-up (destination) perspective.

Data Collection and the Research Sample

As regards the sampling technique employed, the specific locations and survey sites on the three islands were selected as characteristic of the range of different conditions pertaining to tourism and culture on the islands. This approach, based on the axes of tourism and urbanity, was decided upon in order to ensure a certain degree of representativeness vis-à-vis these conditions, in each case. More specifically, locations of both high and low tourist concentration/ development were surveyed, which included both urban and rural/ small town locations, since urbanity also correlates with cultural heritage, cultural activities and certainly cultural tourism. For the island of Andros, these locations were: a) the heavily touristic town of Batsi, b) the capital of the island Chora, and c) the less touristic fishing village of Korthi. 49 resident responses and 47 business representatives' responses were altogether collected in Andros. For the island of Syros, residents were surveyed in a) the highly touristic seaside resort of Galissas, b) the capital city of Hermoupolis, and c) smaller inland towns/villages. In total, 44 resident responses and 9 business representatives' responses were obtained in Syros. Accordingly, for the island of Santorini, onsite survey was conducted in a) the heavily touristic town of Oia, b) the capital of the island Fira (Thera), and c) the less touristic and more residential town of Emporio. 40 resident responses and 30 business representatives' responses were collected there, in total.

Our survey respondents were approached in person in central squares, local shops, cafes, bus/port stations, playgrounds, and seaside promenades. Questionnaires were administered in written and/or electronic form, using Google forms, whereas some respondents preferred on-the-spot completion in the form of an interview, while the researcher read and filled out the questionnaire for them. Additionally, intensive follow-up and e-mail reminders were employed to urge or remind our study participants to deliver the questionnaires. It must be noted that, due to the COVID-19 pandemic, a considerable number of residents avoided contact with our research team members, or refused to fill out printed questionnaires, in order to avoid human contact and exchange of materials (i.e., pens, printed materials, etc.). Finally, the data collected in the form of questionnaire responses were organized, analyzed (both quantitatively and qualitatively) and manipulated with the aid of the SPSS, using descriptive statistics.

A brief description of our interviewees' profiles (Tables 1 and 2) and of the character of local tourism is followed by the findings pertaining to questions of the two questionnaires. The latter comprise 12 closed-ended questions in total (divided into the two questionnaire groups) and 2 open-ended ones, the following: 1) "How does the area profit from cultural tourism?" (open-ended question addressed to the residents) and 2) 'What could be improved in the area – and by whom – in order to improve the experience both for tourists and for your business?' (businesses' open-ended question).

The following section of the chapter presents the survey outputs from the three Cycladic island destinations, ordered by respondent group (residents and business representatives), under which all pertinent questions of the two questionnaires have been encompassed. Pertinent questions represent questions targeting respondents' perceptions, understandings, opinions and attitudes regarding the ways in which currently-practiced cultural tourism may or may not contribute to the viability of local culture and of local cultural tourism in the future, from various perspectives, including economic, social, cultural and environmental parameters and considerations. Nonetheless, additional basic questions about local cultural tourism were included in this study, in order to frame the analysis of cultural and overall sustainability, in its proper context.

Due to their close proximity to and easy accessibility from the large urban metropolis of Athens, the Cyclades tend to attract a very high share of domestic tourism (Prokopiou et al., 2018; Sarantakou &

Table 1. Residents' sample profile

Features	Characteristics of the Residents' Profile Sample			
Gender	males	females	other	
	31%	68%	1%	
Age	20-30	30-50	50-60	over 60
	15%	62%	10%	13%
Country of origin	Greece	Europe	USA	
	97%	2%	1%	
Educational level	Secondary education	High School graduates	Vocational education	University/Master/ PhD graduates
	3%	18%	17%	62%
Profession	Professional	Clerical support/ Elementary worker	Manager	Other
	57%	31%	10%	2%
Household income	Less than €10,000	€11,000 - €20,000	€20,000 - €40,000	More than €40,000
	23%	50%	21%	6%
Household composition	Single person	Couple	Three or four persons	More than 5 persons
	24%	21%	44%	11%

Tsartas, 2015). The majority of our respondents were professionals belonging to low and medium-level economic status categories, and tended to be highly educated (at least, vocational or undergraduate-degree categories). The majority of visitors on the three islands were domestic tourists, while the second most important visitor group were international tourists arriving from the EU, followed by international tourists arriving from outside the EU. Andros and Syros generally tend to attract a great number of domestic tourists, followed by EU tourists. It is only the island of Santorini that seems to attract the majority of its clientele from international visitors, outside the EU. As a rule, Cyclades tourists tend to be professionals, mostly higher-educated, belonging to low and medium-level economic status categories (Tsartas et al., 2020).

As regards the businesses' sample profile, half (48%) of the businesses participating in the questionnaire survey, on all three islands, were hotels and tourism resorts, followed by leisure centers and visitor attraction establishments (27%), catering (9%) and 'other' types of businesses (16%) including tourist shops, travel agencies, museums/ arts galleries, etc. The majority of tourists attracted to these businesses were couples or lone travelers, and secondarily families. Special-interest tourism emerged as the third most important type of tourism for the surveyed businesses, after conventional 3Ss tourism and cultural tourism, i.e. in Andros, it referred mostly to hiking tourism and outdoor activities, while in Santorini it referred to a broad variety of activities with an emphasis on gastronomy/wine tourism.

Generally speaking, the Cycladic islands feature small- and medium-scale tourism, as opposed to 'industrial tourism', meaning that they are not as heavily reliant on mass/ package tourism (with the exception, perhaps of Mykonos and Santorini), a trend also reflected in the locally supplied types of accommodation (Sarantakou & Tsartas, 2015; Sarantakou, 2017; Oikonomou, 2000). The latter include family operations, to a very high degree—some of which may also not be formal/ institutionalized. These trends are especially evident in our business sample profile, where 55% of businesses were reportedly

Table 2. Businesses' sample profile

Features	Characteristics of the Business Profile Sample			
Type of business	Accommodation	Visitor Attraction	Restaurants	Other
	48%	27%	9%	16%
Average number of visitors per year	<1.000	1.000-5.000	5.000-10.000	10.000-50.000
	53%	29%	12%	6%
Total visitor capacity	<1.000	1.000-5.000	5.000-10.000	10.000-50.000
	58%	13%	9%	20%
Type of Accommodation	4* & 5* Hotels	2* & 3* Hotels	1* Hotels/guesthouse	Airbnb type of Vacation rentals
	20%	24%	40%	16%
Number of beds in accommodation	<10	11-20	21-50	51-100
	28%	50%	16%	6%
Estimated occupancy	2-4 nights	5 - 7 nights	8 nights	
	70%	30%	0%	
Business ownership	Family business	Private business	International chain	Public sector/ co operation
	55%	39%	3%	3%
Years of establishment	>10 years	5 10 years	3-5 years	Last 3 years
	60%	19%	8%	13%
Busiest times of business	Summer	Spring	Autumn	School holidays/Christmas
	99%	52%	36%	11%
Origins of visitors	Tourists from the European Union	International tourists from outside of the EU	Domestic tourists	
	28%	12%	40%	

family owned, a very typical trend for the Cyclades tourism industry, and especially so for the islands of Andros and Syros. When destinations, such as Santorini, gain more international popularity and tourism investments, they also tend to attract more privately-owned businesses (39% of our business sample) or to become part of an international hotel chain (3% of our business sample). The growth and development of all other sectors of local/ regional economy in the Cyclades (such as primary-sector activities, services, culture, gastronomy etc.) tend to follow those of tourism, which constitutes the main source of income for the whole region.

Data Presentation and Analysis

Residents' Questionnaire

1. Most and least important local cultural attractions/sites/events according to residents

Most and least important cultural attractions, sites, and events were addressed at the outset of our questionnaire survey, in order to gain an understanding of the significance that the residents placed on their various local cultural assets, in the tourism context. In this regard, 'historical sites and buildings' came

Figure 3. Most important local cultural attractions/sites/events according to residents

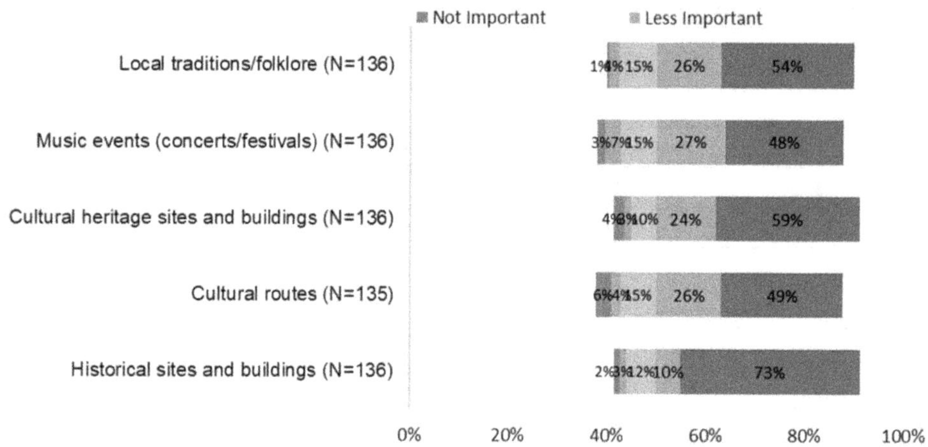

first, as the most significant local cultural attractions category for the islands, according to the majority of the resident respondents, followed by 'Local traditions/folklore' and 'Cultural heritage sites and buildings', and, to a lesser degree, 'Museums' and 'Galleries' (Figure 3). The islands of Andros, Santorini and Syros all boast of important archaeological sites of global significance, such the world-famous Akrotiri site in Santorini, or the site of the Ancient city of Palaiopolis in Andros. Syros also features a very rich heritage of industrial cultural sites (including those pertaining to the long tradition of ship-building on the island) and the famous neoclassical-style buildings of the 19th and early 20th centuries in Hermoupolis. Residents of all three islands, therefore, seemed to acknowledge the importance of these cultural sites and attractions for their islands, at least those that have already received national or international attention. Interestingly, in many cases, residents admitted that they had never visited the museums in their region.

In contrast, the majority of the resident respondents seemed to consider 'Health sites' and 'Sport events' as least important local cultural attractions, followed by 'Dance events' 'Restaurants, food festivals' and 'Townscapes' (Figure 4). These results may simply reflect the residents' own relationship with these attractions and perhaps the latter's inclusion in the residents' local everyday life and mundane routines. For instance, although gastronomy is often associated with culture, it is possible that it does not rate highly in their responses, since most residents tend to eat at home. Dance events often tend to be of significance only for those specific groups active in them, while townscapes may also be deemed not important, due to an overall low level of landscape awareness and appreciation in contemporary Greek society (Pavlis & Terkenli, 2017).

2. Residents' perception of the number of tourists at destination

A large number of respondents (40%) seemed to be at a neutral level as concerned their feelings about the number of tourists visiting their area (Figure 5). Nonetheless, most residents participating in our research survey expressed negative feelings about the incoming numbers of tourists (31% replied that is high and 15% said that it is very high). Therefore, it may be deduced that locals' sentiments about the level of tourist inflow are rather negative, with Santorini seemingly facing possible 'overtourism' and tourism sustainability issues much more intensely than the other two islands. On the other hand, as also highlighted by the findings pertaining to other questions, local residents of our case study tended

Figure 4. Least important local cultural attractions/sites/events according to residents

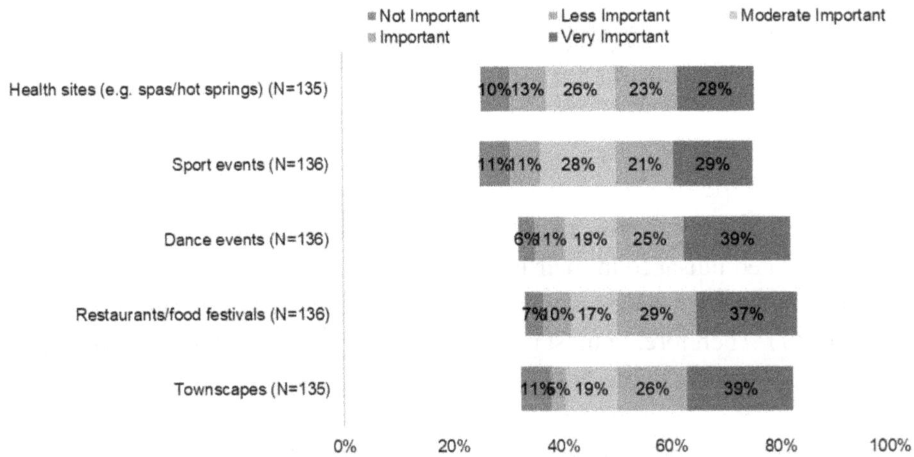

to be generally positive towards tourism, as the latter constitutes the basis of their islands' economies and livelihood.

Figure 5. Residents' perception of the number of tourists at destination

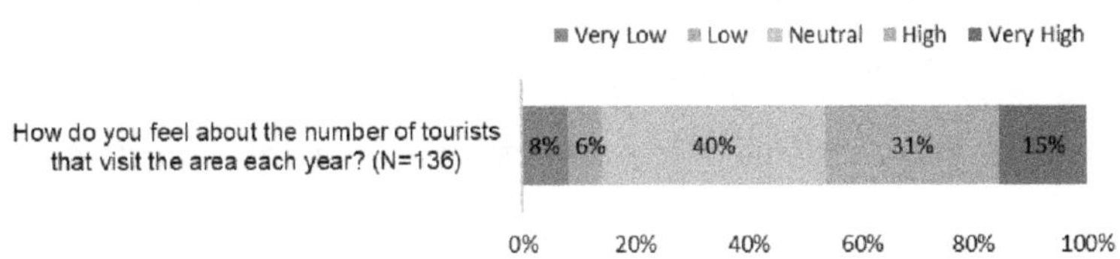

3. Tourism impact on local tradition, according to residents

As expected, the vast majority of our case study resident respondents (73%) stated that tourism also has an impact on local traditions (Figure 6). Most of the respondents rated this impact as neutral (41%), 12% rated it as negative and 7% as very negative, while 27% rated it as positive and 13% as very positive. Thus, overall, the impact of tourism on local traditions was considered to be positive, with lots of space for improvement.

Figure 6. Residents' opinions regarding tourism impacts on local tradition

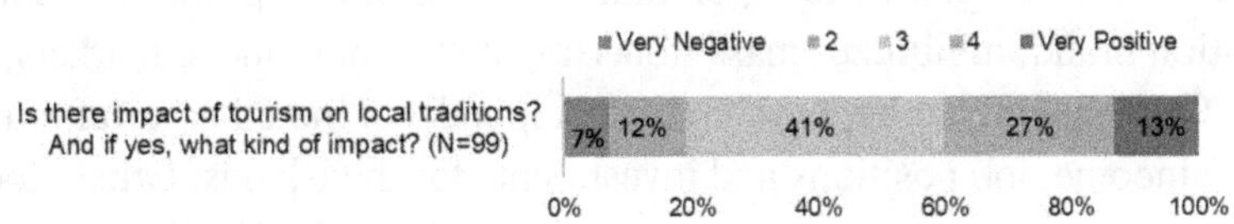

4. Tourism-related nuisance on local daily life, according to residents

Figure 7. Residents' opinions regarding tourism-related nuisance on their daily life

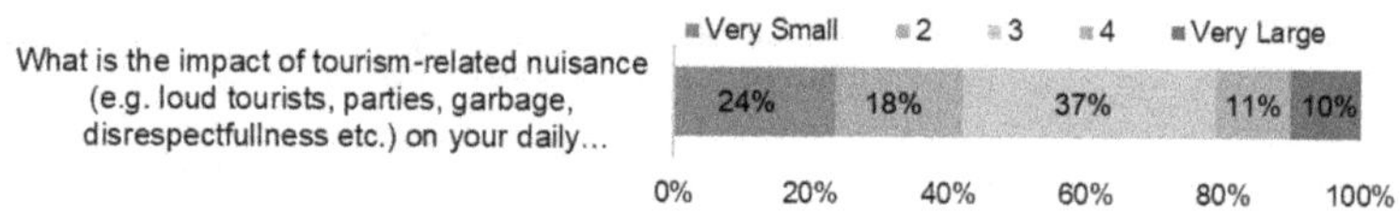

Likewise, these findings show that the majority of resident respondents (37%) declared to be neutral as concerns tourism-related nuisance in their island communities, followed by 24% of our respondent sample who claimed that this nuisance was very small, while 18% of them claimed that such nuisance was just small (Figure 7). Therefore, in most residents' opinion, tourism did not seem to present a serious problem to the locals' everyday life.

5. Perception of the kind of impact that an increase of cultural tourism has on the area (e.g., infrastructure, jobs, quality of life)

Coming now to questions pertaining to cultural tourism in specific, 57% of respondents seemed to believe that cultural tourism could have a very positive impact on their islands, as it could improve not only the economic, but also the social and cultural aspects of their lives (Figure 8). Additionally, 26% of them rated cultural tourism as quite positive for their place. In sum, the respondents' opinion regarding the prospects of cultural tourism for their area was overwhelmingly positive.

Figure 8. Residents' perception of the kind of impact that an increase of cultural tourism has on the area

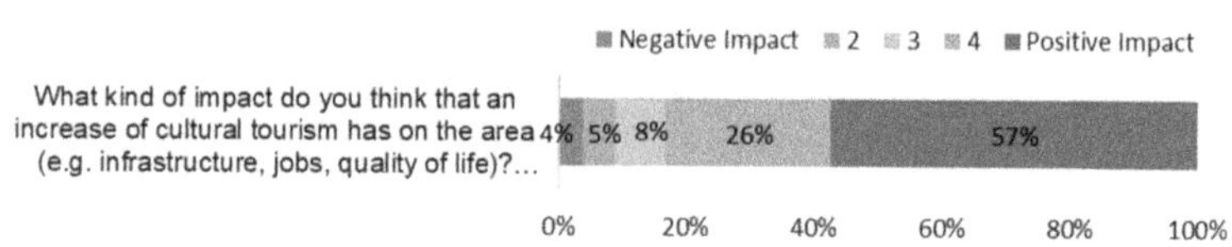

6. Profit from cultural tourism, according to residents (open-ended question)

Out of the 113 residents who took part in our survey, only 81 respondents contributed to this open-ended question, by offering their comments on how their region could generally profit from cultural tourism. Most of them acknowledged the importance of their area's historical and cultural heritage for tourism and expressed their support for further development by both local and national authorities and relevant stakeholders. They all seemed to agree that investing in and promoting cultural tourism can improve the destination brand, minimize 'mass' tourism, attract more niche markets, as well as extend the tourism season. Consequently, a large number of respondents linked cultural tourism to economic benefits of increased income, job positions and investments for the islands. Others focused on the benefits that cultural tourism may impact on the everyday life and standard of living of the local societies. Resident respondents also stated that they wished to feel more connected to their heritage and build closer bonds with it, through cultural activities and socializing. In general, all respondents tended to agree that

Figure 9. Cultural themes offered by local tourism- and culture-related businesses

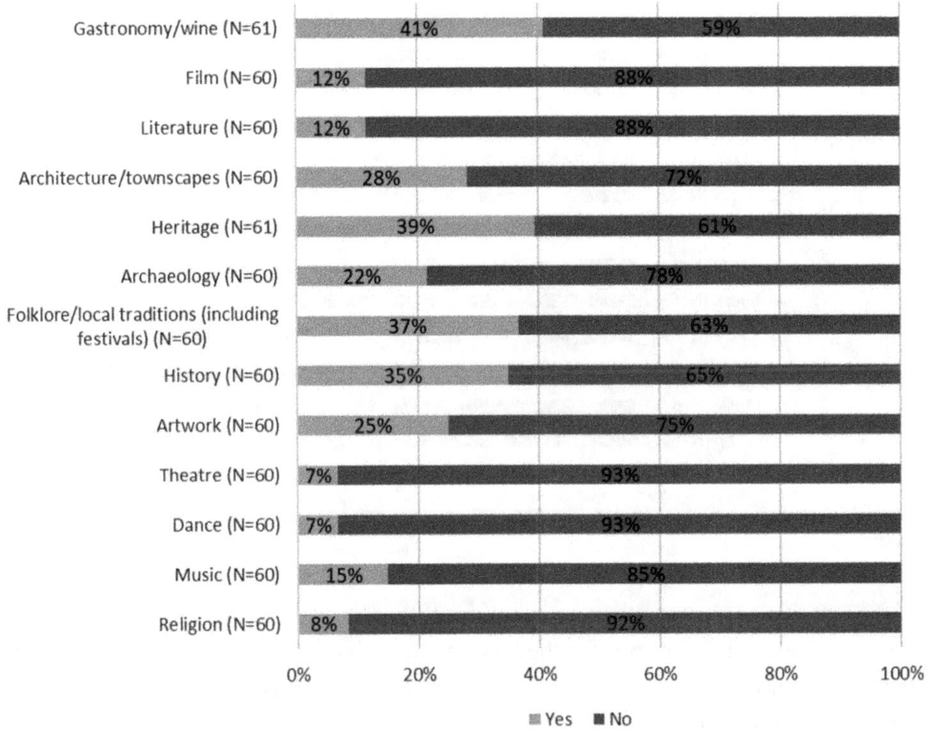

cultural tourism is the most obvious and promising means to upgrade the Cycladic tourism product and to raise both the economic revenues and the cultural standards of their islands.

Businesses' Questionnaire

1. Cultural themes offered by local tourism- and culture-related businesses

To the question as to which cultural services were offered by their businesses (Figure 9), 41% of our respondents replied 'Gastronomy/wine', their type of business being some sort of restaurant/bar. 39% replied that they offered 'Heritage', if their type of business was some local attraction or a museum, while 37% replied 'Folklore/local traditions', since many of the businesses surveyed were tourist souvenir shops. The categories of 'History', 'Architecture' and 'Art' followed next, and finally came all the rest of the categories, which tended to be those more specifically or largely related to some sort of art. This finding is also in accordance with the researchers' own overall survey experience, knowledge of the destination and field impressions.

2. Wished-for development of cultural themes a) for the business and b) for the destination area

This question consists of two parts, which will be addressed separately: a) Wished-for development of cultural themes for the business and b) Wished-for development of cultural themes for the destination area. Beginning with the former question (Figure 10), acknowledging the significance of this series of

Figure 10. Wished-for development of cultural themes for the business (a).

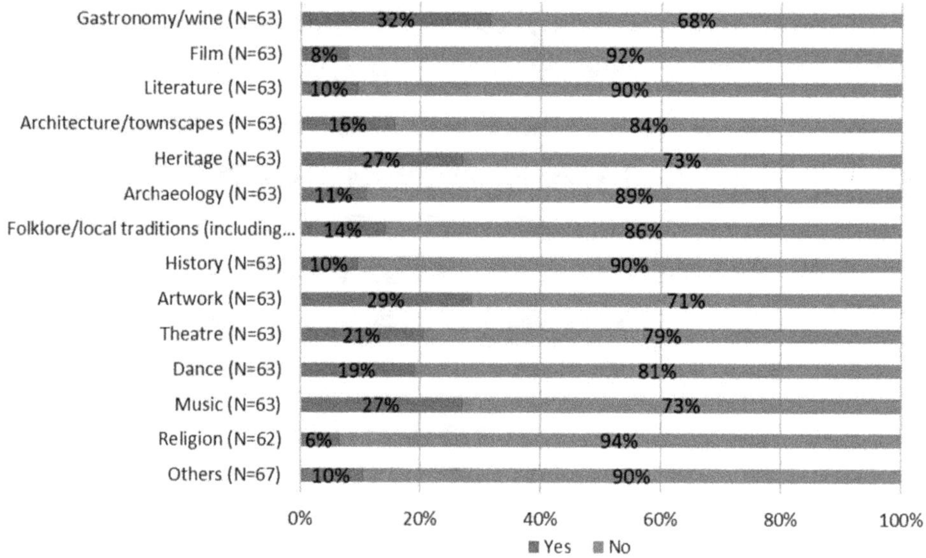

cultural themes/ services and attractions for the business (and indirectly also for the destination at large), the cultural categories most cited here as areas of wished-for business development in the future, were those pertaining to history/heritage/folklore, art and gastronomy. However, the vast majority of business representatives replied with 'No', indicating that their businesses did not seem interested in such future development. 32% were reportedly interested in developing 'Gastronomy and wine', 29% were interested in developing 'Artwork', while 27% were interested in developing 'Heritage' and 'Music', respectively. It may be suggested here that businesses might possibly not be aware that they could develop such cultural themes further in their business, or that they did not have the necessary know-how or tools to enrich their tourism product with cultural attributes and themes/ services. However, the time at which this field survey took place must also be taken into consideration (COVID-19 pandemic era), as great insecurity and uncertainty about the future reigned amongst almost all individuals and businesses interviewed. This fact has been especially relevant in locations highly dependent on tourism such as the Cyclades, since the tourism industry has been marred by some of the most volatile and precarious prospects in the global and local economic sector.

As regards the second part of this question (Wished-for development of cultural themes for the destination area), similarly to the answers received to the previous question, the majority of our respondents replied negatively, indicating that businesses did not especially support or express a strong intention for the development of further cultural themes and services in their area, for the reasons cited above (Figure 11). Accordingly, only 40% of the business representatives surveyed wished for the further development of 'Artwork', 38% seemed to be interested in the further development of 'Folklore/local traditions', 33% of 'Dance', and 32% replied that they wished for the further development of 'Heritage' and 'Architecture respectively at their specific tourist destination area.

3. Feelings about number of tourists visiting business every year

Figure 11. Wished-for development of cultural themes for the destination area (b).

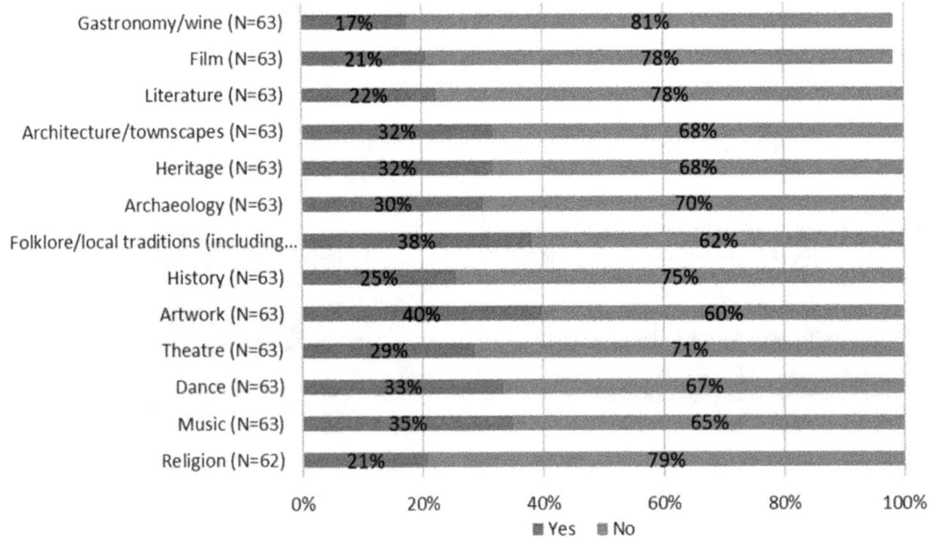

The general picture of the results here follows a normal curve with an emphasis on the higher levels of measurement, namely on respondents' feelings that too many tourists visit the destination and the business every year (Figure 12). Specifically, on a scale of 1 (downward trend) to 5 (upward trend), 45% of business representatives replied that they felt neutral about the number of tourists visiting their businesses every year, whereas 36% said that they felt that they attracted relatively high numbers of tourists and 9% said that this number was one 'too many'. Although, according to some researchers, 'overtourism' seems to be a problem mostly referring to specific Cycladic destinations such as the very touristic island of Santorini, there generally seems to exist a certain widespread feeling of dismay with regard to the large tourism inflows, especially due to the phenomenon's intense seasonality.

Figure 12. Business's feelings about number of tourists visiting business every year.

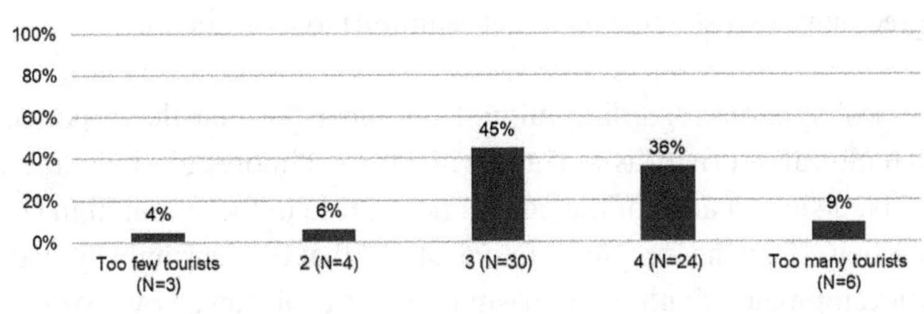

4. Statements regarding the business's visitors

As regards businesses' perceptions of cultural tourists' trends and behaviors, the cultural appeal of our three islands as tourism destinations seemed to run high, according to the responses of business representatives participating in the field survey. Specifically, 55% of them seemed to agree strongly and an additional 24% of them seemed to agree that 'It is important for my visitors to get a taste of the local

Figure 13. Business's statements regarding the business's visitors.

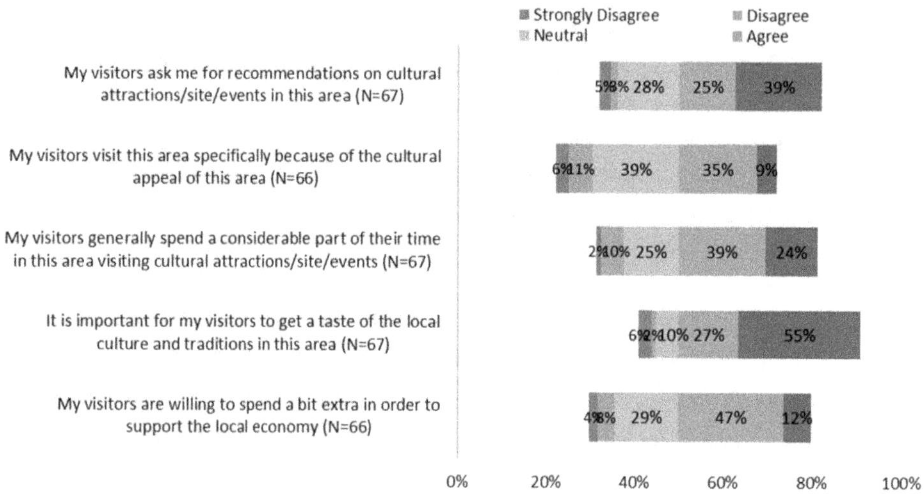

culture and traditions in this area', despite the fact that cultural tourism is not the primary motive for tourist visits to these islands (Figure 13). The visitors' interest in cultural attractions was reported by the business representatives to be strong by 25% and very strong by 39%, in conjunction with the statement 'My visitors ask me for recommendations on cultural attractions/site/events in this area'. Furthermore, 24% of our respondents declared that they strongly agreed and an additional 39% said that they agreed with the statement that 'My visitors generally spend a considerable part of their time in this area visiting cultural attractions/site/events.'

Despite the fact that these visitors, according to the business representatives, do not visit these islands for their cultural attractions by and large, and are rather disinclined to pay more for them in order to support the local economy, it appears that businesses value the cultural heritage and other aspects of their destinations and acknowledge the importance and the interest of visitors for such attractions, sites and events on the islands.

5. Businesses' agreement with statements about (cultural) tourism in area

As regards businesses' opinions regarding cultural tourism in the area, the respondents' stated interest in and concern about (cultural) tourism, as well as the inclusion of more cultural attractions in the tourism product and the perceived significance of the government's role in these, run high (Figure 14). Specifically, 66% of our interviewed business representatives stated that they strongly agreed that 'Government should invest in the development of cultural tourism in the area', followed by 64% of them also strongly agreeing that 'Government should help businesses with developing and increasing cultural tourism in this area'. 54% of them also stated that they strongly agreed with the statement that 'it is important that cultural tourism is present in this area' and 51% also strongly conceded with the high potential of cultural tourism in the area. Most responses to the other questions/statements were fairly evenly distributed among the different answer categories, with the exception of responses vis-à-vis the statement 'Cultural tourism is well developed in this area', where most of our respondents had a rather negative opinion, thus reinforcing previous findings. In general, the majority of business representatives state that they

Figure 14. Businesses' agreement with statements about (cultural) tourism in area.

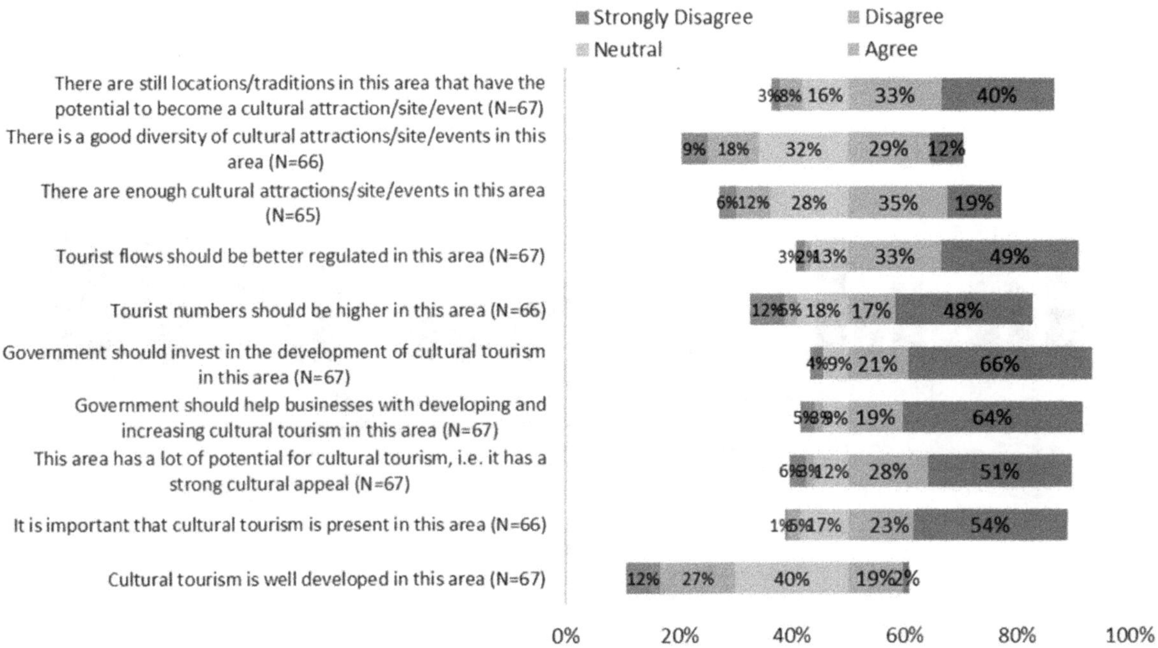

agree or strongly agree with the importance of developing and support the islands' unfulfilled potential for cultural tourism growth for the benefit of tourism and economy.

6. Three factors that could provide most economic improvement in the area

When asked to provide three factors which, according to their opinion, would lead to local economic improvement, business representative respondents emphasized basic infrastructures as the most significant factor for the improvement of the standard of living for both tourists and the islands' permanent population (Figure 15). They specifically rated 'Infrastructure (such as improvements in the road network)' as the first most important aspect for economic improvement by 57%, followed by 'Information provision to tourists' by 43%, and then several other provisions relating to tourism promotion and attraction. The second most important factor, according to their opinion, was 'Public Transport' (72%), followed by 'Events' (44%) and 'Attractions' (24%), very much in line and in agreement with their first reportedly most important factor. The third most important factor, as offered by the business representatives, was a series of aspects mainly pertaining to the tourism sector, including the categories 'Other' (75%), 'Gastronomy' (67%), 'Publicity/marketing' (48%) and 'Attractions' (48%).

Generally speaking, as regards the three most significant factors towards local economic improvement, in all three of our survey research sites, the respondents mostly favored basic infrastructures which tend to improve both the standard of living of the permanent local/ regional population and also contribute to the enhancement of tourist attraction. It is also evident that promoting the islands further or developing their cultural activities and attractions was broadly considered by tourism professionals to play a very important role towards the area's economic improvement.

Figure 15. Businesses' responses regarding three factors which may lead to the destination's economic improvement

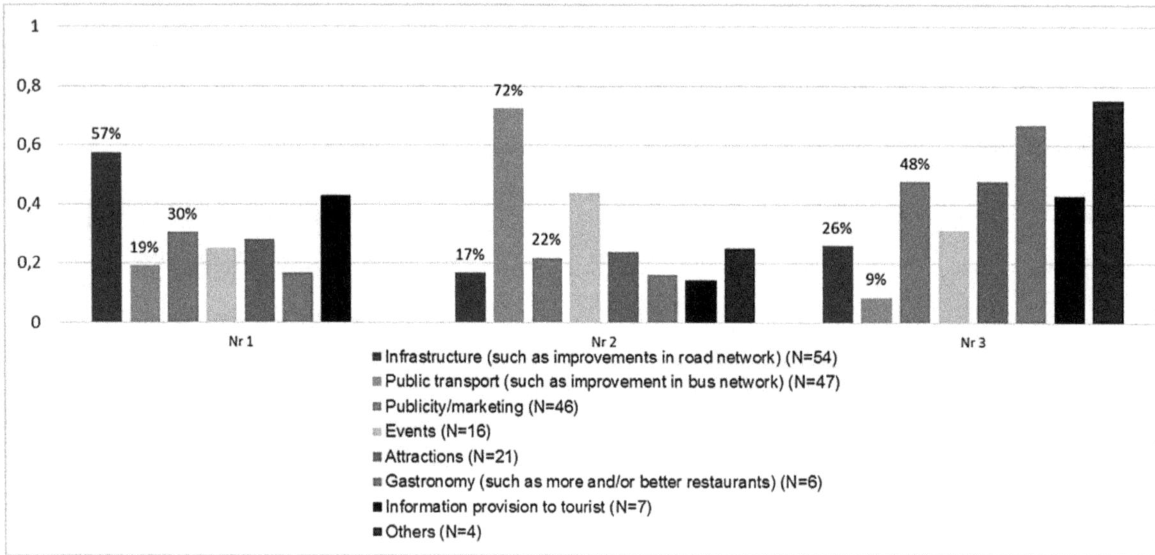

7. Statements regarding cooperation between business, tourist businesses in general, and the cultural sector

Responses to this question indicated that cooperation between interviewed businesses and tourism-related businesses in general with the cultural sector is very low, although there seems to be general agreement that much potential lies in such cooperation (Figure 16). The majority of respondents strongly agreed (25%) or agreed (45%) with the statement that 'I see potential in more cooperation between tourist businesses in this area and the cultural sector', followed by those who strongly agreed (24%) or agreed (43%) with the statement that 'I see potential in more cooperation between my business in this area and the cultural sector'. Cumulatively, these are quite high percentages, emphasizing a widely held understanding of the significance and potential of such cooperation. Moreover, only 22% agreed and 2% strongly agreed with the statement that 'There is cooperation between tourist businesses in this area and the cultural sector' of the island destination, pointing to this cooperation deficiency.

8. Factors of general improvement in the area (open-ended question)

Finally, as regards the responses received from businesses vis-à-vis what can be improved on their island in order to benefit tourists' and businesses' experience, these findings reinforced previous ones of closed-ended questions. On all three islands, the majority of our respondents stressed the importance of improving road and transport infrastructure, just as they had also done in response to other similar questions above, although out of the 67 businesses of our survey sample, only 40 of them contributed to this open-ended question. More specific responses referred to building new road connections and to improving road maintenance. Reportedly, all three islands seemed to offer satisfactory public transport facilities during the summer months, but residents and visitors alike complained about much more

Figure 16. Statements about cooperation between, tourist businesses and the cultural sector.

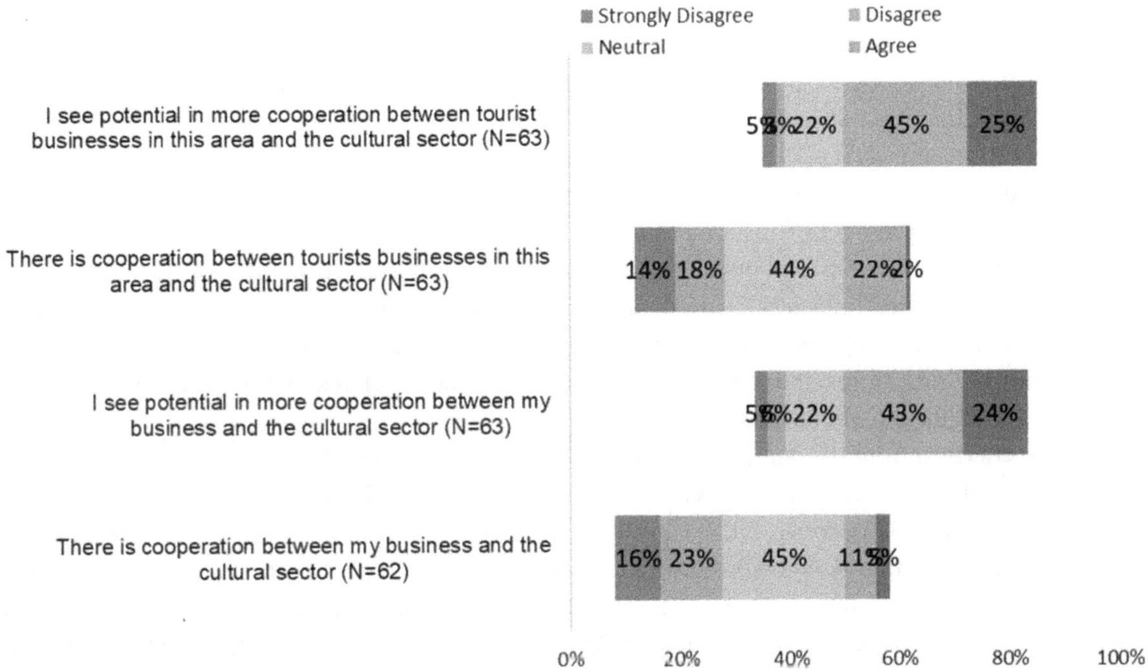

limited choices during off-season times, due to the severe phenomenon of seasonality, typical of Greek island tourism. Additionally, responses stressed the need and significance of more and more frequent ferry connections among the islands and not just between the individual islands and the two main ports of Athens (Piraeus and Rafina). Several other responses addressed the role that Municipalities tend to play and need to play, with regard to improved cleanliness, accessibility, car parking spaces and public area Wi-Fi and lighting. There was a general call for Municipalities to be more assertive and supportive as concerns the organization of cultural events and activities, by providing infrastructure, staff and/or funding to artists and/ or cultural organizations.

The importance of communicating and promoting the natural and cultural beauties of the islands was also stressed by our respondents, by focusing on the islands' cultural heritage, modern arts, cultural events and sports, wine and gastronomy, which, according to their opinions, offer a lot of potential for further cultural tourism growth. The importance of supporting tourism businesses and entrepreneurship was often mentioned in connection with various forms of financial support, vocational staff training, promotion of synergies between private and public organizations, reduced taxes, loans for investment, etc., in the tourism sector. Finally, a number of respondents stressed the value and significance of protecting the physical environment, local character, architecture and heritage of the islands, by imposing more restrictions on hotel construction and other tourism infrastructure development, as well as on mass/ package tourism, in favor of milder, more sustainable forms of tourism and generally local/regional sustainable development. Specifically, for the case of Santorini, most of the answers referred to solutions to problems of 'overtourism' and to the importance of controlling and restricting the number of cruise-ship arrivals—a measure that has been long considered as key to curbing such phenomena on the island.

DISCUSSION AND CONCLUSION

The purpose of this case-study analysis was to investigate cultural tourism perceptions, practices, concerns and prospects, among local residents and business representatives in the Cyclades, and to draw their implications for local cultural sustainability. This exploratory study attempted to achieve this, with a help of a questionnaire survey (including both quantitative and qualitative variables) focusing on local culture, tourism and cultural tourism. Furthermore, the questions addressed to our survey respondents were pertinent to the prospect of local cultural and overall sustainability from all four perspectives/ aspects/ 'pillars' of the concept and practice of sustainability, such as the economic, the social, the environmental, but they predominantly focused on the cultural aspects of destination sustainability. These selected questions, framed by some additional basic background questions concerning Cycladic cultural tourism, aimed at bringing together residents' and businesses' perspectives to this bottom-up endeavor to configure future local (cultural) sustainability.

The results derived from our questionnaire survey point to a very broad acknowledgment of the significant role of culture in local tourism development, as well as that of the implications/repercussions of tourism for local culture (Coccossis, 2011; Coccossis, 2019; Tsartas, 2003; Weaver, 2006; d'Hauteserre, 2017; Cousin, 2008; Girard & Nijkamp, 2009), for both categories of our survey participants. Furthermore, the vast majority of the two categories of respondents tended to agree that cultural tourism is the most obvious and promising means to upgrade the Cycladic tourism product, promote the island destination, boost local/ regional economies and raise the cultural standards of their islands—in short, enhance destination (cultural) sustainability.

Specifically, and in accordance with other similar research findings (Tsartas et al., 2020; Coccossis, 2006; d'Hauteserre, 2017; Liu, 2003), local residents seemed to acknowledge that tourism generally has a positive impact on their islands, but with much space for improvement, e.g. curbing the numbers of incoming tourists and improving tourism impacts on local culture, despite the fact that their opinions were rather spread-out between positive and negative opinions, as regards the latter. They declared that they largely did not mind tourism-related nuisance to their island communities, and acknowledged the crucial role (cultural) tourism plays in the local economy and overall development. Similarly positive findings, upholding the significance of culture for tourism, were obtained from the best part of the surveyed businesses in the study area. Nonetheless, it is important to note that, according to the business representatives, these visitors, by and large, did not visit the three islands primarily for their cultural attractions (but rather for 3Ss tourism), and were consequently rather disinclined to pay more for them, in order to support the local economy. Mirroring similar answers received from residents, businesses seemed to value the cultural heritage and other aspects of their destinations and acknowledged the visitors' stated importance of and interest in such attractions, sites and events at these island destinations.

Taking in consideration that the majority of visitors of the Cycladic islands are domestic tourists, the received answers were in accordance with how Greek tourists choose to spend their holidays (Sarantakou & Tsartas, 2015; Tsartas, 2003; INSETE, 2020; Berg & Edelheim, 2012). For example, local traditions and religious feasts/festivals are very popular events to attend, for all ages and types of Greek and foreign (especially repeat) tourists. These celebrations include food and wine consumption, live local music and dance and usually take place outdoors, during the summer months. Taken all together, such cultural attractions become a most significant pole of tourist attraction for these islands, constituting a competitive advantage vis-à-vis other destinations (Berg & Edelheim, 2012; Coccossis et al., 2011). Obviously, as the Cycladic islands are famous for their crystal-clear waters and beautiful beaches, enjoying the sea and sun generally remains the most important visitation motive, which also generally explains the seasonal character of Cycladic tourism (Tsartas et al., 2020; Berg & Edelheim, 2012; Saariner, 2006; Prokopiou et al., 2018). As regards the opposite relationship, namely tourism repercussions on and implications for local culture, on the one hand, answers tended to be more mixed, in line with other tourism research (Tsartas et al., 2020; Coccossis, 2009; d'Hauteserre, 2017). On the other hand, however, locals did not seem to mind the presence of tourism amidst them very much, a finding which is also in line with other relevant research, illustrating the primacy of the economic significance of tourism over other tourism-related concerns for locals, in popular Greek island destinations (Tsartas et al, 2020; Sarantakou & Terkenli, 2019; Terkenli et al., 2007).

Coming to the question of cultural sustainability and the role of tourism therein, local culture—as part of the tourist product—was generally viewed in a very positive light for the area and local community, according to the theoretical/ conceptual model of cultural tourism (Figure 1), which these findings support and fortify. A very significant role was assigned by both residents and local businesses to cultural tourism, as regards the establishment and enhancement of local cultural sustainability, as shown by the aforementioned model, through the acknowledgement of the importance of cultural sites/ attractions for cultural tourism development, and especially of those sites/ attractions that had already received attention by the Greek State and the international community (i.e. historical/ archeological sites/ buildings and local traditions/ folklore). In parallel, there was general agreement among both categories of interviewees about the unfulfilled potential for cultural tourism growth, for the benefit of tourism and the economy more generally, and culture more particularly. This potential was explicitly stated and the need to meet it was emphasized especially by the interviewed business representatives, who also pointed out the lack of general collaboration among all relevant parties (tourism-related businesses, the authorities and the cultural sector). There was an overwhelming call and appeal for governmental tourism-related support, funding and regulation of cultural activities, heritage preservation and promotion and investment in infrastructure (i.e. ports of entry and public transport). Although there was an acknowledgement by the survey participants of the low degree of local cultural development so far, they stated that they did not necessarily want more tourism. Rather, they indicated that they were in favor of further developing cultural tourism, in specific. In this way, they showed that they were cognizant of and sensitive about issues of local cultural sustainability, which they deemed could be further improved with the right measures and especially with infrastructure development, without compromising their cultural integrity, economic independence and socio-economic standard of living. Decidedly, local residents and tourism- or culture-related businesses appeared to have taken the situation in their hands, in the relative absence of the State regarding most aspects of cultural tourism planning and management, after a long period of depression in the country.

Conclusively, the survey findings revealed the high degree of concern of our study participants for the state and potential of their area's overall sustainability—a crucial outcome concerning the Cyclades which are highly dependent on tourism for their economic sustenance and development. Better promotion of the islands and/ or further development of their cultural activities and attractions were broadly thought to play a very important role towards economic improvement, and thus both cultural and overall local sustainability. The findings also respond to the principles central to the role of cultural tourism in fostering overall, as well as culturally, sustainable development, in accordance with Soini & Birkeland's perspective of 'economic viability' (2014) and the ICOMOS International Charter for Cultural Heritage Tourism 2021. The latter advocates, among other things, placing sustainable cultural management at the center of responsible cultural tourism planning and management, through cooperation for heritage conservation among all stakeholders involved in tourism (ICOMOS, 2021). 'Mass' or 3Ss tourism does not seem to respond to local needs and priorities and to fulfill local bottom-up aspirations for overall sustainable development. Admittedly, the sustainability target for these islands ought to be a combination model of conventional and alternative/ special-interest tourism development, taking into account local communities' needs and the particularities of the Cycladic geography, history and culture (Coccossis, 2001; Coccossis & Tsartas, 2019; Prokopiou et al., 2018).

Albeit exploratory in scope, this study aimed at a synthetic, comparative and in-depth probe into the tourism-culture relationship, in three Aegean islands, towards a bottom-up understanding of this relationship and its role in cultural tourism development, and ultimately in local cultural sustainability. However, as this research survey was undertaken in the context of a broader research project (SPOT), with goals supporting and referring to general sustainability, more focused research specifically on the drivers and workings of cultural sustainability ought to be assumed, in order to tackle the latter area of inquiry more effectively. Such research ought to be replicated, for comparative purposes, in time (repeated in regular time-intervals) and space, e.g. additional Aegean/ other islands or mainland locations of Greece and/ or other tourist destinations. It also ought to be extended in scope, content and depth and undertaken using larger and more diverse research samples, including visitor samples. The practical contribution of such present and future research refers to the generation and analysis of such knowledge, views, prospects etc., for purposes of appropriation from all sides involved—including stakeholders, professionals, and decision-makers—towards more sustainable tourism and cultural destination planning, management, development, preservation, consumption, and, not the least, towards local/ regional life improvement.

ACKNOWLEDGMENT

The authors would like to thank Mr George Tataris who created the case study map and Professor Nikolaos Soulakellis, Head of the Cartography and Geoinformatics Laboratory, Department of Geography, University of the Aegean, where the map was produced.

This research is part of the EU SPOT project, funded by the European Union's Horizon 2020 program for research and innovation under grant agreement no. 870644.

REFERENCES

Adams, W. (1995). Sustainable development. In Geographies of Global Change: Remapping the World in the Late Twentieth Century. Blackwell.

Auclair, E., & Fairclough, G. (2015). Living between past and future: An introduction to heritage and cultural sustainability. In E. Auclair & G. Fairclough (Eds.), *Theory and practice in heritage and sustainability* (pp. 1–22). Routledge. doi:10.4324/9781315771618-1

Berg, I., & Edelheim, J. (2012). The attraction of islands: Travelers and tourists in the Cyclades (Greece) in the twentieth and twenty-first centuries. *Journal of Tourism and Cultural Change*, *10*(1), 84–98. doi:10.1080/14766825.2012.660946

Boström, M. (2012). A missing pillar? Challenges in theorizing and practicing social sustainability: introduction to the special issue. *Sustainability: Science, Practice and Policy*, *8*(1), 3-14.

Coccossis, H. (2001). Sustainable development and tourism in small islands: some lessons from Greece. *Anatolia - An International Journal of Tourism and Hospitality Research*, *12*(1), 53–58.

Coccossis, H. (2006). Tourism, Technological Change and Regional Development in Islands. In M. Giaoutzi & P. Nijkamp (Eds.), *Tourism and Regional Development: New Pathways* (pp. 271–280). Ashgate.

Coccossis, H. (2009). Sustainable Development and Tourism: Opportunities and Threats to Cultural Heritage from Tourism. In L. Fusco Girard & P. Nijkamp (Eds.), *Cultural Tourism and Sustainable Local Development* (pp. 47–56). Ashgate Publishing.

Coccossis, H., Tsartas, P., & Griba, E. (2011). *Special and Alternative forms of Tourism- Demand and Supply of new Tourism Products*. Athens: Kritiki.

Coccossis, H., & Tsartas, P. (2019). *Sustainable Tourism Development* (2nd ed.). Kritiki. (in Greek)

Cousin, S. (2008). UNESCO and the doctrine of cultural tourism. Genealogy of 'good tourism'. *Civilisations*, *57*(1), 41–56. doi:10.4000/civilisations.1541

Culture 21. (2011). *Lobbying for culture as the 4th pillar of sustainable development in the process of the Rio+20 summit*. Agenda 21 for culture. United Cities and Local Governments – Committee on Culture.

d' Hauteserre, A.M. (2017). Cultural tourism in the French Pacific. *Shima, 11*(1), 55–78.

Duxbury, N. (Ed.). (2021). *Cultural Sustainability, Tourism and Development. (Re)articulations in Tourism Contexts*. Routledge. doi:10.4324/9780367201777

Empacher, C., & Wehlung, P. (2002). *Soziale Dimensionen der Nachhaltigkeit*. ISOE-Studientexte.

Gruen, R., Elliott, J., Nolan, M., Lawton, P., Parkhill, A., McLaren, C., & Lavis, J. N. (2008). Sustainability science: An integrated approach for health-program planning. *Lancet*, *372*(9649), 1579–1589. doi:10.1016/S0140-6736(08)61659-1 PMID:18984192

Hawkes, J. (2001). *The fourth pillar of sustainability: Culture's essential role in public planning*. Common Ground Publishing Pty Ltd and Cultural Development Network Victoria.

Helgadóttir, G., Einarsdóttir, A. V., Burns, G. L., Gunnarsdóttir, G. Þ., & Matthíasdóttir, J. M. E. (2019). Social sustainability of tourism in Iceland: A qualitative inquiry. *Scandinavian Journal of Hospitality and Tourism*, *19*(4-5), 4–5, 404–421. doi:10.1080/15022250.2019.1696699

Hughes, E., & Scheyvens, R. (2016). Corporate social responsibility in tourism post-2015. A Development First approach. *Tourism Geographies*, *18*(5), 469–482. doi:10.1080/14616688.2016.1208678

ICOMOS. (2021). *International Charter for Cultural Heritage Tourism 2021: Reinforcing cultural heritage protection and community resilience through responsible and sustainable tourism management*. ICOMOS International Committee on Cultural Tourism.

INSETE. (2020). *Key figures of incoming tourism in Greece*. https://insete.gr/statistika-eiserxomenou-tourismou

James, P., Magee, L., Scerri, A., & Steger, M. (2015). *Urban Sustainability in Theory and Practice: Circles of Sustainability*. Routledge.

Janoušková, S., Hák, T., Nečas, V., & Moldan, B. (2019). Sustainable Development—A Poorly Communicated Concept by Mass Media. Another Challenge for SDGs? *Sustainability, 11*(11), 3181. doi:10.3390u11113181

Katz, C. (1998). Whose nature, whose culture? Private productions of space and the preservation of nature. Remaking Reality: Nature at the End of the Millennium, 46.

Knezevic Cvelbar, L., Antonucci, B., Cutrufo, N., Marongiu, L., Rodrigues, M., & Teoh, T. (2021). Research for TRAN Committee – Relaunching transport and tourism in the EU after COVID-19. Tourism sector, European Parliament, Policy Department for Structural and Cohesion Policies.

Liu, Z. (2003). Sustainable Tourism Development: A Critique. *Journal of Sustainable Tourism, 11*(6), 459–475. doi:10.1080/09669580308667216

Murphy, K. (2012). The social pillar of sustainable development: A literature review and framework for policy analysis. *Sustainability: Science, Practice and Policy, 8*(12), 15–29.

Oikonomou, D. (2000). A System of spatial planning: Greek reality and international experience. Review of Social Researches (Epitheorissi Koinonikon Erenvon), 101(101), 3.

Pavlis, E., & Terkenli, T. (2017). Landscape values and the question of cultural sustainability: Exploring an uncomfortable relationship in the case of Greece. *Norsk Geografisk Tidsskrift, 71*(3), 168–188. doi: 10.1080/00291951.2017.1345977

Prokopiou, D. G., Mavridoglou, G., Manologlou, S., & Tselentis, B. S. (2018). Tourism development of the Cyclades islands: economic, social and carrying capacity assessment and consequences. WIT Transactions on Ecology and the Environment, 217.

Saarinen, J. (2006). Traditions of sustainability in tourism studies. *Annals of Tourism Research, 33*(4), 1121–1140. doi:10.1016/j.annals.2006.06.007

Sarantakou, E., & Terkenli, T. S. (2019). Non-institutionalized forms of tourism accommodation and overtourism impacts on the landscape: The case of Santorini, Greece. *Tourism Planning & Development, 16*(4), 411–433. doi:10.1080/21568316.2019.1569122

Sarantakou, E., & Tsartas, P. (2015). A critical approach to the new framework for creating tourism investment during the current period of economic crisis 2010-2014. *Greek Economic Outlook, 26*, 46–55.

Saunders, F., Gilek, M., Ikauniece, A., Tafon, R. V., Gee, K., & Zaucha, J. (2020). Theorizing Social Sustainability and Justice in Marine Spatial Planning: Democracy, Diversity, and Equity. *Sustainability, 12*(6), 2560. doi:10.3390u12062560

Sawyer, S., & Tapia, A. (2005). The sociotechnical nature of mobile computing work: Evidence from a study of policing in the United States. *International Journal of Technology and Human Interaction, 1*(3), 1–14. doi:10.4018/jthi.2005070101

Shaw, G., & Williams, A. M. (1994). *Critical issues in tourism: A geographical perspective*. Basil, Blackwell.

Sheppard, E., Porter, P. W., Faust, D. R., & Nagar, R. (2009). *A World of Difference: Encountering and Contesting Development* (2nd ed.). Guilford Press.

Sneddon, C. S. (2000). 'Sustainability' in ecological economics, ecology and livelihoods: A review. *Progress in Human Geography*, *24*(4), 521–549. doi:10.1191/030913200100189076

Soini, K., & Birkeland, I. (2014). Exploring the scientific discourse on cultural sustainability. *Geoforum*, *51*, 213–223. doi:10.1016/j.geoforum.2013.12.001

Statista. (2021). *Number of foreign arrivals in hotel establishments on the Cyclades Islands in Greece from 2010 to 2018*. https://www.statista.com/statistics/881851/number-foreign-arrivals-cyclades-hotels

Su, X. (2018). Reconstructing Tradition: Heritage Authentication and Tourism-Related Commodification of the Ancient City of Pingyao. *Sustainability*, *10*(3), 670. doi:10.3390u10030670

Terkenli, T. S. (2000). Landscapes of tourism: a cultural geographic perspective. In H. Briassoulis & J. Van Der Straaten (Eds.), *Tourism and the Environment: Regional, Economic, Cultural and Policy Issues* (2nd ed., pp. 179–202). Kluwer Academic Publishers. doi:10.1007/978-94-015-9584-1_11

Terkenli, T. S., Bellas, M. L., & Dudley-Jenkins, L. (2007, June-December). Tourism impacts on local life: Socio-cultural continuity and change in Crete, Aegean. *The Geographical Journal*, *16*(1-2), 37–52.

Thimm, T. (2019). Cultural sustainability – a framework for Aboriginal tourism in British Columbia. *Journal of Heritage Tourism*, *14*(3), 205–218. doi:10.1080/1743873X.2018.1484469

Throsby, D. (2003). Cultural sustainability. In R. Towse (Ed.), *A handbook of cultural economics* (pp. 183–186). Edward Elgar. doi:10.4337/9781781008003.00028

Throsby, D. (2008). Linking ecological and cultural sustainability. *The International Journal of Diversity in Organisations, Communities and Nations*, *8*(1), 15–20. doi:10.18848/1447-9532/CGP/v08i01/39529

Tsartas, P. (2003). Tourism Development in Greek Insular and Coastal Areas: Sociocultural Changes and Crucial Policy Issues. *Journal of Sustainable Tourism*, *11*(2&3), 116–132. doi:10.1080/09669580308667199

Tsartas, P., Zagotsi, S., & Kyriakaki, A. (2020). *Τουρίστες, Ταξίδια, Τόποι. Κοινωνιολογικές Προσεγγίσεις στον Τουρισμό [Tourists, Travels, Places. Sociological Perspectives to Tourism]*. Kritiki Publications.

United Nations. (2015). Transforming our World: the 2030 Agenda for Sustainable Development. Resolution adopted by the General Assembly on 25 September 2015. New York: United Nations.

United Nations Brundtland Commission. (1987). *Our Common Future*. Oxford University Press.

United Nations World Tourism Organization. (2018). *Report on Tourism and Culture Synergies*. UNWTO.

Weaver, D. B. (2006). *Sustainable Tourism: Theory and Practice*. Elsevier Butterworth-Heinemann.

Williams, R. (1958). *Culture and Society*. Chatto & Windus.

World Population Review. (2021). *Most Visited Countries*. https://worldpopulationreview.com/country-rankings/most-visited-countries

WTTC. (2020). *Travel and Tourism-Global Economic Impact & Trends 2020*. WTTC.

Chapter 16
Favela Tour Experience:
The Impacts in the Host Communities

Natalia Gonçalves da Silva
University of the Algarve, Portugal

Cláudia Ribeiro de Almeida
https://orcid.org/0000-0002-0790-3152
CinTurs, University of the Algarve, Portugal

ABSTRACT

Tourism performed in slums has been raising the interest of researchers and generating controversial opinions regarding its objectives. For this reason, the present study aims to verify the impacts of the slum tours in the host communities by analyzing the pre-concepts and ideas associated with this tourism activity through the collection of data attained by the application of a survey, comparing it with the results obtained during the visits to the slums of Rocinha and Santa Marta. For its validity, the research performed by different authors are used as a guidance reference allowing the reader to have a broader perspective of the phenomena here presented. As for the obtained results, the study shows that the benefits of the tours for those communities can be divided in tangible (offers economic advantages) and intangible (offers change in perspective and reduction of the prejudice).

INTRODUCTION

From the prehistory to tourism take-off era, as discussed by Lickorish and Jenkins (1997), tourism has suffered several modifications. Initially, just practiced by a restricted part of the society (rich, educated and/or by people belonging to the aristocracy), it has evolved into a business that englobes a great number of adepts.

With the technological development and the changes on the needs of the population, the creation of new tourism experiences was required to satisfy the most adverse wishes and curiosities, providing an authentic sensation in an (un)known environment.

DOI: 10.4018/978-1-7998-9217-5.ch016

"Talaya (apud Ramiro, 2004) shows that the behavior of the tourist has followed new trends bringing major changes, and the main ones are the most active and versatile entertainment, the search for experiences, fragmentation of the trips that are increasingly more frequent and shorter, more direct sales channels with the increasing use of the internet, greater demand for customized products defined by the very tourists, greater need for information, and segmenting markets and destinations. (Rezende, 2014, p. 358)"

It is in this scenario that the conception of a new form of tourism has arisen: the reality tourism.

It appears as a response to the demands of tourists seeking for innovative tours that challenge and change their perception of the world, by encouraging the visitors to reconsider their prejudices/pre-concepts related to places affected by social inequality and/or disasters. It is divided in two main categories: dark tourism – focused on visiting places in which occurred natural disasters and/or historical places famous for several deaths; and social tourism – focused on visiting to places with social disadvantages (Freire-Medeiros, 2014; Nisbett, 2017).

As part of the second category, focus of this paper, the practice of tourism in places filled with poverty are not new, dating back to the English Victorian Period (Frenzel et al., 2015; Tavares et al., 2018). However, it was only a century later, in the early nineties, that this type of tourism gained a greater attention by the world population.

The slums in India, townships in South Africa and favelas in Rio de Janeiro started presenting a higher transit of people coming from different areas of the globe interested to see the life in these communities, being exposed to a reality very different to the one they are used to (Freire-Medeiros, 2006, 2008, 2014; Frenzel et al., 2015, Tavares et al., 2018). Freire-Medeiros (2007) states that this type of practice summarizes the premises of the two types of reality tours – it allows an altruistic and politically correct engagement as well as motivates the feeling of adventure and enchantment, since it is an authentic-exotic-risky-tragic experience.

Frenzel et al. (2015, p. 237) defines it as a "mass tourism phenomenon occurring only in a few destinations and a niche form of tourism in a growing number of other destinations (…)". In the city of Rio de Janeiro, the crescent movement of tourists in slums became a topic in several studies, generating questionings such regarding the usage of poverty as a form of tourism and its outcomes for the local host community (Freire-Medeiros, 2014).

Considering the popularity of tourism trend mentioned above, this present research has the objective to explore the experiences provided by the "Favela Tour" and their impacts in the host communities. For this, this study has been divided into different sections to provide a global understanding of the following topics: history of tourism (from its creation to its new trends), history of the slums in Brazil, and development of the slum tourism (pre-concepts, tourism process and its effects in two Brazilian communities).

For the maturing of the investigation, the slums of Rocinha and Santa Marta, both located in the city of Rio de Janeiro, were visited and used as objects of study, due to their good reputation in tourism sector. In addition to it, an online survey has been implemented to demonstrate the image pre-established and compare the answers to the reality observed in tourism experience.

The data collected through the tours in those areas and the survey were then analyzed, providing information to verify if this type of tourism can be considered positive or negative to the host community, as it will be shown on the next sections.

LITERATURE REVIEW

Reality Tourism: Social Tourism

Reality tourism, as described by Freire-Medeiros (2014), is a new type of tourism that aims to explore places affected by natural causes, human intervention (e.g., wars) and/or that present great social inequality. It is divided in two categories (dark and social tourism), however for the purpose of this study only the second one will be taken in consideration.

The social tourism explores the possibility to provide, for tourists, experiences in places that suffer of social and economic disadvantages. Most times, the places visited are neglected by society and by politicians, attracting even more the attention of tourists in search for an authentic, close to reality type of tour.

For this reason, the social tourism experience can be described as an important practice as they may be presented as periods of escape for tourists, and catalysts for change for the communities and their inhabitants. As examples of this type of tours in this segment, we can observe the ones practiced in several parts of the world (Brazil, India, Indonesia, South Africa, etc.) known as slum tours.

Slum Tours

"A Key frame [in the reality tourism segment] is the attempt to see "reality", a promise of authenticity that runs through many forms of tourism (MacCannell, 1976) and is highly prevalent in slum tourism (Dyson, 2012; Meschkank, 2011) (Frenzel et al., 2015, p.241)."

According to the definitions given by Oxford and Cambridge dictionaries, slum is a poor and crowded area around the city, normally considered as a dirty and untidy environment, surrounded by houses in bad condition. They comply with UN definition that categorize slums as "a place where people have insecure residential status. This means that they do not hold a legal title to their property or any legal right to the land that it sits on. Slums are characterized by inadequate access to safe water and sanitation, poorly built housing and overcrowding".

It is given the circumstances/characteristic of these areas, that they have been attracting, for many years, the eyes of curious tourists searching for a more meaningful, authentic and interesting tourist experience.

The first slum tours, as demonstrated in the figure above, have started in the nineteenth century in England, during the Victorian Era. In this period, according to Koven (apud Dürr and Jaffe, 2012), some privileged adventurers would leave their safe environment to explore districts that were considered poor, dangerous and perhaps with a lack of morality to a certain extent this relates to the studies of Spampinato (apud Silva and Barbosa, 2015), who presumes that tourism practiced in the slums can be considered as a desire to know the exotic.

It is possible to notice that the recent practice of this type of tourism does not differ to the way it has been practiced in the past. However, in comparison to the XIX century, its popularity has strongly increased. The reason to it lies in the 1990's event named Rio-92, a UN conference in Rio de Janeiro that raised the attention of the world to the surrounding areas of the city known as favelas (Freire-Medeiros, 2007, 2008, 2014; Tavares et al., 2018).

The idea of visiting areas affected by social and economic inequality extended to several places around the globe. Yearly, places such as Dharavi (India); Soweto (South Africa); Rocinha, Providência, Alemão and Santa Marta (Brazil); Cova da Moura (Portugal) receive a great number of tourists coming along with tourism agencies, NGOs or local guides (Freire-Medeiros, 2007, Urry apud Rhodes, 2016).

Those tours, according to Ma and Sanyal "can provide a way of challenging the stigmatic representations of the 'slum' by educating tourists about its reality". The main objectives proposed in the XX and XXI centuries are to allow visitors to deconstruct the idea of how life and inhabitants are in those environments, while learning about the local culture and assisting the community in question (Rhodes, 2016, p.7).

This cultural exchange – focus of this study – will be discussed in further detail on the next sections, where it will be explained the origin of the slums, worldwide and in Rio de Janeiro, the practice of tourism activity known as Favela Tours, the visits to Rocinha and Santa Marta and its impacts.

Slums

According to studies performed by Nisbett (2017), the number of people living in areas of precarious conditions keeps increasing. Its growth is causally related to the industrialization and the neoliberal capitalism. For this reason, up to now, a countless number of people constantly migrate to areas where there is a higher number of employments.

As a consequence of this expansion of the urban space, inevitably, these migrants are obliged to move into overcrowded areas with poor sanitation, lack of clean water, deficient health and education services. In 2013, the UN stated that this was the current situation of thirty-three per cent of the world's urban population – one billion people are living in slums (Nisbett, 2017).

In Brazil, according to IBGE Census of 2010, the number of slums corresponded to a total of 15868 composed by over eleven million inhabitants. A major part of these communities is located in the Southeast Region.

The increase in the concentration of slums in the region are related to a migratory movement of people, coming from other regions in the country, in pursuit for a better life condition; Nonetheless; studies performed by Freire-Medeiros (2009) and Tavares et al. (2018) show that the first creation of this type of community in Brazil has emerged by the end of XIX century.

Favelas in Brazil: Historical Context

The term favela is related to the popular name given to a plant known as *Cnidoscolus quercifolius* native of some Brazilian areas and it is related to the history of the occupancy of Rio de Janeiro's hills.

Despite serving as a refuge for slaves freed, it was in the end of the 1890s that it gained the significance as we know today. After the military campaign known as Guerra de Canudos finally reached its end, the numerous soldiers returning from it were offered a temporary permission to stay/settle in the hills of Providência and Santo Antonio, both located in Rio de Janeiro's central zones. As they moved to Providência hill, its name was replaced by Favela hill, in reference to the plant that existed in abundance in the semi-arid areas where they lived during the war (Tavares et al., 2018; Freire-Medeiros, 2009).

As Rio de Janeiro started developing and suffering several renovations, people started commuting to other hills around the city of Rio. In 1920, according to Abreu, twenty-six favelas were identified. Due to its growth, twenty years later the government created social housing areas for the working-class with the objective of imposing corrective actions to the residents of the slums, known as workers' housing communities (parques proletários) (Freire-Medeiros, 2009).

According to Burgos, the intensity of the disciplinary actions in the favelas oscillated for twenty years (end of the 40s and beginning of the 60s), receiving great amount of repression at certain periods and some periods of tolerance – when near election periods. However, they were ignored by the former favela residents and the slums kept growing (Freire-Medeiros, 2009).

By the 1950s-decade, seven percent of the population of the city were residents of the favela. It caused a high visibility to this area, attracting the attention of the government, which considered it to become a great problem in the city, fearing that their residents could be tempted by communist ideas (Freire-Medeiros, 2009).

In 1960s, the population in the favelas located in the North and South zone increased exponentially and the Census pointed the existence of 147 slums with a population of over 300,000 (Freire-Medeiros, 2009). Nevertheless, it was during the same period (military coup of 64) that the conflicts in those areas intensified, and several favelas were removed, and their population sent to housing complexes (Valladares apud Freire-Medeiros, 2009).

These commuting movement of people in and out of the favelas, forced by the public power did not stop their expansion again and by the years of 1968 to 1973, they increased 74 per cent.

It has been only in 1992 that Rio de Janeiro state created a plan/program for preservation and upgrading of the slums known under the policy of Favela-Bairro (Tavares et al., 2018).

According to Riley et al. (apud Freire-Medeiros, 2009, p.581), "This highly acclaimed program not only includes sanitation systems and other basic infrastructure but emphasizes the importance of integrating the favelas both spatially (through street connections with the surrounding neighborhoods) and socially (constructing buildings for the operation of social projects). With the financial support of the Inter-American Development Bank, it was launched in 1994 by the Housing Department of the municipal government of Rio de Janeiro. At first, it aimed at medium-sized favelas (which make up nearly one-third of the favelas in Rio), but has been also taken to both large and small favelas."

In addition to it, Valladares states that after the 2000s, the favelas did not present a generic settlement form anymore and that they became more heterogenous – socially and/or economically. However, until the present date the favelas and their residents continue to suffer from a social stigma, which relates them to places full of violence and permeated by the drug culture (Freire-Medeiros, 2009).

Slum Tours in Rio de Janeiro

The favelas of Rio de Janeiro have gained popularity, along the years, due to the increasingly media exposure: from the participation in important movies (e.g., Hulk, Fast and Furious, Cidade de Deus, and others) and clips to the constant violence in the surrounding areas, the media has continuously reinforced the idea that these communities have always been poverty redoubts.

In agreement to this idea, Urry (apud Freire-Medeiros, 2009, p.582) states that: "[The] choice of a certain destination by the tourist/consumer [...] consists of a dialogue with the images of a given locale carried by several media products, images which create an interpretative and behavioral frame for the tourist. Images of a place that circulate through new agencies, films, - national or foreign, fictional or documental – photographs, magazines, travel guides, advertising campaign and so on help build expectations and desirability."

In the case of Rio de Janeiro's slums, while tourists grow their interest in the favelas, the image associated to the life in it causes even more prejudice, as it shows a place ruled by drugs and criminality.

It is in this context that, since Rio-92, residents and tourism companies have been trying to change this image by inviting tourists to visit those communities. The visits, known as Favela Tours, are extremely important to the social/economic development of these places, as well as a great manner to demystify the prejudice tied to it, as reinforced by Tavares et al. (2018, p. 168): "With this cultural diversity, tourism in favelas has been interpreted as a real possibility of economic development, cultural appreciation and combating the stigmas associated with these spaces."

Nowadays, there are several companies responsible for performing tours in the favelas (Favela Tour, Favela Walking Tour, Favela Adventure, TouRio, and many others) and some of them use local/resident guides during the visits. However, the most important factor is that all of them have as a common objective to show the communities as symbols of the Brazilian culture – vivid and creative places (Tavares et al., 2018).

As a form of showing the local culture, the guides integrate tourists with slums' residents, providing unique experiences, such as: teaching the history of the place, sharing the local food, showing the local art (handicraft, graffiti, music, capoeira performances, etc.), visiting alleys and other activities, since it differs from tour to tour.

During the Favela Tour in Santa Marta, Elias (the guide) shows the tourists a mosaic produced by the residents, describing the work developed with kids in the slum. To Zaoual (apud Tavares, 2018)., these types of dialogues between visitors and hosts can change tourism structure, once it produces an intercultural acknowledgement.

This exchange of culture, knowledge and life experience provided by the tour in Rio de Janeiro's slums presents to be complex and it continues to yield various studies aiming to understand its characteristics, verifying the advantages and disadvantages of it in a sociological perspective.

Rocinha: History and Information

"Versatile, multicolored, and with a soul of its own, Rocinha is like that [...] considered one of the most urbanized favelas in Rio de Janeiro, Rocinha has several peculiar points of attractiveness, not to mention the privileged location: green forest, smooth-surfaced hills, the beach and a strong feeling of community. (Lilian Sá apud Freire-Medeiros, 2009, p. 583)"

Rocinha is located between the neighborhoods of Gávea, Vidigal and São Conrado, in the Southern Zone of Rio de Janeiro's capital. According to Rawet, it is considered the biggest slum in Latin America with an official total population of 69,161 inhabitants – data provided by IBGE Census of 2010. However, to the president of UPMRR, unofficially, the total population of this favela is of 180,000 to 220,000 people and the difference on the numbers are related to the difficulty of reaching and mapping some areas in Rocinha (Rodrigues, 2015).

The community size, the complexity its structure and the heterogeneity of the population in Rocinha can be directly connected to its origin.

As states Alberto Beltrão (2020), the Favela Walking Tour's guide, Rocinha is, nowadays, "a city within another city", but it was not always like this.

From the 1920s to the beginning of the 1930s, the area where this favela is now located used to be part of a farm called Fazenda Quebra Cangalha, which was then divided in small land plots of 270 meters square by a company named Castro Guidon. In that period, those pieces of land, were destined to be occupied by people with a low income –foreign migrants, industry workers and national migrants affected by the coffee crises in 1930 – and were transformed in small plantations, creating a new mean livelihood (Rodrigues, 2015).

According to Klintowitz (apud Rodrigues, 2015), the origin of the favela's name is directly related to this period of occupation. The author declares that when sellers were questioned about the origin of the products, their answer was that it came from a "Rocinha" located in Gávea.

In 1935, eighty plots of land had already been sold, driven by the installation of electric power systems in the main road od Rocinha, when this urban allotment did not receive the legal regularization, leaving the company Castro Guidon to bankruptcy. It was in this period that the slumming process of Rocinha started.

The subsequent years, the occupation of the community were as it follows: until the 1940s, the residents were concentrated in the bottom of the hill, along the Estrada da Gávea (main road of the community) and in the land plots located in the first street of Rocinha known as Caminho dos Boiadeiros in the 1950s, the community started receiving migrants coming from the Northeastern region of Brazil in search for a better life condition; in the 1960s the population of Rocinha increased 200% in comparison to the initial occupation, however it suffered three attempts of clearance during the military dictatorial period; during the 1970s and 1980s, the favela started creating a new infrastructure that contributed the social-spatial organization of the territory, acquiring the status of neighborhood in 1986.

According to Rodrigues (2015), nowadays, despite Rocinha has suffered a process of development and verticalization, due to the increase of the population, it still presents problems related to basic sanitation, water supply and violence. This is the main reason why tourists are attracted to this area and it has become a profitable business for numerous operators in tourism sector.

Tourism in Rocinha

Tourism in Rocinha started around 1990s, initially performed by three companies: Favela Tour, Jeep Tours and Exotic Tours. However, it has been only in 2006, due to the law of number 779/2006 – created by the Rio de Janeiro's town councilwoman Liliam Sá – that this slum officially became a tourism site (Freire-Medeiros, 2007, 2009; Bottino, 2016).

For this reason, the tours performed in Favela da Rocinha gained even more visibility, rising tourist's attention and interest, boosting the emergence of several others tourism companies – such as Favela Walking Tour, one of the companies chosen in this present study.

The tours generally consist in a visit to a local house to appreciate the sight – tourists can perceive the landscape of Rocinha, verifying the contrast between São Conrado's high society and the slums that surround it. After that, tourists are invited to discover the alleys that connect the top of the hill with the main commercial area, known as Caminho dos Boiadeiros. Each tour differs in some points, some are made in Jeeps, others on foot; some also stop in NGOs or local bars, where the tourist can have contact with the local community.

Santa Marta: History and Information

Santa Marta slum is located in Dona Marta's hill, in the neighborhood of Botafogo, Southern zone of the city of Rio de Janeiro. According to IBGE's census, the population of Santa Marta in 2010 was of 3980 inhabitants, but unofficially, according to the guide José Elias Pereira Duarte and the Residents Association, there are around 5500 people living in the community (Silva, 2015).

Nowadays, the slum of Santa Marta is the unique slum in Rio de Janeiro, where a wall has been put in place, to avoid a rapid growth in the population and the spreading of residences in the surrounding areas. However, it was not always like this.

The hill of Dona Marta received its first occupation in the decade of 1920s by the construction workers responsible for the expansion of Santo Inácio Jesuit´s school as states Elias, the tour guide: "These workers didn't have money to rent houses, so they started going up the hill of Dona Marta and building small wooden shacks [...] But things here became more expensive every day, you know?! Look at that plot, how much do you think it costs? [...] This small land costs over 40000. Can you imagine that price for something in the slum? Plots are expensive here, so only people that already lived here remain (some even moved out). I think it is also because Santa Marta cannot grow further than the walls, so lots of people go to other slums in the city."

With the development of the city, the real estate price in the Southern zone started increasing each day more and by the year of 1950, the slum of Santa Marta presented a total population of 1632 inhabitants – which is considered small in comparison to slums such as Rocinha, that received an immigration boom between the years of 1950s and 1960s (Barbosa, 2015; Rodrigues, 2015).

Nonetheless, despite being considered a smalls-sized community, Santa Marta became an extremely popular slum in the city of Rio de Janeiro and its popularity is linked to the events described below.

In 1986, the responsible (chief) of the drug trafficking in Santa Marta died, generating a lot of tension in the community. As a consequence of it, the control of such activity was then passed through the hands of several representatives, leading to what is known as Dona Marta hill's Battle – an armed conflict with huge exposition in the media (Barbosa, 2015).

Such exposition generated an international visibility, receiving in 1996 Michael Jackson and his team to produce the videoclip of the song "They don't care about us" directed by Spike Lee. For Elias, the tour guide, "it was a unique moment in our community" as described below: "It was a unique moment in our community. It was the first time our community had seen a helicopter so close! [...] They went down the soccer field and then he exited from it. Everyone was clapping while that little man exited the helicopter. He was so friendly, greeted all the people. [...] It was very important for us to have such visibility, because slums are always forgotten places."

Many other celebrities have visited the community since then, but with the inauguration of the Pacifying Police Unit (UPP) and the funicular railway named as *Plano Inclinado*, the number of people interested in visiting the community have increased.

Tourism in Santa Marta

In the year of 2010, according to Elias Duarte – Santa Marta's tour guide – and Barbosa (2015), the slum was "incorporated in the strategies of the public administration" in a project known as Rio Top Tour. This project aimed to create conditions to receive tourists in the slum, by generating development in the community. It goes in encounter to the objectives proposed by the resident's association: water, health, recycling, education and equality for everyone.

Tourism in Santa Marta is of great importance for the community, since part of the money raised in this activity is destined to the association and then redirected to the needs of its residents.

"Here in Santa Marta, the individual interest should never take precedence over the collective interests. [...] The association works on behalf of the residents, so it means that we all work for the community's benefit [...] You will never see anyone here struggling with medicaments or starving, because part of the gain we have with tourism is given to the association and they [the association] will help the person in need. (Elias Duarte, 2020)"

During the tour in the community, tourists pass in front of the association, where the tour guides explain the principles that guides it. Visits to other places include Space Michael Jackson – where there is a real-size statue of the artist and a mosaic produced by Romero Britto, a visit to the UPP area, a walk through the alleys. The rest of the tour differs from the tour guide and the company chosen.

Recently, there are twelve licensed guides in Santa Marta – born and/or raised in the community and their pick-up point is at the kiosk in the entrance of the community. The tours are done on foot, since cars cannot enter the community, due to the inexistence of roads for vehicles.

EMPIRICAL RESEARCH

Research Methodology

Given the development of new tourism activities and the changes of tourists' needs and desires, this present study aims to describe tourism practiced in two communities (slums) of Rio de Janeiro – Rocinha and Santa Marta, focusing on showing the impacts caused by it. For this reason, qualitative research has been performed using 3 categories of investigation define by Marshall and Rossman (apud Sousa and Baptista, 2011): exploratory, explanatory, and descriptive studies.

The first has the objective to recognize a given situation/reality and raise hypothesis to understand it. The second, explains the forces that create the object of study and identify the causes of it. Lastly, the third study describes the functioning and structure of object of study -the slum tour, in this research.

To obtain theoretical validity, a procedure of triangulation is utilized. According to Denzin and Lincoln (apud Sousa and Baptista, 2011), it means the usage of a combination of investigation methods, such as: data triangulation – use of a range of different sources, theory triangulation – use of different theories to interpret the given data and, method triangulation – use of different methods to study the given problem.

For a more precise result, the information collected is composed by primary and secondary data. In the study to be observed, the primary data refers to the conception and application of surveys with open and closed questions (mixed questionnaire) and semi-structure interviews, as well as studies based on observation. The secondary data refers to the information obtained in studies proposed by other researchers, books, newspapers and magazines and websites, which can be found throughout the entire research. (Guba and Lincoln, Marshall and Rossman, Ketele, apud Sousa and Baptista, 2011).

Lastly, with the results obtained with the analysis of the sampling techniques (not-casual, not-probabilistic and not-aleatory), it is possible to verify tangible and intangible impacts provided by tourism performed in communities of Rio de Janeiro. The following chapters intend to explain the discussed topics in detail.

Slum Tour: Survey – Pre-Concepts of Tourism Activity

Prior to verifying the functioning of the slum tours, a survey has been developed in this research to understand the concepts and the association that people have about this type of tourism. For this reason, mixed questionnaires containing eleven open and closed questions has been created in three languages (English, French and Portuguese) and answered by eighty-four people using Google forms.

Table 1. Questions and Objectives

Questions	Objectives
Age	To verify the age of the participants.
When you hear the expression Favela Tour, what is the first word that comes to your mind?	To verify the image perceived by the participants.
Have you ever taken part in tours performed in the slums/ communities?	To verify the number of participants who attended to this type of tourism activity.
In affirmative case, how was your experience?	To verify and compare tourism experience in the slums.
In negative case, would you have an interest to visit those communities in the future?	To verify the future intention of the participants.
Briefly justify your answer to the question above.	To establish the reason why the participants would or would not go to those communities.
Do you believe this type of tourism benefits the communities by any means?	To verify if the participants believe on the benefits of the Favela Tour.
In case of affirmative response, how does this type of tourism can benefit the communities?	To justify and exemplify the benefits given by the tours held in the communities.
In negative case, justify your answer.	To verify the reason why participants do not believe the activity to be beneficial.
Would you like to receive the result of this study once it has been concluded?	To verify the interest on receiving the result of the present study.
In affirmative case, leave your email down below.	To send the dissertation to the electronic address provided.

Source: Silva (2021)

The questions that have been presented in the survey and their objectives are shown in Table 1.

The methods of research describing the techniques used to analyze the data extracted and the results obtained are described on the subsections.

Population Sample

In order to obtain the needed responses, the survey was shared in three languages on Facebook using a non-probabilistic sampling method named convenience. It means that the population used for it has been not strictly selected by its profile nor it has been required a pre-selection prior to its application. The data collected has been acquired by answers given by participants of the community "Mochileiros" (backpackers) and by residents of Europe and South America with an age variation between 18 to 67+ years old, as presented on the Table 2.

According to the data obtained, it was possible to verify that the major responses were given by participants between 25 to 31 years old (37%), followed by people aging between 18 to 24 (19%) and 32 to 38 years old (15%).

In addition to it, the percentage of the participants aging between 39 to 66 years old are shown as it follows: 39 to 45 (10%), 46 to 52 (6%), 53 to 59 (11%) and 60 to 66 (2%).

Table 2. Age of the participants

Age	Number of Participants
18-24	16
25-31	31
32-38	13
39-45	8
46-52	5
53-59	9
60-66	2
67+	0
Total	84

Source: Silva (2021)

Participation and Intention

Following-up the survey analysis, the participants were asked to answer if any of them have taken part in tours performed in the slums. The answer to the proposed question were then transformed in the subsequent Table 3.

Table 3. Participation in Slum Tours

Have You Ever Taken Part in Tours Performed in the Slums/Communities?	Number of People
Yes	14
No	70
Total	84

Source: Silva (2021)

It has been observed that 83% percent of the contributors of the survey, an equivalent of 70 people, have not been to tours hosted in slums. However, when asked about the intention of visiting such places, it was possible to verify that 47% would like to visit such places in the future (Table 4.).

Table 4. Future intentions of participating in Slum Tours

Intention of Visiting a Slum in the Future	Answers (%)
Yes	47
No	51.5
No response	1.5

Source: Silva (2021)

Table 5. Slum Tour as a benefit to the host community

Do You Believe This Type of Tourism Benefits the Communities by Any Means?	Answers (%)
Yes	63
No	37

Source: Silva (2021)

The table above also shows that 51.5% of the participants do not have the intention of performing tours in those types of communities. The main reasons given lies on the benefits and harms that this type of tourism bring to the host community, as it is demonstrated in the next subchapter.

Slum Tour: Positive and Negative Points

To be able to verify the opinion of the participants regarding tourism hosted in slums and its benefits, it has been proposed the following question: *Do you believe this type of tourism benefits the communities by any means?*

According to the answer given by each of them, they were asked to justify their answers. The data obtained states that 63% of the people who answered the survey believe that the slum tour can be beneficial to the host communities. On the contrary, 37% of the participants does not believe or is uncertain that it can bring any benefit, as demonstrated on the Table 5.

The analysis also shows that the most common justifications for the negative response to the question proposed are the following:

For the participants who considered it beneficial to their inhabitants, however, the main justifications were that this activity can provide jobs, change of tourists' perception, help to the local economy, improvement in the security and life condition of the residents, visibility and recognition, exchange of cultural knowledge, recreational activities, investments in education, etc.

The quotes exhibited above show some of the responses given by contributors who believe that the tours performed in slums are beneficial to the host community. Those answers meet the objective of the present study and will be further discussed and verified in the study cases carried on the communities of Rocinha and Santa Marta.

Nonetheless, it is possible to observe throughout this analysis that a percentage of the thirty-six participants, who have not been to a slum tour and would not have any intention of taking part on this type of tourism attraction, still believe it is beneficial to the community itself.

Lastly, the subchapter of the present study aimed to verify if the image perceived by people who participated of the survey regarding the tours in slum corresponds to the reality witnessed during the tours in the slums of Rio de Janeiro.

Favela Tour: Main Image Perceived

The second question proposed in the survey referred to the image people have when hearing the expression Favela Tour. The most common words verified were related to the condition of life of the communities' inhabitants (insecurity and poverty), followed by the idea of a tour that aims to exploit its local citizens (circus, human zoo, abuse, etc).

Table 6. Negative response and justification

Negative Response Justification	Explanation	Example
Comparison to a Safari Tour	Participants believe the tour to be similar to the activity described.	"Tour is similar to Safari [...]"
Comparison to a Zoo Visit	Participants believe the tour to be similar to the activity described.	"No, because it is extremely uncomfortable to see a lot of rich (mostly) tourists going to visit the favela as if it was a 'zoo'."
Comparison to an Ikea Showroom	Participants believe the tour to be similar to the activity described.	"The places of life are not display stands like the Ikea salons ..."
Poverty Exploitation	Participants believe the tour to be responsible for exploiting fragile/poor communities.	"Exploitation of communities by thrill seeking tourists and benefits limited to tour organizers."
Enrichment of Tour Organizers	Participants believe the tour generates income for tourism company/travel agency.	"I think that only one who wins are the ones that have thought about this type of tourism and have nothing to do with the community."
Unnecessary Inhabitants' Exposition	Participants believe the tour to be responsible for exposing the inhabitants of the communities.	"Dirty and disrespectful, tourists could annoy locals." "Exposure to people who are in unfavorable situations."
Interference in the Residents' Privacy	Participants believe the tour interferes in the privacy of the residents.	"Maybe it's a lack of privacy for them."
Maintenance of Poverty	Participants believe the tour does not stimulate the improvement of the communities nor the change in the residents' lives.	"It enhances the maintenance of the favela, since it has to be abolished (the favela, not who lives in it!), People should live better."
Idealization of Poverty	Participants believe the tour romanticizes the reality lived in the communities.	"This type of tour, romanticizes life in these communities and masks their reality."
Enhancement of Social Inequality	Participants believe the tour enhances the social inequality between the visitors and the residents.	"It only accentuates the great social difference that exists between the poor and the rich."

Source: Silva (2021)

In addition to it, it has also been observed that people favorable to this type of tourism sees it as a form of become solidary to the local inhabitants and learn from its culture

It is possible to conclude that the data obtained in this part of the study will be used to validate the information collected during the visits to the two slums in Rio de Janeiro that will be discussed in the following chapter.

Favela Tour Experience: Study Case

"If you had a flow in the 19th century from the rich parts of the city to the poor, to those territories where the poor were located; we now have a flow from the Rich North to the Global South. [...] It is this idea that for you to be able to really know the metropolises of the Global South, you need to know their territories of poverty. It is as if it [the city] was, and that is very present in the speech of tourists, as if the real city was there. (Freire-Medeiros apud Graziano, 2013)"

The history of tourism has shown that the practitioners of this activity have searched for a diversity of activities that would fulfill their different needs/wishes. Studies performed by Pine ang Gilmore (1999)

Table 7. Positive response and justification

Positive Response Justification	Explanation	Example
Financial Aid	Participants believe that the tour can generate income, that can be used by the residents of the community to improve their lives.	"As long as it is done by the community itself, it can generate resources for it, thus helping in the daily lives of the residents."
Development of New Business	Participants believe that the tour can be responsible for the development of new business.	"Maybe by helping poor communities to start local businesses and earn money with tourism."
Generation of Employment	Participants believe that the tour can generate and/or increase employment.	"Employment within the community."
Recognition and Visibility	Participants believe that the tour is responsible for bringing awareness about the slums.	"The community being recognized by others."
Demystification and Reduction of Prejudice	Participants believe that the tour allows them to change the perception about the life and the people in slums.	"I think it helps to open the favela to the world and to change the perception everybody has from people who is living in the favelas."
Social Development	Participants believe that the tour can be responsible for bringing social development to the communities.	"In a way it can help residents to overcome some obstacles. For example: basic sanitation, water, electricity, gas, etc."
Security	Participants believe that the tour can be responsible for the improvement in security.	"Another point, if there are many visitors, security will be increased and therefore life there will be less dangerous."

Source: Silva (2021)

and Larsen et al. (2008) propose that the success of tourism experiences are connected to its originality, providing an emotional sensation through the involvement in the activity. (Lut, 2013)

The visits performed in the communities of Rio de Janeiro are examples of memorable experiences currently searched by tourists worldwide. Its popularity is connected to its originality; the capacity of providing social engagement through the exchange of culture and knowledge, and its social/economic role.

Objectives of the Tour in Santa Marta

According to the definitions given by Lankford, Grybovych and Lankford (2017, p.3) the form of tourism practiced in Santa Marta can be defined as a community-based tourism, as it is "organized by the community for its overall benefit". This can be verified by observing the main objectives proposed during the described guided tour.

In the beginning of the tour in Santa Marta, Elias Duarte – the tour guide of this present study – explains that tourism in the community has been only created in the year of 2010, much later in comparison to Rocinha which started receiving tourists around the nineties. However, it is possible to determine that each of tourism activities performed in those two communities, presents different purposes.

Along the years, Santa Marta has created a form of tourism that aims to aggregate the services developed by the local inhabitants in tourism activities. They believe by presenting the life in the slum to tourists, it can not only reduce the stigma associated with the place but help the residents and small businesses to flourish.

During the three-hour tour, the group is informed of the history of the place and the activities that are hosted there to raise money, which is then reinvested in the life of the community under the supervision of the Association of the Residents of Santa Marta.

To Elias Duarte (2020): "The individual interest should never take precedence over the collective interests. In our community all the gains we have with tourism will be divided and used to improve the life of all. We are a collective and need to keep thinking that way."

The speech of the tour guide goes in accordance with the studies developed by Russel (apud by Lankford, Grybovych and Lankford, 2017) that notes that the criteria for a community-based tourism are the following: tourism activity should have the participation and support of local people, it should provide economic benefits to its inhabitants, and it must protect the local's cultural identity and environment.

To conclude, tourism developed in Santa Marta presents a tangible and an intangible impact to the local inhabits. This is because, the community can use the resources obtained in this activity to improve their quality of life and create new careers and opportunities in the future. However, this does not exclude the fact that tourism activities hosted there can also decrease the prejudice associated with the slums and its residents and provide cultural exchanges.

Objectives of the Tour in Rocinha

According to Freire-Medeiros (apud Menezes, 2014), the community of Rocinha has a very consolidated tourism activity and it has been occurring since the 1990s, after the event Eco-92. This activity has been attracting over 3000 tourists per month, roughly over 30000 visitors a year.

As a consequence of it, different companies have been created to attend the high demand of people wishing to visit the community for several reasons: curiosity, search for something new/unique, academic research and others.

The company chosen for this present study is named Favela Walking Tour and it operates in the community of Rocinha, performing three to five-hour tours where it "aims to educate you about life in a Favela which is such a big part of the culture of Rio de Janeiro. In addition to that, see how residents live and learn about the difficulties Rocinha – the largest Favela in Latin America.", as described on their website.

During the tour, the visitors are invited to explore the community alleys and to purchase food and beverage from the local commerce. However, what are the impacts of these types of tours in the hosting community? According to Alberto Beltrão, the tours provide an intangible impact as shown below.

"In our tours, the tourist really gets to know the favela and the real life in Rocinha. They visit alleys, see the houses in the communities, eat açaí, go to the bar, but most importantly, they meet people. [...] This tour is all about people, to show that people here are like people anywhere. They work, they shop, and they live. [...]"

Questioned if the tour aimed to demystify the preconception tied to the life in the communities, the tour guide replied: "Definitely, the tour shows to gringos [name used to describe foreigners] that not everyone in the slums have a relation with the criminality as shown in the media. People here are hardworking, but above all, they are friendly and happy people. It is important to show that despite the poor situation that many of these people live, they survive here with a smile on their face. [...] I always hope that by the end of the tour, tourists will have good things to talk about Rocinha. I want people to exit here with a different perspective than the one they arrived with. (Alberto Beltrão, 2020)."

Table 8. Main benefits given by the Favela Tour

Benefits of the Favela Tour According to the Survey Responses
Financial Aid
Social Development
Reduction of Prejudice
Visibility and Recognition
Cultural Exchange

Source: Silva (2021)

As tourism experience provided by the company Favela Walking Tour, explores the reduction and the demystification of prejudice, nothing much is done in relation to the improvement of the community's condition. It means that the money paid for the tour is used solely for the company's interest, which generates the questioning whether if companies that perform this type of tours have any intention to see changes in the quality of life of the inhabitants in the community.

FUTURE RESEARCH DIRECTIONS

The present study suggests performing visits to other communities located in Brazil or in other parts of the world, to verify if the benefits encountered in Rocinha and Santa Marta can be found elsewhere. The results obtained, may be able provide a better understanding of how the life in societies affected by social inequality is and how tourism sector has adapted to provide new experiences to tourists.

CONCLUSION

Observing the data collected in the proposed survey and the visits conducted in the communities of Rocinha and Santa Marta, it was possible to verify some similarities between the responses given by certain participants and the explanations given during the experience offered by the tours hosted in the slums.

The seventh question asked in the survey aimed to know if the people responding to it believed that the Favela Tour could be beneficial to the local inhabitants. sixty-three percent of the participants responded positively, using the justifications presented on Table 8.

The benefits suggested above were verified on the tours performed in the two communities of Rio de Janeiro, Rocinha and Santa Marta. The visit to the first one consisted in giving visibility and recognition of the slum by explaining its history and showing the daily life of their residents. This way, according to Alberto Beltrão, tour guide in Favela Walking Tour, the company was able to achieve its main objective: to reduce the prejudice people [tourists, in this case] have in relation to the life in the slums.

It is also possible to infer that the tour hosted in that community benefited few local businesses, as for example: souvenirs stores, the bar and an açaí shop. Nevertheless, there is not enough evidence to state that it has an effective impact in the social development.

In contrast to it, the tour held in Santa Marta has shown to be partly responsible for bringing new assets to the community. According to Elias Duarte, tour guide in TouRio, the visibility and recogni-

Table 9. Favela Tour – negative image perceived

Favela Tour: Negative Image Perceived				
Madness	Human Zoo	Unhealthy	Poor	Exploitation
Danger	Graffiti	Robbery	Insecurity	Criminal
Poverty	Shacks	Mess	Dangerous	Violence

Source: Silva (2021)

tion gained through the visit of tourists and celebrities have brought social development to the slum. Infrastructure works, the opening of new businesses and new investments in cultural and educational activities allowed the community to improve its social and economic situation.

Equally important, tourism activities in Santa Marta, are responsible for granting cultural exchanges between the residents and its visitors, enabling the reduction of prejudice and the change of the pre-established image associated with life and/or tour in this type of environment, as described in the survey and shown on Table 9.

The same topic has been raised in the documentary Tem Gringo no Morro (2017), demonstrating that the pre-concept associated to the slums are diluted and modified to a more optimistic impression of the people and the place.

Overall, the results obtained in the present study has suggested that the impacts caused by tourism in Rio de Janeiro's communities have been quite positive in terms of social development, cultural exchange and reduction of prejudice. For a better understanding, these outcomes have been classified in two categories, as shown in the Table 10.

Table 10. Tangible and Intangible Impacts

Tangible Impacts	Intangible Impacts
Social development	Decrease of prejudice
Financial aid	Change in the perceived image of slums
	Cultural exchange
	Visibility and Recognition

Source: Silva (2021)

The tangible impacts refer to the visible achievements acquired by the community as a consequence of tourism activity. These outcomes are directly related to the intangible impacts, once it is due to the visibility generated by the tours that it is possible to generate assets and use them for the improvement of the community.

In Rocinha, the tour aimed to demystify the image associated to the slums, producing intangible impacts to the community, as for example: reducing the prejudice through cultural exchange, leading to an increase of its visibility and recognition.

In Santa Marta, tourism activity is responsible for producing tangible and intangible impacts to the community and its residents. As an economic activity, it is accountable for bringing financial aid to local business and entrepreneurs by presenting their space/work to tourists. The places affected by it are restaurants and bars, souvenirs stores, grocery shops, and the people affected by it are local artisans and artists. In addition to it, other tangible impacts are the development in infrastructure, educational and recreational activities inside the communities.

Moreover, it is also possible to infer that the tangible impacts are only feasible, given the intangible impacts caused by the tour in Santa Marta, as also observed in Rocinha.

In the light of those results, it is possible to conclude that the Favela Tour experiences provided in the communities of Rio de Janeiro, are into a certain extent beneficial to their residents. The benefits are either obtained by generating visibility and recognition of the slums and, as a consequence, changing the perceived image by people from outside of it, or by economic and socially assisting the community, through financial aid and social development.

ACKNOWLEDGMENT

This paper is financed by National Funds provided by FCT- Foundation for Science and Technology through project UIDB/04020/2020.

REFERENCES

Barbosa, L. J., & Silva, M. B. (2015). Tourism in favelas: A practice of belonging and sociocultural recognition [Turismo em favelas: uma prática de pertencimento e reconhecimento sociocultural]. *Nuevamérica*, 146.

Bottino, C. M. M. (2016). *Who discovered the favela? The trajectory of tourism agencies in favelas cariocas* [Quem descobriu a favela? A trajetória das agências de turismo nas favelas cariocas] [Master Thesis]. FGV, Rio de Janeiro, Brazil. http://hdl.handle.net/10438/17655

Freire-Medeiros, B. (2007). The Favela that can be seen and sold: Reflections and Controversies around a Tourist Destination [A Favela que se vê e que se vende: Reflexões e polêmicas em torno de um destino turístico]. *Revista Brasileira de Ciências Sociais*, *22*. doi:10.1590/S0102-69092007000300006

Freire-Medeiros, B. (2008). Selling the Favela: Thoughts and Polemics about a Tourist Destination. *Revista Brasileira de Ciências Sociais*, *4*, 1–26. http://socialsciences.scielo.org/pdf/s_rbcsoc/v4nse/scs_01.pdf

Freire-Medeiros, B. (2009). *Gringo na Laje: Production, circulation and consumption of the tourist favela* [Gringo na laje: Produção, circulação e consumo da favela turística]. FGV.

Frenzel, F., Koens, K., Steinbrink, M., & Rogerson, C. M. (2015). Slum Tourism: State of the art. *Tourism Review International*, *18*(4), 237–252. doi:10.3727/154427215X14230549904017

Lickorish, L. J., & Jenkins, C. L. (1997). *An Introduction to tourism*. Butterworth-Heinemann.

Nisbett, M. (2017). Empowering the empowered? Slum tourism and the depoliticization of poverty. *Geoforum*, *85*, 37–45. doi:10.1016/j.geoforum.2017.07.007

Rezende, R. (2014). Community tourism and favela-tour as expressions of the new dynamics of tourist consumption [Turismo comunitário e favela-tour como expressões das novas dinâmicas do consumo turístico]. *Revista Turismo Em Análise, 25*(2), 354–372. doi:10.11606/issn.1984-4867.v25i2p354-372

Rhodes, R. (2016). *Reality tours and travel: A Community perception on slum tourism in Dharavi.* Durham University.

Silva, N. (2021). *Favela tour experience. The impacts in the host communities* [Master Thesis in Management]. University of Algarve, Faro, Portugal.

Tavares, F., Irving, M., Oliveira, E., & Madureira, M. (2018). Slums – Places to visit or to avoid? The contradictions in the media on tours in the favelas of Rio de Janeiro. *Revista Verso e Reverso, 32*(81), 168–186. doi:10.4013/ver.2018.32.81.01

Chapter 17
Challenges in Safeguarding Traditional Festivals Facing Tourism Pressure:
A Case Study of Macao SAR

Kate Chi Wo Kwan
https://orcid.org/0000-0003-1432-0382
The University of Hong Kong, Hong Kong

ABSTRACT

This chapter reports on the findings of a study that investigated the challenges in safeguarding two of Macao's and China's nationally recognized intangible cultural heritage festivals which are facing tourism pressures, namely 1) Macao Feast of the Drunken Dragon and the Lion Dance Gala and 2) The Beliefs and Customs of Na Tcha. Based on the findings, some recommendations are formulated that adopt a community-based approach to safeguard traditional festivals. Face-to-face interviews were conducted with festival organizers and volunteers who empathize with the values of the traditional festival customs and their related cultural spaces.

INTRODUCTION

In 2003, UNESCO put forward guidelines for the Convention for the Safeguarding of the Intangible Cultural Heritage in order to implement and enhance awareness in safeguarding living heritage (UNESCO, 2011). Intangible Cultural Heritage (ICH) includes a variety of living expressions inherited from generation to generation such as traditional customs, rituals, festivals, craftsmanship, and other social practices that carry the shared beliefs and heritage values. In particular, traditional festivals are living heritage that foster social cohesion and also cultural tourism products that may attract tourists. However, it also found that rapid tourism growth generates various social impacts on collective lifestyles and cultural expressions (Ap, 1990). Traditional festivals tend to be transformed into tourist attractions to promote a destination's unique culture and image, but its interpretation and commercialization place the safeguarding process

DOI: 10.4018/978-1-7998-9217-5.ch017

at risk and may significantly impact its authenticity and integrity (Laing, 2018). Consequently, many studies have discussed the conflicts that may arise between traditional festivals and tourism development (McKercher et al., 2002; Huibin et al., 2013; Kaplan, 2015; Guo et al., 2017; Imon, 2017; Su, 2019; Mokgachane et al., 2021). For instance, rapid tourism growth may lead to failure in understanding the meaning of traditional festivals, and the spirit of a place (Imon, 2017). It was found by Xing et al. (2013) that different objectives from the organizers, government agencies, and other stakeholders are one of the crucial factors that influence the transmission of traditional festivals. Thus, it must be recognized that to protect live history and its socio-cultural significance, inherit cultural diversity, sustain cultural authenticity, and advocate community engagement are some of the essential components that can ensure the sustainability of traditional festivals. In addition, Choi et al. (2020) pointed out that, the reasons people participate in traditional cultural events are more socially and culturally driven. Therefore, the organizers and other stakeholders have to carefully consider how to avoid commercializing traditional festivals in order to protect the meaning, as well as its original forms of expression. In contrast to the negative impacts, the United Nations World Tourism Organization (UNWTO) advocates that tourism works closely with countries economically such as increasing job opportunities for locals as to achieve the Sustainable Development: Goal 8: Decent Work and Economic Growth; and Goal 17: Partnerships for the Goals (UNWTO, 2021). As a fact that sustainable tourism requires a long-term commitment to a balance of environmental, economic, and socio-cultural aspects of development. In other words, to protect the past, there is a need to connect with the community. Therefore, the stakeholders should place tourism development appropriately where tourism activities such as tourists' misbehavior and mass tourism pose a threat and may be viewed as hostile to our treasured cultural heritage. Otherwise, an imbalanced situation is likely to arise and further threaten the safeguarding of Intangible Cultural Heritage.

As such, the current study investigates two of Macao's and China's nationally recognized ICH traditional festivals, namely Macao's: 1) Feast of the Drunken Dragon and the Lion Dance Gala; and 2) the Beliefs and Customs of Na Tcha. Based on the findings, some recommendations are formulated which adopt a community-based approach to safeguard traditional festivals which are facing pressures from tourism. As noted by Su et al. (2020) organizers' attitudes directly influence their behavior and the survival of traditional festivals. Thus, this study seeks: 1) to examine the perceptions of the festival organizers and participants toward the sustainability of traditional customs; and 2) to compare how these festivals are interpreted and transmitted by the host communities, as well as by government agencies.

BACKGROUND

There have been few studies focusing on the impacts of the host community in safeguarding traditional festivals under tourism pressure especially the community-based Intangible Cultural Heritage items. To safeguard ICH, maintaining its authenticity and integrity are key factors in preserving ICH as a sustainable tourism resource (Kim et al., 2019). Specifically, to protect the authentic expression of ICH, awareness of the communities' defining values must be conveyed to all key stakeholders. Consequently, the current study will reveal the nature of the relationship in safeguarding ICH progress and tourism development in the Macao Special Administrative Region of China. Case studies of the Macao Feast of the Drunken Dragon and the Lion Dance Gala (as known as The Macao Feast of the Drunken Dragon) which carries the rich history of Macao's fishmonger community, and the Beliefs and Customs of Na Tcha nature (i.e., Na Tcha Festival) that represents the distinctive Macao folklore cultures are presented.

LITERATURE REVIEW

Many papers studied the definitions of folklore beliefs and traditional practices (Rioux, 1950; Ben-Amos, 1971; Motz, 1998; Buccitelli et al., 2016; Dégh, 2020). As Ben-Amos (1971) stated, "...folklore is very much an organic phenomenon in the sense that it is an integral part of culture." (p.4). Gradually, some of the studies reported the interdependence of religious beliefs and their associated traditional events (McKercher et al., 2006; Crespi-Vallbona, 2007; Kuuder et al., 2012; Lee et al., 2017). Kuuder et al. (2012) found that festivals are seasonal occasions that provide people mental support and bring prosperity to the societies through prayers and ritual ceremonies. Hence, this chapter seeks: 1) to introduce the concept of the expressions of folklore beliefs in traditional festivals; 2) to review the notion of socio-cultural values and traditional festivals; and 3) to be more precise, the importance of safeguarding Intangible Cultural Heritage.

EXPRESSIONS OF FOLKLORE BELIEFS IN TRADITIONAL FESTIVALS

Several anthropology studies have revealed there is an interconnection between folklore and human philosophy (Smith, 1959; Bronner, 1984; Bauman, 1992; Motz, 1998; Sassatelli, 2015). Conveniently, present-day cultural activities are based on the creation of the ancestors' ways of living, and this kind of creativity is a fundamental part to deliver their beliefs and traditional myths (Boas, 1916). However, it was noted by Motz (1998) "The term belief is, in or society, an exoteric term like superstition which calls into question its own validity: we usually describe our own beliefs as knowledge." (p.340). In fact, the generic category of traditional ritual ceremonies contains folklore and belief, as well as features ancestors' intellective way of living. The addition and diversification of religious ritual activities constitute the large-scale traditional festivals that people celebrate at present. Nowadays, people attend traditional festivals exhibited by their ancestors as they believe that traditional events bring them prosperity and good fortune. Festival attendees respect the habitual cultures who also wish to pass to the next generation shared traditions. Furthermore, Cialdini et al. (1998) found that customs, traditions, values are similar to norms that are jointly negotiated rules to shape social behavior and maintain human relationships. This idea has likewise been claimed that human social behavior develops reasonably from human beliefs as people associate objects or events to convey their faith and mindset (Ajzen, 1991). That is, common values are derived from religious practices and gradually permeated into the daily thinking and moral ideas of human beings, as Getz (1998) noted that people participate in traditional events which create a cohesive society, and their participation simultaneously cultivates social norms and reduces conflicts.

For instance, the Lantern Lighting Festival (Yeondeunghoe) was originally a ceremony to celebrate Buddha's birthday in Korea. Presently, it is celebrated as a national spring festival that unifies the general public with its profound Buddhist history. Days before the gala starts, Koreans will gather and make lanterns which is a lantern ritual that has been widely practiced among Buddhists. On the eighth day of the fourth month of the lunar calendar, all the lanterns are lit up, and a series of traditional activities such as religious ceremonies are launched during the Korean Lantern Lighting Festival. This festival plays an important role in Korean society as it is believed that the festival grants them the courage to cope with problems (UNESCO, 2021). Another example is the Dragon Boat Festival in China; it is a time-honored custom among various Chinese ethnic groups. This event has transformed with differing practices that vary from region to region, but it still contains a number of shared features which enable the sustainability of its tradition. As noted by UNESCO (2021) the shared features comprise the dragon boat ceremony, rice dumpling making; and families embracing the traditional customs. Both cases also serve to demonstrate that the expressions of folklore beliefs in traditional festive events form social/cultural patterns by a group of individuals which slowly evolve into a common belief within society.

Festivalgoers get involved in different forms of cultural events due to different interests and purposes. Nonetheless, the social norms and shared beliefs are the most important factors that motivate people to be part of a traditional event; the traditional event also generates a platform for people to interact with one and the other. In other words, traditional festivals are also meant to unify people all together and formulate their shared sense of values. Although religious ceremonies were created for worship activities, yet the meanings of religious practices have been rooted in people's daily life and their mindset. As time went by, religious traditional festivals symbolize various values of folk and folklore, the correlation between traditional practices and community groups irreplaceably serves to understand the nature of religious activities, as well as people's socio-cultural identities.

SOCIO-CULTURAL VALUES AND TRADITIONAL FESTIVALS

Numerous studies have reported the socio-cultural significance of traditional festivals (Derrett, 2003; Small et al., 2005; Arcodia et al., 2007; Brown et al., 2011; Kwan, 2021). For instance, Briggs (1992) indicated that "...among different cultures, sometimes in the name of evolution, culture was the creative expressions of a particular society through its symbols, literature, art, and music and, for some, its institutions and the values and experiences that shape them." (pp.4). Traditional festivals constitute a socio-cultural pluralistic system and also significantly serve to build a harmonious society. People as the center that creates and sustains cultures, community groups are the first and foremost segment that can largely understand the shared socio-cultural identities among societies. According to He (2007) Chinese covey their socio-cultural values and ideological achievements through traditional festivals such as the Chinese Lunar Year, which is particularly advocating mutual respect as a shared Chinese national virtue.

Besides, a study by Hauptfleisch (2006) found that some festivals were evolved from particular beliefs and cultures, and the others were introduced by locals and migrants. Kwan (2021) for example, found that historically, the Macao Feast of the Drunken Dragon and the Lion Dance was celebrated in Zhongshan, Zhuhai, and Macao SAR, China by its respective individuals. Due to the Sino-Japanese War and Second World War, many of the villagers who worked as fish traders escaped from wars and moved to Macao. Since then, the fish traders combined with their local counterparts to celebrate this unique traditional custom year by year, which symbols a distinctive culture and the socio-cultural identities of Macao's fishmonger community. However, according to Brown et al. (2011) people started to question and explore the socio-cultural values of traditional festivals. It is because many traditional events are currently driven by the increased market demand and tourism activity (Arcodia et al., 2007; Kwan, 2021). Despite tourism activities are able to benefit the community groups for wider promotion in their traditional items, but the conflicts that touristification might generate are uncountable.

In fact, understanding the socio-cultural values of traditional festivals is an important aspect to sustain human beliefs, the shared values and the collective memories of the community groups. As noted by UNESCO (2011) that through learning and understanding traditional social practices, people are encouraged to help every individual to feel part of the society no matter race, gender, or nationality they are. In addition, UNESCO (2015) stated on the "Intangible Cultural Heritage and Sustainable Development" that "Numerous social practices, from small gatherings to large-scale celebrations and commemorations, strengthen social links and communities' social cohesion by shaping the shared identity of those who practice them." (pp.5). Similarly, relationship and knowledge may be able to enhance between family members or friends through the participation of traditional cultural events.

SAFEGUARDING TRADITIONAL FESTIVALS IN MACAO

As mentioned above, festivals are defined as traditional ceremonies with rich cultural meanings that are linked to a community. Festivals are documenting and expressing the community's beliefs, their collective memories, and socio-cultural identities that simply unify the people and grant them a sense of community pride. Festivals are not intended to generate consumption privately or publicly (Getz, 1998). The Macao Special Administrative Region (Macao SAR) is located on the southern coast of the People's Republic of China. It was a small fishing village when the Portuguese Government settled there in 1553. Four centuries of Portuguese colonial history had shaped Macao into a multicultural city. In 2005, the Historic Centre of Macao was inscribed on the UNESCO World Heritage List. This list is composed of both Chinese and Western historic sites and buildings. Macao has a total of twenty items inscribed in Macao's Intangible Cultural Heritage List, with eleven of them were inscribed on the National Intangible Cultural Heritage List (Cultural Affairs Bureau, Macao SAR, 2021). Both historic buildings and intangible culture feature Macao's cultural uniqueness and inclusiveness. The glory of cultural heritage has offered Macao significant economic growth. As Macao tourism grows, traditional festivals tend to develop as tourism attraction products, especially cultural performances catering for leisure activity needs and traditional festivals have become one of the significant demands that sustain tourism development. Consequently, there have been concerns about the impacts of rapid tourism growth and the need to safeguard traditional festivals. That is, tourism and its promotion of these festivals threaten and have placed at risk the authenticity and integrity of the festivals with the commercialization of traditional festivals which, in turn, may lead to a low-quality interpretation of their cultural meaning and its associated cultural space by the key which may lead to a transformation of heritage values.

Macao's government agencies have been putting steady efforts to promote Macao and diversify its tourism activities with its policy to develop Macao as 'World Centre of Tourism and Leisure' (Yankholmes, 2018). As such, traditional festivals have been largely utilized to rebrand Macao's city image and to attract more tourists. Kwan (2021) for example, found that inconsistencies in the interpretation of Intangible Cultural Heritage festivals have mainly arisen due to the intervention of tourism development and the government often neglecting to consider the communities' needs. The Macao Government Tourism Office's (MGTO) official website contains incorrect and/or an absence of information for both the Macao Feast of the Drunken Dragon and the Lion Dance Gala, as well as the Beliefs and Customs of Na Tch (see Figures. 1 and 2). This issue serves to aggravate and place in danger the safeguarding process for Intangible Cultural Heritage. In addition, the MGTO's inconsistency of ICH interpretation also reflects the government's misunderstanding of the core values and historic background of these intangible cultural festivals.

As the United Nations Environment Programme and the United Nations World Tourism Organization (2005) advocated that sustainable tourism should "Respect the socio-cultural authenticity of host communities, conserve their built and living cultural heritage and traditional values, and contribute to inter-cultural understanding and tolerance." (p.11). Yet, heritage conservation tends to concentrate on

Figure 1. Incorrect and misleading information about the Macao Feast of the Drunken Dragon and the Lion Dance Gala provided by the Macao Government Tourism Office such as: 1) The launching date of the Feast is on the eighth day of the fourth month of the Chinese lunar calendar; 2) The Feast is organized by the Macao Fishmonger Community not a fisherman's association; 3) The Feast starts from the Taoist Opening Ceremony in the morning at the Kuan Tai Temple, and then men perform the drunken dragon dance with a wooden head, as well as tail of a dragon to convey the fishmongers' dragon beliefs; 4) Along with the Drunken Dragon and the Lion Dance Parade, the drunken dragon dance performers pretend to be drunk and do not drink until they drop; 5) The fishmongers have inherited the drunken dragon custom since the Ming Dynasty to show the core spirit of this cultural event, namely: 'selflessness', dedication', 'gratitude' and 'unity' not to honor a man and give him courage; and 6) Venue- local wet markets and the Kuan Tai Temple

Source: Macao Government Tourism Office, 2021; and the author

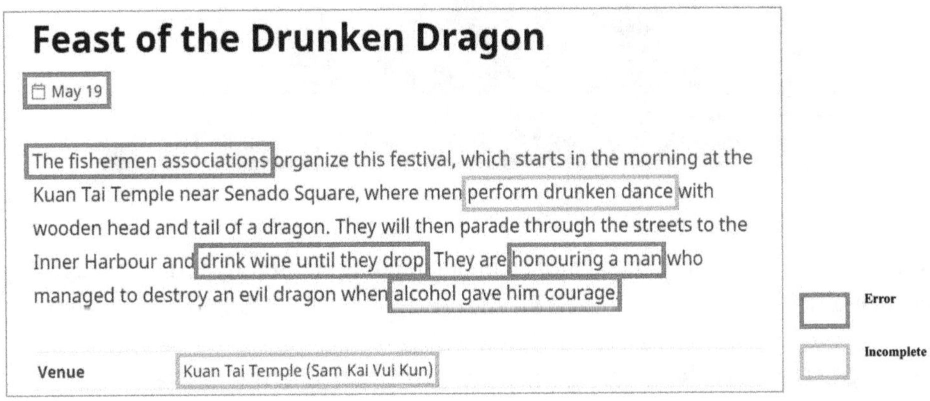

the achievement in conserving the historic buildings, but it rarely focuses on the safeguarding of traditional practices which, in turn, the missing attention and consideration may arise more problems than solutions to protect Intangible Cultural Heritage (Imon, 2017). Decision-makers ought to measure the customization and interpretation are well conserved the meaning, as well as the core spirit of traditional festivals. Therefore, there is a need to establish a platform for stakeholders to liaise and exchange ideas to better present and safeguard ICH. As Tilden (2009) argued that "Interpretation is a mission-based communication process that forges emotional and intellectual connections between in the interests of the audience and the meanings inherent in the resource." (p.16).

HISTORY OF THE MACAO FEAST OF THE DRUNKEN DRAGON AND THE LION DANCE GALA AND THE MACAO BELIEFS AND CUSTOMS OF NA TCHA

The Macao Feast of the Drunken Dragon and the Lion Dance Gala, as well as the Macao Beliefs and Customs of Na Tcha, are both inscribed on the Macao's and China's Nationally Intangible Cultural Heritage List together with the Macao Beliefs and Customs of A-Ma. But, compared to the Macao Beliefs and Customs of A-Ma, which is much more widespread among Chinese society, these two traditional events also receive less academic attention. The following section will cover the evolution of both chosen

Figure 2. Macao Government Tourism Office shows the Macao Beliefs and Customs of Na Tcha on its website where the feast is listed under the category 'Event Calendar' without any supporting information being provided. A video briefly introduces the historical background of the Na Tcha Temple, where the feast takes place, but there is little or no information introducing the traditional customs of this festival.
Source: Macao Government Tourism Office, 2021

case studies for readers to provide context and an understanding of their historical background and their connected communities.

Macao Feast of the Drunken Dragon and the Lion Dance Gala

Historically, the Feast of the Drunken Dragon can be traced back to the Ming Dynasty. According to The Journal of the Xiangshan Country (Qianlong, 1736-1796; Daoguang 1821 -1851), "On the eighth day of the fourth month of the lunar calendar, monks purify all of the Buddha statues, carve wooden dragons in temples, villagers raise flags and beat drums in the village, performing drunken dragon dance to cure of plague, sating dragon transfer". Legend has it that there was a pandemic that happened in a village located in Xingshan (currently Zhongshan, Zhuhai, and Macao SAR). An intoxicated monk killed a huge snake and its blood full into the river. The villagers made cakes from plants along the river to help them recover from the plague pandemic and believed the drunken dragon belief (i.e., drunken dragon dance) brought them blessings. Years later, some of the villagers from other cities in Guangdong province moved to Macao and they cooperated with the local fish traders to carry on the drunken dragon customs. Gradually, this traditional drunken dragon custom has undergone various changes in the past and has now become a hallmark event in Macao (Kwan, 2021). Presently, this annual event is not only a festival that unifies the fishmonger community. It is a three-day event that contains a wide range of social practices, rituals, and performances organized by the Macao Fishmongers' Association, it is a traditional gala that conveys the spirit of 'selflessness', 'dedication', 'gratitude', and 'unity' among the community for the general public (Kwan, 2019).

Kwan (2019) also found that this cultural event is composed of three major components, namely: 1) Traditional Gathering Dinner 'Poon Choi'; 2) The Drunken Dragon and the Lion Dance Parade; 3) Free Longevity Rice Distribution (see Figures. 3, 4 and 5). Among them, the 'Poon Choi' dinner is a private event for the fishmongers and their family members. The Parade and Rice Distribution start right after

Figure 3. Traditional Gathering Dinner 'Poon Choi' and the signature snack of this cultural festival of the fishmonger community – Indian Pluchea Cake
Source: The author, 2018

the Taoist Music Opening Ceremony. The Taoist ceremony is held at the Kuan Tai Temple (also known as Sam Kai Vui Kun Temple) and local wet markets are the main locations where the Macao Feast of the Drunken Dragon and the Lion Dance Gala takes place (see Figure. 6). This festival demonstrates how tangible and intangible cultural heritage always works together in the host community as it works to hold the fishmongers' annual festival event in the traditional local markets.

Macao Beliefs and Customs of Na Tcha

In a wider context, Na Tcha is a Chinese mythological character in the form of a child. Traditionally, people believed Na Tcha is a patron saint who blesses children to grow up healthily. Moreover, Na Tcha possesses the power to eliminate disasters such as pestilence and children's illness. According to the Na Tcha legend in Macao, one of the legends that happened in Mount Hill during the Qing Dynasty, there was a child dressed up with a belly knot, played games with children, and led them all to stand on a boulder. Although the hill was steep, no accidents occurred. After that, the children recalled that it was a child like Na Tcha who saved them from any accident. Another myth marked that there was a serious disease during the 1900s in Macao. One of the villagers who lived in a walled village, Pátio do Espinho, close to the Ruins of Saint Paul, dreamed of Na Tcha who told him that spring on the right side of Ruins of Saint Paul could help people to recover from the illnesses. The villagers built a Na Tcha Temple there to express their belief and gratitude to Na Tcha. Since then, the Mount Hill's Na Tcha Temple Association and the Ruins of Saint Paul's Temple Association have been organizing the Na Tcha Festival regularly on the eighteenth day of the fifth month of the lunar calendar at their respective temples (Wu, 2014).

With a history of more than three hundred years, the Macao Beliefs and Customs of Na Tcha has developed its unique features. Its legend, the birth of Na Tcha, and traditional ceremonies are clearly different from those in neighboring provinces in China. For example, people in Sichuan province believe Na Tcha was born and raised there as he left a huge rock that represents a 'backbone' of the third son of

Figure 4. Drunken dragon dance performed at one of the signature locations of this cultural festival – the Red Market
Source: The author, 2018

the Dragon King who was killed by Na Tcha (Wu, 2014). Although the celebration was suspended for 30 years due to political reasons, this specific custom and belief were fortunately preserved. The Macao Beliefs and Customs of Na Tcha are composed of different unique elements such as the parade with the children dressed as legendary and modern heroes suspended above the crowd on poles and paper fans, the firecracker-grabbing contest, free peaceful rice distribution, and Chinese opera performance (see Figures. 7 and 8). The Macao Beliefs and Customs of Na Tcha embody the uniqueness of the local folklore belief and its social value (Cultural Affairs Bureau, Macao SAR, 2021).

Figure 5. 'Longevity rice' is free to give out to society
Source: The author, 2018

Figure 6. The Drunken Dragon and Lion Dance Parade proceeds to several local markets, and these markets are the representative organizers of the Macao Feast of the Drunken Dragon
Source: Kwan, 2021

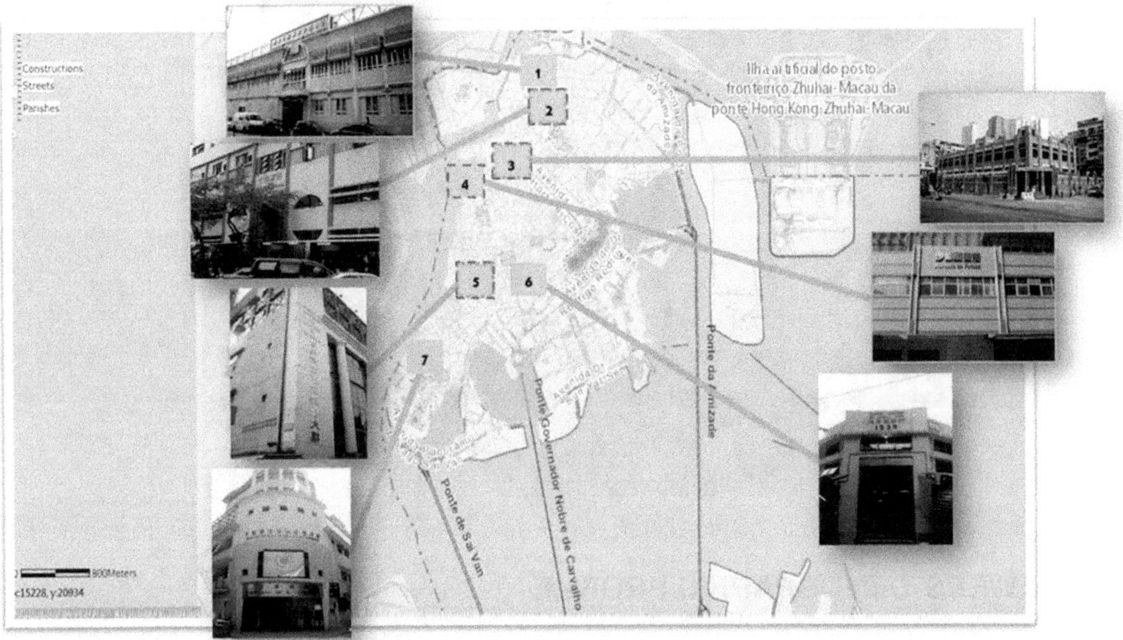

1. Tamagnini Barbosa Municipal Market Complex
2. Bairro Iao Hon Municipal Market
3. Almirante Lacerda Municipal Market (Red Market)
4. Patane Market
5. Mercado de S. Domingos Municipal Complex
6. Horta da Mitra Municipal Market
7. S. Lourenço Municipal Market Complex

Drunken Dragon and Lion Dance Parade
Free Longevity Rice Distribution
Dragon Boat Rice Traditional Gathering Dinner

RESEARCH METHODS

The reported study adopted a qualitative research approach involving 16 festivalgoer interviewees who have been engaging in the festivals. Face-to-face semi-structured interviews were conducted via snowball sampling. Following consent oral given to conduct the interviews, a brief introduction was also given as explaining the purpose of the interview. The interviews were digitally recorded with the interviewees' permission. In particular, due to the long-time and direct engagement, the author will add her previous experiences of both case studies, in order to provide plenty of understanding and possible explanations.

Findings toward the heritage values, the associated cultural spaces, and the challenges that tourism development has been generated by both cultural events were found via primary sources and secondary data. Chinese respondents conducted all the data were done a full English transcript based on the voice records. Content analysis involving a manual coding system was used and significant quotes from different interviewees were assembled into categories, sub-themes, and common themes. In addition, a frequency analysis of similar responses was identified and presented.

Figure 7. One of the ritual ceremonies launched a day before the Na Tcha Festival starts. The picture took at Mount Hill's Na Tcha Temple
Source: The author, 2021

INTERVIEWEES' DEMOGRAPHIC PROFILE

Respondents from three generations and aged between 21-90 years were interviewed. All were local citizens with most being students, fish traders, and housewives. Furthermore, as a study objective to explore the perceptions and attitudes of the core festivals' participants, all respondents had participated in the Macao Feast of the Drunken Dragon and the Na Tcha Festival ranging from 4 to 70 years. A summary of the interviewees' profile is presented in Table 1 – Interviewee Profile.

In order to explore the impacts of the organizers and participants toward the sustainability of the festivals and investigate how these festivals are interpreted and transmitted by the host communities, as well as by government agencies; the interviewees were chosen based on their roles and engagement in the respective festivals. Additionally, all of the senior and middle-aged interviewees were the key members in preparing and upholding two of Macao's and China's nationally recognized intangible cultural heritage festivals for years.

The Problems of Misunderstanding Heritage Values of the Festivals

This study examines the heritage values of the festivals and seeks to develop a bottom-up approach to better safeguard Intangible Cultural Heritage. Accordingly, this will involve an examination of the cultural features of these two festivals. Firstly, the author will identify the related socio-cultural and spiritual values based upon the definition of heritage value. It aims to find out the elements in interpreting these two festivals and maintaining their authenticity and integrity. As Tilden (2009) noted, interpretation ought to present a whole rather than a part, it transmits the social, cultural, and spiritual meaning that lies behind an event and also provokes feeling in an individual.

In accordance with The Burra Charter (2013) "Practice Note: Understanding and Assessing Cultural Heritage", social value refers to "the associations that a place has for a particular community or cultural

Figure 8. Ritual ceremony at the Ruins of Saint Paul's Temple on the eighteenth day of the fifth month of the Chinese lunar calendar
Source: The author, 2021

group and the social or cultural meanings that it holds for them." (p.4). Spiritual value are "the intangible values and meanings embodied in or evoked by a place which gives it importance in the spiritual identity, or the traditional knowledge, art, and practices of a cultural group. …Spiritual value may also be expressed through cultural practices and related places." (p.4).

Macao Feast of the Drunken Dragon and the Lion Dance Gala

The Kuan Tai Temple (also known as Sam Kai Vui Kun Temple) is used as a meeting venue for local commercial representatives which is also one of the UNESCO cultural heritage-listed sites of "The Historic Centre of Macao". More importantly, for decades it represents a time capsule of a traditional fishmonger's cultural event as a gathering place and an event venue. Furthermore, the local wet markets which are irreplaceable tangible elements, embody the collective memory and social connection shared by the fishmongers as well as locals. Spiritually, the Macao Feast of the Drunken Dragon and the Lion Dance Gala symbolizes the shared social identity and cultural belief of the fish trading community,

Table 1. Interviewee Profile

The Macao Feast of the Drunken Dragon and the Lion Dance Gala Interviewees				
	Gender	Age (Years)	Occupation	Remark
I-1	Male	71-80	Fish trader	-
I-2	Male	81-90	Fish trader	-
I-3	Male	21-30	Student	The youngest inheritor of the Drunken Dragon Dance
I-4	Male	21-30	Student	-
I-5	Male	21-30	Student	-
I-6	Male	51-60	Fish trader	Chairman, Fishmongers' Association (Red Market) & rewarded the Outstanding Inheritor Award
I-7	Female	81-90	Retired	-
I-8	Female	51-60	Fish trader	-
I-9	Female	21-30	Insurance Agent	-
The Macao Belief and Customs of Na Tcha Interviewees				
I-10	Female	41-50	Housewife	-
I-11	Female	41-50	Salesperson	-
I-12	Male	41-50	Service Industry	-
I-13	Female	41-50	Housewife	-
I-14	Female	41-50	Housewife	-
I-15	Female	51-60	Manager of local garment factory	-
I-16	Male	61-70	Chairman, Ruins of Saint Paul's Na Tcha Temple Association	National's Representative Inheritor of the Beliefs and Customs of Na Tcha

Source: (The author, 2021)

which is socially expressed through the unique drunken dragon custom of traditional gathering dinner, the drunken dragon and the lion dance parade, and free 'longevity rice' distribution, as well as its related cultural spaces to demonstrate the core meaning of this fishmonger's annual event: 'selflessness', dedication', 'gratitude' and 'unity'.

Macao Beliefs and Customs of Na Tcha

The Na Tcha Festival is a shared belief among numerous ethnic Chinese communities. Its abundant history has also influenced two groups in the Macao community. Its social and cultural significance strongly exemplifies specific groups of inhabitants' religious beliefs during the Qing Dynasty from1900s to the present day. Additionally, its value also lies in its associated traditional temples, Mount Hill's Na Tcha Temple and the Ruins of Saint Paul's Na Tcha Temple, which are the carrier of the spirit that upholds one of Macao's representative intangible cultural festivals. Meanwhile, the Ruins of Saint Paul's Na Tcha Temple is also one of the historical buildings in the UNESCO cultural heritage-listed sites of "The Historic Centre of Macao". This irrefutably demonstrates the interconnection between a traditional festival and its cultural space.

Other than cultural space, the social connection between the community members is also an important aspect to better understand the festivals' heritage values. As one of the interviewees (I-2) mentioned:

"The Macao Feast of the Drunken Dragon is certainly a shared tradition for us. I still remember at the very beginning, fishmongers raised money for the gathering dinner that took place in front of the Sam Kai Vui Kun Temple. Several fishmongers danced with the wooden dragons in hand as dragon worship. It is an annual event full of festive and delightful atmosphere.".

Other than the associated temple, the local markets are also placed to build harmonious human interactions and develop the social fabric between fish vendors and customers. Therefore, the Macao Feast of the Drunken Dragon has been attracting the general public to be part of this event and this nationally recognized ICH traditional festival. According to the interviewees (I-1) and (I-8):

"Plenty of the participants are the families' members, friends, customers of the fishmongers, we much appreciate their kindness from the volunteers to complete the launching of this annual event.".

"As a volunteer, I am delighted to see people coming for the 'longevity rice' no matter how the weather is, in a way, their engagement is sort of great support for us (host community). Meanwhile, I feel blessed to see people holding a box of 'longevity rice' with their smiley faces."

In terms of the Na Tcha Festival, interviewee I-15 claimed:

"It is not easy to organize such a hallmark event as large as the Na Tcha Festival. I am very much appreciative of the opportunity to be a member of this community. I also found so much pleasure to share the meanings of the Na Tcha Festival with people who are not familiar with this cultural event, which represents traditional Chinese virtues, namely: filial piety and loyalty.".

Based on the author's observation, some of the interviewees (I-10, I-13, I-15) were quite emotional during their interviews. The relationship between the festivalgoers is irreplaceable, their enthusiasm is very catching and moving. Another interviewee (I-12) added:

"It will be a pity that if there is no one continue launching the Na Tcha Festival. I am happy to put all my efforts to sustain this tradition and its spirit. ...My son has been participating in the Na Tcha Festival for years.".

Last, but not least, interviewee I-15 added:

"Like the Macao Feast of the Drunken Dragon and the Lion Dance Gala, the Na Tcha Festival also represents local custom. These religious festivals not only gain blessings for us and our families but also the entire Macao society. I believe my engagement is charity work, therefore, I would love to join the Macao Feast of the Drunken Dragon as a volunteer to support this cultural event of the fishmonger community.".

The statement on "Intangible Cultural Heritage and Sustainable Development" by UNESCO (2015) stated that "Intangible Cultural Heritage can help strengthen social cohesion and inclusion." (pp.5). The influence of the Macao Feast of the Drunken Dragon and the Na Tcha Festival were clearly identified with people encouraging their family members and friends to take part in traditional festivals whereby, people can get spiritual rest and strengthen social connections through their participation in traditional festivals. In other words, traditional festivals can help festivalgoers build and maintain their social relationships, like the power of social norms clarify human attitudes and their behaviors that are expected by those who are in the same social circle.

CHALLENGES IN SAFEGUARDING THE FESTIVALS

This section also seeks to discover and examine the challenges that these specific cultural events are facing. Lastly, suggestions are made to address the issues through a bottom-up approach from an analysis of the two festival cases.

- Issue 1: Impacts of the top-down approach in safeguarding community-based festivals

The two selected festivals commemorate the histories and collective memories of specific community groups. These festivals also shape social and cultural identities that connect people and motive them to continue the traditions. Mahika (2011) noted that although the organizers 'and festivalgoers' motivations are driven primarily by social and psychological mechanisms, the cultural space is also a critical piece in upholding the heritage values of living heritage.

As the chairman of the Fishmongers' Association (Red Market) (I-6) stated:

"There are several vital meanings of the Macao Feast of the Drunken Dragon, namely: 1) Sodality, as the Feast gathers all the fishmongers together intending to organize this annual event; 2) Gather the fishmonger community and the general public as to make a closer relationship with the local society; 3) We hope through this cultural event to share with the general public about the history of our fish trading industry, which includes our social status and shared cultures.".

Yet he also argued that:

"Unfortunately, I found that the government agencies intend to transform our time-honored event into a tourism product through incomplete interpretation, incorrect information.".

Redpath et al. (2013) argued, the top-down approach is able to guarantee processes at all local levels are fair, all-encompassing, and not subject to corruption and control by professionals. However, this research study also found that several legislators proposed to develop one of the signature local markets, Red Market, and transform it into a trendy seafood market as means to diversify Macao's tourism attractions (Kwan, 2021). It is thought-provoking that possibly Red Market may become another renovated Hong Kong Central Market, which only left a new modern market that neglects to preserve the local culture and its cultural space where memories were formed (see Figure. 9).

Figure 9. The interior of Hong Kong's Central Market before and after its restoration
Source: HONG KONG FREE PRESS, 2021

In contrast to the Macao Feast of the Drunken Dragon, the existing Na Tcha Exhibition Room, situated next to the Ruins of Saint Paul's Na Tcha Temple to showcase the history and some tangible elements of the Na Tcha Festival. As interviewee (I-16) claimed, the exhibition room is run by the Ruins of Saint Paul's Na Tcha Temple Association and the Macao Cultural Affairs Bureau, while Mount Hill's Na Tcha Temple Association has zero responsibility for it. He also said that:

"Although the tourism office issued different types of local tours to stimulate local tourism during the covid outbreak, there is a lack of cultural tours focusing on the promotion of religious culture. …I would suggest the tourism office offer tours that mainly share local beliefs and it would be great to include the Na Tcha Exhibition Room, with the Mount Hill's Na Tcha Temple Association and the Ruins of Saint Paul's Na Tcha Temple Association together to promote our Na Tcha custom. Additionally, I wished the government will soon put a forward guideline to revitalize Pátio do Espinho (a village connected to the Na Tcha community) since it is an important place that connects the Na Tcha Festival.".

Therefore, the author suggests establishing an Intangible Cultural Heritage Exhibition Centre for the locals and tourists, where both can help the host community group to share their traditions, and also help government agencies to diversify Macao's tourism activities. Furthermore, based on the author's previous experiences, it was difficult to accommodate large groups of people at the small exhibition rooms of the Macao Fishmongers' Association and the Na Tcha Associations. This issue has raised more problems especially as the Macao Fishmongers' Association exhibition room is a private venue that often welcomes outsiders to visit. The proposed Intangible Cultural Heritage Exhibition Centre (the ICH Centre) by the author aims to strengthen the recognition of intangible heritage, specific groups of community, individuals, as well as the associated place or spaces of ICH. As such, research, documentation, and the digital database will be developed through collaboration with relevant key stakeholders and government agencies. The ICH Centre is an ICH Museum seeks to increase the awareness of ICH and encourage the new generation's participation in traditional events. In addition, the organizers as gatekeepers will be

Figure 10. Sam Tung Uk Museum was a walled house and currently is served as a cultural centre that presenting the Intangible Cultural Heritage items of Hong Kong. Visitors are able to learn the unique folklore customs and other intangible legacies through different forms of informative and innovative activities such as videos watching, ritual music listening and various educational games. Additionally, visitors are able to reserve a guided tour for an in-depth field study of the conservation project of the Sam Tung Uk Village and the traditional cultural features of the Sam Tund Uk Museum.
Source: The author, 2021

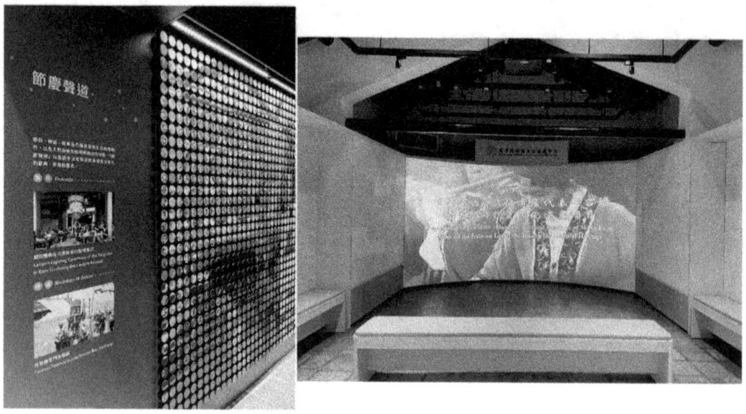

able to host tours and interact with the ICH Centre visitors. Such interactions will also serve as a tool to encourage the visitors' participation in Intangible Cultural Heritage activities, according to Denes et al. (2013), the Intangible Cultural Heritage Museum is able to offer visitors a complete understanding of living heritage and its cultural significance. Accordingly, the proposed Intangible Cultural Heritage Exhibition Centre will be a hub for the host communities to showcase and share their traditions, as well as a museum for the general public to study intangible heritage through authentic interpretation and their interactions with the core community (see Figure. 10).

- Issue 2: Relationship between the host community and other stakeholders

As discussed earlier, the issues and threats caused by touristification that have been identified clearly indicate there is a lack of attention and concern in the festivals' interpretation among the government authorities and a less than equal partnership between the host community and other stakeholders. In regard to these festival case studies, the Macao Feast of the Drunken Dragon and Na Tcha Festival both feature legends that enable their respective ancestors to gain blessings through Drunken Dragon and Na Tcha worship. Despite both of the legend stories have no medical evidence proofed the ritual ceremonies helped the believers to get rid of illness, the believers keep their faith that directly and collectively inherit the shared traditions of each specific community group. The roles of different government departments, for example, the promotions from Macao Government Tourism Office neither mention nor provide accurate information on its online platform. In addition, incomplete interpretations from Macao Cultural

Figure 11. The government authorities and other event managers positioned the drunken dragon dance as a dancing performance. Frankly, it is a ceremony that coveys and worships drunken dragon belief. However, the drunken dragon team was invited as a performing team in the Macao International Parade and showcased at different sites of "The Historic Centre of Macao" such as the Ruins of Saint Paul's
Source: Macao Cultural Affairs Bureau, 2019

Affairs Bureau threaten the original expression and meaning of traditional festivals, which directly harms the sustainability of intangible cultural heritage festivals (see Figure. 11).

As one of the youth drunken dragon performers (interviewee I-4) argued that:

"The venue should remain its original setting. We should be aware and respect the traditions that our ancestors have passed on to us.".

Indeed, stakeholders should pay more attention in interpreting the intangible festivals, especially as these festivals are distinguished by their profound histories with the fundamental part of their tangible buildings and its surrounding residential areas that completely display the living habits, economic, social, and environmental conditions in the city during the old days. To address such issue, the author suggests that the festivals' organizers (Macao Fishmongers' Association; The Mount Hill's Na Tcha Temple Association; and The Ruins of Saint Paul's Na Tcha Temple Association) select their respective ambassadors (or curators) and to:

- establish a non-governmental organization to communicate, update and liaise with the government agencies;
- have the ambassadors regularly communicate with government agencies and cooperate with them.

For example, launch public workshops to promote the significance of traditional events in order to increase greater social awareness in the safeguarding of Intangible Cultural Heritage. A study by Giglitto (2017) found that empowerment refers to the attempt to use bottom-up management in preserving cultural

Figure 12. The conceptual framework of the Safeguarding Intangible Cultural Heritage Association
Source: The author, 2021

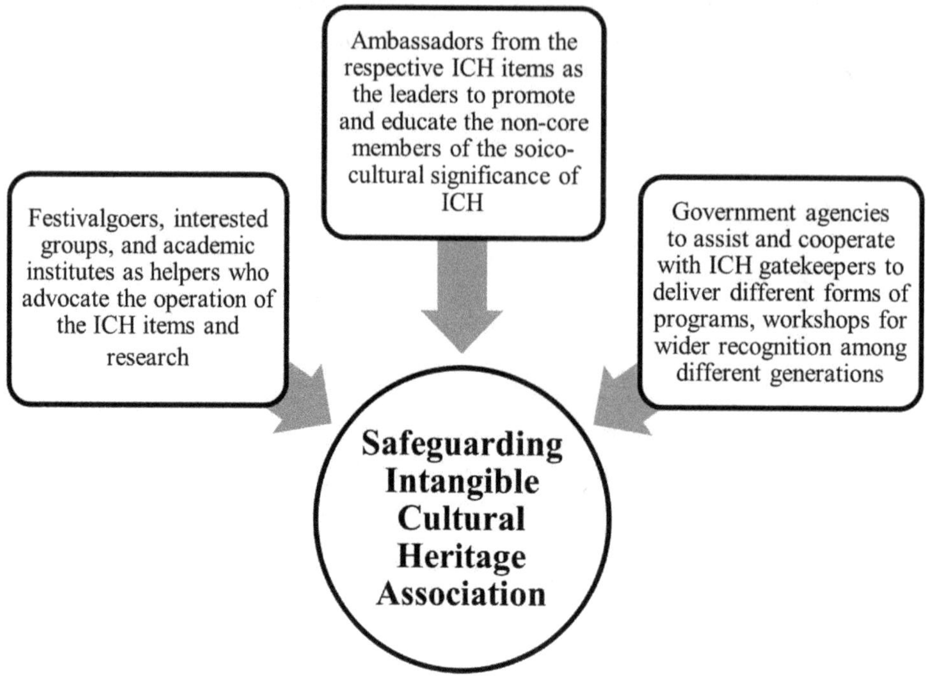

heritage, which is a factor to enhance community and social cohesion, similar to community involvement in tourism management while encouraging the creation of collective memory. Furthermore, the aim of this strategy seeks to address and achieve the United Nation's Sustainable Development Goals 8 and 17. A summary of the concept is presented in Figure 12 (see Figure.12).

More importantly, the festival believers' superstition sincerely links to the nature of religious activities which they believe drunken dragon worship and Na Tcha belief bring them prosperity, respectively. It exemplifies the harmonious relationship between humanity and nature. Also, the local community believes in the magic of longevity rice which will give them a healthy life, etc. The influences of folklore beliefs and traditional festivals are no longer on a micro-level, as a fact that both festivals are recognized at the national level having been listed as China's Nation Intangible Cultural Heritage. Therefore, the safeguarding process of both selected festivals should firstly deliver the values ascribed by the key communities (namely, the Macao Fishmongers' Association; The Mount Hill's Na Tcha Temple Association; and The Ruins of Saint Paul's Na Tcha Temple Association) to emphasize the social and cultural significance of these festivals. Using one of the major venues of the Macao Feast of the Drunken Dragon as an example, Red Market has remained without any changes of its fabric and functions to the present day. The market itself is a tangible heritage that has witnessed the evolution of living heritage. Although the entire Red Market building retains its traditional character and atmosphere, the related government offices have demonstrated a lack of effort in either correctly interpreting or understanding the local cultures of Macao. For example, the Macao Government Tourism Office has also indicated a failure to convey the dynamic, transformative spirit, living culture, and cultural space of local Intangible Cultural Heritage in providing incorrect information of this particular event. The cooperation between the core festival community groups and other stakeholders is a vital factor in safeguarding traditional festivals. As the chairman of the Fishmongers' Association (Red Market) (I-6) said:

"Frankly, we (the fishmongers) don't have regular meetings with the government authorities. I have been wondering why the government has no interest in pursuing our concerns and needs.".

Furthermore, recent research on the Macao Feast of the Drunken Dragon and the fishmonger community by Kwan (2021) found that many tourists learned about the Drunken Dragon Feast from their friends and the online platforms, which means, the gatekeepers should ensure the accuracy of the information and interpretation about the Feast. Otherwise, people who have not experienced the Feast may not understand the authentic meanings and its cultural significance.

Even though the government authorities made several mistakes in promoting these traditional festivals, some interviewees claimed the government funding is relatively important in launching a large-scale event such as the Macao Feast of the Drunken Dragon and the Lion Dance Gala. For instance, a youth drunken dragon performer (interviewee I-3) said:

"The government funding assists to invite the overseas teams to perform in the Feast, and this novel element helps to attract more younger generations to the Macao Feast of the Drunken Dragon.".

One senior festivalgoer (I-7) said, *"I am very grateful to the government support in upholding this cultural event for years."*. While a middle-aged participant (I-8) added, *"The industry can't hold this festival without government funding since the fundraising from the fishmongers is not enough to hold such a hallmark event."*. Their comments show the nature of the harmonious connection between the fishmonger community and government agencies. It clearly shows that the government is generous in providing financial support for local cultural events. That is to say, the concerns toward safeguarding traditional festivals are the interpretation of the festivals and communication between the host community and government authorities.

FUTURE RESEARCH DIRECTIONS

Since the author proposed two solutions in order to generate more benefits in safeguarding Intangible Cultural Heritage in Macao, both of them are communities-led tasks while some unfair conditions may arise, especially during the negotiations among the different parties. Thereby, future research should investigate the nature of the relationship between different Intangible Cultural Heritage stakeholder organizations. To build up a more effective bottom-up approach from core members, there is a need to establish a systematic mechanism to safeguard the history, buildings, spaces, and their related time-honored cultures. Meanwhile, this study collected data from two stakeholder groups and future studies should include other stakeholders such as the general public, government agencies and event managers.

CONCLUSION

The global outbreak of the novel coronavirus pandemic has put travel activities at a standstill which has substantially influenced the celebration of cultural events due to social prevention and travel restriction measures. Travelers have a desperate need to discover and travel the world again. In accordance with the issues and challenges discussed above, some alarming impacts on the sustainability of traditional ceremonies such as the inconsistency and inaccuracy of its original interpretation have put the safeguarding of living heritage in a more difficult situation. Due to the COVID-19 pandemic, only ritual ceremonies were held in 2020 and 2021. With cooperation and positive thinking, there is an opportunity for host communities and government agencies to further discuss how to present and interpret the traditional festivals to the general public and tourists. When the pandemic is over, the establishment of the proposed ICH Centre will be able to present accurate information and interactive tours. Other than interpretation, the associated cultural spaces of these traditional cultural events are also an important aspect that requires attention and consideration. The festivals and the venues constitute people's daily interactions with affiliated temples of Na Tcha Festival representing human settlement dating back in the early 1900s and the local wet markets which also exemplify the rich history of the Macao fish trading industry. These cultural spaces carry and encapsulate human daily interactions but also their decades-long beliefs; the spirit of these public places should be better conserved and pertained to the community. As stated by Kilipiris (2005), "Local involvement and shared decision making is crucial to sustainable development. …Local involvement is essential to tourism: local peoples, their culture, environment, way of life, and traditions are important factors that attract tourists to a location. The needs and aspirations of local peoples must therefore be fully supported. Carefully developed tourism can provide real economic, environmental, and cultural benefits to the community. In turn, genuine community involvement can enrich the tourism experience and product." (p.32-33). To conclude, the author suggests that a bottom-up approach provides a better strategy for safeguarding community-based festivals, which can also serve as a strategy for maintaining a mutually beneficial relationship between the community and government agencies, and collectively achieve the Sustainable Development Goals.

ACKNOWLEDGMENT

I would like to express my gratitude to the interviewees who agreed to be interviewed and also offered uncountable, meaningful information in the current research. Without their generously assistant, this research will never exist. I am very much appreciated by each of them.

To my respected Dr. John Ap, thank you so much for your excellent advice and patient teaching. With your help and encouragement, this research has made it to a good standard. Meanwhile, thank you to Dr. Sharif Shams Imon for your useful advice. A big thank you to both of you!

This research received no specific grant from any funding agency in the public, commercial, or not-for-profit sectors.

REFERENCES

Ajzen, I. (1991). The theory of planned behavior. *Organizational Behavior and Human Decision Processes*, *50*(2), 179–211. doi:10.1016/0749-5978(91)90020-T

Ap, J. (1990). Residents' perceptions research on the social impacts of tourism. *Annals of Tourism Research*, *17*(4), 610–616. doi:10.1016/0160-7383(90)90032-M

Arcodia, C., & Whitford, M. (2007). Festival attendance and the development of social capital. In *Journal of convention & event tourism* (pp. 1–18). Taylor & Francis Group.

Australia, I. C. O. M. O. S. (2013). *Practice Note: Understanding and Assessing Cultural Heritage*. https://australia.icomos.org/wp-content/uploads/Practice-Note_Understanding-and-assessing-cultural-significance.pdf

Bauman, R. (Ed.). (1992). *Folklore, cultural performances, and popular entertainments: A communications-centered handbook*. Oxford University Press.

Ben-Amos, D. (1971). Toward a definition of folklore in context. *Journal of American Folklore*, *84*(331), 3–15. doi:10.2307/539729

Boas, F. (1916). The origin of totemism. *American Anthropologist*, *18*(3), 319–326. doi:10.1525/aa.1916.18.3.02a00020

Briggs, A. (Ed.). (1992). *Folklore, cultural performances, and popular entertainments: A communications-centered handbook*. Oxford University Press.

Bronner, S. J. (1984). The Early Movements of Anthropology and Their Folkloristic Relationships. *Folklore*, *95*(1), 57–73. doi:10.1080/0015587X.1984.9716297

Brown, S. J. W., & Trimboli, D. (2011). The real "worth" of festivals: challenges for measuring socio-cultural impacts. *Asia Pacific Journal of Arts and Cultural Management, 8*(1).

Buccitelli, A. B., & Schmitt, C. R. (2016). Everyday Practice and Tradition. *Cultural Analysis*, *15*, 1–5.

Choi, S. H., Imon, S. S., & Couto, U. (2020). Negotiating with Authenticity to Ensure Sustainability of Traditional Cultural Festivals: Residents' Perspectives about Macao's Drunken Dragon Festival. *Sustainability*, *12*(3), 885. doi:10.3390u12030885

Cialdini, R. B., & Trost, M. R. (1998). Social influence: Social norms, conformity and compliance. In D. T. Gilbert, S. T. Fiske, & G. Lindzey (Eds.), *The handbook of social psychology* (pp. 151–192). McGraw-Hill.

Crespi-Vallbona, M., & Richards, G. (2007). The meaning of cultural festivals: Stakeholder perspectives in Catalunya. *International Journal of Cultural Policy*, *13*(1), 103–122. doi:10.1080/10286630701201830

Cultural Affairs Bureau. Macao SAR. (2019). *Macao International Parade*. https://www.icm.gov.mo/macaoparade/9/en/

Cultural Affairs Bureau. Macao SAR. (2021). *Belief and Customs of Na Tcha*. https://www.culturalheritage.mo/en/detail/133/1

Cultural Affairs Bureau. Macao SAR. (2021). *Festival of the Drunken Dragon*. https://www.cultural-heritage.mo/en/detail/135/1

Dégh, L. (2020). The "Belief Legend" in Modern Society: Form, Function, and Relationship to Other Genres. In American Folk Legend (pp. 55-68). University of California Press.

Denes, A., Koanantakool, P. C., Davis, P., Kreps, C., Hennessy, K., Alivizatou, M., & Stefano, M. L. (2013). Critical reflections on safeguarding culture: The intangible cultural heritage and museums field school in Lamphun, Thailand. *Heritage & Society*, *6*(1), 4–23. doi:10.1179/2159032X13Z.0000000004

Derrett, R. (2003). Festivals & regional destinations: How festivals demonstrate a sense of community & place. *Rural Society*, *13*(1), 35–53. doi:10.5172/rsj.351.13.1.35

Getz, D. (1998). 24 Event tourism and the authenticity dilemma. In Global tourism (pp. 409-426). Reed Elsevier Plc Group.

Giglitto, D. (2017). Community empowerment through the management of intangible cultural heritage in the Isle of Jura, Scotland. *Imperial Journal of Interdisciplinary Research*, *3*(5), 567–578.

Guo, X., Pikas, B., & Lee, T. (2017). Study of Intangible Chinese Cultural Heritage Protection and Tourism Development. The Case of Intangible Cultural Heritage Protection of the Hezhe Nationality. *Journal of Marketing Development and Competitiveness*, *11*(2), 39–45.

Hauptfleisch, T. (2006). Eventifying identity: Festivals in South Africa and the search for cultural identity. *NTQ. New Theatre Quarterly*, *22*(2), 181–198. doi:10.1017/S0266464X0600039X

He, X. L. (2007). *Religlionary Faith and Ethnic Culture (I)*. Social Sciences Academic Press.

Hong Kong Free Press. (2021). *A gentrified mess: How Hong Kong's Central Market restoration tarnished an architectural gem*. https://hongkongfp.com/2021/09/11/a-gentrified-mess-how-hong-kongs-central-market-restoration-tarnished-an-architectural-gem/

Huibin, X., Marzuki, A., & Razak, A. A. (2013). Conceptualizing a sustainable development model for cultural heritage tourism in Asia. *Theoretical and Empirical Researches in Urban Management*, *8*(1), 51–66.

Imon, S. S. (2017). Cultural heritage management under tourism pressure. *Worldwide Hospitality and Tourism Themes*, *9*(3), 335–348. doi:10.1108/WHATT-02-2017-0007

Kaplan, G. G. (2015). The global burden of IBD: from 2015 to 2025. *Nature Reviews. Gastroenterology & Hepatology*, *12*(12), 720–727. doi:10.1038/nrgastro.2015.150 PMID:26323879

Kilipiris, F. (2005). Sustainable tourism development and local community involvement. *Tourism and Hospitality Management*, *11*(2), 27–39. doi:10.20867/thm.11.2.3

Kim, S., Whitford, M., & Arcodia, C. (2019). Development of intangible cultural heritage as a sustainable tourism resource: The intangible cultural heritage practitioners' perspectives. *Journal of Heritage Tourism*, *14*(5-6), 422–435. doi:10.1080/1743873X.2018.1561703

Kuuder, C. J., Adongo, R., & Abanga, J. (2012). The socio-cultural significance of the Kakube Festival of the Dagara of Nandom Traditional Area in Ghana. *Ghana Journal of Development Studies*, *9*(2), 103–124. doi:10.4314/gjds.v9i2.6

Kwan, C. W. K. (2019). The impact of festival participants' attitude on the sustainability of traditional festivals: A case study of the Feast of the Drunken Dragon and the Lion Dance Gala of Macao. In *Proceedings of the Tourism Education Student Summit, Macao SAR, China* (pp. 162-168). Macao Institute for Tourism Studies, Macao SAR, China.

Kwan, C. W. K. (2021). *The Living Drunken Dragon: Macao's Fishmonger Community and the Feast of the Drunken Dragon and the Lion Dance Gala* [Unpublished master's dissertation]. The University of Hong Kong, Pokfulam, Hong Kong SAR, China.

Laing, J. (2018). Festival and event tourism research: Current and future perspectives. *Tourism Management Perspectives*, *25*, 165–168. doi:10.1016/j.tmp.2017.11.024

Lee, T. H., & Chang, P. S. (2017). Examining the relationships among festivalscape, experiences, and identity: Evidence from two Taiwanese aboriginal festivals. *Leisure Studies*, *36*(4), 453–467. doi:10.1080/02614367.2016.1190857

Macao Government Tourism Office. (2021). *Feast of the Drunken Dragon*. https://www.macaotourism.gov.mo/en/events/calendar/feast-of-the-drunken-dragon

Macao Government Tourism Office. (2021). *Na Tcha Temple*. https://www.macaotourism.gov.mo/en/sightseeing/temples/na-tcha-temple

Mahika, E. C. (2011). Current trends in tourist motivation. *Cactus Tourism Journal*, *2*(2), 15–24.

McKercher, B., & Du Cros, H. (2002). *Cultural tourism: The partnership between tourism and cultural heritage management*. Routledge.

McKercher, B., Mei, W. S., & Tse, T. S. (2006). Are short duration cultural festivals tourist attractions? *Journal of Sustainable Tourism*, *14*(1), 55–66. doi:10.1080/09669580608668591

Mokgachane, T., Basupi, B., & Lenao, M. (2021). Implications of cultural commodification on the authenticity of iKalanga music: A case of Domboshaba traditional music festival in Botswana. *Journal of Tourism and Cultural Change*, *19*(2), 153–165. doi:10.1080/14766825.2019.1700989

Motz, M. (1998). The practice of belief. *Journal of American Folklore*, *111*(441), 339–355. doi:10.2307/541314

Rioux, M. (1950). Folk and Folklore. *Journal of American Folklore*, *63*(248), 192–198. doi:10.2307/537160

Sassatelli, M. (2015). *Festivals, urbanity and the public sphere: reflections on European Festivals. In Focus on Festivals. Contemporary European Case Studies and perspectives*. Goodfellow Publishers.

Small, K., Edwards, D., & Sheridan, L. (2005). A flexible framework for evaluating the socio-cultural impacts of a (small) festival. *International Journal of Event Management Research*, *1*(1), 66-77.

Smith, M. W. (1959). The importance of folklore studies to anthropology. *Folklore*, *70*(1), 300–312. doi:10.1080/0015587X.1959.9717162

Su, J. (2019). Understanding the changing Intangible Cultural Heritage in tourism commodification: The music players' perspective from Lijiang, China. *Journal of Tourism and Cultural Change, 17*(3), 247–268. doi:10.1080/14766825.2018.1427102

Su, X., Li, X., Wu, Y., & Yao, L. (2020). How is intangible cultural heritage valued in the eyes of inheritors? Scale development and validation. *Journal of Hospitality & Tourism Research (Washington, D.C.), 44*(5), 806–834. doi:10.1177/1096348020914691

The United Nations Educational, Scientific and Cultural Organization. (2011). *What is Intangible Cultural Heritage?* https://ich.unesco.org/en/what-is-intangible-heritage-00003

The United Nations Educational, Scientific and Cultural Organization. (2015). *Intangible cultural heritage and sustainable development.* https://ich.unesco.org/doc/src/34299-EN.pdf

The United Nations Educational, Scientific and Cultural Organization. (2021). *Text of the Convention for the Safeguarding of the Intangible Cultural Heritage.* https://ich.unesco.org/en/convention#art2

The United Nations Educational, Scientific and Cultural Organization. (2021). *Yeondeunghoe, lantern lighting festival in the Republic of Korea.* https://ich.unesco.org/en/RL/yeondeunghoe-lantern-lighting-festival-in-the-republic-of-korea-00882

The United Nations Educational, Scientific and Cultural Organization. (2021). *Dragon Boat festival.* https://ich.unesco.org/en/RL/dragon-boat-festival-00225

The United Nations Environment Programme and the United Nations World Tourism Organization. (2005). *Making Tourism More Sustainable: A Guide for Policy Makers.* The United Nations Environment Programme and the United Nations World Tourism Organization.

The United Nations World Tourism Organization. (2021). *Tourism 4 SDGS.* https://www.unwto.org/tourism4sdgs

Tilden, F. (2009). *Interpreting our heritage.* University of North Carolina Press.

Wu, K.L. (2014). *Macao Belief and Customs of Na Tcha.* Joint Publishing.

Yankholmes, A. (2018). Stakeholders' Opinions on How to Brand and Diversify Macau's Tourism Product. In *Workshop Proceedings of the Inclusive Innovation for Enhanced Local Experience in Tourism, Phuket, Thailand* (pp. 11-15). University of Birmingham.

Chapter 18
Recognition of Tangible and Intangible Cultural Heritage for the Sustainability of Tourism:
Turkey (Türkiye) Case

Nil Sonuç
https://orcid.org/0000-0002-7572-9192
İzmir Katip Çelebi University, Turkey

ABSTRACT

This chapter is designed to portray how sustainability for cultural heritage tourism management can be ensured through recognition of the tangible and intangible cultural heritages. Creating a bridge among people, the recognition of the intangible and tangible cultural heritages brings their revaluation for more sustainable tourism and opens the way to their acknowledgement in tourism industry practices. The chapter gives an overall idea of the tangible, intangible, and mixed cultural heritages in Turkey, representing one of the most prominent destinations in the world for the richness of its cultural heritage. Academic studies and the reports related to cultural heritage recognition practices in Turkey and the world are revised to bring recommendations and to develop a base for the discussion of practical implications for sustainable cultural heritage tourism management in Turkey. The study aims to be inspirational for both theoreticians and practitioners to create brighter cooperative projects.

INTRODUCTION

This chapter aims to reflect the dynamics of national and international recognition of cultural heritage resulting in a positive basis for the sustainability of tourism in Turkey. The question asked to reach this objective is as follows: How can cultural heritage management tourism be sustainable through the national and international recognition of cultural heritages in Turkey? The study searched for and analysed academic literature in Turkey and in the world to unearth the trends in cultural heritage management and their relationship with sustainability and tourism. In addition, document analysis is used. The data selected

DOI: 10.4018/978-1-7998-9217-5.ch018

through the official online documents of the Ministry of Culture and Tourism and videos on its official website, WTO's and UNESCO's official documents and official web sites are reviewed and analysed to provide information concerning the national and international recognition efforts of cultural heritage and its impacts on the sustainable cultural heritage management. Additionally, reliable news data on the web is also used to support the analysis used for the methodological basis of this chapter. Finally, the chapter has not only provided some recommendations for a more sustainable cultural heritage tourism management in Turkey, but it has also highlighted the need for more multidisciplinary researches for a more sustainable cultural heritage tourism management in different locations of the world.

Cultural heritage tourism is a selected area of differentiation of tourism products in Turkey and the World. To avoid the negative impacts of mass tourism, which is based on the "sun-sea-sand" triangle, and to make tourism more sustainable, cultural heritage has been an inevitable target of strategy for sustainable tourism in Turkey. Turkey is a country situated on the Anatolian Peninsula on the Commercial Routes of Silk, Spice and Royal Roads and located on land that has been inhabited uninterruptedly during the historical periods by many civilizations such as Hittites, Lydians, Phrygians, Ionians, Seljuks, Byzantines, Ottomans who left behind universally invaluable tangible and intangible heritages.

Cultural heritage tourism is mentioned besides cultural tourism and is often considered as a subset of cultural tourism (Kaminski et al., 2014). In UNWTO General Assembly in China, cultural tourism is defined as "a type of tourism activity in which the visitor's essential motivation is to learn, discover, experience and consume the tangible and intangible cultural attractions/products in a tourism destination. These attractions/products relate to a set of distinctive material, intellectual, spiritual and emotional features of a society that encompasses arts and architecture, historical and cultural heritage, culinary heritage, literature, music, creative industries and the living cultures with their lifestyles, value systems, beliefs and traditions." (World Tourism Organization, 2018, p. 11).

Cultural tourism is estimated to generate 40% of world tourism revenues (UNESCO, 2020). Cultural tourism gained momentum in the 1990s with the emergence of sub-sectors such as heritage tourism, arts tourism, gastronomic tourism, film tourism and creative tourism (UNESCO, 2020). Nearly all these sub-sectors are included or conceptualised in the intangible or tangible heritages. Additionally, it would not be wrong if cultural heritage tourism were mentioned very synonymously to cultural tourism.

Since antiquity, people have visited shrines, monuments, and historical places for many different motivations. Discovering other cultures and interacting with them has reawakened and appears as a rapidly increasing trend today. The motivations of the domestic tourists may be to learn their roots and pride, while international motivations may result in higher tolerance and understanding among different cultures (UNESCO, 2020).

The craftsmanship of various arts (Pelit & Türkoğlu, 2019, Alagöz et al., 2018), visiting of war places and war monuments (Akbulut & Ekin, 2018), literature and literary heritage (Aliağaoğlu & Narlı, 2012), gastronomy museums (Ağcakaya & Can, 2019) are mentioned as diversified products that appear in cultural heritage tourism market. Today's existing cultures and folkways are included in cultural heritage tourism too, as they are becoming inheritances from the past (Timothy & Nyaupane, 2009, World Tourism Organization, 2018).

Concerning the sustainability of the cultural heritage tourism management, it is indicated that much of the rhetoric, initiatives and advice offered are still in their infancy and it is therefore too soon to tell the extent of stakeholders' engagement and consequently whether coherent implementation will happen (Kaminski et al., 2014, p. 9). Cultural heritage tourism management is a field that automatically needs a sustainable perspective, that is why it cannot be considered apart from sustainability. Resource protection in terms of preservation and revaluation of the heritages, participation-coordination and cooperation of all the stakeholders, sharing both benefits and burdens are some important pillars of this strategic sustainable heritage tourism management perspective.

On the international institutional level, UNESCO verifies the undeniable sustainability dimension trying a new approach with its "World Heritage and Sustainable Tourism Programme". It is based on dialogue and stakeholder cooperation where planning for tourism and heritage management is integrated at a destination level, the natural and cultural assets are valued and protected, and appropriate tourism is developed. UNESCO World Heritage and Sustainable Tourism Programme has created a project called the "Silk Road Project" joining 34 member countries in Europe (including Turkey), Asia, the Middle East the and the Far East together to realize sustainable tourism in the Silk Road Corridor (UNESCO, 2021a).

If a sustainability strategy is operational in tourism, this will benefit simultaneously the protection and revaluation of these cultural heritages. In Turkey, the government authorities have contributed to this aim more concretely especially since the Turkey Tourism Strategy, Action Plan 2007-2013 was accepted. Before this concrete step at the national level, many other measures were taken by the coordination of the Ministry of Culture and Tourism and the local authorities such as municipalities to inscribe the intangible and tangible heritages to the UNESCO World Heritage lists since the 1980s. After UNESCO declared the acceptance of the Convention Concerning the Protection of the World Cultural and Natural Heritage in 1972, it was officially accepted in Turkey in 1983. These are the proofs that show the recognition efforts of Turkey for the protection of its cultural heritages on the national and international level and that enable their revaluation for tourism.

If the development of cultural heritage tourism is not based on sustainable planning and management principles, it may have negative impacts on the cultural heritage resources such as deterioration, destruction, alienation or the loss of authenticity (Baykan, 2007). This conflict in the relationship between tourism and cultural heritage management can be returned into an advantageous win-win relationship by the cautious calculation of the carrying capacities of the cultural heritage tourism destinations and preference of correct management styles (Arslan Kalay & Yenişehirlioğlu, 2017).

The chapter flow includes an introduction with conceptualisations related to cultural heritage and cultural heritage tourism management followed by the cultural heritage tourism management in the World with sustainability aspect. The chapter then continues with the cultural heritage tourism management in Turkish Tourism, followed by the current state of national and international recognition of cultural heritage serving the purpose of tourism in Turkey. Finally, the chapter includes the recommendations for a more sustainable cultural heritage tourism management in Turkey and ends with suggestions of related future research directions.

CULTURAL HERITAGE TOURISM MANAGEMENT

Conceptualisations Related to Cultural Heritage and Cultural Heritage Tourism

Before analyzing cultural heritage and its close relationship with tourism, it is better to review some conceptualisations related to cultural heritage and tourism.

Cultural heritage is defined as a treasure strengthening the sentiments of unity and solidarity and supporting the moral values among the members of society, and telling them their common history and additionally comprising a basis for the current world and giving a reference for creating the future (Ünal,

2014). Showing the depth of a nation in the civilization, cultural heritage is also one of the basic elements that enlighten the cultural and intellectual evolution of the nation (Jassam, 2019).

The Convention Concerning the Protection of the World Cultural and Natural Heritage was accepted at the General Conference of UNESCO in 1972. Two basic factors have increased the subsistence of this convention worldwide: One is the rising acceptance of the 'common heritage of the humanity' concept meaning that natural or human-made unique works of art are far beyond belonging to only one country. The second factor is that, in the 1970s, the international community was eager to mobilize for the protection of the archaeological sites or protected areas (Akipek, 2001).

Another important step of UNESCO for the international acknowledgement of intangible heritages is the 2003 Convention for the Safeguarding of the Intangible Cultural Heritage. Following this important step, it is recognized that the interrelationship of the tangible and intangible heritage is essential for understanding the meaning and importance of each other. The commonality between the 1972 and 2003 Convention is the *universal* value of all heritages and some legal issues related to their international cooperative mechanisms, administrative structures and educational aspects. The difference between them is signified by *safeguarding* of intangible or living heritages, which means that the control to use and exploit is given to the holder of the intangible heritages (Bouchenaki, 2003).

Even if there is a universal categorization of cultural heritage such as tangible and intangible as seen in UNESCO World Heritage Lists, sub-categories may vary according to varying features of different countries and localizations. For the case of tangible heritages movable and unmovable are the basic categories for cultural heritages (Asatekin 2004, Ünal 2014). As mentioned above, there are also the "mixed heritages" having both natural and cultural heritages in one area or location.

Natural assets and the natural environment formed by them that need protection have a definition that is restricted to "rare", "privileged natural properties and beauties"; thus, "to be rare" is the most important input. This rareness can be described as a geological formation, a special type of flora or a biogenetic area where a special creature lives (Ministry of Culture, 1990 as cited in Asatekin, 2004).

For the intangible heritages, there are more categories such as traditions and verbalization types including languages, theatre and show performances, arts and traditions, social practices, ceremonies, festivities, practices related to nature and the universe, traditional handicrafts skills. Intangible heritages are not only traditions inherited from the past (such as customs and traditions, rituals, art, music, dance, theatre, fairy tale) but they are also contemporary and living heritages comprising the urban or rural practices of different groups (such as culinary arts, food and beverages) (Jassam, 2019).

A culturally authentic resource is both culturally and economically valuable, thus the interests in the preservation of cultural heritage may be dependent on the economic value of cultural heritage as a tourism resource (Cohen, 2002).

Halewood and Hannam (2001, p. 567) describe three authenticity views for heritage tourism: *Landscapes of nostalgia* which gives stability and security instead of the instability of the contemporary era, *staged authenticity* which is a created authenticity that makes tourists experiences genuine and *commodification* which leads to standardization of the touristic resources becoming a commodity.

Chabbra (2010, p.17) defines cultural heritage tourism as "a phenomenon that focuses on the management of past, inheritance, and authenticity to enhance participation and satisfy consumer motivations by evoking nostalgic emotions; its underlying purpose is to stimulate monetary benefits for its various constituencies and custodian of heritage such as the museums, historic houses, festivals, heritage hotels and other stakeholders".

People and natural conditions are among the factors that threaten cultural and natural assets or heritages. (Ministry of Culture, 1990, as cited in Asatekin, 2004).

Natural disasters or climate change are not considered as threatening factors here, even if nowadays they too are mentioned among the harm resulting from human intervention. Most importantly, historical artefacts smuggling and illegal auctions are two basic factors that harm the tangible heritages. Illegal treasure hunting, illegal excavations and other illegal methods and activities seriously damage cultural heritage sites. This kind of illegal activity may even lead to the loss of the artefacts forever (Akkuş &Efe, 2015, p. 125). That is why this situation requires urgent legal action from the related authorities.

Harmon (2007) has argued that people can find a way to preserve natural and cultural resources at the same time even if the culture scientists or practitioners are completely different from biologists. He argues that only by being aware of the plurality of the world and accepting that epistemological varieties are richness, and by creating an integrated perspective it can be possible.

Sustainable Cultural Heritage Tourism Management in the World

When the academic literature is searched for sustainable cultural heritage tourism management, it is seen that conceptualisations for the theoretical and practical implications in this field are still in the developing phase thus, far from being mature. Researches trying to conceptualise and create a framework for a sustainable cultural heritage tourism management in different destinations emphasize the collaborative involvement of all the related shareholders such as public-private partnership (Gilmore et al., 2007); tourism operators and practitioners, tourists, authorities and local communities (Huibin et al., 2012; Arthur & Mensah, 2006; Vargas-Hernández, 2012). A conceptual model to realize a sustainable cultural heritage tourism management suggested by Li & Hunter (2015) includes the leading elements of organizational motivation, community empowerment and involvement of community stakeholders. A study searched for an answer to the question of how to negotiate the concept of heritage among community stakeholders to use it as a necessary and strategic method in the development of cultural tourism and it suggested that critical ethnography should not be undermined during attempts to understand the sociology of tourism (Santa & Tiatco, 2019).

Ngamsomsuke et al. (2011) searched for sustainable cultural heritage tourism indicators in their study in one of the five heritage sites of Thailand, Ayutthaya. Economic activities, social supports, management of cultural heritage sites, surrounding environment of cultural heritage sites were four basic indicators in turn by weight. Convenient access to cultural heritage sites; services system and capacity at airports, railways and bus terminals; tourism facilities support; tourism service supports; tourism leisure supports were the 5 sub-indicators of economic activities. Tourists' satisfaction with cultural heritage sites, tourists' attitude towards cultural heritage sites, public awareness prints of possible criminality at cultural heritage sites, public prints of regulation at cultural heritage sites resulted as the 4 sub-indicators of social supports. Cultural heritage site preservation condition, maintenance of cultural heritage sites, cultural heritage site attracting promotion, facilities and services provided at cultural heritage sites, knowledge and beliefs received from cultural heritage sites visit, advanced information technology for accessing cultural heritage sites, major site narration are the 7 sub-indicators of management of cultural heritage sites. And finally, level of air pollution at the cultural heritage site, garbage disposal and control at cultural heritage sites, the overall architectural character of location surrounding cultural heritage sites, urban design surrounding cultural heritage sites are 4 sub-indicators of the surrounding environment of cultural heritage sites.

Nocca (2017) analyzed 40 case studies on heritage management from the sustainability perspective considering the multidimensional impacts of preservation and regeneration of cultural heritages. This study suggested that, while trying to create a matrix of sustainable cultural heritage management there is a need for a circular tourism system operating as a circular economy that balance the short and long-term impacts of climate change, social, environmental, economic and well-being dimensions and that also comprises the synergy of the collaborators/actors. Especially improving the well-being of the local community is emphasized regarding Rypkema's saying "if you do it for the locals, the tourists will come; if you do it for the tourists, only the tourists will come" (Rypkema, 2003 and 2010 as cited in Nocca, 2017, p. 17). This saying explains briefly the meaning of the strategic sustainability approach with all its dimensions. It means that by doing it for the locals you do the right thing, the visit of the tourists will later automatically follow.

Apart from the supply side, demand-side studies are also needed especially for sustainable promotion of the cultural heritages in the tourism industry. In Korea for example, Park et al (2019) examined the role of the authenticity of the heritage for the intention to revisit and they suggested that constructive authenticity can strongly contribute to the satisfaction of heritage tourists when intangible tourism resources become tangible.

Intangible cultural heritage is also a growing field of study. A very valuable study was realised from the practitioner's -living human treasures'- perspective in South Korea. The findings of this study suggested that the development of means intangible cultural heritage as a sustainable tourism resource requires a positive and constructive symbiotic relationship between safeguarding intangible cultural heritage authenticity and increasing the socioeconomic value of intangible cultural heritage (Kim et al., 2019).

Researching the consumer trends and making a proper cultural heritage marketing plan has also been of great importance. Chabbra (2010) points out some trends in heritage tourism such as growing demand, increasing competition, tourists' preference for optimal effective use of time, growing demand for authenticity, ethical consumption and volunteering, continued relationship with politics, experience economy, resistance to marketing by both consumers and conventional suppliers of heritage, multiculturalism especially as a result of internet and heritage economics.

Finally, to achieve a more holistic and sustainable approach to tourism management and marketing that provides economic benefit to the local economy and sustainable marketing outputs at a World heritage site, Gilmore et al. (2007, p. 262) remark the need for a continuous collaborative, interactive and hands-on management with relevant expertise to ensure the unique environment is respected while making the sites' offerings attractive to the diversity of tourists who visit.

CULTURAL HERITAGE TOURISM IN TURKEY

Before Covid 19 Pandemic, Turkey was among the top ten countries in terms of the number of visitors. Additionally, the tourism revenues of Turkey were among the top 15 to 20 countries of the World. Cultural and natural heritage is a reason of attraction not only for international tourists but also for national tourists.

Before the pandemic, the number of international tourist arrivals to Turkey in 2019 was over 45 million. (Ministry of Culture and Tourism, 2021a). After the pandemic, world tourism has witnessed a recession of 85% (UNWTO), while this percentage was lower around 69% to 70% in Turkey (Kalyoncuoğlu, 2021). With the "Safe Tourism" Certification Program of Türkiye Tourism Promotion and Development Agency (TGA), which is an organization working in coordination with the Turkish Ministry of Culture and Tourism, the international arrivals were 12.7 million in 2020 in Turkey, citizens living abroad (more than 3 million) excluded (Ministry of Culture and Tourism, 2021a). In addition to its Safe Tourism Program, which makes tourists feel safe, Turkey keeps its popularity in terms of cultural and natural attractiveness for international tourists.

In addition, the limitation of international mobility has turned into an advantage for domestic tourism in Turkey, increased the visits of domestic tourists to natural and cultural heritages of Turkey.

This means that intangible or tangible natural and cultural heritages form an undeniable resource and attraction for both domestic and international tourism in Turkey.

Possessing an enormous part of World cultural heritage, the evaluation of this heritage through tourism has gained importance in Turkey. The basic aim should be to preserve this rich cultural heritage concordantly while benefiting from it and keep a balance of usage-preservation-conservation (Emekli, 2005).

Sustainable Cultural Heritage Tourism Management in Turkey

The main tourism authority in Turkey called the Ministry of Culture and Tourism considers that the protection of the cultural heritage is the responsibility of all the people. However, each individual in society might not have the same level of awareness for this responsibility. A guidebook prepared by the Ministry of Culture and Tourism in Turkey emphasizes the importance of the involvement of all the citizens in this issue. The guidebook highlights the value of the natural and cultural heritages in certain dimensions such as *historical, documentary, aesthetic, architectural, indigenous, scarce, economic and holistic value* (Asatekin, 2004). Additionally, the guidebook emphasizes that tourism must be a tool, not the basic aim to shortcut for economic purposes (Asatekin, 2004). That is to say, the protection and correct restoration of the cultural heritages must be the first aim. Then tourism will follow naturally.

One of the undesired examples of wrong strategies of the past is given as the wrong restorations (Aydın & Güner, 2013) and another is the mass tourism's concrete hotel buildings harming the environment in Antalya similar to 1970s Spain (Asatekin, 2004).

Since 2004, the awareness on the level of the tourism stakeholders concerning the value of cultural heritage has increased. Efforts are made for more sustainable tourism at least for the inscription and protection of the cultural heritage as it is seen, on the UNESCO Heritage Lists. Local and National governmental bodies are involved in the process.

About the revaluation of the cultural heritage, Turkey has a high potential relating the historical geography to cultural tourism (Gümüşçü, 2018). Digitalisation and models of open access for the cultural heritages are also considered alternative ways of protection as well as reachability (Öztemiz, 2016). "Benefiting the local economy, keeping the originality of the heritage, elaborate service, providing experience" are among the suggestions for developing sustainable tourism products using intangible cultural heritages (Türker & Çelik, 2012, p. 94, 95).

Academics also try to create the linkage among tourism, cultural heritage and the revaluation of cultural heritage via tourism. The importance of the correct understanding and application of sustainable cultural heritage management for the protection, preservation and revalorization of tangible heritages for tourism, specifically "the Göbeklitepe" is emphasized (Karapınar & Barakazı, 2017). About the intangible cultural heritage valorization, craftsmanship and its revaluation by the tourism has been studied (Öter 2010, Arıoğlu & Aydoğdu Atasoy 2015, Alagöz et al. 2018). A study made in a destination of Turkey on the tentative list of UNESCO "Beypazarı" shows that the revalorization of the tangible and intangible heritages can contribute positively to the local socio-economic life (Öter & Ünal, 2011).

Figure 1. Map of Turkey (Türkiye) located on the connection points of Europe Asia and Africa
Source: Adapted from Google Maps

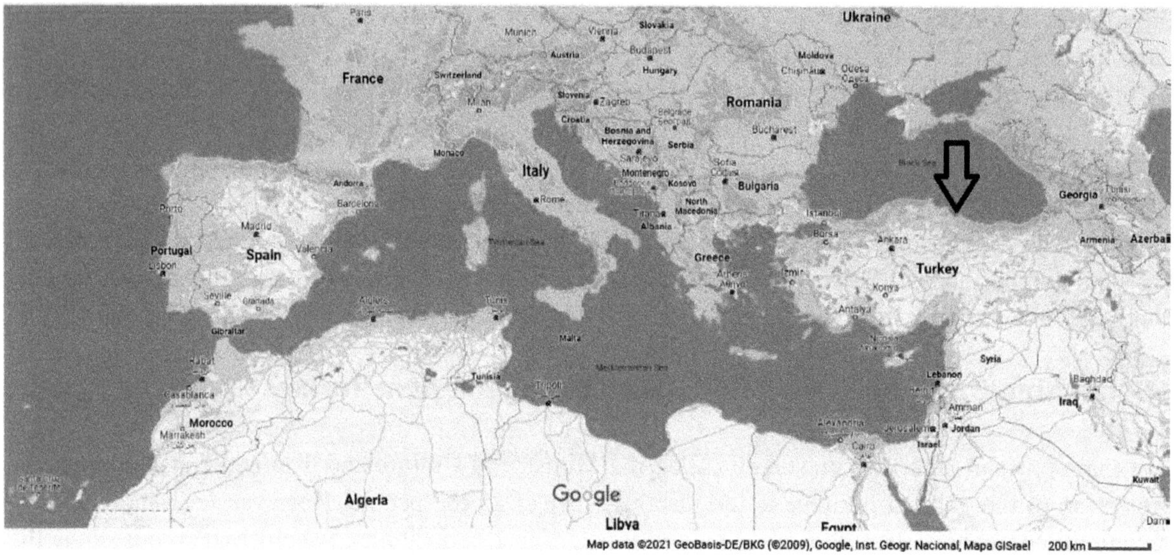

Apart from the protection and revaluation, there are the problems of risk management (Aydın & Güner, 2013) and efficient and up-to-date marketing strategies, which are prevailing the cultural heritage management for sustainable tourism.

Issues related to risk management for cultural heritage tourism include the fragile nature of tourism and cultural heritages in reaction to natural disasters, terror attacks, wars, political instabilities and internal disturbances (Okuyucu & Somuncu, 2013).

The sustainable marketing strategies include creating mindful visitors, stakeholder collaboration, economic viability, leveraging technology, emerging new middle class (Chhabra, 2010) and many more dimensions. The promotional efforts of the local government tourism offices on their websites may not be enough for generation Z focusing more on social media and having preferences more on their channels of communication.

National and International Recognition of Tangible and Intangible Cultural Heritages of Turkey

Possessing two of the Seven Wonders of the Ancient World, *the Temple of Artemis* in Ephesus and *the Mausoleum of Halicarnassus* in Bodrum, Turkey is located on the very important connection points of three continents: Europe, Asia and Africa (Figure 1). Throughout history, this geographical region has been on the cultural crossroads that witnessed the reign of many civilisations. As a result of its uninterrupted inhabitance, this land has tremendously rich tangible and intangible cultural heritage transferred from generation to generation.

In recent years, Turkey has become very active in the recognition of its cultural heritages on the national and international scale. The role of the national authorities especially the Ministry of Culture and Tourism is undeniable in this respect, even if there is a lot more to do in this long path of sustainable cultural heritage tourism management.

Table 1. Turkey (Türkiye) on UNESCO Intangible Heritage List (21 Intangible Heritages)

UNESCO Intangible Heritage (21) - Turkey (Türkiye)	Date of Inscription
Arts of Meddah, public storytellers	2008
Mevlevi Sema Ceremony	2008
Minstrelsy (Aşıklık) Tradition	2009
Karagöz	2009
Nevruz *	2009
Traditional Sohbet Meetings	2010
Semah, Alevi Bektaşi Ritual	2010
Kırkpınar Oil Wrestling Festival	2010
Ceremonial Keşkek Tradition	2011
Mesir Macunu Festival	2012
Turkish Coffee and Culture Tradition	2013
Ebru: Turkish Art of Marbling	2014
Flatbread Making and Sharing Culture, Lavash, Katyrma, Jupka, Yufka*	2016
Traditional Craftsmanship of Çini-Making	2016
Spring Celebration: Hıdrellez*	2017
Whistled Language**	2017
Heritage of Dede Korkut-Korkut Ata, Epic Culture, Folk Tales and Music*	2018
Traditional Turkish Archery	2019
Art of Miniature*	2020
Traditional Art and Strategy Game: Mangala/Göçürme*	2020
Hüsn-i Hat, Traditional Calligraphy in İslamic Art in Turkey	2021
** Common File together with one/more countries* *** Intangible heritage requiring urgent protection*	

Source: Compiled from ich.unesco.org, unesco.org.tr and aregem.ktb.gov.tr

The successful efforts of Turkish authorities including the Ministry of Culture and Tourism of Turkey on the national level to inscribe countless tangible and intangible cultural heritages of this land on the UNESCO World Heritage List on the International level makes Turkey a prominent country on the way to achieving the sustainability of its tangible and intangible cultural heritages. It is a well-known fact that the inscription process of UNESCO Heritage lists requires a wide array of actors and finally the inscription is a result of serious teamwork with high responsibility. In addition, if the sustainability for the protection and preservation of the heritages is not provided, the inscribed assets may be removed as seen in Liverpool, Dresden and other similar examples. Turkey is reported as the top five countries inscribing its intangible cultural heritages on the UNESCO list (Ministry of Culture and Tourism, 2021b; AvrupaGazete, 2021) which increases its visibility among its rivals. Turkey has recorded 21 intangible heritages between 2008 and 2021 (UNESCO, 2021b; Ministry of Culture and Tourism, 2021d). (Table 1) and continues to inscribe new heritages to the list each year.

Among these heritages that are on the list, many of them are practised around Turkey and some other countries that are recorded to UNESCO Heritage List with a common file with Turkey. As an example of intangible cultural heritage, "Spring Celebration" is practised in many different countries with many different formats, so is "Flatbread making and sharing culture".

Especially in Turkey, some of the intangible heritages listed in Table 1 also have a specific period and place of celebration. Mevlevi Sema Ceremony is practised in its true manner in Konya each year between December 7th-17th as the "Şeb-i Arus" meaning "wedding" which signifies the death of Mevlana but metaphorically meaning, the reach of Mevlana Celaleddin-i Rumi to God. It is a ceremonial organization receiving international visitors each year.

Another example, "Mesir Macunu" is a festival organized each year in Manisa, the city located in the west of Turkey in the Aegean Region, during which a very rich flavoured dessert with 40 different spices is sprinkled onto people, which is a tradition dating back to Ottoman period.

Kırkpınar Oil Wrestling Festival traditionally takes place each year in Edirne, a city located on the west country border of Turkey. Wrestling is a traditional sport, but the distinctive feature of this Kırkpınar Oil Wrestling Festival is the wrestlers wear special clothing and their body is covered with oil which makes the bout more difficult. Other specific features of this unique festival include the prize for the winner, the characteristics of the referee and the area the festival practised including the spectators.

It is impossible to give all the details of all the heritages (be tangible or intangible) in this context. Each has a potential content and perspective for separate academic or field study. Table 1 shows only the intangible heritages recorded to UNESCO Intangible Heritage list.

Many other intangible heritages have not been recorded on this list yet but efforts are there to archive them before they are forgotten. For this purpose, the Turkish Ministry of Culture and Tourism works in collaboration with UNESCO for keeping sustainable the intangible cultural heritages by the *"Yaşayan İnsan Hazineleri"* (Living Human Treasures) Program (Ministry of Culture and Tourism, 2021c). Ministry of Culture and Tourism records the living people who are the latest representatives of the intangible cultural heritages such as arts, craftsmanship, folk music and so on and record them by taking their videos and creating an archive by listing them (Ministry of Culture and Tourism, 2021c). This is a practice proving that Turkey is a good example for its efforts of providing sustainability in intangible heritage and is creating a good resource for humanity.

In addition to the Living Human Treasure Program, in the framework of UNESCO and Ministry of Culture and Tourism studies, Turkey also continues to provide a full Inventory of National Intangible Cultural Heritages which are at the same time candidates to inscribe for the UNESCO heritage list. They are, "Tea Culture", "Traditional Mey Craftsmanship and Performance" (Mey or Balaban is a musical instrument played in folk music), "Art of Hat" (A traditional art of writing), "Sericulture and Silk Weaving", "The Telling Tradition of Nasreddin Hodja Anecdotes", "Art of Mother Pearl Inlay", "Ahlat Stonework" (Ministry of Culture and Tourism, 2021d).

Much earlier in time of approval and inscription, tangible heritages are also the attraction points for both national and international tourists. Turkey has 19 (tangible) heritages on the list, 2 of which are mixed (having both natural and cultural heritage status). Table 2 includes these 19 Heritages of Turkey with their date of inscription to UNESCO Heritages Lists and their city of location in Turkey.

According to the last revision in April 2021, there are 84 more sites of Turkey on the UNESCO Tentative Lists, of which 3 are natural, 4 are mixed (both cultural and natural) heritages (UNESCO Tentative Lists - Turkey, UNESCO Türkiye Milli Komisyonu). It is a fact that Turkey still has many tangible, intangible, natural, cultural and mixed heritages that are not on the tentative list. The efforts of the local, national and related national representatives of international authorities continue to improve the inscription process.

Table 2. Turkey (Türkiye) on UNESCO Heritage List (19 Tangible Heritages)

Cultural (17) Heritages of Turkey (Türkiye)	**Date of Inscription**	**City of Location**
Great Mosque and Hospital of Divriği	1985	Sivas
Historic Areas of Istanbul	1985	İstanbul
Hattusha: the Hittite Capital	1986	Çorum
Nemrut Dağ	1987	Adıyaman
Xanthos-Letoon	1988	Antalya and Muğla
City of Safranbolu	1994	Karabük
Archaeological Site of Troy	1998	Çanakkale
Selimiye Mosque and its Social Complex	2011	Edirne
Neolithic Site of Çatalhöyük	2012	Konya
Bursa and Cumalıkızık: the Birth of the Ottoman Empire	2014	Bursa
Pergamon and its Multi-Layered Cultural Landscape	2014	İzmir
Diyarbakır Fortress and Hevsel Gardens Cultural Landscape	2015	Diyarbakır
Ephesus	2015	İzmir
Archaeological Site of Ani	2016	Kars
Aphrodisias	2017	Aydın
Göbekli Tepe	2018	Şanlıurfa
Arslantepe Mound	2021	Malatya
Mixed (Both Natural and Cultural) (2) Heritages of Turkey (Türkiye)	**Date of Inscription**	**City of Location**
Göreme National Park and the Rock Sites of Cappadocia	1985	Nevşehir
Hierapolis-Pamukkale	1988	Denizli

Source: Compiled from unesco.org.tr and http://whc.unesco.org

In addition to the UNESCO Cultural Heritage inventory, the Turkish Ministry of Culture and Tourism tries to create an inventory of intangible heritages. The National Intangible Cultural Heritage Inventory of Turkey list includes 35 different elements. These 35 elements all have different sub-elements varying from 1 to 57 sub-elements for each. In total, excluding the repetitions in different destinations, sub-elements total up to 297. Including the repetitions which mean similar sub-elements in different locations included the total is 1264 sub-elements. Let's look at these 35 elements with some (not all are given) of their sub-elements here (Ministry of Culture and Tourism, 2021e):

1. Turkish Culinary Culture / Traditional Producing of Food and Beverage and Social Practices: *Coffee Culture, Aşure (Noah's Pudding) Tradition, Traditional Simit (Bagel) Making, Traditional Boza Making, Şerbet (Sherbet) Culture, Tarhana Making and Tradition, Yayık (Churn) Tradition, Cellar and Storehouse Tradition, Turkish Delight Culture, Künefe (Kunefe) Culture, Höşmerim Culture, Traditional Local Methods of Collecting and Using Edible Grass and Spice, Tea (Çay) Culture, Köfte (Meatball) Culture, Kebap Culture, Yoğurt Culture.*
2. Traditional Healing and Cleansing Practices, Production of Care Products: *Hot Spring Culture, Traditional Herbal Healing Products Producing, Soap Producing, Turkish Bath Culture*

3. Traditional Livestock Raising and Social Practices: *Silk and Sericulture, Apiculture and Honey Culture, Aviculture, Traditional Horse Breeding*
4. Agricultural and Horticultural Practice and Traditions: *Rose Culture and Producing of Rose Products, Olive and Olive Oil Culture, Traditional Craftsmanship of Tree Grafting, Viticulture and Gardening Culture*
5. Traditional Sports: *Javelin, Wrestling, Traditional Turkish Archery, Game of Guhşegi, Game of Matrak, Game of Tomak, Traditional Game of Tulum, Traditional Game of Wisdom Mangala/ Göçürme, Traditional Strategy Game Peçiç*
6. Traditional Theatrical Arts and Plays
7. Traditional Folk Dances: *Sword Shield Dance/Game, Spoon Folk Dances, Horon, Halay*
8. Folk Music and Tradition of Singing Folk Songs: *Singing Heroic Songs (Türkü)*
9. Tradition of Minstrelsy
10. Tradition of Elegising
11. Tradition of Narration and Traditional Rhetorics: *Tradition of Telling Sagas, Epics*
12. Traditional Festivals: *Oil Wrestling, Mesir, Kermess, Revelry*
13. Traditions of Summer Pasture and Winter Quarters
14. Beliefs, Ceremonies and Traditions Related to the Calendar
15. Traditional Pedlar's Trade and Culture
16. Traditional Communication Methods and Practices
17. Commemorative Ceremony Traditions for Important Dates and Personalities
18. Traditions and Practices Related to Beliefs and Faith
19. Traditions and Practices Related to Death: *Halva*
20. Wedding Traditions: He*nna Night, Dowery Tradition*
21. Prenatal, Postnatal and Childhood Traditions: *Preparing Cradle*
22. Traditions of Sendoff and Greeting: *To Haj, to military service*
23. Traditions of Guest Welcoming and Hospitality
24. Traditions of Fellowship and Kindredness
25. Traditions of Social Welfare and Solidarity and Charity
26. Social Practices Related to Cultural Places
27. Traditional Meetings and Organizations: *Ahilik*
28. Traditional Child Games and Toys
29. Traditional Musical Instrument Playing and Producing
30. Embroidery, Sewing and Knitting Traditions
31. Traditional Clothes Producing and Dressing Techniques
32. Art and Traditions of Weaving: *Carpet, Kilim, Fabric, Objects*
33. Traditional Architecture and Construction Craftsmanship: *Stone, Brick, Wood, Glass, Metal*
34. Traditional Handicraft Craftsmanship: *Art of Kazaziye, Shipbuilding*
35. Traditional Turkish Decorative Arts and Handicrafts: *Çini, Hat, Miniature, Kündekari*

The inscription of the cultural heritages on the UNESCO heritage list is believed to be an important instrument for protection, inspection and provision of necessary promotional studies for the sustainable management of cultural heritages. However, a concrete mechanism to measure the short and/or long-term impacts of the cultural heritage revalorization on the numbers of tourist arrivals or tourism revenues in

Figure 2. UNESCO Cultural Heritages of Turkey (Türkiye) Shown on the Map
Source: Adapted from Republic of Turkey Ministry of Culture and Tourism (2014)

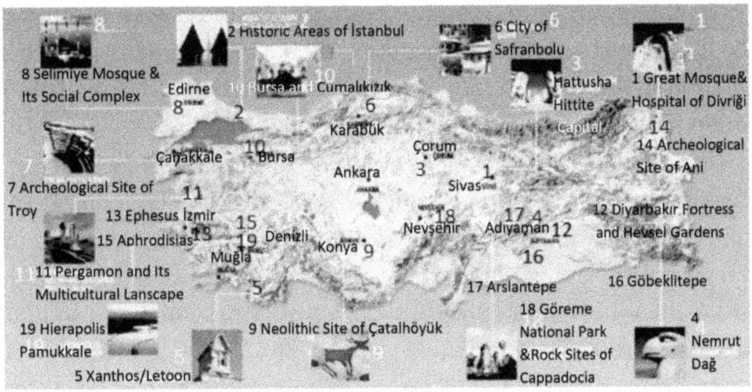

Turkey yet needs to be developed (Yenişehirlioğlu & Arslan Kalay, 2017). In this way, another important source of data for improving the sustainability in cultural heritage tourism management will be gained.

CONCLUSION

Conclusion and Recommendations for Sustainable Cultural Heritage Tourism Management in Turkey

As Turkey has a huge legacy of tangible and intangible cultural heritage background, it is a very attractive destination for international tourism. The national and international efforts of recognition of cultural heritages bring a busy agenda for Turkish tourism authorities and professionals to make tourism more sustainable.

The location of the heritages on the Turkey map numbered from 1 to 19 according to the order given in Table 2 can be seen in Figure 2. The map is taken from a promotional document of the Ministry of Culture and Tourism dated back to 2014 but, a more up-to-date version of the map with the latest 6 heritages inscribed to the UNESCO heritage list in the following years, from 2015 till 2021 could not be reached, unfortunately. That is why the white numbers and little pictures show only 13 of the heritages in the original version. The numbers from 13 to 19 are added by the author to adjust the map and to give an idea to the reader where the rest of the heritages are located currently on the map. That is why the black numbers from 1 to 19 on the map in Figure 2 show the locations of all the 19 heritages.

The distance between the locations as seen on the map are quite long. To give an idea: from Edirne (8) to Kars (14) (West to East) is 1706 km and from İstanbul (2) to Antalya (5) (North to South) is 718 km (calculated at https://della.eu/en-tr/distance/). When the tours of the travel agencies in Turkey with the content of UNESCO heritage sites are searched for on the Internet it can be seen that 14 to 20-day tours include too many different regions from north to south and east to west squeezed in just one tour. Examples include ambertravel.com, mywayturkey.com. It might seem advantageous from the eye of the visitor to visit many different heritages at once; but, it's a tiring trip and in addition, it means hundreds of kilometres to go each day which is not also consistent with the basic principles of sustainability.

However, sustainable cultural heritage tourism is not just seeing and leaving the heritage site. If the local people do not get their share from the economic profit, then it will produce depriving results for the destination. The planning and programming of tours should have a completely different aspect. The professionals should be aware of the fact that the tangible and intangible cultural heritages support each other at a destination in Turkey and they should re-organize the tourism policies and operational programs and marketing plans according to this fact.

As a recommendation, synthesis of each spot could be designed from a sustainable marketing perspective as "a city-destination" or "a heritage regional area" that can be focused as an attraction point for visiting at once for its tangible and intangible cultural heritage assets. As an example, 'Aegean Tour' can be the preferred region for a tour including Pergamon and Ephesus as UNESCO Heritage Sites and gastronomy tour in the Kemeraltı Bazaar in the city-centre of İzmir which is also in the UNESCO tentative list as "The Historical Port City of İzmir".

A study offering a "spatial planning model" for the sustainability of Turkish Tourism suggests a spatial environmental and architectural design planning model that integrates many different stakeholders and institutions which is a must for sustainable tourism in Turkey (Dede & Ayten, 2012). In addition to spatial planning, other perspectives such as geographical aspects in terms of the division of thematic zones for intangible, tangible, and mixed cultural heritage should be integrated to create a more specific sustainable tourism policy and implications in Turkey for cultural heritage management.

Another point to consider is the demand studies. A segmentation study should be made for both national and international tourists and products or packages should be prepared accordingly.

A good example of promotional activity has been announced as "UNESCO Touristica Photo Safari" tours that will simultaneously serve to create a unique archive of 17 of the 19 cultural heritages of Turkey. The reason why the two heritages are not included is Arslantepe in Malatya has been inscribed to list very recently and unfortunately, Divriği Ulu Camii in Sivas is excluded as it is in restoration. The photo safari is a social responsibility project sponsored by Touristica and prepared in coordination with Türkiye Tourism Promotion and Development Agency (TGA) and many NGOs from Turkish tourism practitioners such as the Turkish Federation of Hoteliers (TÜROFED), Turkish Promotional Platform (TUTAP), Tourism Writers and Journalists (TUYED). The duration of the project is 8 months beginning on September 22nd in 2021 in Cappadocia and ending in May 2022 in İstanbul. This long-term project will increase the sustainability of tourism as it will put the cultural heritage at the centre of the tours and additionally, it neutralises the seasonality (DHA, 2021).

Tourism Strategy of Turkey 2023- Action Plan 2007-2013 (Ministry of Culture and Tourism, 2007) makes a different categorization for Turkish Tourism and is an interesting policy document that is overviewed and studied by academics from various aspects (Paksoy 2009, Ozturk & Van Niekerk 2014, Demirkol et al 2015, Bulin 2015, Uygun & Ekiz 2016, Çavuşoğlu 2017, Karamustafa & Yılmaz 2020, Sonuç 2020). However, this tourism policy document has only been developed with an action plan of 2007-2013, which was not a sufficient period to realize fully its planned sustainable tourism objectives. Tourism Strategy of Turkey document has put enormous importance on operational sustainability of Turkish Tourism with the diversification of tourism products. In this document, Turkey was divided into several thematic "tourism development zones" and "corridors" comprising cultural heritages as two prominent elements. For instance, most of these zones include archaeological areas that are at the same time cultural heritage sites combined with another theme such as thermal tourism. The policy document defines "Tourism cities" and "Ecotourism zones" as well.

The policy document may have some strong aspects as it emphasizes the sustainable evolution of tourism in Turkey. However, theoretical aspects need deeper analysis to be applicable in the practical world. Sustainability in cultural heritage tourism may be possible if the practical and theoretical aspects can be adjusted from a holistic and multidisciplinary perspective. This policy document that has terminated its literal life in 2013, still has importance on sustainable development of Turkish tourism as it has given place to the ever-widest sustainability suggestions for Turkish tourism. That is why, with a deeper analysis of its strong and weak points, taking into consideration the changing global and national tourism environment, it is high time to create a more convenient version of the action plan. The new version should analyse the natural and cultural assets of this land correctly and most importantly, it should adapt the tourism planning and policy implications for the managerial and promotional aspects prioritizing the protective measures for preservation and conservation of the cultural and natural assets.

Another important issue in the protection of cultural heritage is the illegal activities that harm or damage cultural heritages. The cultural heritages of Turkey have always attracted citizens and people from abroad and have been subjected to smuggling, illegal excavations and treasure hunting many times and many valuable heritages were stolen and taken abroad (Akkuş&Efe, 2015, p. 107). The archaeological sites, ruins, tumuluses, mounds that have not been examined by the archaeological excavations yet, have become an illegal livelihood for the smugglers and treasure hunters; by this way, it can be concluded that; with the richness of its cultural heritage Turkey is among the countries where most of the illegal excavations are made (Akkuş&Efe 2015, p. 121-122). Despite the efforts of the Ministry of Culture and Tourism that has brought back many artefacts from abroad (Ministry of Culture and Tourism, 2021f), there are still missing artefacts the number of which is not known (Ministry of Culture and Tourism, 2021g).

FUTURE RESEARCH DIRECTIONS

Both for management and marketing purposes, "sustainable cultural heritage tourism" can be considered as a developing area of study both in Turkey and in the World with high potential of promising multidisciplinary studies.

Cultural heritage studies comprise many disciplines such as history, geography, biology, historical geography, archaeology, sociology, anthropology, psychology, behavioural sciences, folklore and not limited to that. When joined with tourism and sustainability, the field enlarges and becomes more holistic including sustainable management, tourism business, management and marketing disciplines even engineering.

Each tangible and intangible (and mixed) heritage in Turkey and different locations of the World also deserve to be studied and researched from different perspectives with different research methods to enrich academic literature in tourism and cultural heritage tourism management. The supply and demand-side studies are also needed to reach the objective of sustainable cultural heritage tourism management.

Multi-national comparative studies can also result in interesting findings. Finally, there is a need for participative and collaborative insight not only in the practical, industrial or managerial area of cultural heritage tourism but also in academic researches. Additionally, a sharing perspective is needed between these two actors: Industry and academy in succeeding a sustainable cultural heritage tourism management structure.

ACKNOWLEDGMENT

This research received no specific grant from any funding agency in the public, commercial, or not-for-profit sectors.

REFERENCES

Ağcakaya, H., & Can, İ. I. (2019). Somut Olmayan Kültürel Miras Kapsamında Mutfak Kültürünün Sürdürülebilirliği: Türkiye'deki Gastronomi Müzeleri Örneği [Sustainability of Culinary Culture in the Context of Cultural Heritge: An Example for Gastronomy Museums in Turkey]. *Gastroia: Journal of Gastronomy and Travel Research, 3*(4), 788–804.

Akbulut, O., & Ekin, Y. (2018). Kültürel Miras Turizmi Olarak Savaş Alanları Turizmi: Türkiye'de Yer Alan Savaş Anıtlarının Coğrafi Bilgi Sistemleri Analizi [War Areas Tourism as Cultural Heritage Tourism: The Analysis of Turkish War Monuments with GIS]. *Hitit Üniversitesi Sosyal Bilimler Enstitüsü Dergisi, 11*(1), 395–420.

Akipek, S. (2001). Dünya Kültürel ve Doğal Mirasının Korunmasına Dair Sözleşmenin Değerlendirilmesi [The Evaluation of the Convention Concerning the Protection of the World Cultural and Natural Heritage]. *Ankara Üniversitesi Hukuk Fakültesi Dergisi, 50*(4), 13–39.

Akkuş, Z., & Efe, T. (2015). Türkiye'de Defineciliğin Kültürel Mirasın Korunması Bağlamında Değerlendirilmesi [The Evaluation of Treasure Hunting in Turkey in the Context of Protection of Cultural Heritage]. *Birey ve Toplum Sosyal Bilimler Dergisi, 5*(2), 105–132.

Alagöz, G., Çalık, İ., & Güneş, E. (2018). Kültürel Miras Turizmi Açısından Erzincan Bakır İşleme Sanatının Mevcut Durumu ve Sürdürülebilirliği [Current State of the Erzincan Art of Chalcography from the Perspective of Cultural Heritage Tourism and Its Sustainability]. *Dumlupınar Üniversitesi Sosyal Bilimler Dergisi, 55*, 174–191.

Aliağaoğlu, A., & Narlı, M. (2012). Edebi Miras Turizmi ve Türkiye'de Edebi Mekanlar [Literary Heritage Tourism and Literature Related Places in Turkey]. *Bilig, 60*, 1–18.

Arıoğlu, İ. E., & Aydoğdu Atasoy, Ö. (2015). Somut Olmayan Miras Kapsamında Geleneksel El Sanatları ve Kültür ve Turizm Bakanlığı [Traditional Handicrafts in the Context of Intangible Cultural Heritage and Ministry of Culture and Tourism]. *Turkish Studies: International Periodical for Languages. Literature and History of Turkish or Turkic., 10*(16), 109–126.

Arslan Kalay, H., & Yenişehirlioğlu, E. (2017). *Kültürel Miras Yönetiminde Bir Çatışma Unsuru: Turizm* [A Conflict in Cultural Heritage Management: Tourism] [Paper presentation]. *International Congress on Cultural Heritage and Tourism (ICCHT-2017).*

Arthur, S. N. A., & Mensah, J. V. (2006). Urban management and heritage tourism for sustainable development: The case of Elmina Cultural Heritage and Management Programme in Ghana. *Management of Environmental Quality, 17*(3), 299–312. doi:10.1108/14777830610658719

Asatekin, N. G. (2004). Kültür ve Doğa Varlıklarımız, Neyi Niçin Nasıl Korumalıyız? [Culture and Nature Assets, What and How to Protect?] T.C. Kültür ve Turizm Bakanlığı Yayın No: 3016, DÖSİMM Basımevi, Ankara.

AvrupaGazete. (2021). *Türkiye'nin Geleneksel Ahlat Taş İşçiliği, İpek Böcekçiliği ve Sedef Kakma Sanatı UNESCO adayı* [Turkey's Traditional Sericulture and Ahlat Stonework]. https://www.avrupagazete.co.uk/kultur-sanat/turkiye-nin-geleneksel-ahlat-tas-isciligi-ipek-bocekciligi-h348175.html

Aydın, M., & Güner, S. (2013). Kültürel Mirasta Risk Yönetimi [Risk Management in Cultural Heritage]. *Batman University Journal of Life Sciences*, *3*(1), 70–81.

Baykan, E. (2007). Kültür Turizmi ve Turizmin Kültürel Varlıklar Üzerindeki Etkileri [Effects of Cultural Tourism on Cultural Assets]. *Gazi Üniversitesi Ticaret ve Turizm Eğitim Fakültesi Dergisi*, *2*, 30–49.

Bouchenaki, M. (2003) *The Interdependency of the Tangible and Intangible Cultural Heritage.* In 14th ICOMOS General Assembly and International Symposium: Place, memory, meaning: preserving intangible values in monuments and sites, 27 – 31 Oct 2003, Victoria Falls, Zimbabwe.

Bulin, D. (2015). Policies and Strategies for Tourism Development–Some Lessons for Romania. *Knowledge Horizons-Economics*, *7*(2), 77–81.

Çavuşoğlu, P. P. (2017). Evaluation of Tourism Strategy 2023 within Perspective of Transportation Strategy. *Journal of Education & Social Policy*, *4*(3), 61–68.

Chhabra, D. (2010). *Sustainable Marketing of Cultural and Heritage Tourism*. Routledge. doi:10.4324/9780203855416

Cohen, E. (2002). Authenticity, Equity and Sustainability in Tourism. *Journal of Sustainable Tourism*, *4*(10), 267–276. doi:10.1080/09669580208667167

Dede, O. M., & Ayten, A. M. (2012). The Role of Spatial Planning for Sustainable Tourism Development: A Theoretical Model for Turkey. *Tourism: An International Interdisciplinary Journal*, *60*(4), 431–445.

Demirkol, Ö., Yıldırım, M., & Gökkaya, S. (2015). *Recent Developments in Turkish Tourism Policy within the context of Accession to the European Union* [Paper presentation]. *2nd International Congress of Tourism & Management Researches.*

DHA. (2021). *UNESCO'nun 17 Kültürel Mirasında Foto Safari Başlıyor* [Photo Safari Starts in 17 Cultural Heritage of UNESCO]. https://www.cnnturk.com/seyahat/unesconun-17-kulturel-mirasinda-foto-safari-basliyor

Emekli, G. (2005). Avrupa Birliği'nde Turizm Politikaları ve Türkiye'de Kültürel Turizm [The Policies of Tourism in European Union and Cultural Tourism in Türkiye]. *Ege Coğrafya Dergisi*, *14*, 99–107.

Gilmore, A., Carson, D., & Ascenção, M. (2007). Sustainable Tourism Marketing at a World Heritage Site. *Journal of Strategic Marketing*, *15*(2-3), 253–264. doi:10.1080/09652540701318930

Gümüşçü, O. (2018). Tarihi Coğrafya ve Kültürel Miras [Historical Geography and Cultural Heritage]. *Erdem*, *75*, 99–120.

Halewood, C., & Hannam, K. (2001). Viking Heritage Tourism: Authenticity and Commodification. *Annals of Tourism Research*, *28*(3), 565–580. doi:10.1016/S0160-7383(00)00076-1

Harmon, D. (2007). A Bridge over the Chasm: Finding Ways to Achieve Integrated Natural and Cultural Heritage Conservation. *International Journal of Heritage Studies*, *13*(4-5), 380–392. doi:10.1080/13527250701351098

Huibin, X., Marzuki, A., & Razak, A. A. (2012). Protective Development of Cultural Heritage Tourism: The case of Lijiang, China. *Theoretical and Empirical Researches in Urban Management*, *7*(1), 39–54.

Jassam, O. (2019). Küresel Kültürel ve Doğal Mirası Tehdit Eden Risk Yönetimi [Risk Management Threatening the Global Cultural and Natural Heritage]. *5. Uluslararası Öğrenci Sempozyumu Bildiriler Kitabı -2- Din-Tarih-Sosyoloji*, 75-89.

Kalyoncuoğlu, Y. (2021). *Türkiye Salgın Koşullarına Rağmen 2020'de ve 2021'in İlk İki Ayında 17 Milyondan Fazla Turist Ağırladı* [Turkey has Received more than 17 Million the First Two Months of 2020 and 2021 Despite the Conditions of the Pandemic]. https://www.aa.com.tr/tr/turkiye/turkiye-salgin-kosullarina-ragmen-2020de-ve-2021in-ilk-iki-ayinda-17-milyondan-fazla-turist-agirladi/2213246

Kaminski, J., Benson, A. M., & Arnold, D. (2014). *Contemporary Issues in Cultural Heritage Tourism*. Routledge.

Karamustafa, K., & Yılmaz, M. (2020). Destination Life Cycle: A Conceptual Approach Based on the Turkish Experience. *Journal of Tourism and Gastronomy Studies, Special Issue*, *8*(4), 439–451. doi:10.21325/jotags.2020.696

Karapınar, E. & Barakazı, M. (2017). Kültürel Miras Turizminin Sürdürülebilir Turizm Açısından Değerlendirilmesi [Evaluation of Cultural Heritage Tourism from Aspect of Sustainable Tourism: Göbeklitepe Ruins]. *Güncel Turizm Araştırmaları Dergisi, 1*, 5-18.

Kim, S., Whitford, M., & Arcodia, C. (2019). Development of Intangible Cultural Heritage as a Sustainable Tourism Resource: The Intangible Cultural Heritage Practitioners' Perspectives. *Journal of Heritage Tourism*, *14*(5-6), 422–435. doi:10.1080/1743873X.2018.1561703

Li, Y., & Hunter, C. (2015). Community Involvement for Sustainable Heritage Tourism: A Conceptual Model. *Journal of Cultural Heritage Management and Sustainable Development*, *5*(3), 248–262. doi:10.1108/JCHMSD-08-2014-0027

Ministry of Culture and Tourism. (2007). *Tourism Strategy of Turkey 2023 - Action Plan 2007-2013*. https://www.ktb.gov.tr/Eklenti/43537,turkeytourismstrategy2023pdf.pdf?0&_tag1=796689BB12A540BE0672E65E48D10C07D6DAE291

Ministry of Culture and Tourism. (2021a). *Republic of Turkey: General Directorate of Investment and Business, Turizm İstatistikleri* [Tourism Statistics]. https://yigm.ktb.gov.tr/TR-9851/turizm-istatistikleri.html

Ministry of Culture and Tourism. (2021b). *Republic of Turkey: General Directorate of Research and Training; video: "Turkey On UNESCO Intangible Cultural Heritage Lists"*. https://aregem.ktb.gov.tr/TR-268296/unesco-somut-olmayan-kulturel-miras-listelerinde-turkiy-.html#

Ministry of Culture and Tourism. (2021c). *Republic of Turkey: General Directorate of Research and Training; Yaşayan İnsan Hazineleri* [Living Human Treasures]. https://aregem.ktb.gov.tr/TR-167971/yasayan-insan-hazineleri.html

Ministry of Culture and Tourism. (2021d). *Republic of Turkey: General Directorate of Research and Training. Somut Olmayan Kültürel Miras Ulusal Envanteri* [National Inventory of Intangible Cultural Heritage]. https://aregem.ktb.gov.tr/TR-159257/somut-olmayan-kulturel-miras-ulusal-envanteri.html

Ministry of Culture and Tourism. (2021e). *Republic of Turkey: General Directorate of Research and Training. Somut Olmayan Kültürel Miras Türkiye Ulusal Envanteri* [National Inventory of Intangible Cultural Heritage of Turkey]. https://aregem.ktb.gov.tr/TR-279417/somut-olm-kult-miras-turkiye-ulusal-envanteri.html

Ministry of Culture and Tourism. (2021f). *Republic of Turkey: General Directorate of Cultural Heritage and Museums. Yurt Dışından İadesi Sağlanan Eserler* [The Artefacts that Have Been Returned from Abroad]. https://kvmgm.ktb.gov.tr/TR-44470/yurt-disindan-iadesi-saglanan-eserler.html

Ministry of Culture and Tourism. (2021g). *Republic of Turkey: Q&As about the Activities to Stop Smuggling*. https://kvmgm.ktb.gov.tr/TR-63359/sikca-sorulan-sorular.html

Ngamsomsuke, W., Hwang, T., & Huang, C. (2011). *Sustainable Cultural Heritage Tourism Indicators* [Paper presentation]. In International Conference on Social Science and Humanity IPEDR.

Nocca, F. (2017). The Role of Cultural Heritage in Sustainable Development: Multidimensional Indicators as Decision-Making Tool. *Sustainability, 9*(10), 1-28.

Okuyucu, A., & Somuncu, M. (2013). *Türkiye'de İnanç Turizmi, Bugünkü Durum, Sorunlar ve Gelecek* [Religious Tourism in Turkey: Today's Situation, Problems and Future] [Paper presentation]. *International Conference on Religious Tourism and Tolerance.*

Öter, Z. (2010). Türk El Sanatlarının Kültür Turizmi Bağlamında Değerlendirilmesi [Valorization of Turkish Handicrafts in Cultural Tourism Context]. *Milli Folklor, 22*(86), 174–185.

Öter, Z., & Ünal, E. (2011). *Somut Olmayan Kültürel Miras ve Turizm Etkileşiminin Beypazarı İlçesinde Yerel Ekonomiye Katkıları* [Paper presentation]. Milletlerarası Türk Halkbilimi Kongresi.

Öztemiz, S. (2016). *Türkiye'de Dijitalleştirilen Kültürel Miras Ürünlerine Açık Erişim: Bir Model Önerisi* [Open Access to Digitalized Cultural Heritage Products: A Model Suggestion] [PhD Thesis]. Hacettepe University, Institute of Social Sciences, Information and Document Management Major Field of Study.

Ozturk, A. B., & Van Niekerk, M. (2014). Volume or Value: A Policy Decision for Turkey' s Tourism Industry. *Journal of Destination Marketing & Management, 3*(4), 193–197. doi:10.1016/j.jdmm.2014.04.001

Paksoy, P. (2009). *The Environmental Focus In Tourism Strategy Of Turkey-2023 Within The Outlook Of Tourism Planning*. Sofia: Surveying Geology & Mining Ecology Management (SGEM). Retrieved from https://www.proquest.com/conference-papers-proceedings/environmental-focus-tourism-strategy-turkey-2023/docview/1285255677/se-2?accountid=141837

Park, E., Choi, B. K., & Lee, T. J. (2019). The Role and Dimensions of Authenticity in Heritage Tourism. *Tourism Management, 74*, 99–109. doi:10.1016/j.tourman.2019.03.001

Pelit, E., & Türkoğlu, T. (2019). Somut Olmayan Kültürel Miras Değerlerinin Turizme Yansımaları: Ebru Sanatı Üzerine Bir İnceleme [Tourism Reflections of Intangible Cultural Heritage Values: A Review on Ebru Art]. *Güncel Turizm Araştırmaları Dergisi*, *3*(1), 101–118. doi:10.32572/guntad.463910

Republic of Turkey Ministry of Culture and Tourism. (2014). *Turkey in World Heritage*. https://kvmgm.ktb.gov.tr/Eklenti/27154,turkeyinworldheritage.pdf?0

Santa, E. D., & Tiatco, A. (2019). Tourism, Heritage and Cultural Performance: Developing a Modality of Heritage Tourism. *Tourism Management Perspectives*, *31*, 301–309. doi:10.1016/j.tmp.2019.06.001

Sonuç, N. (2020). Social Tourism: An Analysis from the Sustainability Perspective-İzmir Case. *International Journal of Contemporary Economics and Administrative Sciences*, *10*(1), 35–52.

TGA (Türkiye Tourism Promotion and Development Agency). (n.d.). https://tga.gov.tr/home/

Timothy, D. J., & Nyaupane, G. N. (2009). *Cultural Heritage and Tourism in the Developing World, A Regional Perspective*. Routledge. doi:10.4324/9780203877753

Türker, A., & Çelik, İ. (2012). Somut Olmayan Kültürel Miras Unsurlarının Turistik Ürün Olarak Geliştirilmesine Yönelik Alternative Öneriler [Alternative Proposals for Tourist Product Development of Intangible Cultural Heritage Elements]. *Yenifikir*, *9*, 86–98.

Ünal, Z. G. (2014). *İSMEP Rehber Kitaplar 6, Kültürel Mirasın Korunması* [Protection of Cultural Heritage]. Beyaz Gemi Sosyal Proje Ajansı, İstanbul Valiliği.

UNESCO. (2020). *Cutting Edge | Bringing Cultural Tourism Back in the Game*. https://en.unesco.org/news/cutting-edge-bringing-cultural-tourism-back-game

UNESCO. (2021a). *UNESCO World Heritage and Sustainable Tourism Programme*. https://whc.unesco.org/en/tourism/

UNESCO. (2021b). *UNESCO Intangible Cultural Heritage, Turkey, Elements on the Lists of Intangible Cultural Heritage*. https://ich.unesco.org/en/state/turkey-TR?info=elements-on-the-lists

UNESCO Tentative Lists - Turkey. (n.d.). http://whc.unesco.org/en/tentativelists/state=tr

UNESCO Türkiye Milli Komisyonu. (n.d.). https://www.unesco.org.tr/Pages/125/122/UNESCO-D%C3%BCnya-Miras%C4%B1-Listesi

UNWTO (United Nations World Tourism Organization). (n.d.). https://www.unwto.org/unwto-tourism-dashboard

Uygun, M., & Ekiz, E. (2016). An Overview of Health Tourism within the Context of Turkey's Tourism Strategy (2023). *Journal of Hospitality and Tourism*, *14*(1), 18–26.

Vargas-Hernández, J. G. (2012). Sustainable Cultural and Heritage Tourism in Regional Development of Southern Jalisco. *World Journal of Entrepreneurship, Management and Sustainable Development*, *8*(2/3), 146–161. doi:10.1108/20425961211247752

World Tourism Organization. (2018). *Tourism and Culture Synergies*. UNWTO., doi:10.18111/9789284418978

Yenişehirlioğlu, E., & Arslan Kalay, H. (2017). *UNESCO Dünya Miras Listesi'nin Kültürel Miras Alanlarındaki Ziyaretçi Sayılarına Etkisi* [The Impact of UNESCO World Heritage List on The Number Of Visitors To Cultural Heritage Sites] [Paper presentation]. *International Congress on Cultural Heritage and Tourism (ICCHT-2017).*

KEY TERMS AND DEFINITIONS

Cultural Heritage: The tangible and intangible assets and common values that create a nation with customs and traditions dating back to history, continuing with current rituals, ceremonies, festivities artistic creations, handicrafts, gastronomy and carrying to the future by transmission from generation to generation.

Cultural Heritage Tourism: The thematic area of tourism that benefits as a resource from tangible, intangible, and mixed cultural heritage resources at a destination.

Intangible Cultural Heritage: The cultural heritage related to the human creation of a nation and having artistic or traditional values such as spring festivals, rituals, ceremonies, folkloric dances, musical instruments and songs played by them, traditional storytelling, handicrafts, gastronomic arts, and traditional recipes and other folkloric elements.

Living Human Treasures: The people, the artists or craftsmen, specifically practitioners that contribute to the creation of cultural heritages by for example making handicrafts, musical instruments and playing at the same time and singing local songs or who create intangible heritages with artistic value. The UNESCO Living Human Treasures program is created to archive information about and short videos of these people while practising when they are still alive.

Mixed (Cultural+Natural) Heritage: Heritage with values of both cultural assets and natural assets around the same area such as Pamukkale or Cappadocia in Turkey. In Pamukkale, travertines are natural pools and there is an archaeological site just beside to visit. Similarly, Cappadocia has "Peribacaları", the natural scenery created by the natural formation of volcanic rocks, and at the same time an important historical background and places to visit for Christianity.

Sustainable Cultural Heritage Tourism: Cultural heritage tourism realised according to the sustainability rules, that is, prioritizing the protection and preservation of the natural and cultural heritage assets, the local economy, the participation of the stakeholders and adopting multidisciplinary, inclusionary, and holistic management and decision-making perspectives.

Tangible Cultural Heritage: The cultural heritages belonging to one land and civilization and having universal value and that have a tangible asset such as archaeological sites, buildings of historical or traditional value, monuments, artistic creations such as sculptures and other tangible resources.

Compilation of References

Aarstad, J., Ness, H., & Haugland, S. (2015). Destination Evolution and Network Dynamics: New Research Agendas. In D. Chambers & T. Rakić (Eds.), *Tourism Research Frontiers: Beyond the Boundaries of Knowledge* (pp. 141–164). Emerald Publishing. doi:10.1108/S1571-504320150000020002

Abatedaga, N., Haiquel, M., & González, V. (2015). ¿Cómo permear la extensión universitaria desde la práctica social? *Actas Del XVII Congreso de La Red de Carreras de Comunicación Social y Periodismo de Argentina.*

Acevedo, M., González, O., Zamudio, L., Abello, R., Camacho, J., Gutiérrez, M., Barreto, E., Ochoa, J., Torres, G., Quintero, M., & Baeza, Y. (2005). Un análisis de la transferencia y apropiación del conocimiento en la investigación de universidades colombianas. *Investigación & Desarrollo, 13*(1), 128–157.

Adams, W. (1995). Sustainable development. In Geographies of Global Change: Remapping the World in the Late Twentieth Century. Blackwell.

Adams, R., Jeanrenaud, S., Bessant, J., Denyer, D., & Overy, P. (2016). Sustainability-oriented Innovation: A Systematic Review. *International Journal of Management Reviews, 18*(2), 180–205. doi:10.1111/ijmr.12068

Agapito, D. (2013). *The role of sensescapes in tourist experiences in rural areas.* Faculdade de Economia - Universidade do Algarve. https://sapientia.ualg.pt/handle/10400.1/7776

Agapito, D. (2020). The senses in tourism design: A bibliometric review. *Annals of Tourism Research, 83*(December). doi:10.1016/j.annals.2020.102934

Agapito, D., Pinto, P., & Mendes, J. (2017). Tourists' memories, sensory impressions and loyalty: In loco and post-visit study in Southwest Portugal. *Tourism Management, 58*, 108–118. doi:10.1016/j.tourman.2016.10.015

Agapito, D., Valle, P., & Mendes, J. (2014). The sensory dimension of tourist experiences: Capturing meaningful sensory-informed themes in Southwest Portugal. *Tourism Management, 42*, 224–237. doi:10.1016/j.tourman.2013.11.011

Ağcakaya, H., & Can, İ. I. (2019). Somut Olmayan Kültürel Miras Kapsamında Mutfak Kültürünün Sürdürülebilirliği: Türkiye'deki Gastronomi Müzeleri Örneği [Sustainability of Culinary Culture in the Context of Cultural Heritge: An Example for Gastronomy Museums in Turkey]. *Gastroia: Journal of Gastronomy and Travel Research, 3*(4), 788–804.

Agyeiwaah, E., McKercher, B., & Suntikul, W. (2017). Identifying core indicators of sustainable tourism: A path forward? *Tourism Management Perspectives, 24*, 26–33. https://doi.org/10.1016/j.tmp.2017.07.005

Ahumada, J., & Miranda, F. (2003). Ciencia, tecnología y Sociedad: Algunas Reflexiones. In Documento preparado para la Organización de los Estados Américanos. doi:10.2307/j.ctvn5tw9m.5

Ajzen, I. (1991). The theory of planned behavior. *Organizational Behavior and Human Decision Processes, 50*(2), 179–211. doi:10.1016/0749-5978(91)90020-T

Akademia e Shkencave e Shqipërisë. (2009). *Fjalor Enciklopedik Shqiptar* [Albanian Encyclopedic Dictionary]. III. Author.

Akbulut, O., & Ekin, Y. (2018). Kültürel Miras Turizmi Olarak Savaş Alanları Turizmi: Türkiye'de Yer Alan Savaş Anıtlarının Coğrafi Bilgi Sistemleri Analizi [War Areas Tourism as Cultural Heritage Tourism: The Analysis of Turkish War Monuments with GIS]. *Hitit Üniversitesi Sosyal Bilimler Enstitüsü Dergisi, 11*(1), 395–420.

Akipek, S. (2001). Dünya Kültürel ve Doğal Mirasının Korunmasına Dair Sözleşmenin Değerlendirilmesi [The Evaluation of the Convention Concerning the Protection of the World Cultural and Natural Heritage]. *Ankara Üniversitesi Hukuk Fakültesi Dergisi, 50*(4), 13–39.

Akkuş, Z., & Efe, T. (2015). Türkiye'de Defineciliğin Kültürel Mirasın Korunması Bağlamında Değerlendirilmesi [The Evaluation of Treasure Hunting in Turkey in the Context of Protection of Cultural Heritage]. *Birey ve Toplum Sosyal Bilimler Dergisi, 5*(2), 105–132.

Aksoy, L., Alkire, L., Choi, S., Kim, P. B., & Zhang, L. (2019). Social innovation in service: A conceptual framework and research agenda. *Journal of Service Management, 30*(3), 429–448. doi:10.1108/JOSM-11-2018-0376

Al-Ababneh, M., & Masadeh, M. (2019). Creative cultural tourism as a new model for cultural tourism. *Journal of Tourism Management Research, 6*(2), 109–118. doi:10.18488/journal.31.2019.62.109.118

Alagöz, G., Çalık, İ., & Güneş, E. (2018). Kültürel Miras Turizmi Açısından Erzincan Bakır İşleme Sanatının Mevcut Durumu ve Sürdürülebilirliği [Current State of the Erzincan Art of Chalcography from the Perspective of Cultural Heritage Tourism and Its Sustainability]. *Dumlupınar Üniversitesi Sosyal Bilimler Dergisi, 55*, 174–191.

Alba-Pagán, E., & Pitarch-Garrido, M. D. (2018). Las propuestas culturales socio-participativas en la ciudad de Valencia. Innovación social y cultura en los barrios: la emergencia necesaria. Actas del Congreso Internacional de Ciencia Regional.

Albareda, L., & Hajikhani, A. (2019). Innovation for Sustainability: Literature Review and Bibliometric Analysis. In N. Bocken, P. Ritala, L. Albareda, & R. Verburg (Eds.), *Innovation for Sustainability. Business Transformations Towards a Better World* (pp. 35–57). Palgrave Macmillan. doi:10.1007/978-3-319-97385-2_3

Algarve Drone. (2021). https://www.dronealgarve.pt/fotos-aereas-algarve/

Aliağaoğlu, A., & Narlı, M. (2012). Edebi Miras Turizmi ve Türkiye'de Edebi Mekanlar [Literary Heritage Tourism and Literature Related Places in Turkey]. *Bilig, 60*, 1–18.

Ali, F., Rasoolimanesh, S. M., Sarstedt, M., Ringle, C. M., & Ryu, K. (2018). An assessment of the use of partial least squares structural equation modeling (PLS-SEM) in hospitality research. *International Journal of Contemporary Hospitality Management, 30*(1), 514–538. doi:10.1108/IJCHM-10-2016-0568

Alkier, R., Milojica, V., & Roblek, V. (2017). Challenges of the social innovation in tourism. *Tourism in South East Europe., 4*, 1–13. doi:10.20867/tosee.04.24

Alsos, G. A., Eide, D., & Madsen, E. L. (2014). Introduction: Innovation in tourism industries. In G. Alsos, D. Eide, & E. L. Madsen (Eds.), *Handbook of Research on Innovation in Tourism Industries* (pp. 1–24). Edward Elgar Publishing.

Álvarez, C. (2008). La etnografía como modelo de investigación en educación. *Gazeta de Antropología, 24*(1). http://hdl.handle.net/10481/6998

Álvarez-García, J., Maldonado-Erazo, C. P., Río-Rama, M. C., & Sánchez-Fernandez, M. D. (2018). Análisis de los Estudios Referentes a los Impactos del Turismo Creativo Indexados en la Base Scopus y WoS. *Revista Portuguesa de Estudos Regionais, 48*, 17-32. https://dialnet.unirioja.es/servlet/articulo?codigo=6438098

Alves, D., Filho, D., & Henrique, A. (2015). O poderoso NVivo: Uma introdução a partir da análise de conteúdo [The powerful NVivo: An introduction from content analysis]. *Revista Política Hoje, 24*(2), 119–134.

Alves, S. M. P. P. (2011). *A Antevisão do Peregrino na Iconografia de São Tiago no Caminho Português de Santiago entre Viseu e Chaves. Subsídios para a criação de uma rota turística* (Master's dissertation). Universidade Católica Portuguesa.

Amaro, S., Antunes, A., & Henriques, C. (2018). A closer look at Santiago de Compostela's pilgrims through the lens of motivations. *Tourism Management, 64*, 271–280. doi:10.1016/j.tourman.2017.09.007

Amir, S., Osman, M. M., Bachok, S., & Ibrahim, M. (2015). Sustaining Local Community Economy Through Tourism: Melaka UNESCO World Heritage City. *Procedia Environmental Sciences, 28*, 443–452. doi:10.1016/j.proenv.2015.07.054

Ap, J. (1990). Residents' perceptions research on the social impacts of tourism. *Annals of Tourism Research, 17*(4), 610–616. doi:10.1016/0160-7383(90)90032-M

Arboleda Jaramillo, C. A., Arias Arciniegas, C. M., Pérez Sánchez, E. O., & Correa Janne, P. (2020). Innovación social como estrategia para fortalecer el turismo rural comunitario en Colombia. *Revista Venezolana de Gerencia, 25*(89), 92–104.

Arcodia, C., & Whitford, M. (2007). Festival attendance and the development of social capital. In *Journal of convention & event tourism* (pp. 1–18). Taylor & Francis Group.

Argueta, A. (2012). El diálogo de saberes, una utopía realista. *Integra Educativa, V*(3), 15–29.

Arıoğlu, İ. E., & Aydoğdu Atasoy, Ö. (2015). Somut Olmayan Miras Kapsamında Geleneksel El Sanatları ve Kültür ve Turizm Bakanlığı [Traditional Handicrafts in the Context of Intangible Cultural Heritage and Ministry of Culture and Tourism]. *Turkish Studies: International Periodical for Languages. Literature and History of Turkish or Turkic., 10*(16), 109–126.

Arnáez, J., Lana-Renault, N., Lasanta, T., Ruiz-Flaño, P., & Castroviejo, J. (2015). Effects of farming terraces on hydrological and geomorphological processes. A review. *Catena, 128*, 122–134. doi:10.1016/j.catena.2015.01.021

Arslan Kalay, H., & Yenişehirlioğlu, E. (2017). *Kültürel Miras Yönetiminde Bir Çatışma Unsuru: Turizm* [A Conflict in Cultural Heritage Management: Tourism] [Paper presentation]. *International Congress on Cultural Heritage and Tourism (ICCHT-2017).*

Artal-Tur, A., Villena-Navarro, M., & Alamá-Sabater, L. (2018). The relationship between cultural tourist behaviour and destination sustainability. *Anatolia, 29*(2), 237–251. doi:10.1080/13032917.2017.1414444

Arthur, S. N. A., & Mensah, J. V. (2006). Urban management and heritage tourism for sustainable development: The case of Elmina Cultural Heritage and Management Programme in Ghana. *Management of Environmental Quality, 17*(3), 299–312. doi:10.1108/14777830610658719

Asatekin, N. G. (2004). Kültür ve Doğa Varlıklarımız, Neyi Niçin Nasıl Korumalıyız? [Culture and Nature Assets, What and How to Protect?] T.C. Kültür ve Turizm Bakanlığı Yayın No: 3016, DÖSİMM Basımevi, Ankara.

Ashworth, G. (1995). Managing the cultural tourism. In G. Ashworth & A. Dietvorst (Eds.), *Tourism and spatial transformation. Implications for policy and planning* (pp. 265–283). CAB International.

Ashworth, G. J. (1995). Heritage, Tourism and Europe: a European Future for a European Past. In D. T. Herbert (Ed.), *Heritage, Tourism and Society* (pp. 68–83). Mansell Publishing Ltd.

Asins, S., & Romero, J. (2010). Gestión, ordenación y protección de las laderas aterrazadas en la Comunidad Valenciana. *Investigaciones Geográficas, 53*(53), 85–98. doi:10.14198/INGEO2010.53.04

Aspridis, G., Rossidis, I., Katsimardos, P., & Bouas, K. (2016). Modulation of conditions and infrastructure for the integration of organizational changes in tourist sector. In *Tourism, culture and heritage in a smart economy*. Springer.

Auclair, E., & Fairclough, G. (2015). Living between past and future: An introduction to heritage and cultural sustainability. In E. Auclair & G. Fairclough (Eds.), *Theory and practice in heritage and sustainability* (pp. 1–22). Routledge. doi:10.4324/9781315771618-1

Australia, I. C. O. M. O. S. (2013). *Practice Note: Understanding and Assessing Cultural Heritage*. https://australia.icomos.org/wp-content/uploads/Practice-Note_Understanding-and-assessing-cultural-significance.pdf

AvrupaGazete. (2021). *Türkiye'nin Geleneksel Ahlat Taş İşçiliği, İpek Böcekçiliği ve Sedef Kakma Sanatı UNESCO adayı* [Turkey's Traditional Sericulture and Ahlat Stonework]. https://www.avrupagazete.co.uk/kultur-sanat/turkiye-nin-geleneksel-ahlat-tas-isciligi-ipek-bocekciligi-h348175.html

Awang, K. W. (2010). Tourism policy development: The Malaysian experience. *Arab World Geographer*, *13*(2), 150–158.

Axon, S. (2020). The socio-cultural dimensions of community-based sustainability: Implications for transformational change. *Journal of Cleaner Production*, *266*, 121933. doi:10.1016/j.jclepro.2020.121933

Aydın, M., & Güner, S. (2013). Kültürel Mirasta Risk Yönetimi [Risk Management in Cultural Heritage]. *Batman University Journal of Life Sciences*, *3*(1), 70–81.

Azevedo, P. (2021). The Ways of Saint James in Trás-os-Montes and Alto Douro as an example of soft tourism in rural areas. *European Countryside*, *13*(2), 314–329. doi:10.2478/euco-2021-0020

Baggio, R. (2020). Tourism destinations: A universality conjecture based on network science. *Annals of Tourism Research*, *82*(102929), 1–13. doi:10.1016/j.annals.2020.102929

Baggio, R., Scott, N., & Cooper, Ch. (2010). Network science. A review focused on tourism. *Annals of Tourism Research*, *37*(3), 802–827. doi:10.1016/j.annals.2010.02.008

Baigorri, A. (2000). La red urbana ibérica. In L. de la Machorra & M. Brandâo (Eds.), *La economía ibérica: una fértil apuesta de futuro* (pp. 261–289). Editora Regional de Extremadura.

Bakas, F. E., Duxbury, N., & Vinagre de Casto, T. (2019). Creative tourism: Catalysing artisan entrepreneur networks in rural Portugal. *International Journal of Entrepreneurial Behaviour & Research*, *25*(4), 731–752. doi:10.1108/IJEBR-03-2018-0177

Bandeira, P. (1999). *Participação, Articulação de Atores Sociais e Desenvolvimento Regional.* Texto para discussão, Instituto de Pesquisa Econômica Aplicada (IPEA), nº 630.

Barbosa, L. J., & Silva, M. B. (2015). Tourism in favelas: A practice of belonging and sociocultural recognition [Turismo em favelas: uma prática de pertencimento e reconhecimento sociocultural]. *Nuevamérica*, 146.

Bardin. (1977). Análise de Conteúdo. Academic Press.

Barra, A. (2018). Crossumer, Prosumer, Fansumer... ¿Qué es todo esto? *Lead Hospitality*. https://www.leadhospitality.es/crossumer-prosumer-fansumer-que-es-todo-esto

Barrionuevo, C. A. (2012). *El territorio como construcción social : una pregunta que importa. El caso de Rincón de Las Perlas*. Universidad Nacional de la Plata.

Bates, F. L. (1960). Institutions, organizations, and communities: A general theory of complex structures. *Pacific Sociological Review*, *3*(2), 59–70. doi:10.2307/1388201

Bauman, R. (Ed.). (1992). *Folklore, cultural performances, and popular entertainments: A communications-centered handbook*. Oxford University Press.

Baute-Plana, A., & Zamora-Fung, R. (2021). Una aproximación a la COVID-19 y sus lecciones. *Educación Médica Superior, 35*. http://www.ems.sld.cu/index.php/ems/article/download/2516/1170

Bayer, J., Fehringer, A., Lehar, G., Jurgeit, F., & Leitner, T. (2017). The relevance of visitor's national park affinity for effective visitor management in protected areas. In J. N. Albrecht (Ed.), *Visitor Management in Tourism Destinations* (pp. 75–90). Cabi. doi:10.1079/9781780647357.0075

Baykan, E. (2007). Kültür Turizmi ve Turizmin Kültürel Varlıklar Üzerindeki Etkileri [Effects of Cultural Tourism on Cultural Assets]. *Gazi Üniversitesi Ticaret ve Turizm Eğitim Fakültesi Dergisi, 2*, 30–49.

Becker, E. (2016). *Overbooked: The exploding business of travel and tourism*. Simon and Schuster.

Belhassen, Y., Caton, K., & Stewart, W. P. (2008). The search for authenticity in the pilgrim experience. *Annals of Tourism Research, 35*(3), 668–689. doi:10.1016/j.annals.2008.03.007

Bell, S., Eason, K., & Frederiksen, P. (2011). *POINT Policy use and influence of indicators*. Project co-funded by the European Commission within the Seventh Framework (2007-2013).

Ben-Amos, D. (1971). Toward a definition of folklore in context. *Journal of American Folklore, 84*(331), 3–15. doi:10.2307/539729

Bench, S., & Day, T. (2010). The user experience of critical care discharge: A meta-synthesis of qualitative research. *International Journal of Nursing Studies, 47*(4), 487–499. https://doi.org/10.1016/j.ijnurstu.2009.11.013

Benjamin, S., Dillette, A., & Alderman, D. H. (2020). "We can't return to normal": Committing to tourism equity in the post-pandemic age. *Tourism Geographies, 22*(3), 476–483. doi:10.1080/14616688.2020.1759130

Berg, I., & Edelheim, J. (2012). The attraction of islands: Travelers and tourists in the Cyclades (Greece) in the twentieth and twenty-first centuries. *Journal of Tourism and Cultural Change, 10*(1), 84–98. doi:10.1080/14766825.2012.660946

Berki, M., & Csapó, J. (2008). The Geographical Basis for the Development of Thematic Routes. In Progress in Geography in the European Capital of Culture 2010. University of Geography, University of Pécs.

Bermúdez, O. M., Mayorga, M. L., Jacanamijoy, B., Seygundiba, A., & Fajardo, T. (2005). *El diálogo de saberes y la educacíon ambiental*. Academic Press.

Berti, E. (2015). Defining the Cultural Routes of the Council of Europe. *Cultural Routes Management: from theory to practice; Step-by-step guide to the Council of Europe Cultural Routes.*

Berti, E. (2012). *Itinerari culturali del Consiglio d'Europa: tra ricerce di identità e progetto di paesaggio*. Firenze University Press. doi:10.36253/978-88-6655-142-3

Berti, E. (2013). Cultural Routes of the Council of Europe: New Paradigms for the Territorial Project and Landscape. *Alma Tourism, 7*, 1–10.

Besculides, A., Lee, M. E., & McCormick, P. J. (2002). Resident's perceptions of the cultural benefits of tourism. *Annals of Tourism Research, 29*(2), 303–319. doi:10.1016/S0160-7383(01)00066-4

Bichler, B. F. (2019). Designing tourism governance: The role of local residents. *Journal of Destination Marketing & Management, 100389*. Advance online publication. doi:10.1016/j.jdmm.2019.100389

Bichler, B. F., & Lösch, M. (2019). Collaborative governance in tourism: Empirical insights into a community-oriented destination. *Sustainability*, *11*(23), 6673. doi:10.3390u11236673

Bindé, J. (2005). Hacia las sociedades del conocimiento. In *Informe Mundial*. UNESCO.

Blanco, E., Rey-Maquieira, J., & Lozano, J. (2009). Economic incentives for tourism firms to undertake voluntary environmental management. *Tourism Management*, *30*(1), 112–122. doi:10.1016/j.tourman.2008.04.007

Blapp, M., & Mitas, O. (2018). Creative tourism in Balinese rural communities. *Current Issues in Tourism*, *21*(11), 1285–1311. doi:10.1080/13683500.2017.1358701

Blom, T., Nisson, M., & Santos, X. (2016). The way to Santiago beyond Santiago. Fisterra and the pilgrimage's post-secular meaning. *European Journal of Tourism Research*, *12*, 133–146.

Boas, F. (1916). The origin of totemism. *American Anthropologist*, *18*(3), 319–326. doi:10.1525/aa.1916.18.3.02a00020

Bocken, N., Ritala, P., Albareda, L., & Verburg, R. (2019). *Innovation for Sustainability: Business Transformations Towards a Better World*. Palgrave Macmillan. doi:10.1007/978-3-319-97385-2

Boisier, S. (2007). Territorio, estado y sociedad en chile. La dialéctica de La descentralización: entre la geografía y la gobernabilidad (Thesis). Universidad De Alcalá, Alcalá De Henares.

Boisier, S. (2001). *El lenguaje emergente en el desarrollo territorial*. CATS.

Bonnafous-Boucher, M., & Rendtorff, J. D. (2016). *Stakeholder theory: A model for strategic management*. Springer International Publishing. doi:10.1007/978-3-319-44356-0

Borgatti, S., Everett, M., & Freeman, C. (2002). *Ucinet for Windows: Software for Social Networks Analysis*. Harvard University.

Borgatti, S., Everett, M., & Johnson, J. (2018). *Analyzing Social Networks*. SAGE Publications.

Borgatti, S., Mehra, A., Brass, D., & Labianca, G. (2009). Network Analysis in the Social Sciences. *Science*, *323*(5916), 892–895. doi:10.1126cience.1165821 PMID:19213908

Borges, M. A., Carbone, G., Bushell, R., & Jaeger, T. (2011). *Sustainable tourism and natural World Heritage: Priorities for action. IUCN, Gland, Switzerland*. http://cmsdata.iucn.org/downloads/sustainable_tourism_and_natural_world_heritage_report.pdf

Bosak, K. (2019). The tourism system. In S. F. McCool & K. Bosak (Eds.), *A Research Agenda for Sustainable Tourism* (pp. 14–20). Edward Elgar Publishing. doi:10.4337/9781788117104.00009

Boström, M. (2012). A missing pillar? Challenges in theorizing and practicing social sustainability: introduction to the special issue. *Sustainability: Science, Practice and Policy*, *8*(1), 3-14.

Bottino, C. M. M. (2016). *Who discovered the favela? The trajectory of tourism agencies in favelas cariocas* [Quem descobriu a favela? A trajetória das agências de turismo nas favelas cariocas] [Master Thesis]. FGV, Rio de Janeiro, Brazil. http://hdl.handle.net/10438/17655

Bouchenaki, M. (2003) *The Interdependency of the Tangible and Intangible Cultural Heritage*. In 14th ICOMOS General Assembly and International Symposium: Place, memory, meaning: preserving intangible values in monuments and sites, 27 – 31 Oct 2003, Victoria Falls, Zimbabwe.

Boucher, B. J. (1990). Punta Gorda to create ecotourism trail. *The Belize Review*, *2*(9), 12–13.

Boullón, R. C. (2006). *Planificacion del espácio turístico*. http://prepacihuatlan.sems.udg.mx/sites/default/files/planificaciondelespacioturisticorobertoc.boullon.pdf

Bourdieu, P. (1986). The Forms of Capital. In J. G. Richards (Ed.), *Handbook of Theory and Research for the Sociology of Education* (pp. 241–258). Greenwood Press.

Bowitz, E., & Ibenholt, K. (2009). Economic impacts of cultural heritage – Research and perspectives. *Journal of Cultural Heritage*, *10*, 1–8. https://doi.org/10.1016/j.culher.2008.09.002

Boyd, H., & Charles, A. (2006). Creating community-based indicators to monitor sustainability of local fisheries. *Ocean and Coastal Management*, *49*(5-6), 237–258. doi:10.1016/j.ocecoaman.2006.03.006

Božić, S., & Tomić, N. (2016). Developing the cultural route evaluation model (CREM) and its application on the Trail of Roman Emperors, Serbia. *Tourism Management Perspectives*, *17*, 26–35. doi:10.1016/j.tmp.2015.11.002

Bramwell, B., & Lane, B. (2011). Critical research on the governance of tourism and sustainability. *Journal of Sustainable Tourism*, *19*(4-5), 411–421. doi:10.1080/09669582.2011.580586

Brcar, F., & Colaric-Jakse, L. M. (2017). Evaluation of cultural heritage in Posavje region by tourists and employees. *Challenges of the Future*, *2*(1), 1–15.

Briedenhann, J., & Wickens, E. (2004). Tourism routes as a tool for the economic development of rural areas - vibrant hope or impossible dream? *Tourism Management*, *25*(1), 71–79. doi:10.1016/S0261-5177(03)00063-3

Brito, M. (2009). *Ciudades históricas como destinos patrimoniales. Una mirada comparada*. Junta de Andalucía, Consejería de Cultura.

Bronner, S. J. (1984). The Early Movements of Anthropology and Their Folkloristic Relationships. *Folklore*, *95*(1), 57–73. doi:10.1080/0015587X.1984.9716297

Brønnøysundsavis. (2021). *Ny bobilparkering på Sundsvoll* [New motorhome parking at Sundsvoll]. https://www.banett.no/nyheter/2021/06/28/Ny-bobilparkering-på-Sundsvoll-Har-hatt-besøkende-siden-vi-åpnet-24177

Brown, S. J. W., & Trimboli, D. (2011). The real "worth" of festivals: challenges for measuring socio-cultural impacts. *Asia Pacific Journal of Arts and Cultural Management*, *8*(1).

Buccitelli, A. B., & Schmitt, C. R. (2016). Everyday Practice and Tradition. *Cultural Analysis*, *15*, 1–5.

Buckley, R. (2012). Sustainable tourism: Research and reality. *Annals of Tourism Research*, *39*(2), 528–546. https://doi.org/10.1016/j.annals.2012.02.003

Buitrago, F., & Duque, I. (2013). *La economía naranja. Una oportunidad infinita*. Banco Interamericano de Desarrollo.

Bulin, D. (2015). Policies and Strategies for Tourism Development–Some Lessons for Romania. *Knowledge Horizons-Economics*, *7*(2), 77–81.

Bullvåg, E., Nyvold, C. E., Ovesen, S., Kvarum, M. K., Finsveen, R., & Steffensen, T. (2020). *Indeks Nordland, Nummer 17 - Årgang 2020, En rapport om utviklingen gjennom 2019 og utsiktene for 2020*. Indeks Nordland. https://indeksnordland.no/

Bushati, H. (1998). *Shkodra dhe Motet*. Rozafat.

Butler, G., & Ivanovic, M. (2016). Cultural heritage tourism development in post-apartheid South Africa: Critical issues and challenges. *Cultural Tourism in Southern Africa*, 58-75.

Butler, R. W. (1999). Sustainable tourism: A state-of-the-art review. *Tourism Geographies, 1*(1), 7–25. doi:10.1080/14616689908721291

Byrd, E. T. (2007). Stakeholders in sustainable tourism development and their roles: Applying stakeholder theory to sustainable tourism development. *Tourism Review, 62*(2), 6–13. doi:10.1108/16605370780000309

Cabeça, S. M., Gonçalves, A. R., Marques, J. F., & Tavares, M. (2019). Mapping intangibilities in creative tourism territories through tangible objects: A methodological approach for developing creative tourism offers. *Tourism & Management Studies, 15*(SI, 1SI), 42–49. doi:10.18089/tms.2019.15SI05

Caca, E., Ladias, Ch., & Polo, A. (2016). The Development of tourism in Albania and the importance of cultural tourism. *Tourism and Culture in the Age of Innovation*, 109-119.

Caffyn, A., & Jobbins, G. (2003). Governance capacity and stakeholder interactions in the development and management of coastal tourism: Examples from Morocco and Tunisia. *Journal of Sustainable Tourism, 11*(2-3), 224–245. doi:10.1080/09669580308667204

Camaj, M. (1978). *There before the tribes arrived.* http://www.albanianliterature.net/authors/modern/camaj/camaj_poetry.html

Cámara de Comercio de Duitama. (2020). *Efectos y Expectativas de Recuperación económica frente a la pandemia COVID-19 en el departamento de Boyacá.* https://ccduitama.org.co/documentos/Observatorio/Efectos%20y%20expectativas%20recuperación%20económica%20Boyacá-%20covid.pdf?_t=1595618870

Cámara de Comercio de Duitama. Cámara de Comercio deo Bogotá, & Consejo Provincial de Turismo de Gutiérrez. (2018). *Estudio de Mercados, con enfoque de la demanda turística de producto turístico diseñado para el entorno de la Sierra Nevada del Cocuy, Guican de la Sierra y Chita.* https://ccduitama.org.co/documentos/Observatorio/ESTUDIO%20DE%20MERCADOS%20CON%20ENFOQUE%20DE%20LA%20DEMANDA%20TURSTICA%20DE%20PRODUCTO%20TURSTICO%20DISEADO%20PARA%20EL%20ENTORNO%20DE%20LA%20SIERRA%20NEVADA%20DEL%20COCUY%20GUICAN%20DE%20LA%20SIERRA%20Y%20CHITA.pdf

Câmara Municipal de Olhão. (n.d.). https://www2.cm-olhao.pt/en/conhecer-olhao/geografia-do-concelho

Cammarata, E. B. (2006). El turismo como práctica social y su papel en la apropiación y consolidación del territorio. *América Latina: Cidade, Campo e Turismo*, 351–366.

Campolo, D., Bombino, G., & Meduri, T. (2016). Cultural landscape and cultural routes: Infra- structure role and indigenous knowledge for sustainable development of inland areas. *Procedia: Social and Behavioral Sciences, 223*, 576–582. doi:10.1016/j.sbspro.2016.05.350

Campos, E. (2020). Etnografía. In D. Páramo Morales, S. J. Campo Sierra, & L. M. Maestre Matos (Eds.), *Métodos de investigación cualitativa. Fundamentos y aplicaciones* (pp. 163–203). Editorial UniMagdalena.

Camus, S., Hikkerova, L., & Sahut, J. (2012). Systemic Analysis and Model of Sustainable Tourism Sandra Camus Lubica Hikkerova. *International Journal of Business,* 365–378.

Canclini, N. G. (2006). *Culturas Híbridas: estratégias para entrar e sair da modernidade.* Editora da Universidade de São Paulo.

Cano, A. M. (Ed.). (2018). Diálogo de saberes, memorias y territorios (1a ed.). Sello Editorial Unversidad de Medellín.

Capdepón, F. P. (2014). The literary landscape and its connection with cultural tourism. *Cuadernos de Turismo, 33*, 407–409.

Carayannis, E. G., & Campbell, D. F. (2018). *Smart quintuple helix innovation systems: How social ecology and environmental protection are driving innovation, sustainable development and economic growth.* Springer.

Carbone, F., Corinto, G., & Malek, A. (2016). New Trends of Pilgrimage: Religion and Tourism, Authenticity and Innovation, Development and Intercultural Dialogue: Notes from the Diary of a Pilgrim of Santiago. *AIMS Geosciences, 2*(2), 152–165. doi:10.3934/geosci.2016.2.152

Carden, A. R. (2006). Sign of the times: An examination of the self-concept and emotional motivation of travelers on route 66 (1926-Present). *Journal of Travel & Tourism Marketing, 20*(2), 51–61. doi:10.1300/J073v20n02_04

Cardita, A. (2012). Peregrinação: Possibilidades de compreensão crítica de uma experiência. (Sociologia). *Revista da Faculdade de Letras da Universidade do Porto, 24*, 195–213.

Careto, H., & Lima, S. (2006). *Turismo e Desenvolvimento Sustentável.* GEOTA - Grupo de Estudos de Ordenamento do Território e Ambiente.

Carreira, V., Azeredo, J., González-Rodríguez, M. R., & Díaz-Fernández, M. C. (2021). The Role of Tourism Education in Overtourism and Destination Management. In A. Sharma & A. Hassan (Eds.), *Overtourism as Destination Risk* (pp. 35–48). Emerald Publishing Limited. doi:10.1108/978-1-83909-706-520211003

Carrington, P., Scott, J., & Wasserman, S. (Eds.). (2005). *Models and methods in social network analysis.* Cambridge University Press. doi:10.1017/CBO9780511811395

Carrión, G. C., Nitzl, C., & Roldán, J. L. (2017). Mediation analyses in partial least squares structural equation modeling: Guidelines and empirical examples. In H. Latan & R. Noonan (Eds.), *Partial Least Squares Path Modeling: Basic Concepts, Methodological Issues and Applications* (pp. 173–195). Springer. doi:10.1007/978-3-319-64069-3_8

Carson, S., Hawkes, L., Gislason, K., & Cantrell, K. (2016). Literature, tourism and the city: Writing and cultural change. *Journal of Tourism and Cultural Change, 15*(4), 380–392. doi:10.1080/14766825.2016.1165237

Carter, R. W., & Beeton, R. J. S. (2004). A model of cultural change and tourism. *Asia Pacific Journal of Tourism Research, 9*(4), 423–442. doi:10.1080/1094166042000311282

Carter, R. W., Thok, S., O'Rourke, V., & Pearce, T. (2015). Sustainable tourism and its use as a development strategy in Cambodia: a systematic literature review. *Journal of Sustainable Tourism, 23*(5), 797–818. doi:10.1080/09669582.2014.978787

Carvalho, I., Baptista, M. M., & Costa, C. (2012). Binding culture and tourism through literature: The potential of literary tourism. *International Journal of Hospitality and Tourism, 2*(2), 1–27.

Carvalho, K. D., & Moquete, S. (2011). El turismo en la dinámica territorial ¿Lógica global, desarrollo local? *Estudios y Perspectivas en Turismo, 20*, 441–461.

Casanueva, C., Gallego, A., & García-Sánchez, M. (2016). Social network analysis in tourism. *Current Issues in Tourism, 19*(12), 1190–1209. doi:10.1080/13683500.2014.990422

Casas, L. F., Pineda, M., Andrade, M., & Pulgarin, Y. (2017). *El Alto Ricaurte e Iguaque territorio de patrimonio natural y cultural, acciones para un manejo sostenible.* La Imprenta Ltda. https://www.natura.org.co/publicaciones/alto-ricaurte-e-iguaque-territorio-patrimonio-natural-cultural-acciones-manejo-sostenible/

Castellani, V., & Sala, S. (2010). Sustainable performance index for tourism policy development. *Tourism Management, 31*(6), 871–880. doi:10.1016/j.tourman.2009.10.001

Castillo, J., & Martínez, C. (2014). El Patrimonio agrario: Definición, caracterización y representatividad en el ámbito de la UNESCO. *Boletín de la Asociación de Geógrafos Españoles*, *66*(66), 105–124. doi:10.21138/bage.1782

Castillo-Vergara, M. (2020). La teoría de las N-hélices en los tiempos de hoy. *Journal of Technology Management & Innovation*, *15*(3), 3–5. doi:10.4067/S0718-27242020000300003

Caton, K., & Santos, C. A. (2007). Heritage tourism on Route 66: Deconstructing nostalgia. *Journal of Travel Research*, *45*(4), 371–386. doi:10.1177/0047287507299572

Çavuşoğlu, P. P. (2017). Evaluation of Tourism Strategy 2023 within Perspective of Transportation Strategy. *Journal of Education & Social Policy*, *4*(3), 61–68.

Celata, F., & Sanna, V. S. (2019). A multi-dimensional assessment of the environmental and socioeconomic performance of community-based sustainability initiatives in Europe. *Regional Environmental Change*, *19*(4), 939–952. doi:10.100710113-019-01493-9

Cerin, P. (2006). Bringing economic opportunity into line with environmental influence: A discussion on the Coase theorem and the Porter and van der Linde hypothesis. *Ecological Economics*, *56*(2), 209–225. https://doi.org/10.1016/j.ecolecon.2005.01.016

Cerquetti, M., & Cutrini, E. (2021). The role of social ties for culture-led development in inner areas. The case of the 2016–2017 Central Italy earthquake. *European Planning Studies*, *29*(3), 556–579. doi:10.1080/09654313.2020.1759512

CERTESS. (2020). Retrieved from CERTESS documents: http://certess.culture-routes.lu/documents

Cetin, G., & Bilgihan, A. (2016). Components of cultural tourists' experiences in destinations. *Current Issues in Tourism*, *19*(2), 137–154. doi:10.1080/13683500.2014.994595

Chandralal, L., & Valenzuela, F.-R. (2013). Exploring Memorable Tourism Experiences: Antecedents and Behavioural Outcomes. *Journal of Economics, Business and Management*, *1*(2), 177–181.

Chaparro, F. (n.d.). *Apropiación Social del Conocimiento*. Aprendizaje y Capital Social.

Chaplin, I. (2003). Cultural tourism: The partnership between tourism and cultural heritage management, by Bob McKercher and Hilary du Cros. The Haworth Hospitality Press, 2002. xii + 262 pp. $69.95 (hardback); $39.95 (paperback) ISBN 0-7890-1106-9. *International Journal of Tourism Research*, *5*(3), 245. https://doi.org/10.1002/jtr.423

Chenail, R. J. (2009). Communicating Your Qualitative Research Better. *Family Business Review*, *22*(2), 105–108. https://doi.org/10.1177/0894486509334795

Chen, H., & Rahman, I. (2018). Cultural tourism: An analysis of engagement, cultural contact, memorable tourism experience and destination loyalty. *Tourism Management Perspectives*, *26*, 153–163. doi:10.1016/j.tmp.2017.10.006

Chhabra, D. (2010). *Sustainable Marketing of Cultural and Heritage Tourism*. Routledge. doi:10.4324/9780203855416

Choi, H. C., & Sirakaya, E. (2006). Sustainability indicators for managing community tourism. *Tourism Management*, *27*(6), 1274–1289. doi:10.1016/j.tourman.2005.05.018

Choi, S. H., Imon, S. S., & Couto, U. (2020). Negotiating with Authenticity to Ensure Sustainability of Traditional Cultural Festivals: Residents' Perspectives about Macao's Drunken Dragon Festival. *Sustainability*, *12*(3), 885. doi:10.3390u12030885

Cialdini, R. B., & Trost, M. R. (1998). Social influence: Social norms, conformity and compliance. In D. T. Gilbert, S. T. Fiske, & G. Lindzey (Eds.), *The handbook of social psychology* (pp. 151–192). McGraw-Hill.

Ciro, A., & Toska, M. (2018). Sustainable tourism development in Albania. *Annual Review of Territorial Governance in Albania, I*, 84–93. doi:10.32034/CP-TGAR-I01-06

Cluster Development. (2021). Presentación Pública Cluster Economía Naranja Departamento de Boyacá—Colombia. *Clust Development*. https://www.cluster-development.com

Co Evolve Deliverable Library. (2020). Retrieved from Co Evolve: https://co-evolve.interreg-med.eu/what-we-achieve/deliverable-library/

Coccossis, H. (2001). Sustainable development and tourism in small islands: some lessons from Greece. *Anatolia - An International Journal of Tourism and Hospitality Research, 12*(1), 53–58.

Coccossis, H., Tsartas, P., & Griba, E. (2011). *Special and Alternative forms of Tourism- Demand and Supply of new Tourism Products*. Athens: Kritiki.

Coccossis, H. (2006). Tourism, Technological Change and Regional Development in Islands. In M. Giaoutzi & P. Nijkamp (Eds.), *Tourism and Regional Development: New Pathways* (pp. 271–280). Ashgate.

Coccossis, H. (2008). Cultural heritage, local resources and sustainable tourism. *International Journal of Services Technology and Management, 10*(1), 8–14. doi:10.1504/IJSTM.2008.020340

Coccossis, H. (2009). Sustainable Development and Tourism: Opportunities and Threats to Cultural Heritage from Tourism. In L. Fusco Girard & P. Nijkamp (Eds.), *Cultural Tourism and Sustainable Local Development* (pp. 47–56). Ashgate Publishing.

Coccossis, H., & Koutsopoulou, A. (2020). Measuring and monitoring sustainability of coastal tourism destinations in the Mediterranean. *Tourism: An International Interdisciplinary Journal, 68*(4), 482–498. doi:10.37741/t.68.4.8

Coccossis, H., & Tsartas, P. (2019). *Sustainable Tourism Development* (2nd ed.). Kritiki. (in Greek)

Cohen, A. P. (2013). *Symbolic construction of community*. Routledge. doi:10.4324/9780203131688

Cohen, E. (2002). Authenticity, Equity and Sustainability in Tourism. *Journal of Sustainable Tourism, 4*(10), 267–276. doi:10.1080/09669580208667167

Cohen, E. (2004). *Contemporary Tourism – Diversity and Change*. Elsevier.

Cohen, S. A., Prayag, G., & Moital, M. (2014). Current Issues in Tourism Consumer behaviour in tourism : Concepts, influences and opportunities. *Current Issues in Tourism, 0*(0), 1–38. doi:10.1080/13683500.2013.850064

Colciencias. (2005). Encuentro Internacional de Investigadores "Colombia en la Sociedad del Conocimiento." In *Memorias del Encuentro de Investigadores Colombianos Residentes en el País y en el Exterior*. Panamericana formas e impresos S.A.

Colciencias. (2018). Política de Ética de la Investigación, Bioética e Integridad Científica. In *Documento de Política Nacional de Ciencia, Tecnología e Innovación*. Colciencias.

Coleman, J. (1988). Social Capital in the creation of Human Capital. *American Journal of Sociology, 94*, 95–120. doi:10.1086/228943

Collins-Kreiner, N., & Wall, G. (2015). Tourism and religion: spiritual journeys and their consequences. In The Changing World Religion Map (pp. 689-707). Springer. doi:10.1007/978-94-017-9376-6_34

Collins-Kreiner, N. (2016). The lifecycle of concepts: The case of 'Pilgrimage Tourism'. *Tourism Geographies, 18*(3), 322–334. doi:10.1080/14616688.2016.1155077

Collins-Kreiner, N. (2020). Pilgrimage tourism-past, present and future rejuvenation: A perspective article. *Tourism Review*, *75*(1), 145–148. doi:10.1108/TR-04-2019-0130

Colombo, P., & Gold, P. (2004). Multiple memory systems. *Neurobiology of Learning and Memory*, *82*(3), 169–170. doi:10.1016/j.nlm.2004.07.008 PMID:15464401

Connelly, F. M., & Clandinin, D. J. (1990). Stories of experience and narrative inquiry. *Educational Researcher*, *19*(5), 2–14. doi:10.3102/0013189X019005002

Cooper, C., Scott, N., & Baggio, R. (2009). Network position and perceptions of destination stakeholder importance. *Anatolia*, *20*(1), 33–45. doi:10.1080/13032917.2009.10518893

Cooper, C., Volo, S., Gartner, W. C., & Scott, N. (2018). *The SAGE Handbook of Tourism Management: Aplications of Theories and Concepts to Tourism*. SAGE Publications Ltd.

Corpoboyaca. (2021). https://www.corpoboyaca.gov.co

Correia, A., Oliveira, C., & Pereira, R. (2017). From Emotions to Place Attachment. In A. Correia, M. Kozak, J. Gnoth, & A. Fyall (Eds.), *Co-Creation and Well-Being in Tourism* (pp. 163–177). Springer International Publishing. doi:10.1007/978-3-319-44108-5_13

Costa, C. (2014). Gestão Estratégica do Turismo: Evolução Epistemológica dos Modelos e Paradigmas, e Tendências para o Futuro. In Turismo nos Países Lusófonos: Conhecimento, Estratégia e Territórios (pp. 19-40). Escolar Editora.

Costanza, R. (1996). Ecological Economics: Reintegrating the study of humans and nature. *Nature Sciences Sociétés*, *5*(2), 978–990.

Costanza, R., Cumberland, J. H., Daly, H., Goodland, R., Norgaard, R. B., Kubiszewski, I., & Franco, C. (2014). *An Introduction to Ecological Economics*. CRC Press. doi:10.1201/b17829

Council of Europe. (2000). *European Landscape Convention, CETS No. 176*. Council of Europe.

Council of Europe. (2010). *Resolution CM/Res (2010)53. Establishing an Enlarged Partial Agreement on Cultural Routes (Adopted by the Committee of Ministers at its 1101st meeting of the Ministers' Deputies).* https://search.coe.int/cm/Pages/result_details.aspx?ObjectID=09000016805cdb50

Council of Europe. (2010). Resolution CM/Res(2010)52 on the rules for the award of the "Cultural Route of the Council of Europe" certification.

Council of Europe. (2011). *Impact of European Cultural Routes on SMEs' innovation and competitiveness*. Council of Europe.

Council of Europe. (2013). *Resolution CM/Res(2013)66*. Retrieved from Council of Europe: https://search.coe.int/cm/Pages/result_details.aspx?ObjectId=09000016805c69ac

Council of Europe. (2015). *Cultural routes management: from theory to practice. Step by step guide to the Council of Europe Cultural Routes*. Council of Europe.

Council of Europe. (2020). *Policies for the Danube region: Transnational cultural policies for the Danube Region (EUSDR)*. Council of Europe.

Council of Europe. (2020). Retrieved from Cultural Routes of the Council of Europe programme: https://www.coe.int/en/web/cultural-routes

Council of Europe. (2021). *Cultural Routes of the Council of Europe Programme Activity Report.* https://rm.coe.int/cultural-routes-of-the-council-of-europe-programme-activity-report-202/1680a245ec

Courtney, M. (2013). A review of the literature exploring the occupation of contemporary pilgrimage. *Journal of Occupational Science, 22*(2), 170–182. doi:10.1080/14427591.2013.764816

Cousin, S. (2008). UNESCO and the doctrine of cultural tourism. Genealogy of 'good tourism'. *Civilisations, 57*(1), 41–56. doi:10.4000/civilisations.1541

CPIS. (2021). Available at: https://cpis.utad.pt/

Creative Tourism Network. (2020). *The concept: Creative Friendly by Creative Tourism Network.* Creative Tourism Network. http://www.creativetourismnetwork.org/presentation/?lang=es

Crespi-Vallbona, M., & Richards, G. (2007). The meaning of cultural festivals: Stakeholder perspectives in Catalunya. *International Journal of Cultural Policy, 13*(1), 103–122. doi:10.1080/10286630701201830

Csapo, J. (2012). The Role and Importance of Cultural Tourism in Modern Tourism Industry. In Strategies for Tourism Industry - Micro and Macro Perspectives. doi:10.5772/38693

Cult-Ring Project. (2020). Retrieved from Interreg Europe: https://www.interregeurope.eu/cult-ring/

Cultural Affairs Bureau. Macao SAR. (2019). *Macao International Parade.* https://www.icm.gov.mo/macaoparade/9/en/

Cultural Affairs Bureau. Macao SAR. (2021). *Belief and Customs of Na Tcha.* https://www.culturalheritage.mo/en/detail/133/1

Cultural Affairs Bureau. Macao SAR. (2021). *Festival of the Drunken Dragon.* https://www.culturalheritage.mo/en/detail/135/1

Culture 21. (2011). *Lobbying for culture as the 4th pillar of sustainable development in the process of the Rio+20 summit.* Agenda 21 for culture. United Cities and Local Governments – Committee on Culture.

Cunha, A. de M. R. da. (1995). *Caminhos Portugueses de Peregrinação a Compostela. Itinerários portugueses* (Identificação dos caminhos e coordenação da investigação). Santiago de Compostela/Porto, Xunta de Galicia/Centro Regional de Artes Tradicionais.

Cunha, A. de M. R. da. (2019). Viseu > Vila real > Chaves. Da devoção ao turismo: O Caminho Português Interior de Santiago de Compostela Caminho Português Interior de Santiago de Compostela. In Património cultural jacobeu, turismo e peregrinação: O Caminho Português Interior de Santiago de Compostela (CPIS) (pp. 79-112). PASOS.

Czarniawska, B. (1997). *Narrating the organization. Dramas of institutional identity.* University of Chicago Press.

d' Hauteserre, A.M. (2017). Cultural tourism in the French Pacific. *Shima, 11*(1), 55–78.

Damari, C., & Mansfeld, Y. (2014). Reflections on pilgrims' identity, role and interplay with the pilgrimage environment. *Current Issues in Tourism, 19*(3), 1–24.

de Colombia, C. (2017). *Ley nº 1834—23 May 2017. Fomento de la Economía Creativa Ley Naranja.* https://economianaranja.gov.co/media/5ruhnndp/ley-1834-del-23-de-mayo-de-2017.pdf

de Colombia, C. (2019). *Ley nº 1955—25 May 2019. Plan Nacional de Desarrollo 2018-2022 «Pacto por Colombia, pacto por la equidad».* https://www.funcionpublica.gov.co/eva/gestornormativo/norma.php?i=93970

de Cultura, M., de Comercio, M., Industria y Turismo, & Departamento Nacional de Planeación. (2010). Documento CONPES3659 Política nacional para la promoción de las industrias culturales en Colombia. *Observatorio Cultura y Economía*. https://culturayeconomia.org/blog/documento-conpes-3659-politica-nacional-para-la-promocion-de-las-industrias-culturales-en-colombia/

de Cultura, M., de Comercio, M., Industria y Turismo, & Departamento Nacional de Planeación. (2020a). *¡En 2019 el turismo en Colombia rompió récords!* https://www.mincit.gov.co/prensa/noticias/turismo/en-2019-el-turismo-en-colombia-rompio-records

de Cultura, M., de Comercio, M., Industria y Turismo, & Departamento Nacional de Planeación. (2020b). *Ley Naranja*. https://economianaranja.gov.co/ley-naranja/

DECRA. (2019). *Developing European Cultural Routes for All Erasmus Plus project*. French Federation of European Cultural Routes.

Dedek, P. B. (2007). *Hip to the trip: a cultural history of Route 66*. UNM Press.

Dede, O. M., & Ayten, A. M. (2012). The Role of Spatial Planning for Sustainable Tourism Development: A Theoretical Model for Turkey. *Tourism: An International Interdisciplinary Journal*, *60*(4), 431–445.

Dégh, L. (2020). The "Belief Legend" in Modern Society: Form, Function, and Relationship to Other Genres. In American Folk Legend (pp. 55-68). University of California Press.

Demirkol, Ö., Yıldırım, M., & Gökkaya, S. (2015). *Recent Developments in Turkish Tourism Policy within the context of Accession to the European Union* [Paper presentation]. *2nd International Congress of Tourism & Management Researches*.

Denes, A., Koanantakool, P. C., Davis, P., Kreps, C., Hennessy, K., Alivizatou, M., & Stefano, M. L. (2013). Critical reflections on safeguarding culture: The intangible cultural heritage and museums field school in Lamphun, Thailand. *Heritage & Society*, *6*(1), 4–23. doi:10.1179/2159032X13Z.0000000004

Departamento Administrativo Nacional de Estadística - DANE. (2021). Empleo—Economía Naranja. *DANE - INformación para todos*. https://www.dane.gov.co/files/investigaciones/pib/sateli_cultura/economia-naranja/presentacion-rp-4to-reporte-economia-naranja.pdf

Derrett, R. (2003). Festivals & regional destinations: How festivals demonstrate a sense of community & place. *Rural Society*, *13*(1), 35–53. doi:10.5172/rsj.351.13.1.35

Destimed Deliverable Database. (2020). Retrieved from Destimed: https://destimed.interreg-med.eu/what-we-achieve/deliverable-database/

Destination Røros. (n.d.). *Information about Røros*. Destination Røros. https://www.roros.no/en/ information/

Destinations, G. (2020). Retrieved from Green Destinations Standard: http://greendestinations.org/wp-content/uploads/2017/11/Green-Destinations-Standard-1.4.2.pdf

Devereux, C. (2003). Spirituality, pilgrimage and the road to Santiago: questions for cultural tourism. In *Religious Tourism and Pilgrimage*. Tourism Board Of Leiria/Fátima.

DHA. (2021). *UNESCO'nun 17 Kültürel Mirasında Foto Safari Başlıyor* [Photo Safari Starts in 17 Cultural Heritage of UNESCO]. https://www.cnnturk.com/seyahat/unesconun-17-kulturel-mirasinda-foto-safari-basliyor

Di Giovine, M. A., & Choe, J. (2019). Geographies of religion and spirituality: Pilgrimage beyond the 'officially' sacred. *Tourism Geographies*, *21*(3), 361–383. doi:10.1080/14616688.2019.1625072

Di Giovine, M., & Elsner, J. (2016). Pilgrimage tourism. In J. Jafari & H. Xiao (Eds.), *The Encyclopedia of Tourism* (pp. 722–724). Springer., doi:10.1007/978-3-319-01384-8_146

Dias, J., Correia, A., & Cascais, T. (2017). Traits in Tourists' Experiences: Senses, Emotions and Memories. In A. Correia, M. Kozak, J. Gnoth, & A. Fyall (Eds.), *Co-Creation and Well-Being in Tourism* (pp. 179–194). Springer International Publishing., doi:10.1007/978-3-319-44108-5_14

Díaz Zarco, V., & Castro Morales, F. (2017). El turismo emocional creativo: La ciudad como taller inclusivo. *International Journal of Scientific Management and Tourism, 3*(1), 357–371.

Díaz, M. G., Ortíz, P., & Núñez, I. (2004). *Interculturalidad, saberes campesinos y educación. Un debate con la diversidad cultural.* El Colegio de Tlaxcala, A. C.

Dimitrovski, D., Ioannides, D., & Nikolaou, K. (2020). The tourist-pilgrim continuum in consumer behaviour: The case of international visitors to Kykkos Monastery, Cyprus. *Journal of Tourism and Cultural Change*, 1–19.

Dixit, S. K. (2019). The Routledge handbook of gastronomic tourism (S. K. Dixit, Ed.). Taylor & Francis Group. doi:10.1080/13032917.2019.1643641

Dixon-Woods, M., Cavers, D., Agarwal, S., Annandale, E., Arthur, A., Harvey, J., Hsu, R., Katbamna, S., Olsen, R., Smith, L., Riley, R., & Sutton, A. J. (2006). Conducting a critical interpretive synthesis of the literature on access to healthcare by vulnerable groups. *BMC Medical Research Methodology, 6*(1), 35. https://doi.org/10.1186/1471-2288-6-35

Djellal, F., & Gallouj, F. (2010). The innovation gap and the performance gap in the service economies: A problem for public policy. In F. Gallouj & F. Djellal (Eds.), *The Handbook of Innovation and Services* (pp. 653–675). Edward Elgar Publishing. doi:10.4337/9781849803304.00041

D-maps. (2021). https://d-maps.com/pays.php?num_pay=214&lang=pt

Domínguez-Quintero, A. M., González-Rodríguez, M. R., & Roldán, J. L. (2019). The role of authenticity, experience quality, emotions, and satisfaction in a cultural heritage destination. *Journal of Heritage Tourism, 14*(5/6), 491–505. doi:10.1080/1743873X.2018.1554666

Dragouni, M. (2017). *Sustainable heritage tourism: Towards a community-led approach.* Academic Press.

du Cros, H. (2001). A new model to assist in planning for sustainable cultural heritage tourism. *International Journal of Tourism Research, 3*(2), 165–170. https://doi.org/10.1002/jtr.297

Duncan, O. D. (1961). From Social System to Ecosystem. *Sociological Inquiry, 31*(2), 140–149. doi:10.1111/j.1475-682X.1961.tb00518.x

Durovic, M., & Lovrentjev, S. (2014). Indicators of sustainability in cultural tourism. *The Macrotheme Review, 3*(7), 180–189. https://macrotheme.com/yahoo_site_admin/assets/docs/19MR37mj.202135138.pdf

DuVal, L. (2001). Historic Route 66 is still kicking in Albuquerque. *Arizona Republic, 17*(June), T3.

Duxbury, N., Silva, S., & Vinagre de Castro, T. (2019). Creative Tourism Development in Small Cities and Rural Areas in Portugal: Insights from Start-Up Activities. In *Creating and Managing Experiences in Cultural Tourism* (pp. 291-304). World Scientific Publishing. https://www.worldscientific.com/doi/abs/10.1142/9789813233683_0018

Duxbury, N. (Ed.). (2021). *Cultural Sustainability, Tourism and Development. (Re)articulations in Tourism Contexts.* Routledge. doi:10.4324/9780367201777

Duxbury, N., Bakas, F. E., Vinagre de Castro, T., & Silva, S. (2020). Creative Tourism Development Models towards Sustainable and Regenerative Tourism. *Sustainability, 13*(1), 2. Advance online publication. doi:10.3390u13010002

Dybdahl, L. M. (2019). Business model innovation for sustainability through localism. In N. Bocken, P. Ritala, L. Albareda, & R. Verburg (Eds.), *Innovation for Sustainability. Business Transformations Towards a Better World* (pp. 193–211). Palgrave Macmillan. doi:10.1007/978-3-319-97385-2_11

Ebony. (1990). Georgia's Black History Trail. *Ebony, 45*(4), 167-172.

Eckert, C., Zacher, D., Pechlaner, H., Namberger, P., & Schmude, J. (2019). Strategies and measures directed towards overtourism: A perspective of European DMOs. *International Journal of Tourism Cities*, *5*(4), 639–655. doi:10.1108/IJTC-12-2018-0102

Edgell, D. L. (2019). *Managing Sustainable Tourism: A Legacy for the Future*. Routledge. doi:10.4324/9780429318122

Edwards, J., & Kaimal, G. (2016). Using meta-synthesis to support application of qualitative methods findings in practice: A discussion of meta-ethnography, narrative synthesis, and critical interpretive synthesis: *The Arts in Psychotherapy, 51*. doi:10.1016/j.aip.2016.07.003

Eide, D. (2011). *Turisme basert på fortellinger, bøker og film: en litteratur og case studie*. UiN-rapport nr. 2-2011. University in Nordland.

Eide, D., & Hoarau-Heemstra, H. (2022). Innovation for sustainable destinations: The role of certification and partnership. In I. Booyens & O. Brouder (Eds.), Handbook of Innovation for Sustainable Tourism. Edward Elgar Publishing.

Eide, D., & Mossberg, L. (2019). Toward a framework of experience quality assessment in cultural tourism. In D. A. Jelinčić & Y. Mansfeld (Eds.), *Creating and Managing Experiences in Cultural Tourism* (pp. 101–120). World Scientific. doi:10.1142/9789813233683_0007

Ekman, P. (1992). Are there basic emotions? *Psychological Review*, *99*(3), 550–553. doi:10.1037/0033-295X.99.3.550 PMID:1344638

Elcacho, J. (2020). Efecto secundario de la Pandemia: Canales de Venecia con aguas limpias y peces por el confinamiento del Covid-19. *La Vanguardia - Natural*. https://www.lavanguardia.com/natural/20200317/474234537648/canales-de-venecia-con-aguas-limpias-y-peces-por-el-confinamiento-del-covid-19.html

Elden. (2021). *About Elden*. https://elden-roros.no/utendorsteateret/historiene/

Elías, S. R., & Tuma, C. M. (2009). Turismo y teoría de clusters, las potencialidades de la localidad de Monte Hermoso. *Aportes y Transferencias, 13*(1), 67-92. http://nulan.mdp.edu.ar/1546/

Elias, S., & Barbero, A. C. (2021). Social innovation in a tourist coastal city: A case study in Argentina. *Social Enterprise Journal*, *17*(1), 44–62. doi:10.1108/SEJ-02-2020-0011

Ellis, C. (1997). Evocative autoethnography: Writing emotionally about our lives. In W. G. Tierney & Y. S. Lincoln (Eds.), *Representation and the Text: Re-Framing the Narrative Voice* (pp. 116–139). State University of New York Press.

Ellis, C., & Bochner, A. (2006). Analyzing analytic autoethnography: An autopsy. *Journal of Contemporary Ethnography*, *35*(4), 429–449. doi:10.1177/0891241606286979

Elsie, R. (2004). *Tales from Old Shkodra*. Dukagjini.

Elsie, R. (2005). *Albanian Literature: A Short History*. Palgrave.

Emblematic Deliverable Library. (2020). Retrieved from Emblematic: https://emblematic.interreg-med.eu/what-we-achieve/deliverable-library/

Emekli, G. (2005). Avrupa Birliği'nde Turizm Politikaları ve Türkiye'de Kültürel Turizm [The Policies of Tourism in European Union and Cultural Tourism in Türkiye]. *Ege Coğrafya Dergisi*, *14*, 99–107.

Empacher, C., & Wehlung, P. (2002). *Soziale Dimensionen der Nachhaltigkeit*. ISOE-Studientexte.

Engelmann, J., & Al-Sadi, M. (2019). Business-Driven Ecological Innovations in Green Growth Strategies. In N. Bocken, P. Ritala, L. Albareda, & R. Verburg (Eds.), *Innovation for Sustainability. Business Transformations Towards a Better World* (pp. 97–113). Palgrave Macmillan. doi:10.1007/978-3-319-97385-2_6

Epler Wood, M., Milstein, M., & Ahamed-Broadhurst, K. (2019). *Destinations at Risk: The Invisible Burden of Tourism*. The Travel Foundation. https://www.thetravelfoundation.org.uk/invisible-burden/

Erkoçi, I. (2016). Ismail Kadare as tourist attraction. *DosAlgarves. A Multidisciplinary e-Journal, 28*, 1-12.

Erkoçi, I. (2017). *Stats and Insights*. Mapping Literary Albania. https://wordpress.com/stats/day/mappingliteraryalbania.wordpress.com?startDate=2021-09-19

Erkoçi. I. (2013). A modern-day journey in Byron's footsteps in Albania. *Academic Journal of Interdisciplinary Studies, 2*(4), 97-106.

Esparcia, J., Escribano, J., & Serrano, J. (2016). Una aproximación al enfoque del capital social y a su contribución al estudio de los procesos de desarrollo local. *Investigaciones Regionales—. Journal of Regional Research*, *34*, 49–71.

Espeso-Molinero, P. (2018). Collaborative capacity building as a resilience strategy for tourism development in indigenous Mexico. In J. Cheer & A. A. Lew (Eds.), *Tourism, Resilience, and Sustainability: Adapting to Social, Political and Economic Change* (pp. 184–201). Routledge.

Espinosa, R. A., & Ríos, S. (2017). El diario de campo como instrumento para lograr una práctica reflexiva. *Congreso Nacional de Investigación Educativa - COMIE*, 11.

EU. (2010). *Green Paper - Unlocking the potential of cultural and creative industries* (Vol. 148). EU.

EUROPARC Federation. (2020). Retrieved from European Charter for Sustainable Tourism: https://www.europarc.org/library/europarc-events-and-programmes/european-charter-for-sustainable-tourism/

European Commission. (1999). *European Territorial Strategy*. Author.

European Commission. (2013). The European Tourism Indicator System. Toolkit for Sustainable Destinations. *European Union.*

European Commission. (2016). *The European Tourism Indicator System. ETIS Toolkit for sustainable destination management*. European Commission.

European Commission. (2016). *The European Tourism Indicator System. ETIS Toolkit for sustainable destination management.* Luxembourg: European Union. Retrieved from https://ec.europa.eu/growth/sectors/tourism/offer/sustainable/indicators_en

European Commission. (2018). *A New European Agenda for Culture*. https://eur-lex.europa.eu/legal-content/EN/TXT/PDF/?uri=CELEX:52018DC0267&from=EN

Fals Borda, O. (1981). Investigación participativa y praxis rural nuevos conceptos en educación y desarrollo comunal. Mosca Azul Editores.

Fals, O. (2009). La investigación Acción en Convergencias Disciplinarias. *Revista Paca, 1*, 7–21.

Farid, S. M. (2015). Tourism Management in World Heritage Sites and its Impact on Economic Development in Mali and Ethiopia. *Procedia: Social and Behavioral Sciences, 211*, 595–604. https://doi.org/10.1016/j.sbspro.2015.11.078

Farrell, B. H., & Twining-Ward, L. (2004). Reconceptualizing tourism. *Annals of Tourism Research, 31*(2), 274–295. doi:10.1016/j.annals.2003.12.002

Feather, N. T. (1995). Values, valences, and choice: The influences of values on the perceived attractiveness and choice of alternatives. *Journal of Personality and Social Psychology, 68*(6), 1135–1151. doi:10.1037/0022-3514.68.6.1135

Fernández, L. (2006). Fichas para investigadores ¿Cómo analizar datos cualitativos? *Bulletí LaRecerca, 7*, 1–13.

Fernández, B. M. C., González, R. C. L., & Lopez, L. (2016). Historic city, tourism performance and development: The balance of social behaviours in the city of Santiago de Compostela (Spain). *Tourism and Hospitality Research, 16*(3), 282–293. doi:10.1177/1467358415578473

Ferrari, S., Navarro, V., Mazzoni, E., Albrieu, C., & Biott, J. (2020). Ruta nacional 40 en Santa Cruz (Patagonia, Argentina): Potencialidad turística y propuestas para su planificación responsable. *Revista DELOS, 13*(37), 409–430.

Ferreira, A., Alén, E., Liberato, P., & Liberato, D. (2019). Literary tourism: A cultural trip? *Advances in Tourism. Technology and Smart Systems, 171*, 505–515. doi:10.1007/978-981-15-2024-2_44

Fesenmaier, D. R., & Xiang, Z. (2017). Introduction to tourism design and design science in tourism. *Design Science in Tourism*, 3-16. doi:10.1007/978-3-319-42773-7_1

Fitchett, J., Lindberg, F., & Martin, D. M. (2021). Accumulation by symbolic dispossession: Tourism development in advanced capitalism. *Annals of Tourism Research, 86*, 103072. doi:10.1016/j.annals.2020.103072

Florida, R. (2003). *The Rise of the Creative Class: And How It's Transforming Work, Leisure, Community and Everyday Life.* University of Toronto Press on behalf of Canadian Public Policy. https://www.jstor.org/stable/3552294

Flutur, L., & Nedelea, A. (2018). Cultural Tourism. *Revista De Turism - Studii Si Cercetari In Turism, 0*(25). Retrieved from http://www.revistadeturism.ro/rdt/article/view/382/257

Fondo Mixto de Cultura de Boyacá. (2020). Tunja: Seminario Gobernanza y Patrimonio Cultural: El Papel de la Actividad Turística en el Desarrollo Local. *Fondo Mixto de Cultura de Boyacá.* https://www.fondocultura.org/eventos/?tribe_paged=1&tribe_event_display=list&tribe-bar-date=2020-09-01

Fontan, J. M., Klein, J.-L., Caillouette, J., Doyon, M., Lévesque, B., Tremblay, D.-G., Tremblay, P.-A., & Trudelle, C. (2014). Vers de nouveaux modèles d'action en développement territorial: L'expérimentation à l'échelle locale de la transition vers le «buen vivir». *Economie et Solidarites, 44*(1-2), 84–102.

Fornell, C., & Larcker, D. F. (1981). Evaluating Structural Equation Models with Unobservable Variables and Measurement Error. *JMR, Journal of Marketing Research, 18*(1), 39–50. doi:10.1177/002224378101800104

Forrest, N., & Wiek, A. (2014). Learning from success—Toward evidence-informed sustainability transitions in communities. *Environmental Innovation and Societal Transitions, 12*, 66–88. doi:10.1016/j.eist.2014.01.003

FRANC & IŞTOC. (2002). Cultural tourism and sustainable development. *Museum International, 54*(1–2), 126–130. doi:10.1111/1468-0033.00376

Freeman, L. (2004). *The Development of Social Network Analysis. A Study on the Sociology of Science.* Empirical Press.

Freire-Medeiros, B. (2007). The Favela that can be seen and sold: Reflections and Controversies around a Tourist Destination [A Favela que se vê e que se vende: Reflexões e polêmicas em torno de um destino turístico]. *Revista Brasileira de Ciências Sociais, 22.* doi:10.1590/S0102-69092007000300006

Freire-Medeiros, B. (2008). Selling the Favela: Thoughts and Polemics about a Tourist Destination. *Revista Brasileira de Ciências Sociais, 4*, 1–26. http://socialsciences.scielo.org/pdf/s_rbcsoc/v4nse/scs_01.pdf

Freire-Medeiros, B. (2009). *Gringo na Laje: Production, circulation and consumption of the tourist favela* [Gringo na laje: Produção, circulação e consumo da favela turística]. FGV.

Freire, P. (2005). *Pedagogía del Oprimido* (2nd ed.). Siglo XXI Editores.

Frenzel, F., Koens, K., Steinbrink, M., & Rogerson, C. M. (2015). Slum Tourism: State of the art. *Tourism Review International, 18*(4), 237–252. doi:10.3727/154427215X14230549904017

Fuglsang, L., & Eide, D. (2013). The experience turns as 'bandwagon': Understanding network formation and innovation as practice. *European Urban and Regional Studies, 20*(4), 417–434. doi:10.1177/0969776412448090

Fundação para a Ciência e a Tecnologia. (2019). *Agenda Temática de Investigação e InovaçãoTurismo, Lazer e Hospitalidade* [Thematic Agenda for Research and Innovation Tourism, Leisure and Hospitality]. FCT. . doi:10.34621/fct.edicoes.agendastematicas-2

Fyall, A., Garrod, B., & Wang, Y. (2012). Destination collaboration: A critical review of theoretical approaches to a multi-dimensional phenomenon. *Journal of Destination Marketing & Management, 1*(1-2), 10–26. doi:10.1016/j.jdmm.2012.10.002

Galí, N., Majó, J., & Vidal, D. (2000). Patrimonio Cultural Y Turismo: Nuevos Modelos De Promoción Vía Internet. *Cuadernos de Turismo, 6*(julio-diciembre), 73–88.

Galvani, A., Lew, A. A., & Perez, M. S. (2020). COVID-19 is expanding global consciousness and the sustainability of travel and tourism. *Tourism Geographies, 22*(3), 567–576. doi:10.1080/14616688.2020.1760924

Garcés Feliú, E. (2011). Cultural tourism in fragile institutional contexts: Project for cul-tural routes in Tierra del Fuego. *Journal of Tourism Research, 3*(2), 35–43.

Garg, P., & Pandey, A. (2021). Towards sustainable tourism: An empirical investigation. *Foresight, 23*(2), 188–200. doi:10.1108/FS-04-2020-0042

Garrido, F. (1996). *El análisis de redes en el desarrollo local.* http://pendientedemigracion.ucm.es/info/eurotheo/materiales/hismat/index.html

Gasparini, M. L., & Mariotti, A. (2021). Sustainable tourism indicators as policy making tools: Lessons from ETIS implementation at destination level. *Journal of Sustainable Tourism*, 1–19. doi:10.1080/09669582.2021.1968880

Gazzaniga, M. S., Ivry, R. B., & Mangun, G. R. (2014). *Cognitive Neuroscience: The Biology of the Mind* (A. Javsicas & S. Snavely, Eds.; 4th ed.). W. W. Norton & Company, Inc.

Geiger, B. F., & Werner, K. A.National Voting Rights Walk and Selma Antebellum Historic District. (2009). A guided walking trail to explore the Martin Luther King Jr. National Voting Rights Walk and Selma Antebellum Historical District. *International Journal of Heritage Studies, 15*(5), 467–476. doi:10.1080/13527250903072807

Getz, D. (1998). 24 Event tourism and the authenticity dilemma. In Global tourism (pp. 409-426). Reed Elsevier Plc Group.

Ghetau, L. & Esanu, L.V. (2011). Literary tourism as a promoter of cultural heritage. *Sustainable Development and Planning, 150.*

Giddens, A. (1984). Elements of the theory of structuration. In G. M. Spiegel (Ed.), *Practicing History* (pp. 121–142). Routledge.

Giddens, A., & Giddens, A. (1973). *The class structure of the advanced societies*. Hutchinson.

Giglitto, D. (2017). Community empowerment through the management of intangible cultural heritage in the Isle of Jura, Scotland. *Imperial Journal of Interdisciplinary Research*, *3*(5), 567–578.

Gil, A. C. (2002). *Como elaborar projetos de pesquisa*. Atlas.

Gilmore, A., Carson, D., & Ascenção, M. (2007). Sustainable Tourism Marketing at a World Heritage Site. *Journal of Strategic Marketing*, *15*(2-3), 253–264. doi:10.1080/09652540701318930

Global Sustainable Tourism Council. (2020). Retrieved from Destination Criteria: https://www.gstcouncil.org/wp-content/uploads/GSTC-Destination-Criteria-v2.0.pdf

Goeldner, C., & Ritchie, J. R. (2003). *Tourism, principles, practices, philisophies*. Academic Press.

Gómez-Carreto, T., Zarazúa Escobar, J. A., Guillen Cuevas, L. A. & Castellanos Albores, A. C. (2018). Innovación social, turismo rural y empresas sociales. Evidencias desde el Sur-Sureste de México. *El periplo sustentable, 34*, 44-81.

Gómez, D. (2007). *Ordenación territorial*. Mundi-Prensa.

Gómez, E., Vásquez, G., Betancur, V., Maartínez, D., Ocampo, M., Uribe, E., Soto, M., Ramírez, M., Jafeth, E., Karupia, W., Lectamo, J. N., Medina, B., Atehortúa, O., Valdés, D., & Arcos, A. (2015). *Diálogo de saberes e interculturalidad. Indígenas, afrocolombianos y campesinado en la ciudad de Medellín* (C. Gaviria & M. Orozco, Eds.). Pulso & Letra Editores.

Gómez, L. A. E., Lopez, L., Antelo, M. A. P., & González, R. M. L. (2021). The Timeline of The Way St James: Motivations and Impacts of the Pilgrimage on Personal Lives. In D. Liutikas (Ed.), *Pilgrims: Values and Identities* (pp. 72–82). CABI. doi:10.1079/9781789245653.0072

Gonçalves, A. R., Oliveira, A. P. D., & Santos, C. F. (2015). *Fragmentos para a história do turismo no Algarve* [Fragments for the history of tourism in the Algarve]. Retrieved from https://sapientia.ualg.pt/handle/10400.1/8743

González, A., & Ojeda, A. (2019). Los paisajes de terrazas en las Canarias Orientales (Lanzarote y Fuerteventura): Hacia una tipología actual. *Revista Española de Estudios Agrosociales y Pesqueros*, *253*, 65–92.

González, A., & Ramón, A. (2017). Análisis de la consideración científica e institucional de los paisajes de terrazas agrícolas en Canarias (España). *Revista Española de Estudios Agrosociales y Pesqueros*, *247*, 137–160.

González, I., Thomé, H., & Osorio, R. (2019). Apropiación turística de territorios indígenas: Cambios y continuidades en una comunidad Matlatzinca. *Turismo e Sociedade*, *25*(junio), 113–120. doi:10.18601/01207555.n25.06

Goodey, B. (1975). Urban trails: Origins and opportunities. *The Planner*, *61*, 29–30.

Gössling, S., Scott, D., & Hall, C. M. (2020). Pandemics, tourism and global change: A rapid assessment of COVID-19. *Journal of Sustainable Tourism*, *29*(1), 1–20. doi:10.1080/09669582.2020.1758708

Gouëset, V. (1999). El territorio colombiano y sus márgenes. La difícil tarea de la construcción territorial. *Territorios*, *1*(Enero), 77–94.

Gouthro, M. B. (2010). Qualitative method research and the 'tourism experience': a methodological perspective applied in a heritage setting. In *Tourist Experience* (pp. 221–234). Routledge.

Government of Andalusia. (2006). *Plan de Ordenación del Territorio de Andalucía. Decreto 206/2006, de 28 de noviembre*. https://www.juntadeandalucia.es/export/drupaljda/pota_completo.pdf

Graf, M. (2013). Industrial Cultural Routes. Journal "Energy, Economy, Ecology", 15(2), 328-335.

Grashof, C., & Van Langevelde, F. (2005). Green Veining: Landscape Determinants of Biodiversity in European Agricultural Landscapes. *Landscape Ecology*, *20*(4), 417–439. doi:10.100710980-004-5646-1

Graves, T. D. (1967). Psychological acculturation in a tri-ethnic community. *Southwestern Journal of Anthropology*, *23*(4), 337–350. doi:10.1086outjanth.23.4.3629450

Green Destinations. (2020). Retrieved from Green Destinations Standard: http://greendestinations.org/wp-content/uploads/2017/11/Green-Destinations-Standard-1.4.2.pdf

Grierson, D. (2009). The shift from a mechanistic to an ecological paradigm. *International Journal of Environmental, Cultural, Economic and Social Sustainability*, *5*(5), 197–206. doi:10.18848/1832-2077/CGP/v05i05/54670

Gruen, R., Elliott, J., Nolan, M., Lawton, P., Parkhill, A., McLaren, C., & Lavis, J. N. (2008). Sustainability science: An integrated approach for health-program planning. *Lancet*, *372*(9649), 1579–1589. doi:10.1016/S0140-6736(08)61659-1 PMID:18984192

GSTC. (2020). *Global Sustainable Tourism Council*. Retrieved from Destination Criteria: https://www.gstcouncil.org/wp-content/uploads/GSTC-Destination-Criteria-v2.0.pdf

Guba, E. G., & Lincoln, Y. S. (1994). Competing paradigms in qualitative research. In *Handbook of qualitative research* (pp. 105–117). Sage Publications, Inc.

Guber, R. (2004). *El salvaje metropolitano. Reconstrucción del conocimiento social en el trabajo de campo*. Paidós. www.paidosargentina.com.ar

Guerreiro, M. M., Henriques, C., & Mendes, J. (2019). Cultural and Creative Tourism: The Case of 'Celebrations' in the Algarve Region. *Journal of Spatial and Organizational Dynamics*, *VII*(4), 320–338.

Guilarte, Y. P., & González, R. C. L. (2016). El Perfil Del Visitante En Santiago De Compostela: Tradición Y Actualidad. *Cuadernos de Turismo*, *37*(37), 305–322. doi:10.6018/turismo.37.256251

Gümüşçü, O. (2018). Tarihi Coğrafya ve Kültürel Miras [Historical Geography and Cultural Heritage]. *Erdem*, *75*, 99–120.

Guo, X., Pikas, B., & Lee, T. (2017). Study of Intangible Chinese Cultural Heritage Protection and Tourism Development. The Case of Intangible Cultural Heritage Protection of the Hezhe Nationality. *Journal of Marketing Development and Competitiveness*, *11*(2), 39–45.

Gurira, N. A., & Ngulube, P. (2016). Using Contingency Valuation Approaches to Assess Sustainable Cultural Heritage Tourism Use and Conservation of the Outstanding Universal Values (OUV) at Great Zimbabwe World Heritage Site in Zimbabwe. *Procedia: Social and Behavioral Sciences*, *225*, 291–302. https://doi.org/10.1016/j.sbspro.2016.06.028

Gusmán Correia, M., Lopez, L., Lois González, R., & Santos, X. (2017). The Challenges of the First European Cultural Itinerary: The Way to St. James. *Almatourism - Journal of Tourism, Culture and Territorial Development*, *8*(6), 1–19.

Güzel, F. Ö. (2014). The Dimensions of Tour Experience, Emotional Arousal, and Post-experience Behaviors: A Research on Pamukkale in Turkey. *Procedia: Social and Behavioral Sciences*, *150*, 521–530. doi:10.1016/j.sbspro.2014.09.069

Haase, J., Wiedmann, K. P., & Bettels, J. (2020). Sensory imagery in advertising: How the senses affect perceived product design and consumer attitude. *Journal of Marketing Communications*, *26*(5), 475–487. doi:10.1080/13527266.2018.1518257

Hair, J., Hult, G., Ringle, C., & Sarstedt, M. (2017). *A Primer on Partial Least Squares Structural Equation Modeling (PLS-SEM)* (2nd ed.). Sage Publications., doi:10.1016/j.lrp.2013.01.002

Hair, J., Risher, J., Sarstedt, M., & Ringle, C. (2019). When to use and how to report the results of PLS-SEM. *European Business Review*, *31*(1), 2–24. doi:10.1108/EBR-11-2018-0203

Halewood, C., & Hannam, K. (2001). Viking Heritage Tourism: Authenticity and Commodification. *Annals of Tourism Research*, *28*(3), 565–580. doi:10.1016/S0160-7383(00)00076-1

Hall, C. M. (2011). A typology of governance and its implications for tourism policy analysis. *Journal of Sustainable Tourism*, *19*(4-5), 437–457. doi:10.1080/09669582.2011.570346

Hall, C. M. (2018). Resilience theory and tourism. In J. Saarinen & A. M. Gill (Eds.), *Resilient Destinations and Tourism* (pp. 34–47). Routledge. doi:10.4324/9781315162157-3

Hall, S. (2002). *A identidade cultural na pós-modernidade*. Editora DP&A.

Hamimi Omar. (2013). The Development of Sustainable Cultural Heritage Tourism in Malaysia: Implication for Planning and Management. *Newcascle Thesis*. https://theses.ncl.ac.uk/dspace/bitstream/10443/2121/1/Omar,H.13.pdf

Hanneman, R., & Riddle, M. (2005). *Introduction to social network methods*. University of California.

Harapi, A. (1995). *Valë mbi valë* [Wave after Wave]. Academic Press.

Harmon, D. (2007). A Bridge over the Chasm: Finding Ways to Achieve Integrated Natural and Cultural Heritage Conservation. *International Journal of Heritage Studies*, *13*(4-5), 380–392. doi:10.1080/13527250701351098

Harwood, S., & Eaves, S. (2017). In search of an autoethnographic method. *16th European Conference on Research Methodology for Business and Management Studies*, 145-153, https://www.pure.ed.ac.uk/ws/portalfiles/portal/33004807/ECRM_FINAL_REVdoc x.pdf

Hasani, A., Moghavvemi, S., & Hamzah, A. (2016). The impact of emotional solidarity on residents' attitude and tourism development. *PLoS One*, *11*(6), 1–14. doi:10.1371/journal.pone.0157624 PMID:27341569

Hatipoglu, B., Alvarez, M. D., & Ertuna, B. (2016). Barriers to stakeholder involvement in the planning of sustainable tourism: The case of the Thrace region in Turkey. *Journal of Cleaner Production*, *111*, 306–317. doi:10.1016/j.jclepro.2014.11.059

Hauptfleisch, T. (2006). Eventifying identity: Festivals in South Africa and the search for cultural identity. *NTQ. New Theatre Quarterly*, *22*(2), 181–198. doi:10.1017/S0266464X0600039X

Hawkes, J. (2001). *The fourth pillar of sustainability: Culture's essential role in public planning*. Common Ground Publishing Pty Ltd and Cultural Development Network Victoria.

Haxeltine, A., Avelino, F., Pel, B., Dumitru, A., Kemp, R., Longhurst, N., . . . Wittmayer, J. M. (2016). A framework for transformative social innovation. *TRANSIT Working Paper, 5*, 2-1.

Hayes, D., & MacLeod, N. (2007). Packaging places: Designing heritage trails using an experience economy perspective to maximize visitor engagement. *Journal of Vacation Marketing*, *13*(1), 45–58. doi:10.1177/1356766706071205

Helgadóttir, G., Einarsdóttir, A. V., Burns, G. L., Gunnarsdóttir, G. Þ., & Matthíasdóttir, J. M. E. (2019). Social sustainability of tourism in Iceland: A qualitative inquiry. *Scandinavian Journal of Hospitality and Tourism*, *19*(4-5), 4–5, 404–421. doi:10.1080/15022250.2019.1696699

Henriques.C. (2019). Turismo Cultural como vetor de Sustenabilidade nas Cidades [Cultural Tourism as a vector of Sustainability in Cities]. *Journal of Tourism and Heritage Research, 2,* 327–352.

Henriques, C. (2014). Algarve Turismo Cultural. *Revista Turismo Em Análise*, *25*(2), 497. doi:10.11606/issn.1984-4867.v25i2p497-499

Herbert, D. (2001). Literary places, tourism and the heritage experience. *Annals of Tourism Research*, *28*(2), 312–333. doi:10.1016/S0160-7383(00)00048-7

Hernández, O., Martínez, E., Perdigón, A., Cebrián, J., & Rodríguez, E. (2019). La gestión de los territorios de bancales (caso experimental La Gomera). In *V Congreso Mundial ITLA Re-encantar Bancales* (p. 21). Gobierno de Canarias, Canarias Cultura en Red, Observatorio del Paisaje de Canarias.

Hernández, R., Fernández, C., & Baptista, P. (2010). Metodología de la investigación (5a ed.). McGraw-Hill.

Hernández, R. M. (2014). La investigación cualitativa a través de entrevistas: Su análisis mediante la teoría fundamentada. *Cuestiones Pedagógicas*, *23*, 187–210.

Heslinga, J., Groote, P., & Vanclay, F. (2019). Strengthening governance processes to improve benefit sharing from tourism in protected areas by using stakeholder analysis. *Journal of Sustainable Tourism*, *27*(6), 773–787. doi:10.1080/09669582.2017.1408635

He, X. L. (2007). *Religlionary Faith and Ethnic Culture (I)*. Social Sciences Academic Press.

Hezri, A. A., & Dovers, S. R. (2006). Sustainability indicators, policy and governance: Issues for ecological economics. *Ecological Economics*, *60*(1), 86–99. doi:10.1016/j.ecolecon.2005.11.019

Hidayat, D., & Stoecker, R. (2021). Collective Knowledge Mobilization Through a Community University Partnership. *Journal of Higher Education Outreach & Engagement*, *25*(2).

Higgins-Desbiolles, F. (2020). Socialising tourism for social and ecological justice after COVID-19′. *Tourism Geographies*, *22*(3), 610–623. doi:10.1080/14616688.2020.1757748

Hillman, W., & Radel, K. (2018). *Qualitative Methods in Tourism Research*. Channel View Publications.

Hinojosa, V. (2020). El turismo creativo, solución para reactivar el sector durante la pandemia. *Hosteltur*. https://www.hosteltur.com/139045_el-turismo-creativo-solucion-para-reactivar-el-sector-durante-la-pandemia.html?%20shared&utm_source=twitter&utm_medium=social-web&utm_campaign=139045_el-turismo-creativo-solucion-para-reactivar-el-sector-durante-lapandemia&%20utm_content=economia&utm_term=noticia

Hitrec, T. (1996). Kulturni itinereri: Važan segment europske turistčke ponude. *Turizam*, *44*(3), 47–60.

Hjalager, A. M. (2010). A review of the innovation research in tourism. *Tourism Management*, *31*(1), 1–12. doi:10.1016/j.tourman.2009.08.012

Høegh-Gulberg, O., Eide, D., & Yati, Y. (2022). Towards sustainable tourism through lab-driven innovations: a systematic literature review. In I. Booyens & O. Brouder (Eds.), Handbook of Innovation for Sustainable Tourism. Edward Elgar Publishing.

Høegh-Guldberg, O., Seeler, S., & Eide, D. (2021). Sustainable Visitor Management to Mitigate Overtourism – What, Who, and How. In A. Sharma & H. Azizul (Eds.), *Over-tourism as Destination Risk: Impacts and Solutions* (pp. 167–186). Emerald Publishing Limited. doi:10.1108/978-1-83909-706-520211012

Holgado, D. (2016). Analyzing Social Networks. *Revista Hispana para el Análisis de Redes Sociales*, *27*(2), 141–145. doi:10.5565/rev/redes.637

Holling, C. S., & Goldberg, M. A. (1971). Ecology and Planning. *Journal of the American Institute of Planners*, *37*(4), 221–230. doi:10.1080/01944367108977962

Holman Jones, S. (2013). Introduction: Coming to know autoethnography as more than a method. In Handbook of Autoethnography. Left Coast Press.

Hong Kong Free Press. (2021). *A gentrified mess: How Hong Kong's Central Market restoration tarnished an architectural gem.* https://hongkongfp.com/2021/09/11/a-gentrified-mess-how-hong-kongs-central-market-restoration-tarnished-an-architectural-gem/

Hoppen, A., Brown, L., & Fyall, A. (2014). Literary tourism: Opportunities and challenges for the marketing and branding of destinations? *Journal of Destination Marketing & Management*, *3*(1), 37–47. doi:10.1016/j.jdmm.2013.12.009

Hosany, S., Prayag, G., Deesilatham, S., Caušević, S., & Odeh, K. (2015). Measuring Tourists' Emotional Experiences: Further Validation of the Destination Emotion Scale. *Journal of Travel Research*, *54*(4), 482–495. doi:10.1177/0047287514522878

Hosany, S., Prayag, G., Van Der Veen, R., Huang, S., & Deesilatham, S. (2017). Mediating Effects of Place Attachment and Satisfaction on the Relationship between Tourists' Emotions and Intention to Recommend. *Journal of Travel Research*, *56*(8), 1079–1093. doi:10.1177/0047287516678088

Hospers, G. J. (2002). Industrial heritage tourism and regional restructuring in the European Union. *European Planning Studies*, *10*(3), 397–404. doi:10.1080/09654310220121112

HOSTELTUR. (2020a). *Creative Tourism Network celebra el día del turismo creativo, naranja y de proximidad.* https://www.hosteltur.com/comunidad/nota/022336_el-creativetourismnetwork-celebrara-el-dia-del-turismo-creativo.html

HOSTELTUR. (2020b). *El turismo poscoronavirus: Hacia un «low cost» de precios intermedio.* HOSTELTUR. https://www.hosteltur.com/lat/136500_el-turismo-poscoronavirus-hacia-un-low-cost-de-precios-intermedio.html

Hristov, D., Scott, N., & Michona, S. (2018). Distributed leadership typologies in destination management organisations. *Tourism Management Perspectives*, *28*, 126–143. doi:10.1016/j.tmp.2018.08.003

Hughes, E., & Scheyvens, R. (2016). Corporate social responsibility in tourism post-2015. A Development First approach. *Tourism Geographies*, *18*(5), 469–482. doi:10.1080/14616688.2016.1208678

Hughes, S., Yordi, S., & Besco, L. (2020). The role of pilot projects in urban climate change policy innovation. *Policy Studies Journal: the Journal of the Policy Studies Organization*, *48*(2), 271–297. doi:10.1111/psj.12288

Huibin, X., Marzuki, A., & Razak, A. A. (2012). Protective Development of Cultural Heritage Tourism: The case of Lijiang, China. *Theoretical and Empirical Researches in Urban Management*, *7*(1), 39–54.

Huibin, X., Marzuki, A., & Razak, A. A. (2013). Conceptualizing a sustainable development model for cultural heritage tourism in Asia. *Theoretical and Empirical Researches in Urban Management*, *8*(1), 51–66.

Hunt, C. A., Durham, W. H., Driscoll, L., & Honey, M. (2015). Can ecotourism deliver real economic, social, and environmental benefits? A study of the Osa Peninsula, Costa Rica. *Journal of Sustainable Tourism*, *23*(3), 339–357. https://doi.org/10.1080/09669582.2014.965176

Hunter, C. (1997). Sustainable tourism as an adaptive paradigm. *Annals of Tourism Research*, *24*(4), 850–867. doi:10.1016/S0160-7383(97)00036-4

ICOMOS & WTO. (1993). *Tourism at World Heritage Cultural Sites: The Site Manager's Handbook* (2nd ed.). World Tourism Organization.

ICOMOS. (2008). *Charter On Cultural Routes*. https://www.icomos.org/images/DOCUMENTS/Charters/culturalroutes_e.pdf

ICOMOS. (2021). *International Charter for Cultural Heritage Tourism 2021: Reinforcing cultural heritage protection and community resilience through responsible and sustainable tourism management*. ICOMOS International Committee on Cultural Tourism.

Imon, S. S. (2017). Cultural heritage management under tourism pressure. *Worldwide Hospitality and Tourism Themes*, *9*(3), 335–348. doi:10.1108/WHATT-02-2017-0007

Ingold, T., Hornborg, A., Pálsson, G., Descola, P., Ellen, R. F., Howell, S., Rival, L., Hviding, E., Arhem, K., Hell, B., Knight, J., Papagaroufali, E., Nothnagel, D., Richards, P., & Ruivenkamp, G. (2001). Naturaleza y Sociedad. Perspectivas antropológicas. In P. Descola & G. Pálsson (Eds.), *Naturaleza y sociedad: perspectivas antropológicas*. Siglo XXI Editores.

INSETE. (2020). *Key figures of incoming tourism in Greece*. https://insete.gr/statistika-eiserxomenou-tourismou

Instituto Nacional de Estatística. (2021). *Anuário Estatístico da Região do Algarve*. https://www.ine.pt/xurl/pub/410492449

Instituto Nacional de Estatística. (2021). *População residente* [Resident population]. https://www.ine.pt/scripts/db_censos_2021.html

Interreg Europe. (2020). *Policy brief on Cultural Routes in Europe*. Retrieved from https://www.interregeurope.eu/fileadmin/user_upload/plp_uploads/policy_briefs/2019-03-27_PB_Cultural_routes_final.pdf

Ismagilova, G., Safiullin, L., & Gafurov, I. (2015). Using Historical Heritage as a Factor in Tourism Development. *Procedia: Social and Behavioral Sciences*, *188*(904), 157–162. https://doi.org/10.1016/j.sbspro.2015.03.355

Itchart, L., & Donati, J. I. (2014). Practicas culturales. In Textos Iniciales (3ra ed.). Academic Press.

Ivakhiv, A. (2003). Nature and self in new age peregrimage. *Culture and Religion*, *4*(1), 93–118. doi:10.1080/01438300302812

Jackson, S. A., Haider, W., & Elliot, T. (2003). Resolving inter-group conflict in winter recreation: Chilkoot Trail National Historic Site, British Columbia. *Journal for Nature Conservation*, *11*, 317–323.

Jakobsen, O. D. (2018). *Anarchism and ecological economics: A transformative approach to a sustainable future*. Routledge. doi:10.4324/9780429486876

Jakobsen, O., & Storsletten, V. M. L. (2019). Beyond the green shift–Ecological economics. In J. S. Methi, A. Sergeev, M. Bieńkowska, & B. Nikiforova (Eds.), *Borderology: Cross-disciplinary Insights from the Border Zone* (pp. 173–183). Springer. doi:10.1007/978-3-319-99392-8_13

James, P., Magee, L., Scerri, A., & Steger, M. (2015). *Urban Sustainability in Theory and Practice: Circles of Sustainability*. Routledge.

Janoušková, S., Hák, T., Nečas, V., & Moldan, B. (2019). Sustainable Development—A Poorly Communicated Concept by Mass Media. Another Challenge for SDGs? *Sustainability*, *11*(11), 3181. doi:10.3390u11113181

Jansen-Verbeke, M., Priestley, G. K., & Russo, A. P. (2008). *Cultural resources for tourism: patterns, processes, and policies*. Nova Science Publishers.

Jarvis, D., Stoeckl, N., & Liu, H. (2016). The impact of economic, social and environmental factors on trip satisfaction and the likelihood of visitors returning. *Tourism Management*, *52*, 1–18. doi:10.1016/j.tourman.2015.06.003

Jassam, O. (2019). Küresel Kültürel ve Doğal Mirası Tehdit Eden Risk Yönetimi [Risk Management Threatening the Global Cultural and Natural Heritage]. *5. Uluslararası Öğrenci Sempozyumu Bildiriler Kitabı -2- Din-Tarih-Sosyoloji*, 75-89.

Javaheri, H. (2011). Optimal tourism route in historical urban areas: A case study of Tehran's down town, Iran. *South Asian Journal of Tourism and Heritage*, *4*(1), 34–43.

Jelincic, D. A., & Senkic, M. (2017). Creating a Heritage Tourism Experience: The Power of the Senses. *Etnološka Tribina*, *47*(40), 109–126. doi:10.15378/1848-9540.2017.40.03

Jenkins, I., & Lund, K. A. (2019). *Literary Tourism: Theories, Practice and Case Studies*. CABI. doi:10.1079/9781786394590.0000

Jensen, R., & Lovrod, H. M. (2017). Pilgrimage, past and present. Norwegian perspectives with emphasis on Santiago de Compostela. *Cuadernos de Estudios Gallegos*, *64*(130), 467–489. doi:10.3989/ceg.2017.130.13

Jociles, M. I. (2018). La observación participante en el estudio etnográfico de las prácticas sociales. In Revista Colombiana de Antropologia (Vol. 54, pp. 121–150). Instituto Colombiano de Antropologia e Historia. doi:10.22380/2539472x.386

Johannesson, G. T., Ren, C., & van der Duim, R. (2015). *Tourism encounters and controversies: Ontological politics of tourism development*. Ashgate Publishing, Ltd.

Joijc, S. (2017). Sustainable Tourism in the Albanian Riviera-Cultural Routes Contibution in Sustainable Tourism in the Albanian Riviera. *European Journal of Economics and Business Studies*, *3*(1), 104–113. doi:10.26417/ejes.v7i1.p104-113

Jørgensen, M. T., Vorre, H. A., Sørensen, F., Fuglsang, L., Sundbo, J., & Jensen, J. F. (2021). Collective tourism social entrepreneurship: A means for community mobilization and social transformation. *Annals of Tourism Research*, *88*, 103171. doi:10.1016/j.annals.2021.103171

Justo, D. (2020). La drástica caída de la contaminación permite a la India volver a ver el Himalaya después de varias décadas. *SER - Contaminación*. https://cadenaser.com/ser/2020/04/09/internacional/1586428245_618888.html

Kadare, I. (1991). *Ardhja e Migjenit në Letërsinë Shqipe* [The Emergence of Migjeni in the Albanian Literature]. 8 Nëntori.

Kajan, E. (2014). Community perceptions to place attachment and tourism development in Finnish Lapland. *Tourism Geographies*, *16*(3), 490–511. doi:10.1080/14616688.2014.941916

Kalyoncuoğlu, Y. (2021). *Türkiye Salgın Koşullarına Rağmen 2020'de ve 2021'in İlk İki Ayında 17 Milyondan Fazla Turist Ağırladı* [Turkey has Received more than 17 Million the First Two Months of 2020 and 2021 Despite the Conditions of the Pandemic]. https://www.aa.com.tr/tr/turkiye/turkiye-salgin-kosullarina-ragmen-2020de-ve-2021in-ilk-iki-ayinda-17-milyondan-fazla-turist-agirladi/2213246

Kaminski, J., Benson, A. M., & Arnold, D. (2014). *Contemporary Issues in Cultural Heritage Tourism*. Routledge.

Kammen, C. (1993). Gateway to Freedom: A black history trail. *Heritage*, *9*(4), 16–21.

Kaplan, G. G. (2015). The global burden of IBD: from 2015 to 2025. *Nature Reviews. Gastroenterology & Hepatology*, *12*(12), 720–727. doi:10.1038/nrgastro.2015.150 PMID:26323879

Káposzta, J., Nagy, A., & Nagy, H. (2016). The impact of tourism development policy on the Hungarian regions. *Региональная икономика,* (1), 10-17.

Karamustafa, K., & Ulama, S. (2010). Measuring the seasonality in tourism with the comparison of different methods. *EuroMed Journal of Business*, *5*(2), 191–214. doi:10.1108/14502191011065509

Karamustafa, K., & Yılmaz, M. (2020). Destination Life Cycle: A Conceptual Approach Based on the Turkish Experience. *Journal of Tourism and Gastronomy Studies, Special Issue*, *8*(4), 439–451. doi:10.21325/jotags.2020.696

Karapınar, E. & Barakazı, M. (2017). Kültürel Miras Turizminin Sürdürülebilir Turizm Açısından Değerlendirilmesi [Evaluation of Cultural Heritage Tourism from Aspect of Sustainable Tourism: Göbeklitepe Ruins]. *Güncel Turizm Araştırmaları Dergisi, 1*, 5-18.

Kato, K., & Progano, R. N. (2017). Spiritual (walking) tourism as a foundation for sustainable destination development: Kumano-kodo pilgrimage, Wakayama, Japan. *Tourism Management Perspectives*, *24*, 243–251. doi:10.1016/j.tmp.2017.07.017

Katz, C. (1998). Whose nature, whose culture? Private productions of space and the preservation of nature. Remaking Reality: Nature at the End of the Millennium, 46.

Kerr, R. (2018). *Roads, Tourism and Cultural History*. Channel View Publications.

Khlaikaew, K. (2015). The Cultural Tourism Management under Context of World Heritage Sites: Stakeholders' Opinions between Luang Prabang Communities, Laos and Muang-kao Communities, Sukhothai, Thailand. *Procedia Economics and Finance, 23*(October), 1286–1295. doi:10.1016/s2212-5671(15)00563-8

Khovanova-Rubicondo, K. (2012). Impact of European Cultural Routes on SMEs' innovation and competitiveness – Study. Provisional Edition, Council of Europe.

Kilipiris, F. (2005). Sustainable tourism development and local community involvement. *Tourism and Hospitality Management*, *11*(2), 27–39. doi:10.20867/thm.11.2.3

Kim, M., & Thapa, B. (2018). Perceived value and flow experience: Application in a nature-based tourism context. *Journal of Destination Marketing and Management, 8*(August), 373–384. doi:10.1016/j.jdmm.2017.08.002

Kim, B., Kin, S. S., & King, B. (2016). The sacred and the profane: Identifying pilgrim traveler value orientations using means-end theory. *Tourism Management*, *56*, 142–155. doi:10.1016/j.tourman.2016.04.003

Kim, J.-H. (2010). Determining the factors affecting the memorable nature of travel experiences. *Journal of Travel & Tourism Marketing*, *27*(8), 780–796. doi:10.1080/10548408.2010.526897

Kim, J.-H., & Jang, S. (2016). Memory Retrieval of Cultural Event Experiences: Examining Internal and External Influences. *Journal of Travel Research*, *55*(3), 322–339. doi:10.1177/0047287514553058

Kim, J.-H., Ritchie, J., & McCormick, B. (2012). Development of a scale to measure memorable tourism experiences. *Journal of Travel Research*, *51*(1), 12–25. doi:10.1177/0047287510385467

Kim, J.-H., & Youn, H. (2017). How to Design and Deliver Stories about Tourism Destinations. *Journal of Travel Research*, *56*(6), 808–820. doi:10.1177/0047287516666720

Kim, S., Whitford, M., & Arcodia, C. (2019). Development of intangible cultural heritage as a sustainable tourism resource: The intangible cultural heritage practitioners' perspectives. *Journal of Heritage Tourism*, *14*(5/6), 422–435. doi:10.1080/1743873X.2018.1561703

Kivela, J., & Crotts, J. C. (2009). Understanding Travelers' Experiences of Gastronomy Through Etymology and Narration. *Journal of Hospitality & Tourism Research (Washington, D.C.)*, *33*(2), 161–192. doi:10.1177/1096348008329868

Klein, J.-L., Fontan, J. M., Harrisson, D. & Lévesque, B. (2009). *L'innovation sociale au Québec: un système d'innovation fondé sur la concertation.* Cahier du CRISES, Collection Études théoriques, n. ET0907.

Klein, J.-L., & Pitarch-Garrido, M. D. (2020). La innovación social. In *Manual de Economía Social*. Tirant lo Blanch.

Klewitz, J., & Hansen, E. G. (2014). Sustainability-oriented innovation of SMEs: A systematic review. *Journal of Cleaner Production*, *65*, 57–75. doi:10.1016/j.jclepro.2013.07.017

Knezevic Cvelbar, L., Antonucci, B., Cutrufo, N., Marongiu, L., Rodrigues, M., & Teoh, T. (2021). Research for TRAN Committee – Relaunching transport and tourism in the EU after COVID-19. Tourism sector, European Parliament, Policy Department for Structural and Cohesion Policies.

Knobloch, U., Robertson, K., & Aitken, R. (2017). Experience, Emotion, and Eudaimonia: A Consideration of Tourist Experiences and Well-being. *Journal of Travel Research, 56*(5), 651–662. doi:10.1177/0047287516650937

Knoke, D. (1990). *Political Networks. The Structural Perspective*. Cambridge University Press. doi:10.1017/CBO9780511527548

Knoke, D., & Kuklinski, J. (1982). *Network Analysis*. SAGE Publications.

Knoke, D., & Yang, S. (2008). *Social Network Analysis*. SAGE Publications. doi:10.4135/9781412985864

Koens, K., Postma, A., & Papp, B. (2018). Is overtourism overused? Understanding the impact of tourism in a city context. *Sustainability, 10*(12), 4384. doi:10.3390/su10124384

Koliqi, E. (1933). *Gjurmat e Stinve* [The Traces Left by the Seasons]. Academic Press.

Koliqi, E. (2001). *Pasqyrat e Narcizit*. Camaj-Pipa.

Koliqi, E. (2003). *Vepra, V*. Faik Konica.

Kombëtare e Turizmit, A. (n.d.). *Turizmi në shifra për vitin 2018* [Tourism in Albania in 2018]. https://akt.gov.al/wp-content/uploads/2019/06/STATISTIKA-TURIZMI-2018.pdf

Ko, T. (2005). Development of a tourism sustainability assessment procedure: A conceptual approach. *Tourism Management, 26*, 431–445. https://doi.org/10.1016/j.tourman.2003.12.003

Kouchi, A. N., Nezhad, M. Z., & Kiani, P. (2016). A study of the relationship between the growth in the number of Hajj pilgrims and economic growth in Saudi Arabia. *Journal of Hospitality and Tourism Management, 36*, 103–107. doi:10.1016/j.jhtm.2016.01.008

Krippendorf, J. (1982). Towards new tourism policies: The importance of environmental and sociocultural factors. *Tourism Management, 3*(3), 135–148. doi:10.1016/0261-5177(82)90063-2

Krishna, A. (2012). An integrative review of sensory marketing: Engaging the senses to affect perception, judgment and behavior. In Journal of Consumer Psychology (Vol. 22, Issue 3, pp. 332–351). doi:10.1016/j.jcps.2011.08.003

Krishna, A., & Schwarz, N. (2014). Sensory marketing, embodiment, and grounded cognition: A review and introduction. *Journal of Consumer Psychology, 24*(2), 159–168. doi:10.1016/j.jcps.2013.12.006

Kugapi, O., & Höckert, E. (2020). Affective entanglements with travelling mittens. *Tourism Geographies*, 1–18. doi:10.1080/14616688.2020.1801824

Kuhn, T. S. (1962). *The Structure of Scientific Revolutions*. University of Chicago Press.

Kulla, N. (2010). Shkodra, Mjeda dhe Koliqi. *Gazeta Shekulli*.

Kuuder, C. J., Adongo, R., & Abanga, J. (2012). The socio-cultural significance of the Kakube Festival of the Dagara of Nandom Traditional Area in Ghana. *Ghana Journal of Development Studies, 9*(2), 103–124. doi:10.4314/gjds.v9i2.6

Kuzior, A., Lyulyov, O., Pimonenko, T., Kwilinski, A., & Krawczyk, D. (2021). Post-Industrial Tourism as a Driver of Sustainable Development. *Sustainability, 13*(15), 8145.

Kwan, C. W. K. (2019). The impact of festival participants' attitude on the sustainability of traditional festivals: A case study of the Feast of the Drunken Dragon and the Lion Dance Gala of Macao. In *Proceedings of the Tourism Education Student Summit, Macao SAR, China* (pp. 162-168). Macao Institute for Tourism Studies, Macao SAR, China.

Kwan, C. W. K. (2021). *The Living Drunken Dragon: Macao's Fishmonger Community and the Feast of the Drunken Dragon and the Lion Dance Gala* [Unpublished master's dissertation]. The University of Hong Kong, Pokfulam, Hong Kong SAR, China.

Labadi, S. (2013). *UNESCO, Cultural Heritage, and Outstanding Universal Value: Value-based Analyses of the World Heritage and Intangible Cultural Heritage Conventions*. Rowman & Littlefield.

Ladkin, A., & Martinez Bertramini, A. (2002). Collaborative tourism planning: A case study of Cusco, Peru. *Current Issues in Tourism, 5*(2), 71-93. doi:10.1080/13683500208667909

Laing, J. (2018). Festival and event tourism research: Current and future perspectives. *Tourism Management Perspectives, 25*, 165–168. doi:10.1016/j.tmp.2017.11.024

Larson, C. (2020). Cierres por coronavirus dejan cielos más despejados en China. *APNews*. https://apnews.com/article/e09cc102eb67432fa0993c10381a8951

Lasanta, T., Arnáez, J., Ruiz, P., & Ortigosa, L. (2011). Los bancales en la montaña mediterránea. Un paisaje multifuncional en proceso de degradación. In Territorio, paisaje y patrimonio rural (pp. 198-209). Universidad de Extremadura.

Lasanta, T., Arnaéz, J., Ruiz, P., & Lana, N. (2013). Los bancales en las montañas españolas: Un paisaje abandonado y un recurso potencial. *Boletín de la Asociación de Geógrafos Españoles, 63*, 299–320.

Latour, B. (1996). On actor-network theory: A few clarifications. *Soziale Welt, 4*, 369–381.

Laumann, E., Marsden, P., & Prensky, D. (1983). The Boundary specification problem in network analysis. In R. Burt & M. Minor (Eds.), *Applied Network Analysis: A methodological introduction* (pp. 18–34). SAGE Publications.

Leach, M., Rockström, J., Raskin, P., Scoones, I., Stirling, A., Smith, A., Thompson, J., Millstone, E., Ely, A., Arond, E., Folke, C., Olsson, P., Scoones, I., Stirling, A., Smith, A., Thompson, J., & Olsson, P. (2012). Transforming innovation for sustainability. *Ecology and Society, 17*(2), art11. doi:10.5751/ES-04933-170211

Ledesma, O., & Simancas, M. (2020). El valor de los datos relacionales en la gestión de destinos turísticos. Una aplicación con Análisis de Redes Sociales (ARS). In M. Simancas and P. Peñarrubia (Coords.), El valor de los datos turísticos (pp. 263-286). Tirant Humanidades.

Ledesma, O. (2012). Actores, gobierno en red y destinos turísticos. El estado del arte. *Turismo. Revista de Estudios de Turismo de Canarias y Macaronesia, 3-4*, 23–32.

Ledesma, O., Merinero, R., & Pulido, J. I. (2021). Tourist destination development and social network analysis: What does degree centrality contribute? *International Journal of Tourism Research, 23*(4), 652–666. doi:10.1002/jtr.2432

Lee, T. H., & Chang, P. S. (2017). Examining the relationships among festivalscape, experiences, and identity: Evidence from two Taiwanese aboriginal festivals. *Leisure Studies, 36*(4), 453–467. doi:10.1080/02614367.2016.1190857

Lee, T. H., & Jan, F. (2019). Can community-based tourism contribute to sustainable development? Evidence from residents' perceptions of the sustainability. *Tourism Management, 70*, 368–380. doi:10.1016/j.tourman.2018.09.003

Lehtonen, M., Sébastien, L., & Bauler, T. (2016). The multiple roles of sustainability indicators in informational governance: Between intended use and unanticipated influence. *Current Opinion in Environmental Sustainability, 18*, 1–9. doi:10.1016/j.cosust.2015.05.009

Leopold, A. (1933). The conservation ethic. *Journal of Forestry, 31*(6), 634–643.

Levalle, S. (2013). El pensamiento latinoamericano puesto en práctica: descolonización y construcción dialógica del conocimiento en las experiencias pedagógico-investigativas del Centro Indígena de Investigaciones Interculturales de Tierradentro de Colombia. *X Jornadas de Sociología*, 26.

Lévesque, B., Fontan, J. M. & Klein, J.-L. (2014). *L'innovation sociale: les marches d'une construction théorique et pratique*. Presses de l'Université du Québec.

Lewicka, M., Rowiński, K., Iwańczak, B., Bałaj, B., Kula, A. M., Oleksy, T., Prusik, M., Toruńczyk-Ruiz, S., & Wnuk, A. (2019). On the essentialism of places: Between conservative and progressive meanings. *Journal of Environmental Psychology*, *65*, 101318. doi:10.1016/j.jenvp.2019.101318

Li, Y., & Liu, X. (2019). "Library + Tourism" A new direction for the sustainable development of libraries. *Recruiting and Managing the New Generation of Employees to Attract New Markets and Create New Services*. http://library.ifla.org/id/eprint/2703/1/S09-2019-li-en.pdf

Lickorish, L. J., & Jenkins, C. L. (1997). *An Introduction to tourism*. Butterworth-Heinemann.

Liening, A. (2013). The breakdown of the traditional mechanistic worldview, the development of complexity sciences and the pretence of knowledge in economics. *Modern Economy*, *04*(04), 305–319. doi:10.4236/me.2013.44034

Lillevold, K., & Haarstad, H. (2019). The deep city: Cultural heritage as a resource for sustainable local transformation. *Local Environment*, *24*(4), 329–341. doi:10.1080/13549839.2019.1567481

Lindberg, F., Fitchett, J., & Martin, D. (2019). Investigating sustainable tourism heterogeneity: Competing orders of worth among stakeholders of a Nordic destination. *Journal of Sustainable Tourism*, *27*(8), 1277–1294. doi:10.1080/09669582.2019.1614188

Li, P. Q., & Kovacs, J. F. (2021). Creative tourism and creative spaces in China. *Leisure Studies*, 1–18. doi:10.1080/02614367.2021.1948596

Liu, G., & Chen, J. (2014). A Dynamic Model for Managing Cultural Tourism. *Asia Pacific Journal of Tourism Research*, *20*, 1–15. https://doi.org/10.1080/10941665.2014.904805

Liutikas, D. (2017). The manifestation of values and identity in travelling: The social engagement of pilgrimage. *Tourism Management Perspectives*, *24*, 217–224. doi:10.1016/j.tmp.2017.07.014

Liu, Z. (2003). Sustainable tourism development: A critique. *Journal of Sustainable Tourism*, *11*(6), 459–475. doi:10.1080/09669580308667216

Li, Y., & Hunter, C. (2015). Community Involvement for Sustainable Heritage Tourism: A Conceptual Model. *Journal of Cultural Heritage Management and Sustainable Development*, *5*(3), 248–262. doi:10.1108/JCHMSD-08-2014-0027

Llanos-Hernández, L. (2010). El concepto del territorio y la investigación en las Ciencias Sociales. *Agricultura, Sociedad y Desarrollo*, *7*(Septiembre-diciembre), 207–220.

Logan, W. S. (2002). Vietnam's Highway No.1: Corridor of power and patrimony. *The Historic Environment*, *16*(2), 23–26.

Lopes, J. R., & Pereira, A. M. (2017). Patrimônio cultural, turismo e desenvolvimento local: Estudo de caso da Cidade Velha, ilha de Santiago, Cabo Verde. Sociabilidades Urbanas –. *Revista de Antropología Social*, *1*(2), 45–60.

Lorenzo, J. & Martín, A. (2001). *Aves del archipiélago canario*. Francisco Lemus Editor.

Lor, J. J., Kwa, S., & Donaldson, J. A. (2019). Making ethnic tourism good for the poor. *Annals of Tourism Research*, *76*, 140–152. doi:10.1016/j.annals.2019.03.008

Loulanskia, T., & Loulanski, V. (2011). The sustainable integration of cultural heritage and tourism: A meta-study. *Journal of Sustainable Tourism*, *19*(7), 837–862. https://doi.org/10.1080/09669582.2011.553286

Lozano, M., & Maldonado, O. J. (2010). *Estrategia Nacional de Apropiación Social de la Ciencia, la Tecnología y la Innovación*. Colciencias.

Lozano-Oyola, M., Blancas, F. J., González, M., & Caballero, R. (2012). Sustainable tourism indicators as planning tools in cultural destinations. *Ecological Indicators*, *18*, 659–675. doi:10.1016/j.ecolind.2012.01.014

Lozares, C. (1996). La teoría de redes sociales. *Papers. Revista de Sociologia*, *48*, 103–126. doi:10.5565/rev/papers/v48n0.1814

Lv, X., Li, C., & McCabe, S. (2020). Expanding theory of tourists' destination loyalty: The role of sensory impressions. *Tourism Management*, *77*, 104–126. doi:10.1016/j.tourman.2019.104026

Macao Government Tourism Office. (2021). *Feast of the Drunken Dragon*. https://www.macaotourism.gov.mo/en/events/calendar/feast-of-the-drunken-dragon

Macao Government Tourism Office. (2021). *Na Tcha Temple*. https://www.macaotourism.gov.mo/en/sightseeing/temples/na-tcha-temple

MacLeod, N. (2016). Self-guided trails – a route to more responsible tourism? *Tourism Recreation Research*, *41*(2), 134–144.

MacLeod, N. E. (2004). Seeing place: A critique of the heritage trail as a visual interpretation of place. *LSA Publication*, *84*, 61–78.

Mahika, E. C. (2011). Current trends in tourist motivation. *Cactus Tourism Journal*, *2*(2), 15–24.

Mahoney, C. (1999). Heritage tourism: Strategies for implementation and the model of The Civil War Discovery Trail. *History News, 54*(1), 21–23.

Maldonado, J. E. (2018). Metodología de la Investigación Social. Paradigmas: cuantitativo, sociocrítico, cualitativo, complementario. In Educación. Ediciones de la U.

Manrique-de-Lara, P., & Guerra-Baez, R. (2016). Exploring the Influence of Ethical Climate on Employee Compassion in the Hospitality Industry. *Journal of Business Ethics*, *133*(3), 605–617. https://doi.org/10.1007/s10551-014-2415-1

Manual, O. E. C. D. (2005). *Guidelines for collecting and interpreting innovation data. OECD/Eurostat*. OECD Publishing.

Marín, S. A. (2012). Apropiación social del conocimiento: Una nueva dimensión de los archivos. *Revista Interamericana de Bibliotecología*, *35*(1), 55–62.

Mariotti, A. (2012). Local System, Networks and International Competitiveness: From Cultural Heritage to Cultural Routes. *Almatourism*, *3*(5), 81–95.

Markard, J., Raven, R., & Truffer, B. (2012). Sustainability transitions: An emerging field of research and its prospects. *Research Policy*, *41*(6), 955–967. doi:10.1016/j.respol.2012.02.013

Markwell, K., Stevenson, D., & Rowe, D. (2004). Footsteps and memories: Interpreting an Australian urban landscape through thematic walking tours. *International Journal of Heritage Studies*, *10*(5), 457–473.

Marsden, P. (1990). Network data and measurement. *Annual Review of Sociology*, *16*(1), 435–463. doi:10.1146/annurev.so.16.080190.002251

Martín De la Rosa, B. (2003). Nuevos turistas en busca de un nuevo producto: El patrimonio cultural. *Pasos. Revista de Turismo y Patrimonio Cultural*, *1*(2), 155–160. doi:10.25145/j.pasos.2003.01.015

Martínez, L. A. (2007). La Observación y el Diario de Campo en la Definición de un Tema de Investigación. *Perfiles Libertadores*, *4*(80), 73–80.

Martini, U., Buffa, F., & Notaro, S. (2017). Community Participation, Natural Resource Management and the Creation of Innovative Tourism Products: Evidence from Italian Networks of Reserves in the Alps. *Sustainability*, *9*(12), 2314. doi:10.3390u9122314

Martins, J. C. (2020). Tangible Cultural Heritage Re-Appropriation Towards A New Urban Centrality. A Critical Crossroad In Semi-Peripheral Eastern Riverside Lisbon. *Geography, Environment, Sustainability*, *13*(3), 139–146. doi:10.24057/2071-9388-2020-58

Martín, V. (2009). Aproximación tipológica a los paisajes agrarios actuales de Canarias. *Papeles de Geografía*, *32*, 97–116.

Mason, P. (2020). *Tourism impacts, planning and management*. Routledge. doi:10.4324/9780429273544

Masson, G. (2001, May). *The Chilkoot Trail National Historic Site: A Canadian cultural route management approach* [Paper presentation]. ICOMOS Making Tracks Conference, Alice Springs, Australia.

Mata, R. (2006). Un concepto de paisaje para la gestión sostenible del territorio. In R. Mata & A. Tarroja (Eds.), *El paisaje y la gestión del territorio: criterios paisajísticos en la ordenación del territorio y el urbanismo* (pp. 17–46). Diputación Provincial de Barcelona.

Matías, G., & Pulido, J. I. (2012). Dinámica relacional interorganizacional para el desarrollo turístico. El caso de Villa Gesell y Pinamar (Argentina). *Revista de Estudios Regionales*, *94*, 167–194.

Matos, N. (2014). *The Impacts of Tourism Experiences in the Destination Image. A Marketing Perspective* (PhD Thesis). Faculty of Economics, University of Algarve.

McCool, S. F., & Bosak, K. (2019). *A Research Agenda for Sustainable Tourism*. Edward Elgar Publishing. doi:10.4337/9781788117104

McCormack, P. (2017). Walking the land: Aboriginal trails, cultural landscapes, and archaeological studies for impact assessment. *Archaeologies*.

McDonough, W., & Braungart, M. (2010). *Cradle to cradle: Remaking the way we make things*. North Point Press.

McKercher, B. (2002). Towards a classification of cultural tourists. *International Journal of Tourism Research*, *4*(1), 29–38. doi:10.1002/jtr.346

McKercher, B., & du Cros, H. (2002). *Cultural Tourism: The Partnership between Tourism and Cultural Heritage Management*. Haworth Hospitality Press.

McKercher, B., & Du Cros, H. (2002). *Cultural tourism: The partnership between tourism and cultural heritage management*. Routledge.

McKercher, B., Ho, P. S. Y., & du Cros, H. (2005). Relationship between tourism and cultural heritage management: Evidence from Hong Kong. *Tourism Management*, *26*(4), 539–548. doi:10.1016/j.tourman.2004.02.018

McKercher, B., Mei, W. S., & Tse, T. S. (2006). Are short duration cultural festivals tourist attractions? *Journal of Sustainable Tourism*, *14*(1), 55–66. doi:10.1080/09669580608668591

Meacci, L., & Liberatore, G. (2018). A senses-based model for experiential tourism. *Tourism & Management Studies, 14*(4), 7–14. doi:10.18089/tms.2018.14401

Meadows, D. H., Meadows, D. L., Randers. J., & Behrens, W. W. (1972). *The Limits to Growth*. Earthscan.

Med Cycle Tour Deliverable Library. (2020). Retrieved from Med Cycle Tour: https://medcycletour.interreg-med.eu/what-we-achieve/deliverable-library/

Med Sustainable Tourism Community. (2019). *Ensuring an effective monitoring of tourism sustainability in the Mediterranean Region.* Author.

MED Sustainable Tourism Community. (2019). *Ensuring an effective monitoring of tourism sustainability in the mediterranean region.* Interreg Med programme.

MED Sustainable Tourism Community. (2019). *Policy Recommendations by the MED Sustainable Tourism Community.* Interreg MED programme.

Meillassoux, C. (1981). *Maidens, Meal and Money - Capitalism and the Domestic Community*. Cambridge University Press.

Mendes, J., Guerreiro, M., & Matos, N. (2016). Monitoring the Quality of Tourism Experience. *Proceedings of the Asia Tourism Forum 2016 - the 12Th Biennial Conference of Hospitality and Tourism Industry in Asia, 19*(1), 294–304. 10.2991/atf-16.2016.44

Mendes, J., Valle, P., Guerreiro, M., & Silva, J. (2010). The tourist experience - Exploring the relationship between tourist satisfaction and destination loyalty. *Tourism (Zagreb), 58*(2), 111–126. https://sapientia.ualg.pt/handle/10400.1/4517

Merinero, R., & Lara, F. (2011, May). *Las Ciudades Medias del interior de Andalucía. Caracterización y retos para el desarrollo turístico en un nuevo entorno* [Paper presentation]. 15th Congress of the Spanish Association of Scientific Experts in Tourism, Mijas, Málaga, Spain.

Merinero, R. (2011). Redes de actores y desarrollo turístico. Estudio de casos en Portugal. *Revista de Análisis Turístico, 11*(1), 9–21.

Merinero, R. (2015). La agenda de investigación del turismo mediante el análisis cuantitativo de redes sociales (QSNA). *Cuadernos de Turismo, 36*(36), 269–294. doi:10.6018/turismo.36.230991

Merinero, R., & Pulido, J. I. (2009). Desarrollo turístico y dinámica relacional. Metodología de análisis para la gestión activa de destinos turísticos. *Cuadernos de Turismo, 23*, 173–193.

Merinero, R., & Pulido, J. I. (2016). Analysing relationships in tourism: A review. *Tourism Management, 54*, 122–13. doi:10.1016/j.tourman.2015.10.010

Mesa, R. (2017). *Masterclass Turismo Naranja* [Formación; Youtube]. IMF Busness School. https://www.youtube.com/watch?v=g0rAFPzMis4

Message from the Skies. (n.d.). *The Guardian*. https://www.theguardian.com/stage/2020/jan/03/message-from-the-skies-shorelines-review-edinburgh-venues

Meyer, D. (2004). *Tourism routes and gateways: Examples and a selection of key issues for development of tourism routes and gateways and their potential for pro-poor tourism.* ODI.

Miller, G. (2001). The development of indicators for sustainable tourism: Results of a Delphi survey of tourism researchers. *Tourism Management, 22*(4), 351–362. doi:10.1016/S0261-5177(00)00067-4

Miller, G., & Twining-Ward, L. (2005). *Monitoring for a Sustainable Tourism Transition: The Challenge of Developing and Using Indicators*. CABI. doi:10.1079/9780851990514.0000

Minh, H. V., Lam, T., & Prabhakaran, S. (2021). Perceptions of Key Stakeholders Towards Sustainable Tourism Development: A Case Study in Mekong Delta, Vietnam. *Journal of Asian Finance*, *8*(4), 717–0726. doi:10.13106/jafeb.2021.vol8.no4.0717

Ministerio de Comercio Industria y Turismo y Fondo de Promoción Turística Colombia. (2013). *Guía Turística Boyacá—Colombia*. https://www.mincit.gov.co/cmspages/getfile.aspx?guid=91c401c7-2894-4cd6-ada2-2853a2c8922f

Ministerio de Hacienda y Credito Público. (2020). *Decreto Númro 286 de 26 de Feb de 2020*. República de Colombia. https://economianaranja.gov.co/media/hddp0pcx/decreto-286-de-2020.pdf

Ministero della Cultura. (2015). *Direttiva del Ministro 2016 - Anno dei Cammini d'Italia*. Retrieved from Ministero della Cultura: https://www.beniculturali.it/comunicato/direttiva-del-ministro-2016-anno-dei-cammini-d-italia

Ministria e Turizmit dhe Mjedisit. (2018). *Strategjia Kombëtare për Zhvillimin e Qëndrueshëm të Turizmit 2018-2022* [National Strategy for Sustainable Tourism Development 2018-2022]. https://www.turizmi.gov.al/wp-content/uploads/2018/02/Strategjia-e-Turizmit-18-22-draft-1.pdf

Ministria e Turizmit dhe Mjedisit. (n.d.). *Turizmi në një vështrim* [Tourism at a Glance]. https://turizmi.gov.al/turizmi-at-a-glance/

Ministry of Culture and Tourism. (2007). *Tourism Strategy of Turkey 2023 - Action Plan 2007-2013*. https://www.ktb.gov.tr/Eklenti/43537,turkeytourismstrategy2023pdf.pdf?0&_tag1=796689BB12A540BE0672E65E48D10C07D6DAE291

Ministry of Culture and Tourism. (2021a). *Republic of Turkey: General Directorate of Investment and Business, Turizm İstatistikleri* [Tourism Statistics]. https://yigm.ktb.gov.tr/TR-9851/turizm-istatistikleri.html

Ministry of Culture and Tourism. (2021b). *Republic of Turkey: General Directorate of Research and Training; video: "Turkey On UNESCO Intangible Cultural Heritage Lists"*. https://aregem.ktb.gov.tr/TR-268296/unesco-somut-olmayan-kulturel-miras-listelerinde-turkiy-.html#

Ministry of Culture and Tourism. (2021c). *Republic of Turkey: General Directorate of Research and Training; Yaşayan İnsan Hazineleri* [Living Human Treasures]. https://aregem.ktb.gov.tr/TR-167971/yasayan-insan-hazineleri.html

Ministry of Culture and Tourism. (2021d). *Republic of Turkey: General Directorate of Research and Training. Somut Olmayan Kültürel Miras Ulusal Envanteri* [National Inventory of Intangible Cultural Heritage]. https://aregem.ktb.gov.tr/TR-159257/somut-olmayan-kulturel-miras-ulusal-envanteri.html

Ministry of Culture and Tourism. (2021e). *Republic of Turkey: General Directorate of Research and Training. Somut Olmayan Kültürel Miras Türkiye Ulusal Envanteri* [National Inventory of Intangible Cultural Heritage of Turkey]. https://aregem.ktb.gov.tr/TR-279417/somut-olm-kult-miras-turkiye-ulusal-envanteri.html

Ministry of Culture and Tourism. (2021f). *Republic of Turkey: General Directorate of Cultural Heritage and Museums. Yurt Dışından İadesi Sağlanan Eserler* [The Artefacts that Have Been Returned from Abroad]. https://kvmgm.ktb.gov.tr/TR-44470/yurt-disindan-iadesi-saglanan-eserler.html

Ministry of Culture and Tourism. (2021g). *Republic of Turkey: Q&As about the Activities to Stop Smuggling*. https://kvmgm.ktb.gov.tr/TR-63359/sikca-sorulan-sorular.html

Misra, M. (2011). Shadow of the Silk Road. *The International Journal of Environmental Studies*, *68*(3), 400–404.

Mitomed Plus Deliverables Library. (2020). Retrieved from Mitomed Plus: https://mitomed plus.interreg-med.eu/what-we-achieve/deliverable-library/

Mitomed Plus Deliverables Library. (2020). Retrieved from Mitomed Plus: https://mitomed-plus.interreg-med.eu/what-we-achieve/deliverable-library/

Mokgachane, T., Basupi, B., & Lenao, M. (2021). Implications of cultural commodification on the authenticity of iKalanga music: A case of Domboshaba traditional music festival in Botswana. *Journal of Tourism and Cultural Change*, *19*(2), 153–165. doi:10.1080/14766825.2019.1700989

Mora, D. (2012). Diálogo y transferencia dialéctica de saberes/conocimientos. *Integra Educativa*, *V*(2), 31–75.

Moreira, A. C., Fortes, N., & Santiago, R. (2017). Influence of Sensory Stimuli on Brand Experience, Brand Equity and Purchase Intention. *Journal of Business Economics and Management*, *18*(1), 68–83. doi:10.3846/16111699.2016.1252793

Moreno Pires, S., Fidélis, T., & Ramos, T. B. (2014). Measuring and comparing local sustainable development through common indicators: Constrainst and achievements in practice. *Cities (London, England)*, *39*, 1–9. doi:10.1016/j.cities.2014.02.003

Moscardo, G. (2008). Sustainable tourism innovation: Challenging basic assumptions. *Tourism and Hospitality Research*, *8*(1), 4–13. doi:10.1057/thr.2008.7

Mosedale, J., & Voll, F. (2017). Social Innovations in Tourism: Social Practices Contributing to Social Development. In P. J. Sheldom & R. Danielle (Eds.), *Social Entrepreneurship and Tourism* (pp. 101–115). Springer. doi:10.1007/978-3-319-46518-0_6

Mossberg, L. (2007). A marketing approach to the tourist experience. *Scandinavian Journal of Hospitality and Tourism*, *7*(1), 59–74. doi:10.1080/15022250701231915

Motz, M. (1998). The practice of belief. *Journal of American Folklore*, *111*(441), 339–355. doi:10.2307/541314

Moulaert, F., Martinelli, F., González, S., & Swyngedouw, E. (2007). Social Innovation and Governance in European Cities Urban Development Between Path Dependency and Radical Innovation. *European Urban and Regional Studies*, *14*(3), 195–209. doi:10.1177/0969776407077737

Moulin, C., & Boniface, P. (2001). Routeing heritage for tourism: Making heritage and cultural tourism networks for socio-economic development. *International Journal of Heritage Studies*, *7*(3), 237–248.

Mowinckel, K. M. (2016). El turismo experiencial y creativo: El caso de Cantabria. *International Journal of Scientific Management and Tourism, 2*(3), 195-203. https://dialnet.unirioja.es/servlet/articulo?codigo=5665941

Moyle, B. D., Moyle, C., Bec, A., & Scott, N. (2019). The next frontier in tourism emotion research. *Current Issues in Tourism*, *22*(12), 1393–1399. doi:10.1080/13683500.2017.1388770

Mulgan, G. (2006). The Process of Social Innovation. *Innovations: Technology, Governance, Globalization*, *1*(2), 145–162. doi:10.1162/itgg.2006.1.2.145

Munsters, W. (2021). *The Study of Cultural Tourism in Ten Models*. https://www.academia.edu /48862966/The_Study_of_Cultural_Tourism_in_Ten_Models_Wil_Munsters

Murphy, K. (2012). The social pillar of sustainable development: A literature review and framework for policy analysis. *Sustainability: Science, Practice and Policy*, *8*(12), 15–29.

Murray, M., & Graham, B. (1997). Exploring the dialectics of route-based tourism: The Camino de Santiago. *Tourism Management, 18*, 513–524.

Næss, I. E. (2013). *Øylandet. Dagligliv på Helgelandskysten*. Orkana.

Najmanovich, D. (1995). *El lenguaje de los vínculos. De la independencia absoluta a la Autonomía Relativa*. Academic Press.

Navarro, D. (2010). *Turismo: portal de la diversidad cultural. El turismo receptivo como espacio para el encuentro multicultural*. Academic Press.

Neumann, H. R., Röder, D., & Röder, H. (2020). Diverse and rich fortified cultural heritage of the Iberian Peninsula. Basis for culture tourism with the European Culture Route Fortified Monuments. *Forte Cultura*, 971-976. . doi:10.4995/FORTMED2020.2020.11394

Ngamsomsuke, W., Hwang, T., & Huang, C. (2011). *Sustainable Cultural Heritage Tourism Indicators* [Paper presentation]. In International Conference on Social Science and Humanity IPEDR.

Niavis, S., Papatheochari, T., Psycharis, Y., Rodriguez, J., Font, X., & Codina, A. M. (2019). Conceptualising Tourism Sustainability and Operationalising Its Assessment: Evidence from a Mediterranean Community of Projects. *Sustainability*, *11*(15), 1–18. doi:10.3390u11154042

Nicolini, D., & Monteiro, P. (2017). The practice approach: For a praxeology of organisational and management studies. In A. Langley & H. Tsoukas (Eds.), *The Sage handbook of process organization studies* (pp. 110–126). Sage.

Nilsson, M. (2018). Wanderers in the shadow of the sacred myth: Pilgrims in the 21st century. *Social & Cultural Geography*, *19*(1), 21–38. doi:10.1080/14649365.2016.1249398

Niño, V. M. (2019). Metodología de la Investigación. Diseño, ejecución e informe. In A. Gutiérrez (Ed.), *Educación* (2nd ed.). Ediciones de la U.

Nisbett, M. (2017). Empowering the empowered? Slum tourism and the depoliticization of poverty. *Geoforum*, *85*, 37–45. doi:10.1016/j.geoforum.2017.07.007

Nocca, F. (2017). The Role of Cultural Heritage in Sustainable Development: Multidimensional Indicators as Decision-Making Tool. *Sustainability*, *9*(10), 1-28.

Nocca, F., & Fusco Girard, L. (2018). Towards an integrated evaluation approach for cultural urban landscape conservation/regeneration. *Region*, *5*(1), 33–51. doi:10.18335/region.v5i1.160

Norfolk County Council & University of East Anglia. (2019). *Green Pilgrimage: A report on methodologies to measure the economic, social and environmental impact of pilgrimage*. Author.

Nuñez, J. (2008). *Manual para operaciones básicas con Visone en el análisis de redes sociales para el desarrollo rural*. http://revista-redes.rediris.es/webredes/talleres/MANUAL-VISONE.pdf

O'Brien, R. & Curtis, C. (1979). The Donkin Heritage Trail. *Planning and Building Developments*, *40*, 13–26.

O'Brien, W., & Sarkis, J. (2014). The potential of community-based sustainability projects for deep learning initiatives. *Journal of Cleaner Production*, *62*, 48–61. doi:10.1016/j.jclepro.2013.07.001

O'Dell, T. (2005). Experiencescapes: Blurring borders and testing connections. In T. O'Dell & P. Billing (Eds.), *Experiencescapes -Tourism, Culture, and Economy* (pp. 11–33). Copenhagen Business School Press.

Obidos. (n.d.). https://www.cityofliterature.nl/en/cities/obidos/

Obrador, P., Crang, M., & Travlou, P. (2009). *Taking Mediterranean tourists seriously*. Ashgate.

OECD. (2008). Roles and impact of culture and tourism on attractiveness. In The Impact of Culture on Tourism. OECD.

Oftedal, E. M., Bertella, G., Lanka, S., Grzegorczyk, M., & Molthan-Hill, P. (2021). Perspectives of Sustainability. *Revista de Administração Contemporânea*, *25*(3), 1–8. PMID:34334863

Oh, H., Fiore, A., & Jeoung, M. (2007). Measuring Experience Economy Concepts: Tourism Applications. *Journal of Travel Research*, *46*(2), 119–132. doi:10.1177/0047287507304039

Oikonomou, D. (2000). A System of spatial planning: Greek reality and international experience. Review of Social Researches (Epitheorissi Koinonikon Erenvon), 101(101), 3.

Okuni, P. A., & Rockhold, J. P. (1995). *An approach for community-based sustainability*. Paper presented at the 21st WEDC Conference: Sustainability of Water and Sanitation Systems, Kampala, Uganda.

Okuyucu, A., & Somuncu, M. (2013). *Türkiye'de İnanç Turizmi, Bugünkü Durum, Sorunlar ve Gelecek* [Religious Tourism in Turkey: Today's Situation, Problems and Future] [Paper presentation]. *International Conference on Religious Tourism and Tolerance.*

Olivé, L. (2009). Por una auténtica interculturalidad basada en el reconocimiento de la pluralidad epistemológica. In *Pluralismo epistemológico* (pp. 19–30). Muela del Diablo Editores.

Olivé, L. (2010). Multiculturalidad, interculturalismo y el aprovechamiento social de los conocimientos. *Recerca. Revista de Pensament i Anàlisi*, *10*, 45–66. doi:10.16309/j.cnki.issn.1007-1776.2003.03.004

Olivé, L., Argueta, A., & Puchet, M. (2018). Interdisciplina y transdisciplina frente a los conocimientos tradicionales. *Revista Iberoamericana de Ciencia. Tecnología y Sociedad*, *13*(38), 135–153.

Olsen, D. H., Trono, A., & Fidgeon, P. R. (2018). Pilgrimage Trails and Routes: The Journey from the Past to the Present. In *Religious PIlgrimage Routes and Trails. Sustainable Development and Management.* CABI. doi:10.1079/9781786390271.0001

ONU. (2016). *Agenda 2030 y los Objetivos de Desarrollo Sostenible—Una oportunidad para America Latina y el Caribe.* Publicación de las Naciones Unidas. https://www.cepal.org/sites/default/files/publication/files/40155/S1801141_es.pdf

Opricǎ, M. A. (2015). European Cultural Routes in Romania: Opportunities for Developing Sustainable Tourism and Architectural Heritage. *Acta Technica Napocensis: Civil Engineering & Architecture*, *58*(4), 119–127.

Orbaşli, A. (2000). *Tourists in Historic Towns: Urban Conservation and Heritage Management.* E & FN Spon.

Organización Mundial del Turismo. (2016). Alianza entre turismo y cultura en el Perú – Modelos de colaboración entre turismo, cultura y comunidad. Author.

Ortega Hoyos, A. J., & Marin Verhelst, K. (2019). La innovación social como herramienta para la transformación social de comunidades rurales. *Revista Virtual Universidad Católica del Norte*, (57), 87–99. doi:10.35575/rvucn.n57a7

Ortega-Ibarra, I. H., Ortega-Ibarra, E., & Hernández-Jiménez, A. (2019). La investigación de Acción Participativa: ¿una alternativa de investigación o una estrategia de solución? *Universita Ciencia*, *20*, 61–70.

Ortiz, A. (2015). Enfoques y métodos de investigación en las ciencias sociales y humanas. In *Educación*. Ediciones de la U.

Öter, Z., & Ünal, E. (2011). *Somut Olmayan Kültürel Miras ve Turizm Etkileşiminin Beypazarı İlçesinde Yerel Ekonomiye Katkıları* [Paper presentation]. Milletlerarası Türk Halkbilimi Kongresi.

Öter, Z. (2010). Türk El Sanatlarının Kültür Turizmi Bağlamında Değerlendirilmesi [Valorization of Turkish Handicrafts in Cultural Tourism Context]. *Milli Folklor*, *22*(86), 174–185.

Oviedo, L., de Courcier, S., & Farias, M. (2014). Rise of pilgrims on the Camino to Santiago: Sign of change or religious revival? *Review of Religious Research*, *56*(3), 433–442. doi:10.100713644-013-0131-4

Öztemiz, S. (2016). *Türkiye'de Dijitalleştirilen Kültürel Miras Ürünlerine Açık Erişim: Bir Model Önerisi* [Open Access to Digitalized Cultural Heritage Products: A Model Suggestion] [PhD Thesis]. Hacettepe University, Institute of Social Sciences, Information and Document Management Major Field of Study.

Ozturk, A. B., & Van Niekerk, M. (2014). Volume or Value: A Policy Decision for Turkey' s Tourism Industry. *Journal of Destination Marketing & Management, 3*(4), 193–197. doi:10.1016/j.jdmm.2014.04.001

Pabón, R. (2018). Apropiación social del conocimiento: Una aproximación teórica y perspectivas para Colombia. *Educación y Humanismo, 20*(34), 116–139. doi:10.17081/eduhum.20.34.2629

Padin, C., & Svensson, G. (2013). A multi-layer matrix model of sustainable tourism: Process, measurement areas, gap and reconnection analyses. *European Business Review, 25*(2), 206–216. https://doi.org/10.1108/09555341311302684

Padrón, H. & Hernández, R. (2017). Los puntos de interés turístico: relevancia analítica, propuesta metodológica y caso de estudio. *Pasos. Revista de Turismo y Patrimonio Cultural, 15*(4), 979-1000.

Page, S. J., & Getz, D. (2019). *Event studies: Theory, research and policy for planned events.* Routledge.

Paksoy, P. (2009). *The Environmental Focus In Tourism Strategy Of Turkey-2023 Within The Outlook Of Tourism Planning.* Sofia: Surveying Geology & Mining Ecology Management (SGEM). Retrieved from https://www.proquest.com/conference-papers-proceedings/environmental-focus-tourism-strategy-turkey-2023/docview/1285255677/se-2?accountid=141837

Páramo, D., Campo, S. J., & Maestre, L. M. (2020). Métodos de Investigación Cualitativa. Fundamentos y aplicaciones. In D. Páramo Morales, S. J. Campo Sierra, & L. M. Maestre Matos (Eds.), *Adminsitración y Negocios.* Editorial UniMagdalena.

Pardellas, X. X., & Padín, C. (2014). El Camino De Santiago Portugués En Galicia Y Las Estrategias Turísticas Locales. *Revista de Análisis Turístico, 17*, 51–59.

Parga Dans, E. (2012). *El perfil de la peregrinación en el Camiño Santiago-Fisterra.* https://digital.csic.es/handle/10261/45922

Park, E., Choi, B. K., & Lee, T. J. (2019). The Role and Dimensions of Authenticity in Heritage Tourism. *Tourism Management, 74*, 99–109. doi:10.1016/j.tourman.2019.03.001

Pastoors, S., Scholz, U., Becker, J. H., & Van Dun, R. (2017). *Towards Sustainable Innovation: A five step approach to sustainable change.* Tectum Wissenschaftsverlag.

Pavlis, E., & Terkenli, T. (2017). Landscape values and the question of cultural sustainability: Exploring an uncomfortable relationship in the case of Greece. *Norsk Geografisk Tidsskrift, 71*(3), 168–188. doi:10.1080/00291951.2017.1345977

Pavlovich, K. (2003). The evolution and transformation of a tourism destination network: The Waitomo Caves, New Zealand. *Tourism Management, 24*(2), 203–216. doi:10.1016/S0261-5177(02)00056-0

Pearce, D. (2014). Toward an integrative conceptual framework of destinations. *Journal of Travel Research, 53*(2), 141–153. doi:10.1177/0047287513491334

Pedersen, A. (2002). Managing Tourism at World Heritage Sites: A Practical Manual for World Heritage Site Managers. In World Heritage Manual No. 1. Paris: UNESCO World Heritage Centre.

Pedro, R. M. (2019). *Senses, Emotions and Memories in Tourism* (PhD Thesis). Faculty of Economics, University of Algarve.

Pedro, R. M., Mendes, J., de Matos, N., & Ascenção, M. P. (2020). A Conceptualization of the Senses, Emotions, and Memories in Memorable Tourism Experiences. In P. Pinto & M. Guerreiro (Eds.), *Handbook of Research on Resident and Tourist Perspectives on Travel Destinations* (pp. 208–231). IGI Global. doi:10.4018/978-1-7998-3156-3.ch010

Pelit, E., & Türkoğlu, T. (2019). Somut Olmayan Kültürel Miras Değerlerinin Turizme Yansımaları: Ebru Sanatı Üzerine Bir İnceleme [Tourism Reflections of Intangible Cultural Heritage Values: A Review on Ebru Art]. *Güncel Turizm Araştırmaları Dergisi*, *3*(1), 101–118. doi:10.32572/guntad.463910

Peñarrubia, M. P. (2018). *Los datos estadísticos públicos y su uso en el conocimiento del comportamiento de los turistas en destinos inteligentes* (PhD thesis). Universitat de València.

Peñarrubia, M. P., Simancas, M., & Forgione, G. (2019). Application of geomarketing to coastal tourist areas. *Tourism & Management Studies*, *15*(4), 7–20. doi:10.18089/tms.2019.150401

Pereira Roders, A., & Van Oers, R. (2011). Editorial: Bridging cultural heritage and sustainable development. *Journal of Cultural Heritage Management and Sustainable Development*, *1*(1), 5–14. https://doi.org/10.1108/20441261111129898

Pereiro, X. (2017). Turiperegrinos portugueses no Caminho Português Interior de Santiago de Compostela. *Revista Turismo e Desenvolvimento*, *27*, 413–423.

Pereiro, X. (2018). Ancient ways, new cultural tourism routes: The Inner Portuguese Way to Santiago de Compostela. In E. G. Pellón (Ed.), *Rural Worlds, Social Sustainability and Local Landscapes in Globalisation Era* (pp. 289–315). Case Studies in Southern Europe.

Pereiro, X., & Gomes, C. (2019). Perfil, motivações e experiências dos peregrinos do Caminho Português Interior de Santiago de Compostela. In X. Pereiro (Ed.), *Património cultural jacobeu, turismo e peregrinação: O Caminho Português Interior de Santiago de Compostela (CPIS)* (pp. 79–112). PASOS.

Pérez, J., & Gardey, A. (2013). *Definición de Práctica Social*. https://definicion.de/practica-social/

Pérez, L., & Parra, C. (2004). Paisajes culturales: El parque patrimonial como instrumento de revalorización y revitalización del territorio. *Theoria*, *13*(1), 9–24.

Perkins, D. N., & Salomon, G. (1992). Transfer of learning. In International Encyclopedia of Education (2nd ed., pp. 443–467). Pergamon Press. doi:10.1037/h0071273

Petrović, M. D., Blešić, I., Ivolga, A., & Vujko, A. (2016). Agritourism impact toward locals' attitudes-an evidence from Vojvodina Province (Serbia). *Journal of the Geographical Institute Jovan Cvijic*, *66*(1), 105–123. doi:10.2298/IJGI1601105P

Pham, K., Vogt, C., & Andereck, K. (2021). Overtourism: Sustainable Tourism Plan and Stakeholder Involvement. *Travel and Tourism Research Association: Advancing Tourism Research Globally*, 4.

Pimentel de Oliveira, D. (2020). Tarjeta Turística Safety and Security: El pasaporte para turistas y destinos seguros. *Cuadernos de Turismo, 46*. doi:10.6018/turismo

Pitarch-Garrido, M.-D., Sales-Ten, A., & Salom Carrasco, J. (2018). Políticas de nueva generación para la innovación social: el caso de la ciudad de Valencia. *XVII Congreso Internacional de Investigadores en economía Social: La Economía Social: transformaciones recientes, tendencias y retos de futuro*. http://ciriec.es/eventos/xvii-congreso-internacional-de-investigadores-en-economia-social-y-cooperativa/comunicaciones

Plessis, C. A., & Brandon, P. (2015). An ecological worldview as basis for a regenerative sustainability paradigm for the built environment. *Journal of Cleaner Production*, *109*, 53–61. doi:10.1016/j.jclepro.2014.09.098

Plichta, J. (2019). The co-management and stakeholder's theory as a useful approach to manage the problem of overtourism in historical cities–illustrated with an example of Krakow. *International Journal of Tourism Cities*, *5*(4), 685–699. doi:10.1108/IJTC-12-2018-0107

Pløger, J. (2016). The Evental City. *Space and Culture*, *19*(3), 260–274. doi:10.1177/1206331215595729

Polese, F., Botti, A., Grimaldi, M., Monda, A., & Vesci, M. (2018). Social Innovation in Smart Tourism Ecosystems: How Technology and Institutions Shape Sustainable Value Co-Creation. *Sustainability*, *10*(1), 140. doi:10.3390u10010140

Polo de Lobatón, G., Expósito, M. Y., & Lobatón Polo, D. (2020). Gerencia social: Estrategia de innovación social con enfoque economía solidaria, para el turismo comunitario en las poblaciones palafíticas de la Ciénaga Grande de Santa Marta. *Revista Loginn: Investigación Científica y Tecnológica*, *4*(1), 51–65. doi:10.23850/25907441.3137

Popay, J., Roberts, H., Sowden, A., Petticrew, M., Arai, L., Rodgers, M., & Britten, N. (2006). Narrative Synthesis in Systematic Reviews: A Product from the ESRC Methods Programme. *ESRC Methods Programme, 2006*, 93. http://www.ccsr.ac.uk/methods/publications/documents/Popay.pdf

PORDATA. (2021). Available at: https://www.pordata.pt/

Porter, M. E. (1991). La ventaja competitiva de las naciones. *Harvard Business Review, 1025*. https://d1wqtxts1xzle7.cloudfront.net/53273995/VCN_PORTER-with-cover-page-v2.pdf?Expires=1630409352&Signature=QrDnYFaMwAGNz-6VBZ0rbJOGM8VSOmn8XKFMdtGPEm7yhNpXJUOpybrqklWxlZAwRHqi4uSp1-Hoq34~Lg0Kv3RN1ukWzpMn-MQnG9C8HFHN2O-NCzx22XQc8YwPw-PltVXM4d4WPk~cKoOVB13PqUe6Hu6kbvuEHejleHLLh8ejKuL3mD~chLEa7l3UHLceiBunaFsNuh3t7upB9schzFbdF~C53DhGOWS3AioyzGbwGPbn1uXDA0Oeb84pkGrZZtZRX~fZgQh-9HaNbce-ZK7kIQFaNAdwNN9aaL84IDJ7OKqZA8LR2MIQBlGy5DnfZ8ZrrUP8zClOCOB2QkXZYk47g__&Key-Pair-Id=APKAJLOHF5GGSLRBV4ZA

Procolombia. (2012). Razones de porqué viajar a Colombia. *Visit Colombia*. https://procolombia.co/especial-colombia-ny-2012/visite-colombia/razones-de-por-qué-viajar-colombia

Procopiak, R. K. A., de Almeida, C. R., & Renda, A. I. (2020). Medieval tourism events in the Iberia peninsula. *Ficha Técnica Omnia*, *10*(2), 29–40.

Prokopiou, D. G., Mavridoglou, G., Manologlou, S., & Tselentis, B. S. (2018). Tourism development of the Cyclades islands: economic, social and carrying capacity assessment and consequences. WIT Transactions on Ecology and the Environment, 217.

Puczkó, L., & Rátz, T. (2007). Trailing Goethe, Humbert and Ulysses Tourism: Cultural Routes in Tourism. In G. Richards (Ed.), *Cultural Tourism: Global and Local Perspectives*. Haworth Press.

Pulido, J. I., García, J. A., Casado, J., & Rodríguez, B. (2021). Measuring creativity in urban-cultural Spanish destinations. *Boletín de la Asociación de Geógrafos Españoles*, *89*. Advance online publication. doi:10.21138/bage.3074

Pulido, J. I., & Merinero, R. (2018). Destinations relational dynamic and tourism development. *Journal of Destination Marketing & Management*, *7*, 140–152. doi:10.1016/j.jdmm.2016.09.008

Qiu, Q., Zheng, T., Xiang, Z., & Zhang, M. (2020). Visiting intangible cultural heritage tourism sites: From value cognition to attitude and intention. *Sustainability, Switzerland*, *12*(1), 1–20. doi:10.3390u12010132

Quandt, C., Ferraresi, A., Kudlawicz, C., Martins, J., & Machado, A. (2017). Social innovation practices in the regional tourism industry: Case study of a cooperative in Brazil. *Social Enterprise Journal*, *13*(1), 78–94. doi:10.1108/SEJ-12-2015-0038

Quinteiro, S., & Baleiro, R. (2017). *Literature and Tourism: Fundamental Concepts* (2nd ed.). University of Lisbon.

Quispe, G., Ayaviri, D., & Maldonado, R. (2018). Participación de los actores en el desarrollo local en entornos rurales. *Revista de Ciencias Sociales*, *24*(3). https://www.redalyc.org/jatsRepo/280/28059580008/html/index.html

Quivy, R. (1995). *Manual de Investigação em Ciências Sociais* [Handbook of Social Science Research]. Retrieved from https://tecnologiamidiaeinteracao.files.wordpress.com/2018/09/quivy-manual-investigacao-novo.pdf

Rametsteiner, E., Pülzl, H., Alkan-Olsson, J., & Frederiksen, P. (2011). Sustainability indicator development - Science or political negotiation. *Ecological Indicators*, *11*(1), 61–70. doi:10.1016/j.ecolind.2009.06.009

Ramírez, S. (2011). Nueva sociedad. *Enero-Febrero 2011*. https://nuso.org/articulo/el-giro-de-la-politica-exterior-colombiana/

Ramsay, J., & Truscott, M. C. (2003). Tracking through Australian forests. *The Historic Environment*, *16*(2), 32–38.

Rasoolimanesh, S. M., & Jaafar, M. (2017). Sustainable tourism development and residents' perceptions in World Heritage Site destinations. *Asia Pacific Journal of Tourism Research*, *22*(1), 34–48. doi:10.1080/10941665.2016.1175491

Ratten, V., Braga, V., Álvarez-García, J., & del Rio, M. D. L. C. (2020). Tourism innovation: The role of technology, sustainability and creativity. In V. Ratten, V. Braga, J. Álvarez-García, & M. d. l. C. del Rio (Eds.), Tourism Innovation: Technology, Sustainability and Creativity (pp. 1-13). Routledge.

Raworth, K. (2017). *Doughnut economics: Seven ways to think like a 21st-century economist*. Chelsea Green Publishing.

Reed, B. (2007). Forum: Shifting from 'sustainability' to regeneration. *Building Research and Information*, *35*(6), 674–680. doi:10.1080/09613210701475753

Reijnders, S. (2015). Stories that move: Fiction, imagination, tourism. *European Journal of Cultural Studies*, *19*(6), 672–689. doi:10.1177/1367549415597922

Republic of Turkey Ministry of Culture and Tourism. (2014). *Turkey in World Heritage*. https://kvmgm.ktb.gov.tr/Eklenti/27154,turkeyinworldheritage.pdf?0

Rezende, R. (2014). Community tourism and favela-tour as expressions of the new dynamics of tourist consumption [Turismo comunitário e favela-tour como expressões das novas dinâmicas do consumo turístico]. *Revista Turismo Em Análise*, *25*(2), 354–372. doi:10.11606/issn.1984-4867.v25i2p354-372

Rhodes, R. (2016). *Reality tours and travel: A Community perception on slum tourism in Dharavi*. Durham University.

Ribes, J. F. P., & Baidal, J. I. (2018). Smart sustainability: a new perspective in the sustainable tourism debate. *Investigaciones Regionales-Journal of Regional Research*, (42), 151-170.

Richards, G. (2004). *Towards a sustainable form of cultural tourism?* https://www.academia.edu /2006332/Towards_a_sustainable_form_of_cultural_tourism

Richards, G. (2014). *Tourism trends: The convergence of culture and tourism*. https://www. academia.edu/9491857

Richards, G. (1996). Production and consumption of European Cultural Tourism. *Annals of Tourism Research*, *23*(2), 261–283. doi:10.1016/0160-7383(95)00063-1

Richards, G. (1996). The scope and significance of cultural tourism. In *Cultural tourism in Europe* (pp. 19–45). CAB International.

Richards, G. (2011). Creativity and tourism: The state of the art. *Annals of Tourism Research*, *38*(4), 1225–1253. doi:10.1016/j.annals.2011.07.008

Richards, G. (2011). Cultural tourism trends in Europe: a context for the development of Cultural Routes. In *Impact of European Cultural Routes on SMEs' innovation and competitiveness* (pp. 21–39). Council of Europe.

Richards, G. (2014). Creativity and tourism in the city. *Current Issues in Tourism, 17*(2), 119–144. doi:10.1080/13683500.2013.783794

Richards, G. (2018). Cultural tourism: A review of recent research and trends. *Journal of Hospitality and Tourism Management, 36*, 12–21. doi:10.1016/j.jhtm.2018.03.005

Richards, G. (2021). *Rethinking cultural tourism*. Edward Elgar Publishing. doi:10.4337/9781789905441

Richards, G. (Ed.). (1996). *Cultural Tourism in Europe*. CABI Publishing.

Richards, G., & Munsters, W. (2010). *Cultural Tourism Research Methods*. CABI. doi:10.1079/9781845935184.0000

Richards, G., & Munsters, W. (2021). *The Study of Cultural Tourism in Ten Models*. University of Applied Sciences.

Richards, G., Russo, A. P., & Grossman, M. (2008). *Cultural Corridors in South East Europe: Refinement of concept and development of pilot projects*. Council of Europe.

Richards, G., Van Oss, F., Fernandes, C., Sousa, L., Tzanakaki, K., & Miettenen, S. (1998). *Developing and Marketing Crafts Tourism: The EUROTEX Project*. Atlas. https://d1wqtxts1xzle7.cloudfront.net/37322387/handbook_EUROTEX-with-cover-page-v2.pdf?Expires=1627030970&Signature=gy9-UBaa3YB3CyfIK-5AtU6LkaC1~k-LqqydeSfK273vw-fVEuvHHxsqTUk3NL9K0WAIeav-E0FywJ1cjZVG-sVtxC3RabpWRBKc3O9LPa3GEcCOx6~e8u1Gkxl74zRGdFmWnYWMO~zRrQv9OBRzicex5eyLdwSVEvr7nh-FfRqpb7N9~o-RUOIeg~3hCm1TnqCDgkVJNPGvN5V5WNnNjeExwwFICdajsnk6PMeiEEfqPL2HGCNSzGW8Viwb1svQQM7Q4jklASLTMADKQIDhMQ3B0Bi07KNWtPxZJX8LJ4DV4YojinAEchI8bTUwHyuWJgqFC3eXJp96Hv~a0fxeh4w__&Key-Pair-Id=APKAJLOHF5GGSLRBV4ZA

Richards, G., & Wilson, J. (2006). Developing creativity in tourist experiences: A solution to the serial reproduction of culture? *Tourism Management, 27*(6), 1209–1223. doi:10.1016/j.tourman.2005.06.002

Richins, H. (2009). Environmental, cultural, economic and socio-community sustainability: A framework for sustainable tourism in resort destinations. *Environment, Development and Sustainability, 11*(4), 785–800. doi:10.100710668-008-9143-6

Riley, R. W., & Love, L. L. (2000). The State of Qualitative Tourism Research. *Annals of Tourism Research, 27*(1), 164–187.

Ring, N., Ritchie, K., Mandava, L., & Jepson, R. (2011). *A guide to synthesising qualitative research for researchers undertaking health technology assessments and systematic reviews*. Academic Press.

Ringle, C. M., Wende, S., & Becker, J. M. (2015). *SmartPLS 3.0*. Scientific Research Publishing - SmartPLS GmbH, Boenningstedt. https://www.scirp.org/(S(vtj3fa45qm1ean45vvffcz55))/reference/ReferencesPapers.aspx?ReferenceID=2060940

Ringle, C., Silva, D., & Bido, D. (2014). Structural Equation Modeling with the SmartPls. *Revista Brasileira de Marketing, 13*(2), 54–71. doi:10.5585/remark.v13i2.2717

Rioux, M. (1950). Folk and Folklore. *Journal of American Folklore, 63*(248), 192–198. doi:10.2307/537160

Ritala, P. (2019). Innovation for Sustainability: Scepital, Pragmatic and Idealist Perspectives on the Role of Business as a Driver for Change. In N. Bocken, P. Ritala, L. Albareda, & R. Verburg (Eds.), *Innovation for Sustainability. Business Transformations Towards a Better World* (pp. 21–34). Palgrave Macmillan. doi:10.1007/978-3-319-97385-2_2

Rivas, L. A. (2017). Capitulo 6. La definición de variables o categorías de análisis. In Elaboración de tesis: estructura y metodología (p. 284). Editorial Trillas. doi:10.13140/RG.2.1.3446.6644

Rizo, M. (2004). *Prácticas culturales y redefinición de las identidades de los inmigrantes en El Raval (Barcelona): aportaciones desde la comunicación*. Universitat Autònoma de Barcelona.

Robins, G. (2015). *Doing social network research: Network-based research design for social scientists*. SAGE Publications. doi:10.4135/9781473916753

Roccas, S., & Elster, A. (2014). Values and religiosity. In V. Saroglou (Ed.), *Religion, Personality, and Social Behavior* (pp. 193–212). Taylor & Francis.

Rodríguez, D. (2010). Territorio y territorialidad Nueva categoría de análisis y desarrollo didáctico de la Geografía. *Unipluriversidad, 10*(3), 12.

Romagosa, F. (2020). The COVID-19 crisis: Opportunities for sustainable and proximity tourism. *Tourism Geographies, 22*(3), 690–694. doi:10.1080/14616688.2020.1763447

Romero, L. (2016). Towards a New Valuation of Cultural Terraced Landscapes: The Heritage of Terraces in the Canary Islands (Spain). *Annales, Series Historia et Sociologia*, 26.

Romero, C. (2005). La categorizacón un aspecto crucial en la investigación cualitativa. *Revista de Investigaciones Cesmag, 11*(Jun), 113–118.

Røros Museum. (n.d.). *About the Røros Museum*. Rorosmuseet. https://rorosmuseet.no/en/about-the-museum

Roxas, F., Rivera, J., & Gutierrez, E. (2020). Mapping stakeholders' roles in governing sustainable tourism destinations. *Journal of Hospitality and Tourism Management, 45*, 387–398. doi:10.1016/j.jhtm.2020.09.005

Rúa, C., Sepúlveda, M., Gutiérrez, S., Cárcamo-Vega, J. J., Surco-Luque, J., Campos-Vallette, M., & Pereira, M. (2017). Raman identification of pigments in wall paintings of the Colonial period from Bolivian churches in the Ruta de La Plata. *Conservation Science in Cultural Heritage, 17*, 117–137. doi:10.6092/issn.1973-9494/7945

Rudihartmann. (1986). Tourism, seasonality and social change. *Leisure Studies, 5*(1), 25-33. doi:10.1080/02614368600390021

Ryan, A. W., & Wollan, G. (2013). Festivals, landscapes, and aesthetic engagement: A phenomenological approach to four Norwegian festivals. *Norsk Geografisk Tidsskrift-Norwegian Journal of Geography, 67*(2), 99–112. doi:10.1080/00291951.2013.784352

Rydin, Y. (2007). Indicators as a governmental technology? The lessons of community-based sustainability indicator projects. *Environment and Planning. D, Society & Space, 25*(4), 610–624. doi:10.1068/d72j

Saarinen, J. (2006). Traditions of sustainability in tourism studies. *Annals of Tourism Research, 33*(4), 1121–1140. doi:10.1016/j.annals.2006.06.007

Saarinen, J. (2014). Critical sustainability: Setting the limits to growth and responsibility in tourism. *Sustainability, 6*(1), 1–17. doi:10.3390u6010001

Saarinen, J. (2019). Communities and sustainable tourism development: Community impacts and local benefit creation in tourism. In S. F. McCool & K. Bosak (Eds.), *A research agenda for sustainable tourism* (pp. 206–222). Edward Elgar Publishing. doi:10.4337/9781788117104.00020

Saarinen, J., & Gill, A. M. (2019). *Resilient destinations and tourism. Governance Strategies in the Transition Towards Sustainability in Tourism*. Routledge.

Sabaté, J. (2005). Paisajes culturales, el patrimonio como recurso básico para un nuevo modelo de desarrollo. *URBAN. Revista del Departamento de Urbanística y Ordenación del Territorio, 0*(9), 8–29.

Sanagustín-Fons, M. V., Tobar-Pesántez, L. B., & Ravina-Ripoll, R. (2020). Happiness and cultural tourism: The perspective of civil participation. *Sustainability*, *12*(8), 3465–3485. doi:10.3390u12083465

Sánchez-Sánchez, M. D., De-Pablos-Heredero, C., & Montes-Botella, J. L. (2020). A behaviour model for cultural tourism: Loyalty to destination. *Economic Research Journal*, *0*(0), 1–18. doi:10.1080/1331677X.2020.1838313

Sánchez, V. L., Fernández, M. D., & Lara, J. Á. S. (2016). Economic impact of a religious and tourist event: A holy week celebration. *Tourism Economics*, *23*(6), 1255–1274. doi:10.1177/1354816616675996

Sandelowski, M., & Barroso, J. (2003). Classifying the Findings in Qualitative Studies. *Qualitative Health Research*, *13*, 905–923. https://doi.org/10.1177/1049732303253488

Santa, E. D., & Tiatco, A. (2019). Tourism, Heritage and Cultural Performance: Developing a Modality of Heritage Tourism. *Tourism Management Perspectives*, *31*, 301–309. doi:10.1016/j.tmp.2019.06.001

Santana, M. O. R., & Alves, A. (2019). A paisagem cultural sagrada no Caminho Português Interior de Santiago. In X. Pereiro (Ed.), *Património cultural jacobeu, turismo e peregrinação: O Caminho Português Interior de Santiago de Compostela (CPIS)* (pp. 59–77). PASOS.

Santiago-Jiménez, M. E., & Morfin-Herrera, M. del C. (2011). Diálogo de saberes a través de la actividad turística. *El Periplo Sustentable,* (21), 31–59. doi:10.21854/eps.v0i21.897

Santiago-Jiménez, M. E., & Morfin-Herrera, M. del C. (2016). *Animación Turística para el Diálogo de Saberes*. Fundación Universidad de las Américas.

Santos, G. E. de O. (2000). Importância das peregrinações para o turismo mundial. *Revista Turismo Em Análise*, *11*(2), 38–44. doi:10.11606/issn.1984-4867.v11i2p38-44

Sarantakou, E., & Terkenli, T. S. (2019). Non-institutionalized forms of tourism accommodation and overtourism impacts on the landscape: The case of Santorini, Greece. *Tourism Planning & Development*, *16*(4), 411–433. doi:10.1080/21568316.2019.1569122

Sarantakou, E., & Tsartas, P. (2015). A critical approach to the new framework for creating tourism investment during the current period of economic crisis 2010-2014. *Greek Economic Outlook*, *26*, 46–55.

Sarantou, M., Kugapi, O., & Huhmarniemi, M. (2021). Context mapping for creative tourism. *Annals of Tourism Research*, *86*, 103064. doi:10.1016/j.annals.2020.103064

Saso, E., & Oliver, E. (2003). Las comunidades de aprendizaje: Un modelo de educación dialógica en la sociedad del conocimiento. *Revista Interuniversitaria de Formación Del Profesorado*, *17*(3), 91–103.

Sassatelli, M. (2015). *Festivals, urbanity and the public sphere: reflections on European Festivals. In Focus on Festivals. Contemporary European Case Studies and perspectives*. Goodfellow Publishers.

Saunders, F., Gilek, M., Ikauniece, A., Tafon, R. V., Gee, K., & Zaucha, J. (2020). Theorizing Social Sustainability and Justice in Marine Spatial Planning: Democracy, Diversity, and Equity. *Sustainability*, *12*(6), 2560. doi:10.3390u12062560

Sawyer, S., & Tapia, A. (2005). The sociotechnical nature of mobile computing work: Evidence from a study of policing in the United States. *International Journal of Technology and Human Interaction*, *1*(3), 1–14. doi:10.4018/jthi.2005070101

Schmitz, C. L., Stinson, C. H., & James, C. D. (2010). Community and environmental sustainability: Collaboration and interdisciplinary education. *Critical Social Work, 11*(3).

Schot, J., & Geels, F. W. (2008). Strategic niche management and sustainable innovation journeys: Theory, findings, research agenda, and policy. *Technology Analysis and Strategic Management*, *20*(5), 537–554. doi:10.1080/09537320802292651

Schutt, A. M. (1997). *The Bruce Trail, Ontario Users and their Economic Impact*. Trent University.

Scott, N., Baggio, R., & Cooper, C. (2008). *Network Analysis and Tourism: From Theory to Practice*. Channel View Publications.

Scott, N., Baggio, R., & Cooper, C. (2011). Network analysis methods for modeling tourism inter-organizational systems. *Advances in Culture. Tourism and Hospitality Research*, *5*, 177–222.

Scott, N., Baggio, R., & Cooper, Ch. (2008). *Network analysis and tourism: From theory to practice*. Channel View Publications. doi:10.21832/9781845410896

SENA. (2017). Lo que debes saber sobre la Economía Naranja. *Servicio Nacional de Aprendizaje - SENA*. https://www.sena.edu.co/es-co/formacion/Paginas/economia_naranja.aspx

Severo, M. (2017). European Cultural Routes: Building a Multi-Actor Approach. *Museum International*, *69*(1-2), 136–145. doi:10.1111/muse.12157

Sevigny, R. (1992). Florida's black heritage trail. *Historic Preservation Forum*, *6*(4), 6-9.

Seyfang, G., & Haxeltine, A. (2012). *Growing grassroots innovations: Exploring the role of community-based initiatives in governing sustainable energy transitions*. SAGE Publications.

Seyfi, S., Hall, C. M., & Rasoolimanesh, S. M. (2020). Exploring memorable cultural tourism experiences. *Journal of Heritage Tourism*, *15*(3), 341–357. doi:10.1080/1743873X.2019.1639717

Seyfi, S., Hall, C., & Fagnoni, E. (2018). Managing World Heritage Site stakeholders: A grounded theory paradigm model approach. *Journal of Heritage Tourism*, *14*, 1–17. https://doi.org/10.1080/1743873X.2018.1527340

Shackley, M. (2003). The Frankincense Route: A proposed cultural itinerary for the Middle East. *The Historic Environment*, *16*(2), 12–17.

Shackley, M. (Ed.). (1998). *Visitor Management – Case studies from World Heritage Sites*. Butterworth Heinemann.

Sharareh, A. D., & Badaruddin, M. (2013). Local Perception of Tourism Development: A Conceptual Framework for the Sustainable Cultural Tourism. *J. MGMT & Sustainability*, *31*, 31–40.

Shaw, G., & Williams, A. M. (1994). *Critical issues in tourism: A geographical perspective*. Basil, Blackwell.

Shaw, S., & MacLeod, N. (2000). Creativity and conflict: Cultural tourism in London's city fringe. *Tourism, Culture & Communication*, *2*(3), 165–175.

Sheehan, L., & Ritchie, J. (2005). Destination Stakeholders. Exploring Identity and Salience. *Annals of Tourism Research*, *32*(3), 711–734. doi:10.1016/j.annals.2004.10.013

Sheppard, E., Porter, P. W., Faust, D. R., & Nagar, R. (2009). *A World of Difference: Encountering and Contesting Development* (2nd ed.). Guilford Press.

Shkodra. (n.d.). *Qendra e Studimeve Shqiptare "Ernest Koliqi" Shkodër & Bashkimi Katolik i Publicistëve Shqiptarë*. http://kulturserver-hamburg.de/home/shkodra/shkodra/labeat.html

Shoval, N., Schvimer, Y., & Tamir, M. (2018). Real-Time Measurement of Tourists' Objective and Subjective Emotions in Time and Space. *Journal of Travel Research*, *57*(1), 3–16. doi:10.1177/0047287517691155

Sigala, M. (2020). Tourism and COVID-19: Impacts and implications for advancing and resetting industry and research. *Journal of Business Research*, *117*, 312–321. doi:10.1016/j.jbusres.2020.06.015 PMID:32546875

Silva, N. (2021). *Favela tour experience. The impacts in the host communities* [Master Thesis in Management]. University of Algarve, Faro, Portugal.

Simancas, M. (2020). Parar para repensar: ¿el fin del turismo como lo conocemos? Fortalezas del modelo turístico canario ante el escenario de incertidumbre y desconfianza de la crisis de la COVID-19. In *Turismo pos-COVID-19: reflexiones, retos y oportunidades* (pp. 689–706). Cátedra de Turismo CajaCanarias-Ashotel de la Universidad de La Laguna. doi:10.25145/b.Turismopos-COVID-19.2020

Simonis, U. E. (2011). Greening urban development: On climate change and climate policy. *International Journal of Social Economics*, *38*(11), 919–928. doi:10.1108/03068291111171423

Sinclair-Maragh, G. S., & Gursoy, D. (2015). Imperialism and tourism: The case of developing island countries. *Annals of Tourism Research*, *50*, 143–158. doi:10.1016/j.annals.2014.12.001

Sistema de Información Multiterritorial de Andalucía. (2021). *Andalucía pueblo a pueblo - Fichas municipales*. https://www.juntadeandalucia.es/institutodeestadisticaycartografia/sima/index.htm

Small, K., Edwards, D., & Sheridan, L. (2005). A flexible framework for evaluating the socio-cultural impacts of a (small) festival. *International Journal of Event Management Research, 1*(1), 66-77.

Smeeton. N. C., & Smeeton, E. A. (1985). The Interface Between English and Statistics. *Teaching Statistics, 7*(2), 34–37. doi:https://doi.org/10.1111/j.1467-9639.1985.tb00574.x

Smith, A., Voß, J. P., & Grin, J. (2010). Innovation studies and sustainability transitions: The allure of the multi-level perspective and its challenges. *Research Policy, 39*(4), 435–448. doi:10.1016/j.respol.2010.01.023

Smith, M. K. (2003). *Issues in Cultural Tourism Studies*. Routledge.

Smith, M. W. (1959). The importance of folklore studies to anthropology. *Folklore*, *70*(1), 300–312. doi:10.1080/0015587X.1959.9717162

Smith, V. L., Hetherington, A. Q., & Brumbaugh, M. D. D. (1986). California's Highway 89: A regional tourism model. *Annals of Tourism Research*, *13*, 415–433.

Sneddon, C. S. (2000). 'Sustainability' in ecological economics, ecology and livelihoods: A review. *Progress in Human Geography*, *24*(4), 521–549. doi:10.1191/030913200100189076

Snowball, J. D., & Courtney, S. (2010). Cultural heritage routes in South Africa: Effective tools for heritage conservation and local economic development? *Development Southern Africa*, *27*(4), 563–576.

Soares, J. R. R., Casado-Claro, M. F., Lezcano-González, M. E., Sánchez-Fernández, M. D., Gabriel, L. P. M. C., & Abríl-Sellarés, M. (2021). The Role of the Local Host Community's Involvement in the Development of Tourism: A Case Study of the Residents' Perceptions toward Tourism on the Route of Santiago de Compostela (Spain). *Sustainability*, *13*(17), 9576. doi:10.3390u13179576

Soini, K., Battaglini, E., Birkeland, I. N. D., & Fairclough, G. (2015). Culture in, for and as Sustainable Development. Conclusions from the COST Action IS1007 Investigating Cultural Sustainability. In *Culture in, for and as Sustainable Development*. Retrieved from http://www.culturalsustainability.eu/

Soini, K., & Birkeland, I. (2014). Exploring the scientific discourse on cultural sustainability. *Geoforum*, *51*, 213–223. doi:10.1016/j.geoforum.2013.12.001

Soler, P., Planas, A., Ciraso-Calí, A., & Ribot-Horas, A. (2014). Empoderamiento en la comunidad. El diseño de un sistema abierto de indicadores a partir de procesos de Evaluación Participativa. *Pedagogía Social. Revista Interuniversitaria, 24*, 44–77. doi:10.7179/PSRI_2014.24.03

Sonuç, N. (2020). Social Tourism: An Analysis from the Sustainability Perspective-İzmir Case. *International Journal of Contemporary Economics and Administrative Sciences, 10*(1), 35–52.

Sørensen, F., & Balsby, N. (2020). Brokers and saboteurs: Actor roles in destination innovation network development. *Tourism Planning & Development, 18*(5), 547–572. doi:10.1080/21568316.2020.1830300

Splinder, E. (2013). The history of sustainability. The origins and efects of a popular concept. In I. Jenkins & R. Schröder (Eds.), *Sustainability in tourism: A multidisciplinary approach* (pp. 9–31). Springer Gabler.

Squire, L. (2004). Memory systems of the brain: A brief history and current perspective. *Neurobiology of Learning and Memory, 82*(3), 171–177. doi:10.1016/j.nlm.2004.06.005 PMID:15464402

Squire, S. J. (1994). The Cultural Values of Literary Tourism. *Annals of Tourism Research, 21*(1), 103–120. doi:10.1016/0160-7383(94)90007-8

SSB. (2021). *Alders- og kjønnsfordeling i kommuner, fylker og hele landets befolkning (K) 1986-2021*. Statistics Norway. https://www.ssb.no/statbank/table/07459/

Stake, R. E. (2010). *Qualitative Research: Studying how things work*. The Guilford Press.

Stanfield, J. R. (1983). Toward an Ecological Economics. *International Journal of Social Economics, 10*(5), 27–37. doi:10.1108/eb013942

Star Cities Interreg Europe. (2021). *Sustainable tourism for attractivity of riverside cities*. Retrieved from Interreg Europe Star Cities: https://www.interregeurope.eu/starcities/

Starr, L. J. (2010). The use of autoethnography in educational research: Locating who we are in what we do. *Canadian Journal for New Scholars in Education, 3*(1), 1–9.

Stars and Stories. (n.d.). *List*. https://www.list.co.uk/event/657195-canongate-stars-and-stories/

Statista. (2021). *Number of foreign arrivals in hotel establishments on the Cyclades Islands in Greece from 2010 to 2018*. https://www.statista.com/statistics/881851/number-foreign-arrivals-cyclades-hotels

Stefa, G. (2011). *Oazi i Artë i Gjuhadolit* [Gjuhadol's Golden Oasis]. Gjergj Fishta. Wisegeek. What is Literary Tourism? https://www.wise-geek.com/what-is-literary-tourism.htm

Sthapit, E. (2019). Memories of gastronomic experiences, savoured positive emotions and savouring processes. *Scandinavian Journal of Hospitality and Tourism, 19*(2), 115–139. doi:10.1080/15022250.2017.1402702

Sthapit, E., & Coudounaris, D. (2017). Memorable tourism experiences: Antecedents and outcomes. *Scandinavian Journal of Hospitality and Tourism, 18*(1), 72–94. doi:10.1080/15022250.2017.1287003

Stilgoe, J., Owen, R., & Macnaghten, P. (2013). Developing a framework for responsible innovation. *Research Policy, 42*(9), 1568–1580. doi:10.1016/j.respol.2013.05.008

Stocker, L., & Barnett, K. (1998). The significance and praxis of community-based sustainability projects: Community gardens in western Australia. *Local Environment, 3*(2), 179–189. doi:10.1080/13549839808725556

Stoddart, H., & Rogerson, C. (2009). Tourism routes, local economic promotion and propoor development: The case of the crocodile ramble. *Africa Insight, 38*(4), 10–26.

Stöhr, W., & Taylor, F. (Eds.). (1981). *Development from above or below? The Dialectics of Regional Planning in Developing Countries*. Wiley & Sons Ltd.

Suárez, F. M. (n.d.). *Problemas Sociales y Problemas de Programas Sociales Masivos*. CEPAL.

Su, J. (2019). Understanding the changing Intangible Cultural Heritage in tourism commodification: The music players' perspective from Lijiang, China. *Journal of Tourism and Cultural Change*, *17*(3), 247–268. doi:10.1080/14766825.2018.1427102

Sukanthasirikul, K., & Trongpanich, W. (2016). Cultural Tourism Experience on Customer Satisfaction: Evidence From Thailand. *Journal of Economic and Social Development*, *3*(1), 17–25.

Su, M. M., & Wall, G. (2014). Community participation in tourism at a world heritage site: Mutianyu great wall, Beijing, China. *International Journal of Tourism Research*, *16*(2), 146–156. https://doi.org/10.1002/jtr.1909

Su, M. M., & Wall, G. (2015). Exploring the Shared Use of World Heritage Sites: Residents and Domestic Tourists' Use and Perceptions of the Summer Palace in Beijing. *International Journal of Tourism Research*, *17*(6), 591–601. https://doi.org/10.1002/jtr.2026

Su, X. (2018). Reconstructing Tradition: Heritage Authentication and Tourism-Related Commodification of the Ancient City of Pingyao. *Sustainability*, *10*(3), 670. doi:10.3390u10030670

Su, X., Li, X., Chen, W., & Zeng, T. (2020). Subjective vitality, authenticity experience, and intangible cultural heritage tourism: An empirical study of the puppet show. *Journal of Travel & Tourism Marketing*, *37*(2), 258–271. doi:10.1080/10548408.2020.1740141

Su, X., Li, X., Wu, Y., & Yao, L. (2020). How is intangible cultural heritage valued in the eyes of inheritors? Scale development and validation. *Journal of Hospitality & Tourism Research (Washington, D.C.)*, *44*(5), 806–834. doi:10.1177/1096348020914691

Su, Z., Aaron, J. R., McDowell, W. C., & Lu, D. D. (2019). Sustainable synergies between the cultural and tourism industries: An eciency evaluation perspective. *Sustainability (Switzerland)*, *11*(23), 6607. Advance online publication. doi:10.3390u11236607

Swanson, K. K., & DeVereaux, C. (2012). Culturally Sustainable Entrepreneurship: A Case Study for Hopi Tourism. In K. F. Hyde, C. Ryan, & A. G. Woodside (Eds.), Field Guide to Case Study Research in Tourism, Hospitality and Leisure (Vol. 6, pp. 479–494). Emerald Group Publishing Limited. https://doi.org/10.1108/S1871-3173(2012)0000006029

Swarbrooke, J. (1994). The future of the past: heritage tourism in the 21st century. In A. V. Seaton (Ed.), *Tourism, the State of the Art* (pp. 222–229). John Wiley & Sons.

Tajfel, H. (1981). *Human Groups and Social Categories: Studies in Social Psychology*. Cambridge University Press.

Tang, M. D. (1991). Evaluation of the tourist resources of the northern Silk Road, Xinjiang. *Chinese Journal of Arid Land Research*, *4*(4), 329–337.

Tanguay, G. A., Rajaonson, J., & Therrien, M.-C. (2013). Sustainable tourism indicators: Selection criteria for policy implementation and scientific recognition. *Journal of Sustainable Tourism*, *21*(6), 862–879. doi:10.1080/09669582.2012.742531

Tarolli, P., Preti, F., & Romano, N. (2014). Terraced landscapes: From an old best practice to a potential hazard for soil degradation due tol and abandonment. *Anthropocene*, *6*, 10–25. doi:10.1016/j.ancene.2014.03.002

Tavares, F., Irving, M., Oliveira, E., & Madureira, M. (2018). Slums – Places to visit or to avoid? The contradictions in the media on tours in the favelas of Rio de Janeiro. *Revista Verso e Reverso*, *32*(81), 168–186. doi:10.4013/ver.2018.32.81.01

Taylor, E. (2016). *Culture, tourism and sustainability: an ethnographic study of rural community development in Jamaica* (Unpublished PhD Thesis). Coventry: Coventry University.

Teather, E. K., & Chow, C. S. (2003). Identity and place: The testament of designated heritage in Hong Kong. *International Journal of Heritage Studies*, *9*(2), 93–115.

Temesgen, A. K., Storsletten, V. M., & Jakobsen, O. D. (2019). Circular economy–Reducing symptoms or radical change? *Philosophy of Management*, *19*, 1–20.

Terkenli, T. S. (2000). Landscapes of tourism: a cultural geographic perspective. In H. Briassoulis & J. Van Der Straaten (Eds.), *Tourism and the Environment: Regional, Economic, Cultural and Policy Issues* (2nd ed., pp. 179–202). Kluwer Academic Publishers. doi:10.1007/978-94-015-9584-1_11

Terkenli, T. S., Bellas, M. L., & Dudley-Jenkins, L. (2007, June-December). Tourism impacts on local life: Socio-cultural continuity and change in Crete, Aegean. *The Geographical Journal*, *16*(1-2), 37–52.

Terzić, A., Bjeljac, Ž., & Jovanović, R. (2014b). Protection, revitalization and use of cultural heritage through cultural route formation system: Evaluation of the Roman Emperor's Path master plan. *Kultura*, *143*, 319–335.

Terzić, A., Bjeljac, Ž., Jovičić, A., & Penjišević, I. (2014a). Cultural Route and Ecomuseum Concepts as a Synergy of Nature, Heritage and Community Oriented Sustainable Development. *European Journal of Sustainable Development*, *3*(2), 1–16.

Tettamanti, M., Rognoni, E., Cafiero, R., Costa, T., Galati, D., & Perani, D. (2012). Distinct pathways of neural coupling for different basic emotions. *NeuroImage*, *59*(2), 1804–1817. doi:10.1016/j.neuroimage.2011.08.018 PMID:21888979

Teye, V. B., & Timothy, D. J. (2004). The varied colors of slave heritage in West Africa: White American stakeholders. *Space and Culture*, *7*(2), 145–155.

TGA (Türkiye Tourism Promotion and Development Agency). (n.d.). https://tga.gov.tr/home/

Thanh, V. M., & Bosch, O. J. H. (2010). Systems Thinking Approach as a Unique Tool for Sustainable Tourism Development: A Case Study in the Cat Ba Biosphere Reserve of Vietnam. In *54th Annual Conference of the International Society for the Systems Sciences (ISSS) 2010: Governance for a Resilient Planet, 18-23 July 2010*. Waterloo: ISSS.

The United Nations Educational, Scientific and Cultural Organization. (2011). *What is Intangible Cultural Heritage?* https://ich.unesco.org/en/what-is-intangible-heritage-00003

The United Nations Educational, Scientific and Cultural Organization. (2015). *Intangible cultural heritage and sustainable development.* https://ich.unesco.org/doc/src/34299-EN.pdf

The United Nations Educational, Scientific and Cultural Organization. (2021). *Dragon Boat festival.* https://ich.unesco.org/en/RL/dragon-boat-festival-00225

The United Nations Educational, Scientific and Cultural Organization. (2021). *Text of the Convention for the Safeguarding of the Intangible Cultural Heritage.* https://ich.unesco.org/en/convention#art2

The United Nations Educational, Scientific and Cultural Organization. (2021). *Yeondeunghoe, lantern lighting festival in the Republic of Korea.* https://ich.unesco.org/en/RL/yeondeunghoe-lantern-lighting-festival-in-the-republic-of-korea-00882

The United Nations Environment Programme and the United Nations World Tourism Organization. (2005). *Making Tourism More Sustainable: A Guide for Policy Makers*. The United Nations Environment Programme and the United Nations World Tourism Organization.

The United Nations World Tourism Organization. (2021). *Tourism 4 SDGS*. https://www.unwto.org/tourism4sdgs

Thees, H., Pechlaner, H., Olbrich, N., & Schuhbert, A. (2020). The Living Lab as a tool to promote residents' participation in destination governance. *Sustainability*, *12*(3), 1120. doi:10.3390u12031120

Thimm, T. (2019). Cultural sustainability – a framework for Aboriginal tourism in British Columbia. *Journal of Heritage Tourism*, *14*(3), 205–218. doi:10.1080/1743873X.2018.1484469

Thomas, J., & Harden, A. (2008). Methods for the thematic synthesis of qualitative research in systematic reviews. *BMC Medical Research Methodology*, *8*, 1–10. https://doi.org/10.1186/1471-2288-8-45

Thomson, I. (2004). Bonjour blanc: A journey through Haiti (4th ed., Vol. 1). Random House and Vintage Books.

Thomson, A. X. (2021). Commensality, Communitas and Contestation on the Camino Francés: Making Pilgrims through Meals. In D. Liutikas (Ed.), *Pilgrims: Values and Identities* (pp. 94–104). CABI. doi:10.1079/9781789245653.0094

Throsby, D. (2003). Cultural sustainability. In R. Towse (Ed.), *A handbook of cultural economics* (pp. 183–186). Edward Elgar. doi:10.4337/9781781008003.00028

Throsby, D. (2008). Linking ecological and cultural sustainability. *The International Journal of Diversity in Organisations, Communities and Nations*, *8*(1), 15–20. doi:10.18848/1447-9532/CGP/v08i01/39529

Tilden, F. (2009). *Interpreting our heritage*. University of North Carolina Press.

Timothy, D. J. (2007). Empowerment and stakeholder participation in tourism destination communities. In A. Church & T. Coles (Eds.), *Tourism, Power and Space* (pp. 199–216). Routledge.

Timothy, D. J. (2009). River-based tourism in the USA: Tourism and recreation on the Colorado and Mississippi Rivers. In B. Prideaux & M. Cooper (Eds.), *River Tourism* (pp. 41–54). CABI.

Timothy, D. J. (2014). Contemporary cultural heritage and tourism: Development issues and emerging trends. *Public Archaeology*, *13*(1-3), 1–12.

Timothy, D. J., & Boyd, S. W. (2003). *Heritage Tourism*. Prentice Hall.

Timothy, D. J., & Boyd, S. W. (2006). Heritage tourism in the 21st century: Valued tradi- tions and new perspectives. *Journal of Heritage Tourism*, *1*(1), 1–17.

Timothy, D. J., & Nyaupane, G. N. (2009). *Cultural Heritage and Tourism in the Developing World, A Regional Perspective*. Routledge. doi:10.4324/9780203877753

Timothy, D. J., & Olsen, D. H. (2006). *Tourism, Religion and Spiritual Journeys*. Routledge. doi:10.4324/9780203001073

Timothy, D. J., & Teye, V. B. (2004). American children of the African diaspora: Journeys to the motherland. In T. Coles & D. J. Timothy (Eds.), *Tourism, Diasporas and Space* (pp. 111–123). Routledge.

Timothy, D., & Boyd, S. W. (2014). *Tourism and trails: Cultural, ecological and management issues*. Channel View Publications.

Timur, S., & Getz, D. (2008). A network perspective on managing stakeholders for sustainable urban tourism. *International Journal of Contemporary Hospitality Management*, *20*(4), 445–461. doi:10.1108/09596110810873543

Timur, S., & Getz, D. (2009). Sustainable tourism development: How do destination stakeholders perceive sustainable urban tourism? *Sustainable Development*, *17*(4), 220–232. doi:10.1002d.384

Titterton, M., & Smart, H. (2008). Can participatory research be a route to empowerment? A case study of a disadvantaged Scottish community. *Community Development Journal: An International Forum*, *43*(1), 52–64. https://www.jstor.org/stable/44258060

Tobar, L. E., Castillo, M., Mendoza, R., & Tamayo, A. L. (2015). La visión crítica en el construir del conocimiento turístico. *Journal of Tourism Analysis*, *19*, 33–44.

Todd, R., Talmi, D., Schmitz, T., Susskind, J., & Anderson, A. (2012). Psychophysical and neural evidence for emotion enhanced perceptual vividness. *The Journal of Neuroscience: The Official Journal of the Society for Neuroscience*, *32*(33), 11201–11212. doi:10.1523/JNEUROSCI.0155-12.2012 PMID:22895705

Tolić, I. (2020). Creating and managing experiences in cultural tourism. In Croatian International Relations Review (Vol. 26, Issue 86). Institute for Development and International Relations. doi:10.1142/10809

Torggler, B., & France, C. (2015). *UNESCO's Work on Culture and Sustainable Development Evaluation of a Policy Theme.* https://unesdoc.unesco.org/images/0023/002344/234443e.pdf

Torres-Delgado, A., & López Palomeque, F. (2014). Measuring sustainable tourism at the municipal level. *Annals of Tourism Research*, *49*, 122–137. doi:10.1016/j.annals.2014.09.003

Torres, F. E. J. (2019). *Bem-estar comunitário e visitantes através do Caminho em Santiago. Grandes narrativas, ideias e práticas culturais na cidade*. ANDAVIRA.

Tosun, C. (2000). Limits to community participation in the tourism development process in developing countries. *Tourism Management*, *21*(6), 613–633. doi:10.1016/S0261-5177(00)00009-1

Towner, J. (1995). What is tourism's history? *Tourism Management*, *16*(5), 339–343. https://doi.org/https://doi.org/10.1016/0261-5177(95)00032-J

Troitiño, M. A., & Troitiño, L. (2018). Visión territorial del patrimonio y sostenibilidad del turismo. *Boletín de la Asociación de Geógrafos Españoles*, *78*(78), 212–244. doi:10.21138/bage.2716

Tsartas, P. (2003). Tourism Development in Greek Insular and Coastal Areas: Sociocultural Changes and Crucial Policy Issues. *Journal of Sustainable Tourism*, *11*(2&3), 116–132. doi:10.1080/09669580308667199

Tsartas, P., Zagotsi, S., & Kyriakaki, A. (2020). *Τουρίστες, Ταξίδια, Τόποι. Κοινωνιολογικές Προσεγγίσεις στον Τουρισμό [Tourists, Travels, Places. Sociological Perspectives to Tourism]*. Kritiki Publications.

Tung, V., Lin, P., Zhang, H., & Zhao, A. (2017). A framework of memory management and tourism experiences. *Journal of Travel & Tourism Marketing*, *34*(7), 853–866. doi:10.1080/10548408.2016.1260521

Tung, V., & Ritchie, J. R. B. (2011). Exploring the essence of memorable tourism experiences. *Annals of Tourism Research*, *38*(4), 1367–1386. doi:10.1016/j.annals.2011.03.009

Tura, N., Mortimer, G., & Kutvonen, A. (2019). Exploring the Pitfalls of Systemic Innovations for Sustainability. In N. Bocken, P. Ritala, L. Albareda, & R. Verburg (Eds.), *Innovation for Sustainability* (pp. 157–175). Palgrave Macmillan. doi:10.1007/978-3-319-97385-2_9

Türker, A., & Çelik, İ. (2012). Somut Olmayan Kültürel Miras Unsurlarının Turistik Ürün Olarak Geliştirilmesine Yönelik Alternative Öneriler [Alternative Proposals for Tourist Product Development of Intangible Cultural Heritage Elements]. *Yenifikir*, *9*, 86–98.

Ünal, Z. G. (2014). *İSMEP Rehber Kitaplar 6, Kültürel Mirasın Korunması* [Protection of Cultural Heritage]. Beyaz Gemi Sosyal Proje Ajansı, İstanbul Valiliği.

Underberg-Goode, N. (2014). Cultural heritage tourism on Peru's north coast. *Worldwide Hospitality and Tourism Themes, 6*. doi:10.1108/WHATT-03-2014-0013

UNEP. (2011). *Towards a Green Economy: Pathways to Sustainable Development and Poverty Eradication*. UN environment programme. http://www.unep.org/greeneconomy/Portals/ 88/documents/ger/ger_final_dec 2011/Green%20 EconomyReport_Final Dec_2011.pdf

UNESCO Tentative Lists - Turkey. (n.d.). http://whc.unesco.org/en/tentativelists/state=tr

UNESCO Türkiye Milli Komisyonu. (n.d.). https://www.unesco.org.tr/Pages/125/122/UNESCO-D%C3%BCnya-Miras%C4%B1-Listesi

UNESCO World Heritage Sustainable Tourism Toolkit. (2020). Retrieved from World Heritage UNESCO: http://whc.unesco.org/sustainabletourismtoolkit/

UNESCO. (1992-2010). *Røros mining town and the circumference*. UNESCO World Heritage Centre. http://whc.unesco.org/en/list/55

UNESCO. (2009). *The 2009 UNESCO Framework for Cultural Statistics (FCS)*. UNESCO Institute for Statistics.

UNESCO. (2012). *The Slave Route*. https://en.unesco.org/themes/fostering-rights-inclusion/slave-route

UNESCO. (2020). *Cutting Edge | Bringing Cultural Tourism Back in the Game*. https://en.unesco.org/news/cutting-edge-bringing-cultural-tourism-back-game

UNESCO. (2020). Retrieved from UNESCO Culture for Development Indicators (CDIS): https://en.unesco.org/creativity/sites/creativity/files/cdis_methodology_manual_0_0.pdf

UNESCO. (2021a). *UNESCO World Heritage and Sustainable Tourism Programme*. https://whc.unesco.org/en/tourism/

UNESCO. (2021b). *UNESCO Intangible Cultural Heritage, Turkey, Elements on the Lists of Intangible Cultural Heritage*. https://ich.unesco.org/en/state/turkey-TR?info=elements-on-the-lists

UNESCO. (n.d.). *¿Qué es el patrimonio cultural inmaterial?* Patrimonio Cultural Inmaterial. https://ich.unesco.org/es/que-es-el-patrimonio-inmaterial-00003

United Nations Brundtland Commission. (1987). *Our Common Future*. Oxford University Press.

United Nations World Tourism Organization. (2018). *Report on Tourism and Culture Synergies*. UNWTO.

United Nations. (2015). Transforming our World: the 2030 Agenda for Sustainable Development. Resolution adopted by the General Assembly on 25 September 2015. New York: United Nations.

Universidad De Costa Rica. (1987). *Instructivo para la elaboración del Diario de Campo*. Author.

Università degli Studi di Perugia. (2018). Linee guida per l'adozione di strumenti e modelli condivisi di rilevazione dell'impatto economico, sociale ed ambientale dei flussi turistici relativi ai percorsi previsti dal progetto nazionale dei Cammini d'Italia. Perugia.

University of Maribor. (2021). *Deliverable 3.2.3. BEST MED Open Platform*. Interreg Med Programme.

UNWTO (United Nations World Tourism Organization). (n.d.). https://www.unwto.org/unwto-tourism-dashboard

UNWTO INSTO. (2020). Retrieved from UNWTO INSTO: http://insto.unwto.org/

UNWTO. (2004). *Indicators of Sustainable Development for Tourism Destinations: A Guidebook.* World Tourism Organization.

UNWTO. (2015). *Global Report on Cultural Routes and Itineraries.* UNWTO.

UNWTO. (2018). Report on tourism and culture synergies. In *Tourism and Culture Synergies.* World Tourism Organization (UNWTO). https://www.e-unwto.org/doi/book/10.18111/9789284418978

UNWTO. (2020). *Cultural Tourism and COVID19.* Available on https://www.unwto.org/cultural-tourism-covid-19

UNWTO. (2021, May).. . *World Tourism Barometer, 19*(3). Advance online publication. doi:10.18111/wtobarometereng

Usakli, A., & Kucukergin, K. G. (2018). Using partial least squares structural equation modeling in hospitality and tourism: Do researchers follow practical guidelines? *International Journal of Contemporary Hospitality Management, 30*(11), 3462–3512. doi:10.1108/IJCHM-11-2017-0753

Uygun, M., & Ekiz, E. (2016). An Overview of Health Tourism within the Context of Turkey's Tourism Strategy (2023). *Journal of Hospitality and Tourism, 14*(1), 18–26.

Valeri, M., & Baggio, R. (2021). Social network analysis: Organizational implications in tourism management. *The International Journal of Organizational Analysis, 9*(2), 342–353. doi:10.1108/IJOA-12-2019-1971

Valladares, L., & Olivé, L. (2015). ¿Qué son los conocimientos tradicionales? Apuntes epistemológicos para la interculturalidad. *Cultura y Representaciones Sociales, 19*, 61–101.

Valverde, L. (1993). El Diario de Campo. *Revista de Trabajo Social, 12*(10), 308–319. https://www.binasss.sa.cr/revistas/ts/v18n391993/art1.pdf

Van Buuren, A., & Loorbach, D. (2009). Policy innovation in isolation? Conditions for policy renewal by transition arenas and pilot projects. *Public Management Review, 11*(3), 375–392. doi:10.1080/14719030902798289

Van Der Walt, J. S., Buitendag, A. A., Zaaiman, J. J., & Van Vuuren, J. J. (2009). Community living lab as a collaborative innovation environment. *Issues in Informing Science and Information Technology, 6*(1), 421–436. doi:10.28945/1070

van Popering-Verkerk, J., & van Buuren, A. (2017). Developing collaborative capacity in pilot projects: Lessons from three Dutch flood risk management experiments. *Journal of Cleaner Production, 169*, 225–233. doi:10.1016/j.jclepro.2017.04.141

Vana, M. V., & Malaescu, S. (2016). Cultural thematic travel itineraries: Mediators of success. *Procedure Economics and Finance, 39*, 642–652.

Vargas-Hernández, J. G. (2012). *Heritage Tourism in Regional Development of.* Academic Press.

Vargas-Hernández, J. G. (2012). Sustainable cultural and heritage tourism in regional development of Southern Jalisco. *World Journal of Entrepreneurship, Management and Sustainable Development, 8*(2/3), 146–161. doi:10.1108/20425961211247752

Vargas-Sánchez, A., Porras-Bueno, N., & Plaza-Mejía, M. (2011). Explaining residents' attitudes to tourism: Is a universal model possible? *Annals of Tourism Research, 38*, 460–480. https://doi.org/10.1016/j.annals.2010.10.004

Varotto, M. (2015). Terraced landscapes of the Alps: Decay, rediscovery, revitalization. In T. Tillmann, M. Bueno de Mezquita. In *II Congreso Internacional de Terrazas. Encuentro de culturas y saberes de terrazas del mundo* (pp. 38-48). Cuzco.

Vélez, M. (2007). Papel de la población local en procesos de valoración turística. Caso: Turismo Rural en la Región del Alto Ricaurte (Boyacá): 2002-2003. *Anuario Turismo y Sociedad*, *8*(Noviembre), 144–161.

Vera-Rebollo, J. F., & Ivars-Baidal, J. A. (2020). *El impacto del COVID-19 en la actividad turística de la Comunitat Valenciana* (N.° 1; Prospect Comunitat Valenciana, p. 10). https://prospectcv2030.com/wp-content/uploads/2020/04/INFORME-TURISMO-Y-COVID19-CV.pdf

Verd, J. M., & Martí, J. (1999). Muestreo y recogida de datos en el análisis de redes sociales. *Qüestiió. Quaderns d'Estadística, Sistemes, Informàtica i Investigació Operativ*, *23*(3), 507–524.

Vergori, A. S., & Arima, S. (2020). Cultural and non-cultural tourism: Evidence from Italian experience. *Tourism Management*, *78*, 104058. Advance online publication. doi:10.1016/j.tourman.2019.104058

Vieira, B. (1998). The Construction of National Identity – On Primordialism and Instrumentalism. *Human Affairs*, *8*(1), 29–43.

Viera, A. J., & Garrett, J. M. (2005). Understanding interobserver agreement: The kappa statistic. *Family Medicine*, *37*(5), 360–363.

Vilaça, H. (2010). Pilgrims and pilgrimages: Fátima, Santiago de Compostela and Taizé. *Nordic Journal of Religion and Society*, *23*(2), 137–155. doi:10.18261/ISSN1890-7008-2010-02-03

Villoro, L. (1982). *Creer, Saber y Conocer*. Siglo XXI Editores.

Visit Vega. (2021). *Vega World Heritage Centre*. Vega verdensarv-senter. https://www.visitvega.no/ no/vega-verdensarvsenter

Vitić–Ćetković, A., Jovanović, I., & Potočnik Topler, J. (2020). Literary tourism: The role of Russian 19th century travel literature in the positioning of the smallest European capital – Cetinje. *Annales. Series Historia et Sociologia*, *30*(1), 81–98.

Vlizos, S., & Kosta, E. (2020) Designing Cultural Routes in the region of Sparta, Peloponnese: A Methodological approach. *Proceedings of the 6th UNESCO UNITWIN Conference 2019*, 299-306.

Vreugdenhil, H., Slinger, J., Thissen, W., & Rault, P. K. (2010). Pilot projects in water management. *Ecology and Society*, *15*(3), art13. doi:10.5751/ES-03357-150313

Wachyuni, S. S., & Kusumaningrum, D. A. (2020). The Effect of COVID-19 Pandemic: How are the Future Tourist Behavior? *Journal of Education. Society and Behavioural Science*, *33*(4), 67–76.

Waligo, V. M., Clarke, J., & Hawkins, R. (2013). Implementing sustainable tourism: A multi-stakeholder involvement management framework. *Tourism Management*, *36*, 342–353. doi:10.1016/j.tourman.2012.10.008

Wallis, M. (1992). *Route 66: The Mother Road*. St. Martin's Press.

Walsh, C. (2005). La Interculturalidad en la Educación. In O. Chávez (Ed.), Journal of Chemical Information and Modeling (Vol. 53, Issue 9). UNICEF. doi:10.1017/CBO9781107415324.004

Wang, S., & Hji-Avgoustis, S. (2011). Evaluating costs and benefits of a tourism project: A case study of the Indianapolis Cultural Trail. *Tourism Today (Nicosia)*, *11*, 172–181.

Wan, Y. K. P., & Li, X. (2013). Sustainability of Tourism Development in Macao, China. *International Journal of Tourism Research*, *15*(1), 52–65. https://doi.org/10.1002/jtr.873

Wasserman, S., & Faust, K. (2013). *Análisis de Redes Sociales. Métodos y Aplicaciones*. Centro de Investigaciones Sociológicas.

Wasserman, S., & Galaskiewicz, J. (1994). *Advances in Social Networks Analysis*. SAGE Publications. doi:10.1017/CBO9780511815478

Wattanacharoensil, W., & Sakdiyakorn, M. (2016). The Potential of Floating Markets for Creative Tourism: A Study in Nakhon Pathom Province, Thailand. *Asia Pacific Journal of Tourism Research, 21*(sup1), S3-S29. doi:10.1080/10941665.2014.998250

Weaver, D. B. (2006). *Sustainable Tourism: Theory and Practice*. Elsevier Butterworth-Heinemann.

Weaver, D. B. (2009). Reflections on sustainable tourism and paradigm change. In S. Gössling, C. M. Hall, & D. Weaver (Eds.), *Sustainable tourism futures* (pp. 53–60). Routledge.

Webster, L., & Mertova, P. (2007). *Using narrative inquiry as a research method: An introduction to using critical event narrative analysis in research on learning and teaching*. Routledge. doi:10.4324/9780203946268

Wellman, B. (2007). The network is personal. Introduction to a special issue of Social Networks. *Social Networks, 29*(3), 349–356. doi:10.1016/j.socnet.2007.01.006

Wenger, E. (2003). Communities of Practice and Social Learning systems. In D. Nicolini, S. Gerardi, & D. Yanow (Eds.), *Knowing in organizations: A practice-based approach* (pp. 76–99). ME Sharpe.

Williams, H. A., Yuan, J., & Williams, R. L. Jr. (2019). Attributes of Memorable Gastro-Tourists' Experiences. *Journal of Hospitality & Tourism Research (Washington, D.C.), 43*(3), 327–348. doi:10.1177/1096348018804621

Williams, P. W., & Ponsford, I. F. (2009). Confronting tourism's environmental paradox: Transitioning for sustainable tourism. *Futures, 41*(6), 396–404. doi:10.1016/j.futures.2008.11.019

Williams, R. (1958). *Culture and Society*. Chatto & Windus.

Wironen, M. B., & Erickson, J. D. (2020). A critically modern ecological economics for the Anthropocene. *The Anthropocene Review, 7*(1), 62–76. doi:10.1177/2053019619884485

Wolf, E. (1956). Aspects of Group Relations in a Complex Society: Mexico. *American Anthropologist, 58*(6), 1065–1078. doi:10.1525/aa.1956.58.6.02a00070

Wolf, E. (1972). Ownership and Political Ecology. *Anthropological Quarterly, 45*(3), 201–205. doi:10.2307/3316532

Wong, K. (2013). Partial Least Squares Structural Equation Modeling (PLS-SEM) Techniques Using SmartPLS. *Marketing Bulletin - Technical Note, 24*, 1–32.

Wood, A. F. (2010). Two roads diverge: Route 66, "Route 66," and the mediation of American ruin. *Critical Studies in Media Communication, 27*(1), 67–83. doi:10.1080/15295030903554375

Woolcock, M., & Narayan, D. (2000). Social Capital. Implications for Development Theory, Research and Policy. *The World Bank Research Observer, 15*(2), 225–249. doi:10.1093/wbro/15.2.225

World Population Review. (2021). *Most Visited Countries*. https://worldpopulationreview.com/country-rankings/most-visited-countries

World Tourism Organization. (2017), *General Assembly - Twenty-second session*. UNWTO.

World Tourism Organization. (2018). *Tourism and Culture Synergies*. UNWTO., doi:10.18111/9789284418978

World Tourism Organization. (n.d.). *Sustainable Development*. https://www.unwto.org/sustainable-development

World Tourism Organization. (n.d.). *Tourism and Culture*. https://www.unwto.org/tourism-and-culture

WTO. (2001). *The concept of sustainable tourism.* http://www.worldtourism.org/sustainable/concepts.htm

WTO. (2002). *World tourism stalls in 2001.* https://www.world-tourism.org/

WTTC. (2020). *Travel and Tourism-Global Economic Impact & Trends 2020.* WTTC.

Wu, K.L. (2014). *Macao Belief and Customs of Na Tcha.* Joint Publishing.

Xu, J., Lü, Y., Chen, L., & Liu, Y. (2009). Contribution of tourism development to protected area management: Local stakeholder perspectives. *International Journal of Sustainable Development and World Ecology, 16*(1), 30–36. https://doi.org/10.1080/13504500902757189

Yamashita, S. (2016). In J. Jafari & H. Xiao (Eds.), *Cultural tourism BT - Encyclopedia of Tourism* (pp. 212–214). Springer International Publishing. doi:10.1007/978-3-319-01384-8_45

Yankholmes, A. (2018). Stakeholders' Opinions on How to Brand and Diversify Macau's Tourism Product. In *Workshop Proceedings of the Inclusive Innovation for Enhanced Local Experience in Tourism, Phuket, Thailand* (pp. 11-15). University of Birmingham.

Yankholmes, A. K. B., & Boakye, K. A. (2010). 'Awusai Atso': Community attachment to and use of transatlantic slave trade resources in Danish-Osu, Ghana. *Journal of Heritage Tourism, 5*(1), 49–67.

Yeh, S. S. (2021). Tourism recovery strategy against COVID-19 pandemic. *Tourism Recreation Research, 46*(2), 188–194. doi:10.1080/02508281.2020.1805933

Yenişehirlioğlu, E., & Arslan Kalay, H. (2017). *UNESCO Dünya Miras Listesi'nin Kültürel Miras Alanlarındaki Ziyaretçi Sayılarına Etkisi* [The Impact of UNESCO World Heritage List on The Number Of Visitors To Cultural Heritage Sites] [Paper presentation]. *International Congress on Cultural Heritage and Tourism (ICCHT-2017).*

Yun, H. J., & Zhang, X. (2017). Cultural Conservation and Tourism Development in the Consolidation Stage of the Tourism Area Life Cycle Model. *Tourism Planning & Development, 14*(3), 353–368. https://doi.org/10.1080/21568316.2016.1243147

Zabbini, E. (2012). Cultural routes and intangible heritage. *Almatourism-Journal of Tourism. Culture and Territorial Development, 3*(5), 59–80. doi:10.6092/issn.2036-5195/3188

Zabetta, M. C., Sacerdotti, S. L., & Mauro, S. (2014). Community-based monitoring in tourism sector: An application of the European Tourism Indicators System in the "ATL del Cuneese". *Leadership and Governance for Sustainable Tourism: Proceedings of Summer School, 2014*, 29–36.

Zavratnik, V., Superina, A., & Stojmenova Duh, E. (2019). Living Labs for Rural Areas: Contextualization of Living Lab frameworks, concepts and practices. *Sustainability, 11*(14), 3797. doi:10.3390u11143797

Zehrer, A. (2009). Service experience and service design: Concepts and application in tourism SMEs. *Managing Service Quality, 19*(3), 332–349. doi:10.1108/09604520910955339

Zimmer, L. (2006). Qualitative meta-synthesis: A question of dialoguing with texts. *Journal of Advanced Nursing, 53*(3), 311–318. https://doi.org/https://doi.org/10.1111/j.1365-2648.2006.03721.x

Zizka, L., McGunagle, D. M., & Clark, P. J. (2021). Sustainability in science, technology, engineering and mathematics (STEM) programs: Authentic engagement through a community-based approach. *Journal of Cleaner Production, 279*, 123–715. doi:10.1016/j.jclepro.2020.123715

Zografos, C., & Howarth, R. B. (2010). Deliberative ecological economics for sustainable development. *Sustainability, 2*(11), 3399–3417. doi:10.3390u2113399

About the Contributors

Cláudia Ribeiro de Almeida is Professor of Tourism at University of the Algarve. Tourism Course Director since 2017. Holds a PhD and a Post Doctoral Degree in Tourism (University of Aveiro). Current research interests include Air transport & Tourism dynamics; Second home tourism, Cultural tourism and Decision making process. Member of the Research Unit – CinTurs. Reviewer of many international journals and researcher in several national and international funded projects.

Alexandra R. Gonçalves is Assistant Professor at the University of Algarve, School of Management, Hospitality and Tourism. Integrated Researcher of the Research Centre for Tourism, Sustainability and Well-being (CinTurs) for the areas of Tourism, Sustainability, Territory, Heritage, Museums, Cultural Management and Creative Industries. PhD Degree in Tourism. Master's Degree in Cultural Heritage Management. Post-graduate degree in Cultural Heritage Law. Graduate in Marketing. Regional Director of Culture of the Algarve from December 16, 2013 to December 15, 2018; Councilor of the Municipality of Faro (Oct. 2009-Oct. 2013). Since October 30th, 2019, is the Director of the School of Management, Hospitality and Tourism (University of Algarve). Has several published research in the areas of tourism, cultural experiences, cultural and creative tourism, cultural heritage management and museums, events evaluation. Was the regional researcher responsible for the project funded by the Portuguese Science and Technology Foundation -CREATOUR-Creative Tourism Destination Development in Small Cities and Rural Areas (2016-2020). Currently is the UALG coordinator for the iHERITAGE project, ENI CBC MED Program (2020-2023).

Sílvia Quinteiro is Coordinating Professor at University of the Algarve – School of Management, Hospitality and Tourism, where she teaches since 1994 and, among other positions and functions, was President of the Technical and Scientific Council. She holds a first degree in Modern Languages and Literatures (Portuguese and German Studies) from New University of Lisbon, a Master's and a PhD degree in Comparative Literature from the University of Lisbon. She founded and coordinated the Research Cluster in Literature and Tourism Studies: LIT&TOUR (2012-2020). Her research interests include Comparative Literature, and the relation between Literature and Tourism. She is a full member of CIAC – Centre for Research in Arts and Communication (University of the Algarve) member of the Rede Entremeio: Geography, Tourism and Literature Research Network of the Federal University of Rio de Janeiro and collaborates with project LITESCAPE. PT Literary Atlas of Continental Portugal. She coordinates the Literary Route of the Algarve, and has authored several national and international scientific publications.

Maria Laura Gasparini is a tourism researcher and consultant with a solid background in the private and non-profit sectors gained in Latin America, Europe and Oceania. She holds a Master's Degree in Tourism Economics and Management from the University of Bologna (Italy) and a Degree in Tourism from the National University of Lanús (Buenos Aires, Argentina). She has specialized in the role of sustainability indicators as policy-making tools in European destinations. She is currently Research Fellow and PhD candidate at the Center for Advanced Studies in Tourism (CAST), University of Bologna, being her main research focus on rural development, Cultural Routes of the Council of Europe, sustainability measuring through sustainable tourism indicators.

* * *

Pedro Azevedo is a PhD student in Development, Societies and Territories at the University of Trás-os-Montes and Alto Douro - UTAD (Portugal) and researcher at Research Centre for Transdisciplinary Studies for Development (CETRAD-UTAD).

José Juan Cano Delgado is a PhD in Geography. Deputy Director of the La Laguna World Heritage Cultural Chair of University of La Laguna. He has been a professor of the Degree in Tourism at the University School of Tourism of S / C de Tenerife (EUTUR) since 2018 and professor-tutor at the National University of Distance Education (UNED) since 2005. He is a member of ICOMOS-Spain, International Center for Heritage Conservation -CICOP and the World Network of Magellanic Universities. Its main lines of research are developed in the field of territorial development, heritage, and alternative tourism.

Jorge Cebrián Ramos has a degree in Geography from the University of La Laguna. In addition, he has completed a Master's Degree in Spatial Planning and Environment. His work experience has been developed in the company Gestión y Planeamiento Territorial y Medioambiental, S.A. (GESPLAN) in different areas, such as, strategic environmental evaluation of plans, elaboration of master plans and urban plans. He has also developed his work in the field of rural development.

Sandra Cely Andrade is a PhD in projects, with a master's degree in strategic planning and marketing from the International Ibero-American University, professional and specialist in marketing from the Pedagogical and Technological University of Colombia. Director of Competitiveness of the Chamber of Commerce of Duitama, where she has led the planning and creation of business models for companies in the region. With experience in training at the postgraduate level on issues of identification and formulation of projects and strategic management. With extensive experience in institutional, corporate and commercial planning applied to various sectors of the economy; and with publications of local and regional interest; as extensive experience in the design and use of methodologies applied to business development.

Diana María Dueñas Quintero PhD student in Education Sciences of the UPTC, Master in Administration of the National University of Colombia, Specialization Course in Good Practices of Cooperation Universidad Empresa, Business Administrator. Research, Development and Innovation Management Consultant, researcher in research projects funded by the Ministry of Science, Technology and Innovation. He advises on the construction of public policy guidelines for STI, technological management, value chain, value-added productivity, sectoral technological plans and strategies for the transfer and appropriation of knowledge.

Dorthe Eide is a professor in organization and management at Nord University Business School in Norway. She holds a Dr. Polit degree from the University of Tromsø. Her main context focus is on experience sectors/activities (nature, culture, and meals), regarding experience-based innovation and value co-creation, particularly different innovation process approaches (network, triple helix, rural, lab, testing, cross-sector) and experience design, -quality and -consumption. Increasingly also sustainability, visitor management, and innovation for sustainability. She manages the research group Marketing, Management and Innovation of Experiences (MMIE), teaches at all levels, and has managed several research and innovation projects.

Ilda Erkoçi is a lecturer and researcher at the University of Shkodra in Albania. She is interested in a number of fields concerning languages and literatures, such as Sociolinguistics, Translation, post-colonial studies, and so on, but her main passion remains literature, especially when combined with uncommon fields like tourism. She loves approaching literature from new perspectives, which has led to her long-term research in the field of literary tourism and resulted in a number of publications and conference papers.

Carlos Flores Rodríguez studied the university degree Geography and Territory Ordination at the University of La Laguna (ULL), from 2016-2020. In November 2020, he started to collaborate with the research group of ULL ReinvenTUR: Observatory of the Tourist Renewal. Since May 2021, he started to work in Gesplan, updating the Action Plan (2021-2024) of the Macizo de Anaga Biosphere Reserve. His professional interests are Climate Change, carbon footprint, urban sustainability and urban renewal, community processes and Geographic Information System.

Ruhet Genç is a Professor at Turkish-German University Economics & Administrative Sciences Department. He intends to see all issues from management and strategy point of view. His research interests are Sustainability, Innovation, Gastronomy, Value Creation, Medical Value Tourism, Tourist Behaviour & Quality of Life.

Vasiliki Georgoula is a Senior Researcher and Deputy Team Leader for the Horizon 2020 SPOT Project of the University of the Aegean Team. SPOT is an EU-funded Horizon 2020 SPOT project aims to develop a new approach to understanding and addressing cultural tourism and to promote the development of disadvantaged areas. http://www.spotprojecth2020.eu/ She has been an Academic Fellow and Lecturer at the Agricultural University of Athens, School of Applied Economics and Social Sciences, Department of Cultural and Tourism Administration, Economics and Communication in Central Greece for the Academic years 2019-2020 and 2020-2021. Vasiliki completed her Doctoral Thesis at the University of the Aegean, Department of Geography in 2018. Her work has been presented to international conferences and her published articles focus on the field of cultural and event tourism, cultural heritage, destination management and tourism geography. She holds a BA on European Languages, Society and Culture from Goldsmith's College, University of London and a MS on eTourism from the University of Surrey. Since 2005 she has been involved in vocational training, consultancy and higher education. https://orcid.org/0000-0001-6144-4471.

Natalia Gonçalves da Silva has a bachelor degree in International Trade and a master's in Tourism Management. Her main areas of interest are Reality and Community-based to Tourism and her research mainly focus on the impacts caused by these types of tourism in the host community.

Oswaldo Ledesma González is a PhD Candidate in the Department of Geography and History, University of La Laguna (ULL). His main research interests focus on territorial analysis of tourism, planning and management of tourist destinations, tourism public policy, tourism governance, and Social Network Analysis.

Ramón Granados is a Lecturer and Researcher at the Universidad Pedagógica y Tecnológica de Colombia. Public Accountant, Specialist in Finance and Master in Project Management. Diploma in School Education and Transformative Pedagogy EEPT and tutoring in Virtual Learning Environments. Participation in academic and cultural events of the Pan-American Confederation of Schools of Hotel Management, Gastronomy and Tourism, CONPEHT. Exchange Academic experiences at the University of Valencia, University of Guanajuato, University for International Cooperation of Costa Rica-UCI. He is currently director of the Tourism Observatory Group - OTGUIA of the Pedagogical and Technological University of Colombia and professor of financial mathematics, financial administration, budgeting and project formulation and evaluation. Award 'Recognition of Research and Extension CIFAD' -2017.

Desiderio Gutiérrez Taño is an Assistant Professor of Marketing in the Department of Business Administration and Economic History of the University of La Laguna, and a specialist in Tourist Marketing and Business Model Innovations. He is an Industrial Engineer and PhD in Marketing; he has also worked as a consultant for public and private organisations. His research interests include the attitude of residents towards tourism and entrepreneurship.

Olga Høegh-Guldberg is a postdoctoral researcher in experience-based tourism at Nord University Business School and is part of the Marketing, Management, and Innovation of Experiences research group. She holds a Ph.D. in Public and Private Sector Innovation from Inland Norway University of Applied Sciences. Her main research interests are experience innovation, collaboration and cross-sectoral networking, participative innovation, practice-based approaches, innovation for sustainability, sustainable destination, and visitor management.

Seyedeh Yasamin Hosseini is a tourism consultant and tourism researcher currently working as a tourism specialist for a rail transportation Company, Tehran, Iran. She has over 10 years of experience working with businesses, communities, and universities to leverage tourism for development and sector growth. She has obtained her doctorate degree in Cultural Tourism from the University of Girona, Spain. Yasamin is passionate about improved tourism through strengthening potential with various Digital Marketing tools. Her research interests are in the areas of Cultural Tourism, Family and children Tourism, tourism development & management, rural/urban tourism, tourism marketing, and museums.

Kate Chi Wo Kwan studied BSc Tourism Business Management at the Macao Institute for Tourism Studies. She is a member of China Folklore Society and a committee member of the Macao Fishmonger Association. In particular, she has been participating in the Macao Feast of the Drunken Dragon and the Lion Dance Gala for almost two decades. Recently, Kate has finished her further study at The University of Hong Kong, Faculty of Architecture - MSc Conservation. And currently being an Associate Intern at the PATA Sustainability & Social Reasonability. Her research interests focus on cultural tourism, heritage conservation, intangible cultural heritage, festivals and cultural spaces, urban development as well as sustainable development.

Javier Serrano Lara received his Geography and environmental degree from the University of Valencia in 2013 with extraordinary degree award. From 2018 has a PhD in Geography with international mention and he received the research initiation award for his thesis. Currently he is assistant professor in the geography department and Co-director in the master in development cooperation at the University of Valencia. He researches and writes widely on issues of rural development, social capital, social network analysis, gender approach, etc. He also participates in numerous European and national projects on thematic such as, depopulation, dissemination of science in European universities and gender equity.

Raphael Marinho is a PhD student in Development, Societies and Territories at the University of Trás-os-Montes and Alto Douro - UTAD (Portugal) and researcher at Research Centre for Transdisciplinary Studies for Development (CETRAD-UTAD). Master in Administration by Fluminense Federal University - UFF (Brazil), Degree in Administration by Federal University of Viçosa - UFV (Brazil).

Eduardo Martínez Díaz has a degree in Geography from the University of La Laguna. In addition, he has completed a Master's Degree in TIGs for OT: GIS and remote sensing from the University of Zaragoza. His work experience has been developed in the company Gestión y Planeamiento Territorial y Medioambiental, S.A. (GESPLAN) where he has specialized in the creation and maintenance of spatial databases and development of web applications and dashboards. On the other hand, he has participated in the territorial planning documents. At the same time, he has been part of studies related to forest fire management through satellite remote sensing platforms (Landsat and Sentinel) and LiDAR sensors.

João Carlos Martins holds a Ph.D in Sociology (2015) by Universidade Nova de Lisboa. Has a master in urban Anthropology (2009) by ISCTE-Instituto Universitário de Lisboa and Degree in Sociology (20004) by Universidade Nova de Lisboa Faculdade de Ciências Sociais e Humanas. Actual scientific interests are around urban transformation, particularly on multidimensional questions around Tourism, Participation, Leisure, Cultural Heritage; Creative Cities; Knowledge Cities and Contemporary societies. Works in the area(s) of Social Sciences with emphasis on Sociology, Anthropology and Economic, Cultural and Social Geography with emphasis on Cultural and Urban Studies.

Nelson Matos has a PhD in Tourism from the Faculty of Economics of the Universidade do Algarve. He is also an integrated member of Cinturs - Research Center for Tourism, Sustainability and Well-being, a collaborating member at CiTUR, and an invited marketing professor at the School of Management, Hospitality and Tourism, from the Universidade do Algarve. His research interest are tourism experience, destination marketing and consumer behavior.

Evelina Maziliauske is a Ph.D. candidate at Nord University Business School. She holds Master's Degree in Tourism studies from the University of Queensland, Australia. Her Ph.D. thesis focuses on topics of innovation and sustainability in experience-based cultural tourism. It particularly centers on the holistic approach to sustainability, co-creation in innovation practices and community participation in rural destinations in Norway.

Vânia Mendonça holds a degree in Cultural Heritage, and is a student of the Master degree in Tourism at University of the Algarve, Portugal.

Rui M. Mendonça-Pedro holds a Ph.D. in Tourism from the Faculty of Economics of the University of Algarve, Portugal. Currently he is Assistant Professor at the Instituto Superior Manuel Teixeira Gomes (ISMAT), Lusófona University, Portimão, Portugal. His doctoral thesis developed a SEMs Model (Senses, Emotions & Memories Model) in the context between Tourism – Neurosciences – Psychology. His knowledge focuses on the intersection of Tourism, Marketing and Management. His research interests include Tourists' Behaviours Trends, Tourism Experience Modelling, Experience Design and Experience Management, particularly at the Sensory, Emotional and Memorable levels. He is a member of the Research Centre for Tourism, Sustainability and Well-being (CinTurs), University of Algarve, Faro, Portugal.

Rafael Merinero-Rodríguez is Associate Professor in the Department of Sociology, University of Pablo de Olavide (UPO). His main research interests focus on tourism public policy, destination management, and Social Network Analysis.

Maria Pilar Peñarrubia is a Doctor in Geography. She is a PhD in Geography in the Department of Geography of the University of Valencia. She is member of the research group of the University of La Laguna.ReinvenTUR: Tourist Renovation Observatory. She has been responsible for projects of spatial analysis of markets in different companies and consultants (1998-2012). Her main lines of research are Geomarketing techniques and data source analysis.

Xerardo Pereiro is an Associate Professor with aggregation in Anthropology; Department of Economics, Sociology and Management, DESG, University of Trás-os-Montes e Alto Douro (UTAD), Portugal; Researcher at CRIA (Research Network Centre in Anthropology).

Danielle Pimentel de Oliveira is a PhD in Local Development and International Cooperation, with an International mention in Tourism, from the University of Valencia (2020), the Universidad Jaume I de Castellón y the Universidad Politécnica de Valencia, achieving the maximum award of Cum Laude, for her Thesis. She has a Master's Degree in "Dirección y Planificación del Turismo" from the Universidad de Valencia (2013) and is graduated in Tourism from the Universidad de Valencia. (2011) She has experience in the area of Tourism, where she has carried out scientific work presented at national and international congresses in Spain, Portugal, Canada and Colombia. She has extensive work experience as a translator, copywriter and customer service at destination. In her Curriculum, her academic responsibility and commitment to science stand out, in her capacity to relate theory and practice. She regularly attends seminars, conferences and congresses, in a constant search for professional improvement. Her studies have emphasis on Sustainable Tourism, working mainly on the following topics: sustainability, tourism sustainability, tourism planning and management, governance, supra-municipality. She is the author of several articles, seminars and conferences, published and presented in Spain, Brazil, Peru, Colombia, Portugal and Canada. She has participated as a field researcher in the "Ruta de la Seda" project implemented by UNESCO and the Universidad de Valencia. She has carried out international stays at the University of Algarve - Faro, Portugal, as a field researcher, participating in research groups. At present time, she is doing a post-doctorate at the Instituto Politécnico do Cávado e do Ave, in Portugal, besides being a PhD researcher at the Instituto Interuniversitario de Desarrollo Local - IIDL, being part of the INNODES research group.

María D. Pitarch-Garrido is a senior lecturer in the Department of Geography at the University of Valencia (Spain) and director of the Inter-University Institute for Local Development. She was awarded the extraordinary doctorate prize in 2008 and has been a guest lecturer at the Research Centre for Nature and Mankind (Cuba), the University of Trier (Germany), the University of Grenoble (France), the London School of Economics (United Kingdom) and Massey University (New Zealand), among others. She has participated in research projects financed by the VI and VII EU Framework Programme, the Ministry of Science and Innovation and the regional government. Her main lines of research are framed within the sub-discipline of Economic Geography, with particular attention to the analysis and territorial impact of tourism activities.

Juan Ignacio Pulido-Fernández is Associate Professor in the Department of Economics and Head of the Laboratory of Analysis and Innovation in Tourism, University of Jaén (UJA). His main research interests focus on tourism economics, destination management, sustainability of tourism, tourism impacts, and Social Network Analysis.

Ana Isabel Renda is a Professor at School of Management, Hospitality and Tourism - University of the Algarve. Integrated Member of CiTUR - Center for Tourism Research, Development and Innovation.

Anne Wally Ryan is an associate professor in Geography at Nord University, Faculty of Social Sciences. She holds a Ph.D. from NTNU, Norwegian University of Science and Technology. Her main research topics are related to the phenomenological perspective to experiences, experience-based value creation, contextual issues and place development, and sustainable destination management. She is part of the research group Marketing, Management and Innovation of Experiences (MMIE), has managed several research and development projects, and she has a long-lasting teaching experience.

Omid Salek Farokhi is a tourism and hospitality expert, lecturer, advisor, and researcher, involved in this sphere since 1999. Cultural tourism is his exclusive sphere particularly in museum studies and has published several papers in this respect. He has obtained his Ph.D. in Cultural Tourism from the University of Girona, Spain. Also, He is cooperating and advising in several operational projects about marketing, sustainable tourism, and hospitality.

Milena Santiago-Escobar is a Hotel and Tourism Manager From Externado de Colombia University and has a Master's Degree in Cultural Tourism from the University of Girona (Spain). She is a Doctor in Tourism from the University of Girona and is currently doing a postdoctoral stay at the Regional Management Center for Productivity and Innovation of Boyacá - CREPIB in Colombia. She has carried out research stays in the RES-ADMIN LEUVEN bvba Organization (Belgium), in Sustainable Strategies EKOSOS S.A. (Ecuador) and at the Autonomous University of Bucaramanga (Colombia). She has also taken specialization courses from the World Tourism Organization - UNWTO and the UNWTO.

THEMIS Foundation (UNWTO Academy). Her training as a researcher took place at the University of Girona (Spain) where she was a research assistant at the UNESCO Chair of Cultural Policies and Cooperation at the University of Girona. In addition, she has participated in research projects on various topics such as the sustainability of tourism, the use of virtual, social and complementary currencies in tourism activity, and on the impact of tourism in rural areas in international cooperation projects for development. In addition, she has presented various papers at international scientific conferences such as The t-FORUM Global Conference, The International Conference on Heritage and Sustainable Development, Tourism: International Conference on Global Tourism and Sustainability, and the International Tourism Congress CONGRETUR. His main academic interest is focused on the study of social aspects related to the sustainable development of tourism, international cooperation for development, the Sustainable Development Goals - ODS, the identification of economic, socio-cultural, and environmental impacts caused by tourism and cultural tourism and heritage.

Moisés Simancas Cruz is a Senior Lecturer of Human Geography of the the University of La Laguna. Is a PhD in Geography. He is a coordinator of the research group of the ULL "ReinvenTUR: renovation and tourist innovation", and Deputy Director of the Tourism Chair CajaCanarias-Ashotel-University of La Laguna. The current lines of research are analyses of public policies for renovation of mature tourist areas and tourism territorial planning.

Nil Sonuç is an assistant professor at İzmir Katip Çelebi University, Faculty of Tourism. Bachelor at Dokuz Eylül University, Tourism Management (Education in English & French). 4th year of bachelor & Master at Université de Savoie, in Chambéry/France at "Masters of European Hospitality Management". Internship in Guadeloupe. Master thesis in French on Sustainable Tourism Development in Guadeloupe. PhD in Tourism Management, Dokuz Eylül University, İzmir, Turkey. PhD thesis on Social Tourism in İzmir. Pedagogical Formation at Dokuz Eylul University, for teaching English. Licensed tour guide in French & English. Since 2007 teaches tourism at universities in İzmir & taught at European Universities, Belgium, Germany, Italy, Portugal & Denmark by Erasmus Teaching Exchange. Publications on sustainability and tourism, sustainable tourism marketing, environmental and social sustainability of tourism, sustainable tourism product and planning, accessible tourism, digitalisation and tourism, wellness tourism, gender in tourism guidance, yoga tourism, destinations in Turkey and Europe, slow tourism, western European cuisine, bar management, mythology, tourism and hospitality management, tourism education, cultural heritage and tourism. Research interests: Publishing in French, qualitative research, multidisciplinary tourism studies on neuro-tourism, wellness, well-being, psychology, digitalisation, cultural heritage and inclusion, social tourism, tour guiding, sustainable tourism.

Theano S. Terkenli is a professor and founding member of the first Department of Geography in Greece, University of the Aegean, since 1994. Faculty at the Interdepartmental Graduate Program in Tourism Planning, Administration and Policy, (University of the Aegean), since 1999, and at the International Interdepartmental M.Sc. in Sustainable Tourism Development: Cultural Heritage, Environment, Society (Harokopeion University, University of the Aegean, and Université-Paris I- Panthéon Sorbonne), since 2018. Ph.D. in Geography, University of Minnesota—Twin Cities, USA (1986-1993); M.Sc. in Landscape Architecture, University of Wisconsin—Madison, USA (1983-1986). Research, publishing and teaching specialty and interests in Cultural Geography, Landscape Geography and Critical Perspectives to Tourism. Conference organizer, and/or board member of several international and Greek associations and organizations. Roy Wolfe Award 2017, for outstanding contributions to the field and discipline of recreation, tourism and sports geography, from the American Association of Geographers—Recreation, Tourism and Sport Specialty Group. Books, indicatively: 1) Connections, Mobilities, Urban Prospects and Environmental Threats: the Mediterranean in Transition, editor with Annick Douguedroit and Louis F. Cassar, (Cambridge Scholars Publishing, 2015); 2) Human Geography: Humans, Society and Space, editor with T. Iosifidis and I. Chorianopoulos (Athens: Kritiki, 2007); 3) Landscapes of a New Cultural Economy of Space, editor with Anne - Marie d'Hauteserre, Landscape Series (Dordrecht: Springer, 2006; 4) Tourism Geography, with N. Metaxidis and P. Raftopoulou, (Athens: Educational Books Publishing Organization, 2003). https://orcid.org/0000-0001-6982-6132.

Cristina Valente-Pedro is a PhD. candidate in Tourism at the Faculty of Economics of University of Algarve, and a MSc. in Hospitality Management and Administration in specific branch of Hospitality and Commercial Function by the School of Management, Hospitality and Tourism of University of Algarve. Currently, Cristina's is a SPA Manager at Royal SPA in Vale do Lobo Luxury Resort. She is a National and International SPA expert, and she was an Invited Lecture in the Post-graduation of SPA Operations Management by the School of Management, Hospitality and Tourism of University of Algarve. Her research interests are in SPA Management & Marketing, SPA Experience, and Consumer Behavior in SPA.

Valeria Villalobos is currently a Research Fellow at the Center for Advanced Studies in Tourism (CAST), University of Bologna, where she has participated in research projects related to sustainability monitoring and destination management. She received her Master's Degree in Tourism Economics and Management from the University of Bologna (Italy) and her Bachelor's degree in Tourism Studies from the Universidad de Oriente (Venezuela). Her research interests include sustainability monitoring, forecasting and analysing tourism demand, data science applications for tourism and technology in tourism.

Index

D

E

F

G

H

I

L

T

U

V

W

Recommended Reference Books

IGI Global's reference books are available in three unique pricing formats: Print Only, E-Book Only, or Print + E-Book.

Shipping fees may apply.

www.igi-global.com

ISBN: 9781799872061
EISBN: 9781799872085
© 2021; 298 pp.
List Price: US$ **225**

ISBN: 9781799876113
EISBN: 9781799876120
© 2021; 210 pp.
List Price: US$ **225**

ISBN: 9781799847601
EISBN: 9781799847618
© 2021; 254 pp.
List Price: US$ **225**

ISBN: 9781799869924
EISBN: 9781799869948
© 2021; 522 pp.
List Price: US$ **295**

ISBN: 9781799843993
EISBN: 9781799844006
© 2021; 347 pp.
List Price: US$ **195**

ISBN: 9781799850625
EISBN: 9781799850632
© 2021; 367 pp.
List Price: US$ **215**

Do you want to stay current on the latest research trends, product announcements, news, and special offers?

Join IGI Global's mailing list to receive customized recommendations, exclusive discounts, and more.
Sign up at: **www.igi-global.com/newsletters.**

Publisher of Timely, Peer-Reviewed Inclusive Research Since 1988

IGI Global PUBLISHER of TIMELY KNOWLEDGE

www.igi-global.com Sign up at www.igi-global.com/newsletters facebook.com/igiglobal twitter.com/igiglobal linkedin.com/igiglobal

www.ingramcontent.com/pod-product-compliance
Lightning Source LLC
LaVergne TN
LVHW081249011026
849162LV00006BA/1248
* 9 7 8 1 7 9 9 8 9 2 1 7 5 *